Springer Tracts in Modern Physics

Volume 285

Springer Tracts in Modern Physics provides comprehensive and critical reviews of topics of current interest in physics. The following fields are emphasized:

- Particle and Nuclear Physics
- Condensed Matter Physics
- Light Matter Interaction
- Atomic and Molecular Physics

Suitable reviews of other fields can also be accepted. The Editors encourage prospective authors to correspond with them in advance of submitting a manuscript. For reviews of topics belonging to the above mentioned fields, they should address the responsible Editor as listed in "Contact the Editors".

More information about this series at https://link.springer.com/bookseries/426

Viraht Sahni

Schrödinger Theory of Electrons: Complementary Perspectives

Springer

Viraht Sahni
Brooklyn College and the Graduate Center
of the City University of New York
Brooklyn, NY, USA

ISSN 0081-3869 ISSN 1615-0430 (electronic)
Springer Tracts in Modern Physics
ISBN 978-3-030-97408-4 ISBN 978-3-030-97409-1 (eBook)
https://doi.org/10.1007/978-3-030-97409-1

This Springer imprint is published by the registered company Springer Nature Switzerland AG
The registered company address is: Gewerbestrasse 11, 6330 Cham, Switzerland

A new concept appeared in physics, the most important invention since Newton's time: the field. It needed great scientific imagination to realize that it is not the charges nor the particles but the field in the space between the charges and particles that is essential for the description of the physical phenomenon.

—Einstein and Infeld

The Evolution of Physics: The Growth of Ideas from early Concepts to Relativity and Quanta, Simon and Schuster, New York, 1938

To Vikram,

Through the vicissitudes of life
May your dreams come true

Preface

This book presents a complementary perspective to stationary-state and time-dependent Schrödinger theory of electrons in an electromagnetic field. The wave function solutions of the Schrödinger equation constitute a description of the *many-electron* system. According to Born, the wave functions are afforded the interpretation of being a probability amplitude. As such all observables are expectation values of Hermitian operators taken with respect to the wave function. The new perspective, complementary to Schrödinger theory and derived from it, is that of the *individual* electron in the sea of electrons via its stationary-state and temporal equations of motion—the 'Quantal Newtonian' First and Second Laws. These laws are in terms of 'classical' fields experienced by *each* electron. As in classical physics, these fields pervade all space. The sources of the fields are quantum-mechanical in that they are expectations of Hermitian operators taken with respect to the wave functions. The eigen energies are obtained in terms of the fields. The individual-electron perspective is therefore ensconced within the Born probabilistic interpretation. On the other hand, the fields obey equations of motion that are classical, and this then imparts determinism to them. The new perspective leads to further physical insights, and new properties of the electronic system are revealed. There are, in addition, new mathematical understandings of the Schrödinger equation that emerge as a consequence of the Laws.

Another complementary perspective to Schrödinger theory is its manifestation within a local effective potential energy framework as described by Quantal Density Functional Theory (Q-DFT). Q-DFT provides a rigorous physical explanation of how the electron correlations due to the Pauli principle and Coulomb repulsion, and of the contributions of these correlations to all kinetic properties, are incorporated into a local effective potential. The description of the mapping to such a model system is also in terms of 'classical' fields and quantal sources representative of these correlations. From these fields, the contribution of the correlations to both the local potential and total energy is determined. A significant feature of this complementary description of Schrödinger theory is that it allows for the separation of the correlations due to the Pauli principle and Coulomb repulsion. Thus, it is possible to determine the separate Pauli (exchange) and Coulomb energies, as well as their contributions to the kinetic energy. Another feature of all local effective potential theories is the determination of the Ionization Potential via the model system. Q-DFT also generalizes and broadens our understanding of local effective potential theory. Both time-dependent and stationary-state Q-DFT is described.

The two ground-state theorems of Hohenberg and Kohn provide fundamental new insights into Schrödinger theory. The first theorem proves that all the physics of an electronic system lies within its nondegenerate ground state density. Thus, the wave function is a functional of this density. The second theorem states that the ground state energy, being a functional of the density, is obtained for the true density as determined variationally for arbitrary variations of norm-preserving densities. The focus of the chapter on Density Functional Theory (DFT) is on these Theorems, their extension to the temporal domain, corollaries to the stationary-state and time-dependent theorems, and generalizations arrived at by density-preserving unitary transformations. The Kohn-Sham local effective potential theory versions of DFT, both stationary-state and temporal, are described. In Kohn-Sham DFT, the electron correlations are embedded in energy and action functionals of the density, with the corresponding local potentials defined as their respective functional derivatives. A rigorous physical description of these mathematical entities, and of how the various electron correlations are ensconced within them, is then provided via Q-DFT.

There is an emphasis throughout the book on the correlations between electrons. These constitute descriptions of how the electron spin correlations due to the Pauli principle and those of Coulomb repulsion contribute to the 'Quantal Newtonian' Laws of motion of each electron. The various complementary perspectives, and the descriptions of the electron correlations within them, are then elucidated in an *exact* manner by example. This is made possible because there exist exact analytical solutions to the stationary-state Schrödinger and Schrödinger-Pauli theory equations for two-electron 'artificial atoms'. Such 'atoms' are comprised of a system of two harmonically bound electrons. Analytical solutions also exist when a magnetic field is present. The detailed derivation of the solutions to the corresponding Schrödinger-Pauli equation for this system of interacting electrons in the presence of a magnetic field is provided. The evolution of such 'artificial atoms' perturbed by a time-dependent electric field is then described *exactly* via the Generalized Kohn Theorem. The derivation of the Theorem for arbitrary electron number is also given. To further understanding of electron correlations, the structure of the wave function in dimensions $D \geq 2$ at the coalescence of any two particles of arbitrary charge—the electron-electron and the electron-nucleus coalescence constraints—is derived. It is thus possible to study the wave function structure at and about electron-electron coalescence in the 'artificial atoms'. The exact structure of the wave function in the asymptotic classically forbidden region of natural atoms, and of its relationship to the Ionization Potential, is also derived. A new symmetry property of two-electron systems valid for arbitrary binding potential structure, arbitrary interaction, and arbitrary state is explained. The symmetry then leads to an understanding of the parity of the wave functions of such systems. The symmetry property is also elucidated for the two-electron 'artificial atom' in a magnetic field.

The book is addressed to senior undergraduate and graduate students who have had a semester's course on Quantum Mechanics and Electromagnetic theory, to researchers in the field of electronic structure, and to the general physicist and quantum chemist for whom the new and complementary perspectives on Schrödinger theory will be of interest. It is written in a pedagogical manner with the proofs of the various Laws, Theorems and Corollaries provided in detail.

The book is an evolution of ideas going back to the early 70s, essentially my career at Brooklyn College of the City University of New York. Over these decades, Brooklyn College has afforded me the freedom to pursue these ideas, and for that I am grateful.

I also wish to acknowledge Professors Marlina Slamet, Xiaoyin Pan, and Lou Massa, long-term collaborators who have been unsparing with their time in providing a critique to aspects of this work. For their constructive suggestions and support, my thanks.

Finally, I owe an immeasurable debt of gratitude to my wife Catherine for her fortitude and effort over the years in typing the book whilst I was writing it.

Brooklyn, NY, USA
December 2021

Viraht Sahni

Contents

Chapter 1
Introduction

Abstract The introductory chapter is a summary description of the new complementary perspectives to the Schrödinger theory of electrons in an electromagnetic field. The most fundamental of these is the description of the many-electron system from the perspective of the *individual* electron via its equations of motion – the 'Quantal Newtonian' Second and First Laws descriptive of time-dependent and stationary-state Schrödinger theory, respectively. The Laws, derived from the Schrödinger equation, are in terms of 'classical' fields experienced by *each* electron. The sources of these fields are quantum-mechanical expectation values of Hermitian operators taken with respect to the wave function. Hence, the perspective hews to the Born interpretation. The Laws reveal new physics. Each electron experiences an external electric and Lorentz field, as well as an internal field representative of properties of the system: correlations due to the Pauli principle and Coulomb repulsion; kinetic effects; the electron density; and an internal magnetic component. The response of the electron to these fields is described by the current density field. New mathematical understandings of the Schrödinger equation provide a quantum-mechanical definition of the external binding potential, and thus prove the equation to be intrinsically self-consistent. The perspective is extended to Schrödinger-Pauli theory in which the electron spin moment is explicitly considered. The manifestation of these ideas within a local effective potential theory framework as described via Quantal Density Functional Theory constitutes a second complementary perspective. The theory provides a rigorous physical explanation of how electron correlations due to the Pauli principle and Coulomb repulsion, and the contributions of these correlations to all kinetic properties, are incorporated into a local effective potential. This description too is in terms of 'classical' fields and quantal sources. These new perspectives are explicated by application to interacting electron 'artificial atoms' or semiconductor quantum dots in a magnetic field. The complementary perspective to stationary ground state Schrödinger theory founded in the two theorems of Hohenberg and Kohn—Modern Density Functional Theory—is described. The focus is on these theorems, their extension to the presence of a magnetic field, the extension to the temporal domain, various corollaries, and generalizations via unitary transformations. The local effective potential theory version due to Kohn and Sham is described, and a rigorous physical interpretation of both the stationary-state and temporal versions is provided via Quantal Density Functional Theory. Relevant wave function

V. Sahni, *Schrödinger Theory of Electrons: Complementary Perspectives*,
Springer Tracts in Modern Physics 285,
https://doi.org/10.1007/978-3-030-97409-1_1

properties are derived: electron coalescence constraints; structure in the classically forbidden region of atoms; a new symmetry operation for two-electron systems – the Wave Function identity; derivation of the wave functions for harmonically bound electrons in a magnetic field; and in the added presence of a time-dependent electric field.

This book presents new perspectives [1] on the Schrödinger theory [2] of electrons in the presence of external electric and magnetic fields. The new perspectives are ensconced within the Göettingen-Copenhagen [3] probabilistic interpretation of quantum mechanics: the Born rule that interprets the many-electron wave function as a "probability amplitude"; the Heisenberg uncertainty principle; the information content that the wave function incorporates all the possible information about a quantum state; the wave function collapse, that is, when a measurement is made, the state collapses to an eigenstate of the Hermitian operator associated with the observable being measured; the correspondence principle; and the wave-particle complementarity.

Schrödinger Theory: A Complementary Perspective

The fundamental new idea is the description of the many-electron system from the perspective of the *individual* electron via its corresponding equation of motion. The equations of motion obeyed by each electron—the *'Quantal Newtonian' First and Second Laws*—are derived from the Schrödinger equation, and lie within the probabilistic interpretation of quantum mechanics. These quantal Laws parallel those of Newton's Laws for the individual classical particle amongst particles interacting via Newton's Third Law forces in the presence of an external electromagnetic field. The quantum-mechanical Laws then lead to new physical insights into the electronic structure of matter. Additionally, new mathematical understandings of the nature of the Schrödinger equation emerge.

The *'Quantal Newtonian' Second Law* obeyed by *each* electron in the sea of electrons is a complementary description of time-dependent Schrödinger theory. The time-independent *'Quantal Newtonian' First Law* constitutes a special case, and in turn provides a complementary perspective to stationary-state Schrödinger theory. The 'Quantal Newtonian' Laws are expressed in terms of 'classical' fields experienced by each electron. The sources of these fields are quantal in that they are quantum-mechanical expectation values of Hermitian operators taken with respect to the wave function. It is in this manner that the new perspectives hew to the probabilistic interpretation. The added rationales for the designation 'classical' are that as is the case of classical physics, these fields pervade all space, and are thereby tangible. Further, the fields obey a classical equation of motion. In this context, these fields are deterministic.

The simplest description of the fields experienced by the N electrons is for the stationary-state case when the electrons are subjected to an external static electric field $\mathcal{F}^{\text{ext}}(\mathbf{r}) = -\nabla v(\mathbf{r})/e$, with $v(\mathbf{r})$ a scalar potential. In addition to this field, *each electron* experiences a *conservative* internal field $\mathcal{F}^{\text{int}}(\mathbf{r})$. The internal field is comprised of component fields that are representative of (i) electron correlations due to the Pauli principle [4–6] and Coulomb repulsion; (ii) the electronic density; and (iii) kinetic effects. The *statement of the corresponding 'Quantal Newtonian' First Law*, similar to Newton's First Law, *is that the sum of the external* $\mathcal{F}^{\text{ext}}(\mathbf{r})$ *and internal* $\mathcal{F}^{\text{int}}(\mathbf{r})$ *fields experienced by each electron vanishes.* On summing over all the electrons, the contribution of each component of the internal field $\mathcal{F}^{\text{int}}(\mathbf{r})$ vanishes, thus leading to Ehrenfest's First Law Theorem [7]. This then provides a more fundamental understanding of Ehrenfest's Theorem. Finally, the total energy, and the separate contributions to it from the external and internal fields, is expressed in integral virial form in terms of the individual fields.

In the added presence of a magnetostatic field $\boldsymbol{\mathcal{B}}(\mathbf{r}) = \nabla \times \mathbf{A}(\mathbf{r})$, with $\mathbf{A}(\mathbf{r})$ the vector potential, the resulting modified 'Quantal Newtonian' First Law shows that each electron now experiences an added magnetic field component to the external field $\mathcal{F}^{\text{ext}}(\mathbf{r})$, *viz.*, the Lorentz field. Of course, one expects that this must be the case. The result emerges via the derivation of the Law from the Schrödinger equation. The Lorentz field experienced by each electron is comprised of its paramagnetic and diamagnetic components. What the 'Quantal Newtonian' First Law also reveals, is that there now exists (iv) a magnetic field component contribution to the internal field $\mathcal{F}^{\text{int}}(\mathbf{r})$, referred to as the Internal Magnetic field. There is a paramagnetic and diamagnetic component to this internal field. Once again, the sum of the external $\mathcal{F}^{\text{ext}}(\mathbf{r})$ and internal $\mathcal{F}^{\text{int}}(\mathbf{r})$ fields vanish. And once again, on summing over all the electrons, the contributions of the internal field vanishes leading to Ehrenfest's First Law Theorem for the case when an external static electromagnetic field is present. The contributions to the total energy due to the Lorentz and Internal Magnetic fields can also be expressed in integral virial form in terms of these fields.

In the Schrödinger theory description of electrons in an electromagnetic field, the Hamiltonian is written for spinless electrons, and hence only the interaction of the magnetic field with the orbital angular momentum is considered. In an ad hoc manner, Pauli [8] included an additional term in the Hamiltonian involving the interaction of the magnetic field with the electron spin moment, leading to the Schrödinger-Pauli theory equation. The equation can be derived either via the Feynman kinetic energy operator [9] or as the non-relativistic limit of the Dirac equation [10]. The 'Quantal Newtonian' First Law corresponding to the stationary-state Schrödinger-Pauli theory is once again in terms of all the fields experienced by each electron as described above. The difference in the Law for this case is that now the external Lorentz and Internal Magnetic fields have an added magnetization (spin) component. Ehrenfest's First Law Theorem follows on summing over all the electrons, and the contribution of these additional fields to the energy can be expressed in integral virial form.

The 'Quantal Newtonian' First Law for a system of electrons in an external electromagnetic field thus provides new physical insights into the system. It reveals that in a manner similar to the Coulomb-Lorentz Law of classical physics, each electron in the quantum-mechanical system also experiences the external electric field and a Lorentz field. In the quantum-mechanical case, the Lorentz field is comprised of its paramagnetic, diamagnetic, and magnetization components. What the First Law further reveals is that additionally each electron experiences an internal field. The components of this field are representative of properties of the system: correlations between the electrons; the electron density; the electronic kinetic effects; and an internal magnetic field component that arises due to the external magnetic field.

In the presence of a time-dependent electromagnetic field, there is a 'Quantal Newtonian' Second Law for spinless electrons obeyed by each electron derived from the time-dependent Schrödinger equation, and a corresponding Second Law for electrons with a spin moment obtained from the temporal Schrödinger-Pauli equation. The former constitutes a special case of the latter. The fields experienced by each electron are the same as those of the stationary state case, but are now time-dependent. The temporal external fields experienced by each electron are the Coulomb electric field and a Lorentz field. The components of the temporal internal fields are representative of electron correlations due to the Pauli principle and Coulomb repulsion, the electron density, kinetic effects, and an internal magnetic field. Each electron then responds to the sum of the external and internal fields. The temporal response is in turn represented by a Current Density field. Thus, *the statement of the 'Quantal Newtonian' Second Law is that at each instant of time the sum of the external and internal fields experienced by each electron equals the Current Density response field.* It is therefore possible to study the time evolution of the electronic system via the equation of motion of the individual electron. Further, this description provides an understanding of how the correlations between the electrons and other properties of the system evolve as a function of time. On summing over all the electrons, the contribution of the internal field vanishes, thereby leading to Ehrenfest's Second Law Theorem [7].

In addition to the above described new physical perspective of a quantum-mechanical system, there are mathematical insights into the stationary-state and time-dependent Schrödinger/Schrödinger-Pauli theory equations that are arrived at via the 'Quantal Newtonian' First and Second Laws. The origin of these insights lie in the external electric field, or more to the point, the corresponding scalar potential that appears in the Hamiltonian $\hat{H}$. Thus, restricting this discussion to the stationary-state and spinless electron case, consider the external electric field to be $\mathcal{E}(\mathbf{r}) = -\boldsymbol{\nabla} v(\mathbf{r})/e$, where $v(\mathbf{r})$ is the scalar potential. Traditionally, in studying a physical system, one begins by writing down the Hamiltonian $\hat{H}$ descriptive of the system. In the Hamiltonian is included the known analytical expression for the scalar potential $v(\mathbf{r})$ such as the Coulomb or harmonic potential. This is considered an independent input to the Hamiltonian. (As the system is comprised of electrons, the kinetic and electron-interaction operators are assumed known.) As such the Hamiltonian $\hat{H}$ is completely defined. One then proceeds to solve the corresponding Schrödinger eigenvalue equation for the wave functions and energies of the system.

The insights arrived at via the 'Quantal Newtonian' First Law are for the *general arbitrary* potential $v(\mathbf{r})$ and the corresponding Hamiltonian $\hat{H}$. The First Law proves the following:

(i) The function $v(\mathbf{r})$ has the *physical interpretation of being the work done to bring an electron from some reference point at infinity to its position at* $\mathbf{r}$ *in the force of a conservative field* $\mathcal{F}(\mathbf{r})$. Hence, the function $v(\mathbf{r})$ is a potential energy in the rigorous classical sense.

(ii) The field $\mathcal{F}(\mathbf{r})$ is comprised of the external Lorentz field as well as all the various internal fields experienced by each electron. The field $\mathcal{F}(\mathbf{r})$ is therefore explicitly defined. Via the Lorentz and internal field components, the potential energy $v(\mathbf{r})$ is inherently related to the properties of the system such as the current density, correlations between the electrons, the kinetic and internal magnetic effects, and the electron density. In this manner, the potential $v(\mathbf{r})$ is *not independent* of the physical system.

(iii) The component fields of $\mathcal{F}(\mathbf{r})$ are each determined from quantal sources that are expectation values of Hermitian operators taken with respect to the wave function Ψ. The field $\mathcal{F}(\mathbf{r})$ is thus an *exactly known* functional of the wave function i.e. $\mathcal{F}(\mathbf{r}) = \mathcal{F}[\Psi](\mathbf{r})$. Hence, the potential $v(\mathbf{r})$, and therefore the Hamiltonian $\hat{H}$, too are *exactly known* functionals of the wave function, i.e. $v(\mathbf{r}) = v[\Psi](\mathbf{r})$ and $\hat{H} = \hat{H}[\Psi]$. Furthermore, these functionals are *universal* in that they are valid for *arbitrary* electronic system, and *arbitrary* state of the system whether ground, excited, nondegenerate or degenerate. The known functional dependence of $v(\mathbf{r})$ on the wave function Ψ then constitutes the quantum-mechanical definition of the binding potential $v(\mathbf{r})$. (The fact that the potential $v(\mathbf{r})$ is a functional of the *nondegenerate ground state* of the system was originally proved by Hohenberg and Kohn [11]. However, the explicit dependence of the potential $v(\mathbf{r})$ on the ground state wave function Ψ was not provided.)

(iv) The fact that the Hamiltonian $\hat{H}[\Psi]$ is an *exactly known universal functional* of the wave function Ψ means that the Schrödinger equation can now be generalized to read $\hat{H}[\Psi]\Psi = E[\Psi]\Psi$. The significance of the generalization is that the Schrödinger equation can be solved self-consistently for the determination of the eigenfunctions and energies of the system. This is the case irrespective of whether the potential $v(\mathbf{r})$ is known or unknown. (The latter case corresponds to one in which a new artificial system is created in which the potential $v(\mathbf{r})$ may be initially unknown. This occurred, for example, when 'artificial atoms' or semiconductor quantum dots were originally developed and the binding potential $v(\mathbf{r})$ was not known.) The self-consistency procedure is similar to those of *single-particle-orbital* theories, such as Hartree [12], Hartree-Fock [13, 14], Kohn-Sham [15], and Quantal Density Functional theory [16, 17], in which the corresponding Hamiltonians are *known* functionals of the orbitals ϕ_i, i.e. $\hat{H} = \hat{H}[\phi_i]$. The corresponding theory differential equations, which are of the form $\hat{H}[\phi_i]\phi_i = \epsilon_i\phi_i$, are then solved self-consistently for the orbitals ϕ_i and eigenvalues ϵ_i. One begins with an accurate initial guess for the orbitals ϕ_i, and then solves till self-consistency is achieved. In a similar manner, with an accurate initial guess for the wave function Ψ, one may obtain the solutions to the Schrödinger equation. The difference in this case is that the self-consistency procedure is for the *many-electron* wave function and corresponding energy eigenvalue. We learn, therefore, that Schrödinger theory is intrinsically self-consistent.

(v) Similarly, consider a system of electrons in a time-dependent electromagnetic field in which the electric field is $\boldsymbol{\mathcal{E}}(\mathbf{r}t) = -\boldsymbol{\nabla} v(\mathbf{r}t)/e$ and the magnetic field is $\boldsymbol{\mathcal{B}}(\mathbf{r}t) = \boldsymbol{\nabla} \times \mathbf{A}(\mathbf{r}t)$. Then according to the 'Quantal Newtonian' Second Law, the scalar potential $v(\mathbf{r}t)$ has the rigorous *physical interpretation of being, at each instant of time, the work done to bring an electron from infinity to its position at r in the force of a conservative field* $\boldsymbol{\mathcal{F}}(\mathbf{r}t)$. The field $\boldsymbol{\mathcal{F}}(\mathbf{r}t)$ is comprised of the Lorentz and Internal fields, as well as the response Current Density field. Hence, the potential $v(\mathbf{r}t)$ is inherently related to the properties of the system. As the various components of the field $\boldsymbol{\mathcal{F}}(\mathbf{r}t)$ are derived from quantal sources that are expectation values of Hermitian operators taken with respect to the wave function $\Psi(t)$, the potential $v(\mathbf{r}t)$ is a known universal functional of the wave function $\Psi(t)$, i.e. $v(\mathbf{r}t) = v[\Psi](\mathbf{r}t)$. Hence, the Hamiltonian $\hat{H}(t)$ is a *known universal functional* of the wave function $\Psi(t)$, i.e. $\hat{H}(t) = \hat{H}[\Psi](t)$. The temporal Schrödinger equation may then be generalized to read $\{\hat{H}[\Psi](t)\}\Psi(t) = i\hbar\partial\Psi(t)/\partial t$. Once again, the consequence of the generalization of the Schrödinger equation is that it can be solved self-consistently. So, time-dependent Schrödinger theory too is intrinsically self-consistent.

The Hartree [12] and Hartree-Fock [13, 14] theory descriptions of electronic structure are approximate versions of the Schrödinger and Schrödinger-Pauli theories in terms of electron correlations. In Hartree-Fock theory, the wave function is approximated by a Slater determinant of single-particle orbitals. As such, the wave function accounts solely for electron correlations due to the Pauli principle. In Hartree theory, the wave function is a product of single-particle orbitals, and thus does not satisfy the property of antisymmetry of the wave function in an interchange of its coordinates. In applications of Hartree theory, the Pauli principle is invoked ad hoc in its original manifestation [4] that no two electrons can occupy the same quantum state. Both Hartree and Hartree-Fock theories can also be described from the complementary perspective of quantal sources and fields, and the corresponding 'Quantal Newtonian' Laws. As a consequence, the Hartree and Hartree-Fock theory differential equations may be generalized in the same vein as that of Schrödinger theory.

The new quantal-source—field perspective of the more general Schrödinger-Pauli theory is elucidated by example [18]. The application is to the 2-dimensional spin-correlated first excited triplet 2^3S state of a 2-electron 'artificial atom' or semiconductor quantum dot in a uniform magnetic field. The 'artificial atom' has properties similar to that of a natural atom. The principal difference, other than the dimensionality, is that the binding potential is harmonic. As the 'artificial atom' exists in a semiconductor, the corresponding length scales are also different. For such a quantum dot in its triplet first excited state, there exists a closed-form analytical solution to the stationary-state Schrödinger-Pauli theory equation. As a consequence, it is possible to determine exactly the various quantal sources and the fields they give rise to; the correlations between the electrons due to the Pauli principle and Coulomb repulsion; and to demonstrate the satisfaction of the 'Quantal Newtonian' First Law. Further, it is shown that the energy as determined from these fields is the same as the eigenvalue of the Schrödinger-Pauli equation. An understanding of the time evolution of such a quantum dot on application of an external time-dependent electric field, and the requisite satisfaction of the 'Quantal Newtonian' Second Law, is arrived at via the Generalized Kohn Theorem [19] discussed further in the final subsection of the Introduction.

These ideas which constitute a complement to our understanding of Schrödinger and Schrödinger-Pauli theories are explicated in Chaps. 2, 3, 4 and 5.

Quantal Density Functional Theory: A Local Effective Potential Theory Complement to Schrödinger Theory

Another new perspective on Schrödinger (and Schrödinger-Pauli) theory is Quantal Density Functional Theory (Q-DFT) [15, 16]. Q-DFT is a *local effective potential* description of Schrödinger theory for *arbitrary* state. In such a description, all the many-electron correlations due to the Pauli principle and Coulomb repulsion are incorporated into a *local* (multiplicative) effective electron-interaction potential $v_{ee}(\mathbf{r})$. This potential includes in addition the contributions of these correlations to the kinetic energy, *viz.* the Correlation-Kinetic component. One is no longer then concerned with electrons but rather with an equivalent number of *noninteracting* model fermions with each possessing the *same* potential energy. The total energy is determined via the properties of the model system. As a local effective potential theory, there is commonality between Q-DFT and the work of Slater [20] and of the modern density functional theory (DFT) of Kohn and Sham [15]. However, there exist significant fundamental differences between the earlier work and Q-DFT.

Stationary-state Q-DFT constitutes the mapping from an interacting system of electrons in an external electromagnetic field as described by Schrödinger theory to one of noninteracting fermions experiencing the *same* external field and possessing the *same* density $\rho(\mathbf{r})$ and physical current density $\mathbf{j}(\mathbf{r})$. In Q-DFT, the quantum state of the interacting system could be a *nondegenerate* or *degenerate ground* or *excited* state. The state of the model system to which the mapping is performed is *arbitrary*: the model system could be in a *ground* or *excited* state. The Q-DFT mapping is in terms of 'classical' fields whose sources are quantum-mechanical expectation values of Hermitian operators. In Q-DFT, the effective potential $v_{ee}(\mathbf{r})$ is provided a *rigorous physical interpretation: It is the work done to bring a model fermion from some reference point at infinity to its position at* $\mathbf{r}$ *in the force of a conservative effective field* $\boldsymbol{\mathcal{F}}^{\text{eff}}(\mathbf{r})$. This field is a sum of an Electron-Interaction field representative of electron correlations due to the Pauli principle and Coulomb repulsion, and a Correlation-Kinetic field. The Electron-Interaction field may be further decomposed into its Hartree, Pauli and Coulomb components. Thus, within Q-DFT, in contrast to Schrödinger theory, it is possible to delineate between the correlations due to the Pauli principle and Coulomb repulsion, as well as to account for the Correlation-Kinetic effects. The *separate* contributions of these correlations to both the effective potential $v_{ee}(\mathbf{r})$ and the total energy—the Pauli, Coulomb, and Correlation-Kinetic local potentials and the corresponding energies—are *explicitly* defined. The components of the energy are each expressed in integral virial form in terms of the individual fields. Additionally, the highest occupied eigenvalue of the Q-DFT model system differential equation, irrespective of the state in which it is constructed, corresponds to the negative of the Ionization Potential. Thus, properties of the system beyond those obtained via solution of the Schrödinger equation are determined via the mapping, and as such Q-DFT constitutes a complement to Schrödinger theory.

Q-DFT also extends further our understanding of local effective potential theory: ***(a)*** The local electron-interaction potential $v_{ee}(\mathbf{r})$ which generates the same densities $\{\rho(\mathbf{r}), \mathbf{j}(\mathbf{r})\}$ as that of the interacting system is *not unique*. There could exist an infinite number of such functions; ***(b)*** This non-uniqueness is solely due to Correlation-Kinetic effects. The contribution of correlations due to the Pauli principle and Coulomb repulsion to each potential remains the same; ***(c)*** The model system wave function that reproduces the $\{\rho(\mathbf{r}), \mathbf{j}(\mathbf{r})\}$ is also not unique; ***(d)*** That it is solely the Correlation-Kinetic effects which contribute to the discontinuity in the potential $v_{ee}(\mathbf{r})$ as the electron number passes through an integer value. As the Q-DFT description of local effective potential theory is in terms of fields from which potentials and energies are obtained, it further leads to insights into the physics of both Slater theory [20] and Kohn-Sham (KS) DFT [15]. In particular, it provides a rigorous physical interpretation of the undefined energy functionals of the density of KS-DFT, and of their functional derivatives.

There is a corresponding time-dependent Q-DFT. This constitutes the mapping from a system of electrons in a time-dependent electromagnetic field to one of noninteracting fermions in the same external field and possessing the same time-dependent density $\rho(\mathbf{r}t)$ and physical current density $\mathbf{j}(\mathbf{r}t)$. Once again, the description of the model system is in terms of fields and quantal sources evolving with time. A *rigorous physical definition* is provided for the local electron interaction potential $v_{ee}(\mathbf{r}t)$ in which all the many-body effects are incorporated: *It is the work done, at each instant of time, to move the model fermion from its reference point at infinity to its position at* $\mathbf{r}$ *in the force of a conservative effective field* $\boldsymbol{\mathcal{F}}^{\text{eff}}(\mathbf{r}t)$. As in the stationary-state case, the field $\boldsymbol{\mathcal{F}}^{\text{eff}}(\mathbf{r}t)$ is the sum of an Electron-Interaction and Correlation-Kinetic field. And once again, the Electron-Interaction field may be decomposed into its Hartree, Pauli and Coulomb components. Thus, the time evolution of a physical system in terms of the individual electron correlations is determined.

The ideas underlying the stationary-state Q-DFT mapping are then elucidated by two examples. The first [21] is the mapping from a 2-dimensional 2-electron 'artificial atom' or quantum dot in a uniform magnetic field and in its *ground* state to one of noninteracting fermions also in their *ground* state possessing the same $\{\rho(\mathbf{r}), \mathbf{j}(\mathbf{r})\}$. The second example [22] is the mapping from the 2-electron quantum dot in a uniform magnetic field in its *first excited singlet* 2^1S state to a model system in its *ground* state possessing the same $\{\rho(\mathbf{r}), \mathbf{j}(\mathbf{r})\}$ as that of the excited state. (It could equally well be mapped to a model system in its first excited singlet 2^1S state.) The time evolution of these model systems in the presence of a time-dependent external electric field then follows from the Generalized Kohn Theorem [19] described below.

The principal tenets of Q-DFT together with the examples are described in Chap. 6.

Modern Density Functional Theory

The origins of modern Density Functional Theory (DFT) lie in the work of Thomas [23], Fermi [24], Dirac [25], and von Weizsacker [26]. In each theory, the ground state energy is expressed as an approximate but analytically known energy functional of the density $\rho(\mathbf{r})$. The application of the variational principle for the energy [27], and consequent functional differentiation of the energy functional, then leads to the corresponding approximate Euler-Lagrange equation for the density $\rho(\mathbf{r})$. In each case the Euler-Lagrange equation is solved for the density by introducing a Lagrange multiplier to ensure electron number conservation. With the approximate density $\rho(\mathbf{r})$, the energy is determined from the corresponding approximate functional.

A significant advance in our understanding of the quantum theory of electrons occurred with the advent of the two theorems of Hohenberg and Kohn (HK) [11] which put the above DFT idea on a rigorous mathematical foundation. The First Theorem of HK plots a path from the *nondegenerate ground state density* $\rho(\mathbf{r})$ of a system to its Hamiltonian $\hat{H}$. Solutions of the resulting Schrödinger equation then lead to the eigenfunctions and eigenvalues of the system. The wave functions Ψ of the system are consequently functionals of the ground state density, i.e. $\Psi = \Psi[\rho]$. The energies E and all properties O of the system, which are the expectation values of the Hamiltonian $\hat{H}$ and other Hermitian operators $\hat{O}$ taken with respect to the wave functions, are then *unique* functionals of the *ground state density* $\rho(\mathbf{r})$, i.e. $E = E[\rho]$; $O = O[\rho]$. In essence, all the information of the properties of an electronic system is embedded in its ground state density $\rho(\mathbf{r})$. This is a fundamental step forward in our understanding of quantum mechanics.

Although the First Theorem proves that the energy is a functional of the density, it does not provide the explicit dependence of the wave function, and thus of energy and other properties, on the density. The Second Theorem of HK proves that on application of the variational principle for arbitrary variations of the density $\rho(\mathbf{r}) + \delta\rho(\mathbf{r})$, the ground state energy is obtained for the exact ground state density $\rho(\mathbf{r})$. Unlike the Thomas, Fermi, Dirac, and von Weizsacker theories, the Euler-Lagrange equation for the density cannot be written explicitly because the energy functional of the density is unknown, and the functional derivative in the Euler-Lagrange equation cannot be determined.

The HK path from the nondegenerate ground state density $\rho(\mathbf{r})$ to the Hamiltonian $\hat{H}$ is proved for a system of N electrons in an external electrostatic field $\boldsymbol{\mathcal{E}}(\mathbf{r}) = -\boldsymbol{\nabla} v(\mathbf{r})/e$, with $v(\mathbf{r})$ the scalar binding potential. In the proof, the constraint of *fixed electron number* N is of prime significance. The proof is in two parts. It is first proved that there is a *one-to-one* relationship between the potential $v(\mathbf{r})$ (to within a constant) and the nondegenerate ground state Ψ. Hence, for each ground state Ψ, there exists one and only one potential $v(\mathbf{r})$. (The Theorem does not provide the explicit dependence of $v(\mathbf{r})$ on Ψ. That relationship is provided by the 'Quantal Newtonian' First Law, and is not restricted to the ground state. See Chap. 2.) In the second step, it is proved via a *reductio ad absurdum* argument, that there is a similar *one-to-one* relationship between the ground state Ψ and the nondegenerate ground state density $\rho(\mathbf{r})$. Hence, knowledge of the density $\rho(\mathbf{r})$ uniquely determines the potential $v(\mathbf{r})$ to within a constant. With the kinetic and electron-interaction potential operators of the electrons assumed known, the ground state density $\rho(\mathbf{r})$ determines the Hamiltonian $\hat{H}$ to within a constant. The density $\rho(\mathbf{r})$, a gauge invariant property, is thus referred to as a *basic variable*.

In the added presence of a magnetic field $\boldsymbol{\mathcal{B}}(\mathbf{r}) = \boldsymbol{\nabla} \times \mathbf{A}(\mathbf{r})$, where $\mathbf{A}(\mathbf{r})$ is the vector potential, there is a similar path to the Hamiltonian $\hat{H}$. In this case, it is knowledge of the gauge invariant properties of the nondegenerate ground state density $\rho(\mathbf{r})$ and the physical current density $\mathbf{j}(\mathbf{r})$ that determines the scalar $v(\mathbf{r})$ and vector $\mathbf{A}(\mathbf{r})$ potentials to within a constant and gradient of a scalar function, respectively. With the kinetic and electron-interaction potential operators known, the Hamiltonian $\hat{H}$ is determined, as are the consequent eigenfunctions and eigenvalues. The wave functions are thus functionals of the densities $\{\rho(\mathbf{r}), \mathbf{j}(\mathbf{r})\}$, i.e. $\Psi = \Psi[\rho, \mathbf{j}]$, and thus the energy is a unique functional of these densities; $E = E[\rho, \mathbf{j}]$. In this case, the *basic variables* are the gauge invariant ground state densities $\{\rho(\mathbf{r}), \mathbf{j}(\mathbf{r})\}$. The proof of the *one-to-one* relationship between the densities $\{\rho(\mathbf{r}), \mathbf{j}(\mathbf{r})\}$ and the potentials $\{v(\mathbf{r}), \mathbf{A}(\mathbf{r})\}$ is due to Pan and Sahni [28] and differs from that of the First Theorem of HK. The reason is that the relationship between the potentials $\{v(\mathbf{r}), \mathbf{A}(\mathbf{r})\}$ and the nondegenerate ground state wave function Ψ is not one-to-one but *many-to-one* and can be *infinite-to-one*. This fundamental difference in the physics when a magnetic field is present must be accounted for in any proof. The constraints for spinless electrons are those of fixed electron number N and the orbital angular momentum $\mathbf{L}$. For electrons with a spin moment, there is the added constraint of fixed spin angular momentum $\mathbf{S}$. In a manner similar to the Second Theorem of HK, the corresponding Euler-Lagrange equations with the above constraints together with the satisfaction of the continuity equation may be written.

There are further aspects of the Theorems of Hohenberg and Kohn that are described: (a) The Gunnarsson-Lundqvist Theorem [29] is a restatement of the First Theorem of HK for certain excited states based on their symmetry relative to the ground state. These are states for which there exists a *one-to-one* relationship between the potential $v(\mathbf{r})$ and the excited state density $\rho^e(\mathbf{r})$, and for which the variational principle for the energy [27] is applicable; (b) The First Theorem is *generalized* via *density preserving* unitary transformations. The transformation proves that the wave function Ψ is a functional not only of the density $\rho(\mathbf{r})$ but also of a gauge function $\alpha(\mathbf{R})$, $(\mathbf{R} = \mathbf{r}_1, \ldots, \mathbf{r}_N)$; $\Psi = \Psi[\rho(\mathbf{r}), \alpha(\mathbf{R})]$, and that there could be an infinite number of such gauge functions. The significance of this generalization lies in the fact that the wave function Ψ is *gauge variant*, whereas the density is *gauge invariant*. The wave function written as the functional $\Psi[\rho(\mathbf{r}), \alpha(\mathbf{R})]$ then ensures that the functional is gauge variant. The First Theorem of HK is recovered as a special case for $\alpha(\mathbf{R}) = \alpha$, a constant; (c) According to the First Theorem of HK, knowledge of the ground state density $\rho(\mathbf{r})$ uniquely determines the physical system via its Hamiltonian $\hat{H}$. In a Corollary, it is shown that it is possible to construct an *infinite* set of Hamiltonians $\{\hat{H}\}$ representing *different physical systems* with each possessing the *same* ground state density $\rho(\mathbf{r})$. In this case, the density $\rho(\mathbf{r})$ cannot distinguish between the different physical systems on the basis of the First Theorem. This would appear to be a counter example to the First Theorem. However, it is not because the Hamiltonians $\{\hat{H}\}$ each differ by a constant. A counter example would be one in which the Hamiltonians differ by more than a constant and yet possess the same density $\rho(\mathbf{r})$. Nevertheless, there exist physical systems which cannot be distinguished through the knowledge of the density $\rho(\mathbf{r})$; (d) There are other paths from the ground state density $\rho(\mathbf{r})$ to the Hamiltonian $\hat{H}$, and thus to the ground state wave function Ψ. These paths all require pre-knowledge of the fact that the basic variable is the ground state density $\rho(\mathbf{r})$, and as such are not as fundamental as the First Theorem of HK. For example, in the constrained-search path of Percus [30], Levy [31, 32], and Lieb [33], a search is performed over all functions Ψ_ρ that yield the density $\rho(\mathbf{r})$. The function Ψ_ρ which yields the minimum of the energy is then the true ground state wave function Ψ. Since Ψ cannot be an eigenfunction of more than one $\hat{H}$ with a multiplicative potential $v(\mathbf{r})$, the Hamiltonian is thereby determined. The argument is in general terms, and does not explain how the potential $v(\mathbf{r})$ is to be obtained from Ψ. In this constrained-search path too, the explicit functional dependence of $v(\mathbf{r})$ on Ψ is provided by the 'Quantal Newtonian' First Law.

Kohn-Sham (KS) DFT is founded in the Theorems of Hohenberg-Kohn. As a consequence, KS-DFT is a *ground state* theory. The theory constitutes the mapping from the *ground state* of an interacting system of electrons as described by Schrödinger theory to one of noninteracting fermions also in its *ground state* and possessing the same density $\rho(\mathbf{r})$. It is a local effective potential theory. Thus, without further ado, the differential equation may be written down in terms of the single-particle orbitals ϕ_i and eigenvalues ϵ_i. With the assumption that the noninteracting fermions experience the same external field, the only remaining property required to define the model system within KS-DFT is the definition of the local electron-interaction potential $v_{ee}(\mathbf{r})$ in which all the many-electron effects are incorporated. Employing the fact that the total energy E may be expressed as a functional of the density $\rho(\mathbf{r})$, it is assumed that all the many-body effects are embedded in its electron-interaction energy functional $E_{ee}^{\mathrm{KS}}[\rho]$ component. Then, on application of the variational principle for the energy to the energy functional $E[\rho]$, for arbitrary variations of the density $\delta\rho(\mathbf{r})$, the potential $v_{ee}(\mathbf{r})$ is defined to be the *functional derivative* $\delta E_{ee}^{\mathrm{KS}}[\rho]/\delta\rho(\mathbf{r})$. The orbitals ϕ_i and eigenvalues ϵ_i are determined by self-consistent solution of the single-particle differential equations. The density $\rho(\mathbf{r})$ obtained thereby is then substituted into the energy functional $E[\rho]$ to obtain the energy of the system. (The energy $E[\rho]$ may also be written in terms of the properties of the model system: eigenvalues ϵ_i, the functional $E_{ee}^{\mathrm{KS}}[\rho]$, and its functional derivative $v_{ee}(\mathbf{r})$.) The energy functional $E_{ee}^{\mathrm{KS}}[\rho]$ is *universal* in that it is the same for all electronic systems independent of the external binding potential $v(\mathbf{r})$. When the Schrödinger theory differential equation for the model system is expressed such that the local electron-interaction potential $v_{ee}(\mathbf{r})$ is defined as a functional derivative, it is referred to as the *Kohn-Sham equation.*

The KS-DFT mapping to the model system is thus *mathematical* in construct: The description of the electron correlations is in terms of the universal energy functional of the density $E_{ee}^{\mathrm{KS}}[\rho]$; the description of the corresponding local potential in which these correlations are embedded is defined as the functional derivative $\delta E_{ee}^{\mathrm{KS}}[\rho]/\delta\rho(\mathbf{r})$. The mathematical entities of the functional $E_{ee}^{\mathrm{KS}}[\rho]$ and its functional derivative $v_{ee}(\mathbf{r})$ can be provided a rigorous *physical interpretation* via Q-DFT, and this interpretation is described.

KS-DFT does not explain how the various electron correlations are to be incorporated into the functional $E_{ee}^{\mathrm{KS}}[\rho]$. Hence, neither the energy functional $E_{ee}^{\mathrm{KS}}[\rho]$, nor its functional derivative $v_{ee}(\mathbf{r})$, are known. Therefore, all the physics of the many-electron correlations must be extrinsically incorporated. Since the advent of the work by Kohn and Sham, there has been extensive research into the development of accurate approximations $E_{ee}^{\mathrm{KS-approx}}[\rho]$ to the functional $E_{ee}^{\mathrm{KS}}[\rho]$. This is best described in reviews [34, 35] of the various approximations via their numerous applications. A few general remarks can be made about the approximate functionals $E_{ee}^{\mathrm{KS-approx}}[\rho]$ and their functional derivatives $v_{ee}^{\mathrm{approx}}(\mathbf{r})$. (i) In approximating the functional $E_{ee}^{\mathrm{KS}}[\rho]$, one is in essence approximating the electron-interaction potential operator in the Hamiltonian for the Schrödinger equation of the interacting system. Thus, the original physical system is inherently modified. As a consequence, the rigor of the Second Theorem of HK is lost, and the approximate energy obtained from the model system is not an upper bound. The energy could lie below the true value; (ii) In constructing a $E_{ee}^{\mathrm{KS-approx}}[\rho]$, an assumption is made as to the form of the electron correlations. For example, consider the Local Density Approximation (LDA) [11, 15, 36], which constitutes the leading term in essentially all approximate functionals $E_{ee}^{\mathrm{KS-approx}}[\rho]$ in the literature. In the LDA, it is assumed that the electron correlations of a nonuniform electron density system are those of a uniform electron gas but with a structure defined by the local value of the density. It is further assumed that the corresponding functional derivative is representative of the same correlations. This is not the case. The correct description of the electron correlations within the LDA, which are the same for both the approximate LDA energy functional and of its functional derivative, is obtained via Q-DFT [16]. As explained, there are added correlations beyond those assumed within the framework of KS-DFT.

The (stationary-state) First Theorem of Hohenberg and Kohn has been extended by Runge and Gross (RG) [37] to the case of a system of N electrons in a time-dependent (TD) external field $\mathcal{F}^{\mathrm{ext}}(\mathbf{r}t) = -\nabla v(\mathbf{r}t)/e$. The scalar potential $v(\mathbf{r}t)$ is assumed to be Taylor expandable about some initial time t_0 which is finite. For such a system, RG proved that the *basic variables* were the TD density $\rho(\mathbf{r}t)$ and current density $\mathbf{j}(\mathbf{r}t)$. Thus, although the path to the TD Hamiltonian $\hat{H}(t)$ may be achieved from either $\rho(\mathbf{r}t)$ or $\mathbf{j}(\mathbf{r}t)$, TD-DFT is concerned only with the density $\rho(\mathbf{r}t)$. Beyond proving the *one-to-one* relationship between the potential $v(\mathbf{r}t)$ and the current density $\mathbf{j}(\mathbf{r}t)$, it is further proved that there is a *one-to-one* relationship between the density $\rho(\mathbf{r}t)$ and the potential $v(\mathbf{r}t)$ to within a TD function. Then, as the kinetic and electron-interaction operators are known, knowledge of the density $\rho(\mathbf{r}t)$ determines the Hamiltonian $\hat{H}(t)$ to within a TD function. Thus, the solution $\Psi(t)$ to the TD Schrödinger equation is a functional of the density $\rho(\mathbf{r}t)$, i.e. $\Psi(t) = \Psi[\rho(\mathbf{r}t)]$. The RG Theorem is *generalized* by performing a *density preserving* unitary transformation, whereby it is proved that the wave function $\Psi(t)$ must also be a functional of a gauge function $\alpha(\mathbf{R}t)$, i.e. $\Psi(t) = \Psi[\rho(\mathbf{r}t), \alpha(\mathbf{R}t)]$. (The RG Theorem constitutes a special case when $\alpha(\mathbf{R}t) = \alpha(t)$.) The generalization ensures that the wave function $\Psi(t)$ when expressed as a functional is gauge variant. A Corollary shows that it is possible to construct an *infinite* number of TD Hamiltonians $\{\hat{H}(t)\}$ which represent *different physical systems* but yet possess the *same* density $\rho(\mathbf{r}t)$. Thus, knowledge of the density $\rho(\mathbf{r}t)$ cannot distinguish between the different physical systems on the basis of the RG Theorem. This would appear to violate the RG Theorem. However, this is not the case because the Hamiltonians in the set $\{\hat{H}(t)\}$ each differ by a function of time. A counter example would be one in which Hamiltonians differ by more than a function of time but yet possess the same density $\rho(\mathbf{r}t)$. Nonetheless, there exist physical systems for the which the density $\rho(\mathbf{r}t)$ cannot distinguish between them.

In their original work, RG attempted to derive a Second Theorem to parallel the Second Theorem of HK by constructing an action functional $A[\rho(\mathbf{r}t), \Psi_0]$ of the density $\rho(\mathbf{r}t)$, where Ψ_0 is the initial state of the system. Unfortunately, for arbitrary variations of the density $\delta\rho(\mathbf{r}t)$, the first-order variation $\delta A[\rho(\mathbf{r}t), \Psi_0]$ of this functional does not vanish. There have been other attempts at creating such an action function whose functional derivative then leads to the local electron-interaction potential $v_{ee}(\mathbf{r}t)$ for a system of noninteracting fermions possessing the same density $\rho(\mathbf{r}t)$, and consequently to the TD version of the KS equation. More significantly, employing the 'Quantal Newtonian' Second Law [38], van Leeuven [39] proved the existence of such a model system. This then provides a definition of the potential $v_{ee}(\mathbf{r}t)$ as that function which is obtained by self-consistent solution of the TD KS equations such that the orbitals lead to the density $\rho(\mathbf{r}t)$. A rigorous *physical interpretation* of the potential $v_{ee}(\mathbf{r}t)$ in terms of electron correlations is provided via TD Q-DFT.

The principal ideas underlying and new insights into stationary-state and time-dependent Density Functional Theory are explained in Chap. 7.

Wave Function Properties and Wave Functions

In general, the solutions to the Schrödinger or Schrödinger-Pauli equations cannot be obtained in closed-analytical form, and must be approximated. There are, however, *universal* properties that any approximate wave function must satisfy. These properties are universal in that they are valid for *arbitrary* state of all electronic systems irrespective of the binding potential. Thus, the approximate wave function must be single valued, smooth and bounded; it must be anti-symmetric in an interchange of the electronic coordinates to satisfy the Pauli principle; and be normalized with probability density greater than or equal to zero. There are other *exactly known* properties that a many-electron wave function must satisfy, and these are concerned with when electrons coalesce or when an electron coalesces with the nucleus. At such coalescence, the Schrödinger equation is singular. For the wave function to be a solution and remain bounded, it must satisfy either the electron-electron or the electron-nucleus coalescence constraints. These could be either *cusp* or *node* coalescence constraints. The *exact* general expression for the wave function in dimensions $D \geq 2$ when any two charged particles coalesce is derived. Hence, any approximate wave function must satisfy these constraints. Sum rules for other properties such as the electronic density and the pair-correlation function may consequently be derived. The coalescence constraints also translate to *exact* properties of local effective potential theories. For example, as a consequence of the electron-nucleus constraint, it can be proved that the local electron-interaction potential $v_{ee}(\mathbf{r})$ of either Q-DFT or KS-DFT must be finite at the nucleus. Then, via Q-DFT, the exact analytical structure of $v_{ee}(\mathbf{r})$ near the nucleus, and the contributions of the different correlations to this structure, can be determined.

For electronic systems for which the binding potential $v(\mathbf{r})$ is Coulombic, it is possible to derive the *exact* analytical structure of the wave function in the asymptotic classically forbidden region of finite systems. This structure is valid for arbitrary state. The significance of this lies in the fact that the structure is in terms of the Ionization Potential. Experimental input can thus be employed in the construction of approximate wave functions. Once again, this asymptotic structure of the wave function has implications for local effective potential theories. It helps prove that the highest occupied eigenvalue of such theories is the negative of the Ionization Potential. In the case of Q-DFT, this is the case irrespective of the state of the model system. Further, knowledge of the wave function determines via Q-DFT the separate contributions of electron correlations due to the Pauli principle, Coulomb repulsion and Correlation-Kinetic effects to the asymptotic structure of the local electron-interaction potential $v_{ee}(\mathbf{r})$.

The study of the properties of two interacting fermions has provided an understanding of the physics of diverse many-electron systems. Symmetries, and the conservation laws that result therefrom, also provide an understanding of the physics of a system. A new symmetry property [40] of two interacting fermions bound by a *symmetrical* potential in an external electromagnetic field as defined by the Schrödinger-Pauli theory equation is described. The symmetry operation is such that it leads to the equality of the transformed wave function to the wave function. This equality is referred to as the *Wave Function Identity*. The Identity is valid for *arbitrary* structure of the binding potential, *arbitrary* electron interaction of the form $u(|\mathbf{r} - \mathbf{r}'|)$, *all* bound electronic states, and *arbitrary* dimensionality. The symmetry operation is a two-step process: an interchange of the spatial coordinates of the electrons whilst keeping their spin moments unchanged, followed by an inversion. (The switching of the electrons in the first step differs from that of the Pauli principle.) It is proved that the exact wave function satisfies this Identity. The symmetry operation, and the contrast with the Pauli principle, is then elucidated by application to the first excited singlet 2^1S and triplet 2^3S states of a 2-dimensional 2-electron semiconductor quantum dot or 'artificial atom' in a magnetic field. New physics emerges as a consequence of the symmetry. On application of the permutation operation for fermions to the Identity, it is shown that the *parity* of the singlet states is *even* and that of the triplet states *odd*. As a consequence, it follows that at electron-electron coalescence, the singlet state wave functions satisfy the *cusp* coalescence constraint, and triplet state wave functions the *node* coalescence condition. It is further shown that the *parity* of the singlet state wave functions about *all points of electron-electron coalescence* is *even*, and that of triplet state wave functions *odd*. These properties of parity are also exhibited for the above 'artificial atoms'. Note that the Wave Function Identity and subsequent conclusions on parity, are equally valid for the special cases in which the 2-electron bound system, in both the presence and absence of the magnetic field, are described by the corresponding Schrödinger theory for spinless electrons. The Wave Function Identity is thus a new property that any approximate 2-electron wave function must satisfy.

The examples employed to elucidate the new ideas of the 'Quantal Newtonian' First Law, Q-DFT, and the Wave Function Identity, are those of 2-dimensional 2-electron harmonically bound 'artificial atoms' in a uniform magnetic field. Different 'artificial atoms' are considered depending on their state. These are a ground, and first excited singlet 2^1S and triplet 2^3S states. The reasons for employing such 'artificial atoms' are the following: (a) Due to the development of semiconductor technology, such 'atoms' or quantum dots can be physically created; (b) The physics of such 'atoms' is very similar to those of natural atoms. Only the length scales differ; (c) These are examples of an interacting system of electrons for which closed-form analytical solutions to the corresponding stationary-state Schrödinger-Pauli theory differential equation can be derived; (d) Thus, all the properties discussed are determined *exactly*. They are, furthermore, obtained either in analytical or semi-analytical form; (e) Electron correlations, *viz.* the spin correlations due to the Pauli principle, the correlations due to Coulomb repulsion, and the contributions of these correlations to the kinetic energy, are thereby brought to the fore. Further, as a consequence of the theoretical framework developed, these correlations can be studied individually. A detailed derivation of the general solution to the Schrödinger-Pauli equation due to Taut [41] employing the Frobenius series expansion method is provided. The explicit expressions for the wave functions and eigenenergies of a ground and first excited triplet state are derived.

In order to elucidate by example the temporal aspects of the new concepts described, *e.g.* the 'Quantal Newtonian' Second Law and time-dependent Q-DFT, a *'first principles'* derivation of the Generalized Kohn Theorem (GKT) [19] via the 'interaction' representation is provided. The GKT wave function is the solution to the time-dependent Schrödinger equation for harmonically bound electrons in a magnetic field perturbed by a spatially uniform TD electric field. The GKT wave function is comprised of a phase factor times the unperturbed wave function in which the coordinates of each electron are translated by a TD value that satisfies the corresponding classical equation of motion. Hence, if the unperturbed wave functions are known, the evolution of the TD wave function is determined. In particular, the evolution of properties which are expectations of non-differential Hermitian operators, corresponds to that of the unperturbed system translated by a TD function. As the stationary-state eigenfunctions of two harmonically bound electrons in a magnetic field are known analytically, the evolution of the corresponding TD wave function and properties of the system are also known analytically.

The coalescence properties of the wave function, its asymptotic structure in the classically forbidden region, and the Wave Function Identity are described in Chap. 8. The derivation of the general solution to the Schrödinger-Pauli equation for the 2-electron quantum dot in a magnetic field, and the proof of the Generalized Kohn Theorem, is given in Chap. 9.

The Epilogue is Chap. 10.

References

1. V. Sahni, Int. J. Quantum Chem. **121**, e26556 (2021). (This is a representative work. The references to other original literature is provided as the theory is developed in the chapters to follow)
2. E. Schrödinger, Ann. Physik **79**, 361, 489 (1925); **80**, 437 (1926); **81**, 109 (1926); E. Schrödinger, *Collected Papers on Wave Mechanics*, (American Mathematical Society Chelsea Publishing, Providence, Rhode Island, Reprinted 2014)
3. M.Born, Z. Physik **38**, 803 (1926); W. Heisenberg, Z. Physik **43**, 172 (1927); N. Bohr, Naturwis, **16**, 245 (1928); **17**, 483 (1929); **18**, 73 (1930)
4. W. Pauli, Z. Physik **31**, 765 (1925)
5. P.A.M. Dirac, Proc. Roy. Soc. **A112**, 661 (1926)
6. W. Heisenberg, Z. Physik **38**, 411 (1926)
7. P. Ehrenfest, Z. Physik **45**, 455 (1927)
8. W. Pauli, Z. Physik **43**, 601 (1927)
9. J.J. Sakurai, *Advanced Quantum Mechanics* (Addison-Wesley, Reading, MA, 1967)
10. P.A.M. Dirac, Proc. Roy. Soc. (London) **A117**, 610 (1928)
11. P. Hohenberg, W. Kohn, Phys. Rev. **136**, B864 (1964)
12. D.R. Hartree, Proc. Cambridge Philos. Soc. **24**, 39 (1928); **24**, 111 (1928); **24**, 426 (1928)
13. J.C. Slater, Phys. Rev. **35**, 210 (1930)
14. V. Fock, Z. Phys. **61**, 126 (1930)
15. L.J. Sham, W. Kohn, Phys. Rev. **140**, A1133 (1965)
16. V. Sahni, *Quantal Density Functional Theory*, 2nd edn. (Springer, Berlin, Heidelberg, 2016) (Referred to as ***QDFT***)
17. V. Sahni, *Quantal Density Functional Theory: Approximation Methods and Applications*, (Springer, Berlin, Heidelberg, 2010) (Referred to as ***QDFT2***)
18. M. Slamet, V. Sahni, Chem. Phys. **546**, 111073 (2021)
19. H.-M. Zhu, J.-W. Chen, X.-Y. Pan, V. Sahni, J. Chem. Phys. **140**, 024318 (2014)
20. J.C. Slater, Phys. Rev. **81**, 385 (1951); J.C. Slater, T.M. Wilson, J.H. Wood, Phys. Rev. **179**, 28 (1969)
21. T. Yang, X.-Y. Pan, V. Sahni, Phys. Rev. A **83**, 042518 (2011)
22. M. Slamet, V. Sahni, Comp. Theor. Chem. **1114**, 125 (2017)
23. L.H. Thomas, Proc. Camb. Phil. Soc. **23**, 542 (1927)
24. E. Fermi, Rend. Accad. Naz. Linzei **6**, 602 (1927); Z. Physik **48**, 73 (1928)
25. P.A.M. Dirac, Proc. Camb. Phil. Soc. **26**, 376 (1930)
26. C.F. von Weizsacker, Z. Physik **96**, 431 (1935)
27. B.L. Moiseiwitsch, *Variational Principles* (Wiley, London, 1966)
28. X.-Y. Pan, V. Sahni, J. Chem. Phys. **143**, 174105 (2015)
29. O. Gunnarsson, B. Lundqvist, Phys. Rev. B **13**, 4274 (1976)
30. J. Percus, Int. J. Quantum Chem. **13**, 89 (1978)
31. M. Levy, Proc. Natl. Acad. Sci. USA **76**, 6062 (1979)
32. M. Levy, Int. J. Quantum Chem. **110**, 3140 (2010)
33. E. Lieb, Int. J. Quantum Chem. **24**, 243 (1983)
34. R.O. Jones, Rev. Mod. Phys. **87**, 897 (2015)
35. P. Verma, R.G. Truhlar, Trends Chem. **2**, 302 (2020)
36. R. Gáspár, Acta. Phys. Acad. Sci. Hung **3**, 263 (1954); J. Mol. Struct. (Theochem) **501**, 1 (2000)
37. E. Runge, E.K.U. Gross, Phys. Rev. Lett. **52**, 997 (1984)
38. Z. Qian, V. Sahni, Phys. Lett. A **247**, 303 (1998)
39. R. van Leeuwen, Phys. Rev. Lett. **82**, 3863 (1999)
40. M. Slamet, V. Sahni, Chem. Phys. **556**, 111453 (2022)
41. M. Taut, J. Phys. A **27**, 1045 (1994); *ibid* **27**, 4723 (1994) (Corrigenda); J. Phys. Condens. Matter **12**, 3689 (2000); M. Taut, H. Eschrig, Z. Phys. Chem. **224**, 631 (2010)

Chapter 2
Schrödinger Theory of Electrons: A Complementary Perspective

Abstract The temporal Schrödinger theory of electrons in the presence of an external time-dependent electric field is described from the new perspective of the *individual* electron via its equation of motion, the 'Quantal Newtonian' Second Law. According to the law, each electron experiences not only the external electric field, but also an internal field. The internal field is comprised of a sum of fields, each representative of a specific property of the system: electron correlations due to the Pauli principle and Coulomb repulsion; the electron density; and kinetic effects. In the Law, the response of each electron to all the fields is represented by a field which involves the temporal derivative of the electron current density. The sources of the fields are quantum-mechanical in that they are expectation values of Hermitian operators taken with respect to the wave function. In summing the Second Law over all the electrons, each component of the internal field is shown to vanish, thereby leading to Ehrenfest's theorem. In a similar manner, the equation for the torque and the Integral Virial Theorem can also be derived from the Second Law. The new *individual* electron perspective description of stationary-state Schrödinger theory of electrons in an external static electric field, then constitutes a special case. The corresponding equation of motion of each electron is then the 'Quantal Newtonian' First Law. According to the First Law, each electron experiences both the external and an internal field, the sum of which vanishes. The internal field is once again a sum of fields representative of electron correlations due to the Pauli principle and Coulomb repulsion, the electron density, and kinetic effects. The quantal sources are expectation values of Hermitian operators taken with respect to the stationary-state wave function. From the First Law, the resulting stationary-state Ehrenfest's and Integral Virial theorems and the torque equation are derived. Remarks on the equations of Quantum Fluid Dynamics and the 'Quantal Newtonian' Laws are made.

Introduction

The chapter explains how the Schrödinger theory [1] of electrons in an external time-dependent electric field can be described [2–9] from a 'Newtonian' perspective that is tangible in the classical sense. This description of a quantum system is in terms

V. Sahni, *Schrödinger Theory of Electrons: Complementary Perspectives*,
Springer Tracts in Modern Physics 285,
https://doi.org/10.1007/978-3-030-97409-1_2

of 'classical' fields that arise from quantum-mechanical sources. The terminology 'classical' is employed in the original sense of fields pervading all space. The quantal sources of these fields are defined as expectation values of Hermitian operators taken with respect to the wave function of the system. This then hews to and encompasses the probabilistic nature of quantum mechanics. There is, however, no uncertainty with regard to the structure of the fields. The fields obey a classical equation of motion. In this context, the fields are deterministic. The quantum-mechanical description in terms of 'classical' fields and quantal sources mirrors that of the motion in classical physics for an *individual* particle amongst other such particles interacting via Newton's Third Law forces. For such interacting particles, Newton's Second Law of motion *for each particle* involves both the external and internal forces on the particle and of its response to these forces. The special case of the time-independent equation of motion is described by Newton's First Law for each particle. On summing over all the particles, the contribution of the internal forces vanish leading thereby to Newton's Second and First Laws. In a similar manner, for a system of electrons in a time-dependent external field, there exists a 'Quantal Newtonian' Second Law equation of motion for the *individual electron*. This Law involves both the external and an internal field experienced by the electron, and a corresponding field representing the response of the electron to these fields. The internal field is a sum of fields each representative of an intrinsic property of the system. These properties are electron correlations due to the Pauli Exclusion principle [10] or equivalently the requirement of antisymmetry of the wave function in an interchange of two electrons [11, 12]; electron correlations due to Coulomb repulsion; the electron density; and kinetic effects. The 'Quantal Newtonian' Second Law thus brings to light new physics. Whereas it is evident that the electrons experience the external fields, we now understand that each electron also experiences an internal field which is characteristic of fundamental properties of the system. These fields are intrinsic to the quantal system. Further, in a manner similar to that of classical physics, on summing the 'Quantal Newtonian' Second Law over all the electrons, the contribution of the internal field vanishes, leading thereby to a more physically insightful [13] derivation of Ehrenfest's theorem [14]. Additionally, the total (non-conserved) energy and its components can be expressed in integral virial form in terms of the various individual fields. Stationary state Schrödinger theory of electrons is then described by the 'Quantal Newtonian' First Law [3–9, 15] for *each electron* which constitutes a special case. Finally, as a consequence of the 'Quantal Newtonian' Laws, it is possible to generalize the stationary-state and time-dependent Schrödinger equations and write them in a generalized form that exhibits their intrinsic self-consistent nature [16, 17].

We begin with the traditional description of the Schrödinger theory of electrons in a time-dependent external electric field. Then, in order to describe the physical system via the 'Quantal Newtonian' Second Law for each electron, we define the various quantal sources and the 'classical' fields they give rise to. The (non-conserved) total energy and its components are defined both in terms of the quantal sources as well as in integral virial form in terms of the fields. The 'Quantal Newtonian' Second Law, which is the equation of motion for the individual electron amongst all the electrons present, is then stated and explained. It is comprised of the external field and different internal fields as experienced by each electron, and a field representative of the response of the electron to these fields. The parallelism with Newton's Second Law for an individual classical particle in an external force is noted. The derivation of the 'Quantal Newtonian' Second Law is provided in Appendix A for the more general case of electrons in an external time-dependent electromagnetic field. The Law as described in this chapter constitutes a special case. The various internal fields experienced by each electron are described. It is proved that on summing the 'Quantal Newtonian' Second Law over all the electrons, each internal field component separately vanishes, thus providing a new derivation of Ehrenfest's theorem. The 'Quantal Newtonian' Second Law also leads to the integral virial theorem written in terms of the various fields. The stationary-state Schrödinger theory of electrons is described via the 'Quantal Newtonian' First Law, which constitutes a special case of the Second Law. Finally, Quantum Fluid Dynamics, in which the electrons are treated as a classical liquid, is contrasted with the quantum-source—field description of Schrödinger theory via the 'Quantal Newtonian' Laws.

2.1 Time-Dependent Schrödinger Theory

Consider a system of N electrons in the presence of a time-dependent (TD) conservative external field $\mathcal{F}^{\mathrm{ext}}(\mathbf{r}t)$ such that $\mathcal{F}^{\mathrm{ext}}(\mathbf{r}t) = -\boldsymbol{\nabla} v(\mathbf{r}t)/e$, where $v(\mathbf{r}t)$ is the scalar potential energy of an electron. The TD Schrödinger equation in the Born-Oppenheimer approximation [18] is (in atomic units: $e = \hbar = m = 1$)

$$\hat{H}(t)\Psi(\mathbf{X}t) = i\frac{\partial\Psi(\mathbf{X}t)}{\partial t}, \tag{2.1}$$

where $\Psi(\mathbf{Xt})$ is the wavefunction, $\mathbf{X} = \mathbf{x}_1, \mathbf{x}_2, \ldots, \mathbf{x}_N$, $\mathbf{x} = \mathbf{r}\sigma$, $\mathbf{r}$ and σ are the spatial and spin coordinates. The Hamiltonian operator $\hat{H}(t)$ is a sum of the kinetic energy $\hat{T}$, external potential energy $\hat{V}(t)$, and electron-interaction potential energy $\hat{U}$ operators:

$$\hat{H}(t) = \hat{T} + \hat{V}(t) + \hat{U}, \tag{2.2}$$

where

$$\hat{T} = -\frac{1}{2}\sum_i \nabla_i^2, \tag{2.3}$$

$$\hat{V}(t) = \sum_i v(\mathbf{r}_i t), \tag{2.4}$$

and

$$\hat{U} = \frac{1}{2}\sum_{i,j}{}' \frac{1}{|\mathbf{r}_i - \mathbf{r}_j|}. \tag{2.5}$$

As electrons are fermions, the wave function $\Psi(\mathbf{X}t)$ is antisymmetric in an interchange of the coordinates of the particles including spin, and thus accounts for electron correlations due to the Pauli principle. Due to the electron-interaction term in the Hamiltonian, the wave function also accounts for correlations due to Coulomb repulsion. In the Schrödinger equation as written above, the Hamiltonian is *assumed* known. Also *implicit* in the writing of the Hamiltonian is the fact that the external potential energy function $v(\mathbf{r}t)$ is *path-independent* at each instant of time. (In Chap. 3 it will be proved that **(a)** the potential $v(\mathbf{r}t)$ is directly related to the properties of the quantum system; **(b)** that $v(\mathbf{r}t)$ is a functional of the wave function $\Psi(\mathbf{X}t)$, *i.e.* $v(\mathbf{r}t) = v[\Psi(\mathbf{X}t)](\mathbf{r}t)$; **(c)** that the functional $v[\Psi(\mathbf{X}t)](\mathbf{r}t)$ is *universal*; and **(d)** that the functional $v[\Psi(\mathbf{X}t)](\mathbf{r}t)$ is *path-independent.*)

In quantum mechanics, properties of a system are determined in terms of the position probability density, or equivalently as expectation values of the corresponding operators taken with respect to the wavefunction. These expectations are functions of time since the wavefunction depends upon time, and the spatial and spin coordinates are integrated out. Thus, with $\Psi(\mathbf{X}t) = \Psi(t)$, the (non conserved) energy $E(t)$ is the expectation

$$E(t) = \langle \Psi(t) \mid i\frac{\partial}{\partial t} \mid \Psi(t)\rangle = \langle \Psi(t) \mid \hat{H}(t) \mid \Psi(t)\rangle. \tag{2.6}$$

The energy in turn may be written in terms of its kinetic $T(t)$, external potential $E_{\text{ext}}(t)$, and electron-interaction potential $E_{\text{ee}}(t)$ energy components:

$$E(t) = T(t) + E_{\text{ext}}(t) + E_{\text{ee}}(t), \tag{2.7}$$

where

$$T(t) = \langle \Psi(t) \mid \hat{T} \mid \Psi(t)\rangle, \tag{2.8}$$

$$E_{\text{ext}}(t) = \langle \Psi(t) \mid \hat{V}(t) \mid \Psi(t)\rangle, \tag{2.9}$$

and

$$E_{\text{ee}}(t) = \langle \Psi(t) \mid \hat{U} \mid \Psi(t)\rangle. \tag{2.10}$$

The Schrödinger equation (2.1) is a linear, homogeneous, differential equation that is first-order in time. The wave function (or probability amplitude) $\Psi(\mathbf{X}t)$ of the quantum-mechanical system, with initial condition $\Psi(\mathbf{X}t_0)$, describes the evolution of the N electron system. However, the same physical system can also be described from the perspective of the *individual electron* in the sea of electrons. The corresponding equation of motion for the electron is the 'Quantal Newtonian' Second Law. This 'Newtonian' perspective is in terms of 'classical' fields that pervade all space, and which arise from sources that are quantum-mechanical in nature. These quantal sources are expectation values of Hermitian operators taken with respect to the wave function $\Psi(\mathbf{X}t)$. The fields obtained from these sources are separately representative of the kinetic, external, and electron-interaction components of the physical system. Thus, with each property is associated a 'classical' field.

We next define the various quantal sources, their physical interpretation, and the sum rules satisfied by them.

2.2 Definitions of Quantal Sources

In this section we define the quantum-mechanical sources of the fields intrinsic to the system. These sources are the electronic density $\rho(\mathbf{r}t)$, the spinless single-particle density matrix $\gamma(\mathbf{r}\mathbf{r}'t)$, the pair-correlation density $g(\mathbf{r}\mathbf{r}'t)$ and from it the Fermi-Coulomb hole charge distribution $\rho_{xc}(\mathbf{r}\mathbf{r}'t)$, and the current density $\mathbf{j}(\mathbf{r}t)$. The current density may also be expressed in terms of the density matrix. The sources, defined as expectations of Hermitian operators, are also provided a physical interpretation.

2.2.1 *Electron Density* $\rho(\mathbf{r}t)$

The electron density $\rho(\mathbf{r}t)$ is the expectation value

$$\rho(\mathbf{r}t) = \langle \Psi(t) \mid \hat{\rho}(\mathbf{r}) \mid \Psi(t) \rangle, \tag{2.11}$$

where the Hermitian density operator $\hat{\rho}(\mathbf{r})$ is defined as

$$\hat{\rho}(\mathbf{r}) = \sum_i \delta(\mathbf{r} - \mathbf{r}_i). \tag{2.12}$$

The expectation value of (2.11) may also be written in terms of the wavefunction $\Psi(\mathbf{X}t)$ as

$$\rho(\mathbf{r}t) = N \sum_\sigma \int \Psi^* \left(\mathbf{r}\sigma, \mathbf{X}^{N-1}, t\right) \Psi \left(\mathbf{r}\sigma, \mathbf{X}^{N-1}, t\right) d\mathbf{X}^{N-1}, \tag{2.13}$$

where $\mathbf{X}^{N-1} = \mathbf{x}_2, \mathbf{x}_3, \ldots, \mathbf{x}_N$, $d\mathbf{X}^{N-1} = d\mathbf{x}_2, \ldots, d\mathbf{x}_N$, and $\int d\mathbf{x} = \sum_\sigma \int d\mathbf{r}$. Thus, the electron density $\rho(\mathbf{r}t)$ is N times the probability of an electron being at $\mathbf{r}$ at time t.

The sum rule satisfied by the electron density is

$$\int \rho(\mathbf{r}t) d\mathbf{r} = N. \tag{2.14}$$

Thus, integration of the density at each instant of time gives the electron number N.

The electron density is a *local* or *static* charge distribution in that its structure remains unchanged as a function of electron position at each instant of time.

2.2.2 *Spinless Single-Particle Density Matrix* $\gamma(\mathbf{rr}'t)$

The spinless single-particle density matrix $\gamma(\mathbf{rr}'t)$ is defined as the expectation value of the complex density matrix operator $\hat{\gamma}(\mathbf{rr}')$ [19, 20]:

$$\gamma\left(\mathbf{rr}'t\right) = \langle \Psi(t) \mid \hat{\gamma}\left(\mathbf{rr}'\right) \mid \Psi(t)\rangle, \tag{2.15}$$

$$\hat{\gamma}\left(\mathbf{rr}'\right) = \hat{A} + \mathrm{i}\hat{B}, \tag{2.16}$$

$$\hat{A} = \frac{1}{2}\sum_j \left[\delta\left(\mathbf{r}_j - \mathbf{r}\right) T_j(\mathbf{a}) + \delta\left(\mathbf{r}_j - \mathbf{r}'\right) T_j(-\mathbf{a})\right], \tag{2.17}$$

$$\hat{B} = -\frac{i}{2}\sum_j \left[\delta\left(\mathbf{r}_j - \mathbf{r}\right) T_j(\mathbf{a}) - \delta\left(\mathbf{r}_j - \mathbf{r}'\right) T_j(-\mathbf{a})\right], \tag{2.18}$$

$T_j(\mathbf{a})$ is a translation operator such that $T_j(\mathbf{a})\Psi(\ldots,\mathbf{r}_j,\ldots,t) = \Psi(\ldots,\mathbf{r}_j + \mathbf{a},\ldots,t)$, and $\mathbf{a} = \mathbf{r}' - \mathbf{r}$. The real $\hat{A}$ and imaginary $\hat{B}$ components of the complex operator $\gamma(\mathbf{rr}'t)$ are each Hermitian.

The expectation value of (2.15) for $\gamma(\mathbf{rr}'t)$ may then be expressed in terms of the wavefunction $\Psi(\mathbf{X}t)$ as

$$\gamma(\mathbf{rr}'t) = N\sum_\sigma \int \Psi^*\left(\mathbf{r}\sigma, \mathbf{X}^{N-1}, t\right) \Psi\left(\mathbf{r}'\sigma, \mathbf{X}^{N-1}, t\right) d\mathbf{X}^{N-1}. \tag{2.19}$$

To prove the equivalence, we note that

$$\langle \hat{A}\rangle = \langle \Psi(t)|\hat{A}|\Psi(t)\rangle = \frac{1}{2}\left[\gamma(\mathbf{rr}'t) + \gamma(\mathbf{r}'\mathbf{r}t)\right], \tag{2.20}$$

and since the density matrix is 'Hermitian':

$$\gamma(\mathbf{r}'\mathbf{r}t) = \gamma^\star(\mathbf{rr}'t), \tag{2.21}$$

we have

$$\langle \hat{A}\rangle = \Re\gamma(\mathbf{rr}'t). \tag{2.22}$$

Similarly

$$\langle \hat{B} \rangle = -\frac{i}{2}\left[\gamma(\mathbf{r}\mathbf{r}'t) - \gamma(\mathbf{r}'\mathbf{r}t)\right] \tag{2.23}$$

$$= \Im\gamma(\mathbf{r}\mathbf{r}'t). \tag{2.24}$$

Thus, the single-particle density matrix is the expectation value of a complex operator whose real and imaginary parts are Hermitian operators.

The density matrix $\gamma(\mathbf{r}\mathbf{r}'t)$ is a *nonlocal* or *dynamic* source as it depends on the electron positions at both $\mathbf{r}$ and $\mathbf{r}'$. At each instant of time its structure changes with a change in $\mathbf{r}$ and $\mathbf{r}'$.

The sum rule satisfied by the density matrix $\gamma(\mathbf{r}\mathbf{r}'t)$ is

$$\int \gamma\left(\mathbf{r}\mathbf{r}''t\right)\gamma\left(\mathbf{r}''\mathbf{r}'t\right)d\mathbf{r}'' < \gamma\left(\mathbf{r}\mathbf{r}'t\right). \tag{2.25}$$

As a consequence of the inequality, the density matrix is said to be *non-idempotent*. (This is in contrast to the Dirac density matrix which is the expectation value of the density matrix operator $\hat{\gamma}(\mathbf{r}\mathbf{r}'t)$ taken with respect to a Slater determinant of spin-orbitals.)

The diagonal matrix element of the density matrix is the density: $\gamma(\mathbf{r}\mathbf{r}t) = \rho(\mathbf{r}t)$.

2.2.3 *Pair-Correlation Density* $g(\mathbf{r}\mathbf{r}'t)$, *And Fermi-Coulomb Hole* $\rho_{xc}(\mathbf{r}\mathbf{r}'t)$

The pair-correlation density $g(\mathbf{r}\mathbf{r}'t)$ is a property representative of electron correlations due to the Pauli principle and Coulomb repulsion. At each instant of time, it is the conditional density at $\mathbf{r}'$ of all the other electrons, given that one electron is at $\mathbf{r}$. It is defined as the ratio of the expectations of two Hermitian operators:

$$g(\mathbf{r}\mathbf{r}'t) = \frac{P(\mathbf{r}\mathbf{r}'t)}{\rho(\mathbf{r}t)}, \tag{2.26}$$

with the pair function $P(\mathbf{r}\mathbf{r}'t)$ being the expectation

$$P(\mathbf{r}\mathbf{r}'t) = \langle \Psi(t)|\hat{P}(\mathbf{r}\mathbf{r}')|\Psi(t)\rangle, \tag{2.27}$$

where $\hat{P}(\mathbf{r}\mathbf{r}')$ is the Hermitian pair-correlation operator

$$\hat{P}(\mathbf{r}\mathbf{r}') = \sum_{i,j}{}' \delta(\mathbf{r}_i - \mathbf{r})\delta(\mathbf{r}_j - \mathbf{r}'). \tag{2.28}$$

The pair function $P(\mathbf{r}\mathbf{r}'t)$ is the probability of simultaneously finding electrons at $\mathbf{r}$ and $\mathbf{r}'$ at time t.

The sum rule satisfied by the pair-correlation density for each electron position **r** at time t is

$$\int g(\mathbf{r}\mathbf{r}'t)d\mathbf{r}' = N - 1. \tag{2.29}$$

To prove the sum rule of (2.29) we rewrite the pair function $P(\mathbf{r}\mathbf{r}'t)$ as

$$P(\mathbf{r}\mathbf{r}'t) = \langle \Psi(t)|\sum_{i,j} \delta(\mathbf{r}_i - \mathbf{r})\delta(\mathbf{r}_j - \mathbf{r}')|\Psi(t)\rangle$$

$$-\langle \Psi(t)|\sum_{i} \delta(\mathbf{r}_i - \mathbf{r})\delta(\mathbf{r}_i - \mathbf{r}')|\Psi(t)\rangle \tag{2.30}$$

$$= \langle \Psi(t)|\sum_{i} \delta(\mathbf{r}_i - \mathbf{r})\sum_{j} \delta(\mathbf{r}_j - \mathbf{r}')|\Psi(t)\rangle - \delta(\mathbf{r} - \mathbf{r}')\rho(\mathbf{r}t). \tag{2.31}$$

On integrating:

$$\int P(\mathbf{r}\mathbf{r}'t)d\mathbf{r}' = \langle \Psi(t)|\sum_{i} \delta(\mathbf{r}_i - \mathbf{r})\sum_{j} \int \delta(\mathbf{r}_j - \mathbf{r}')d\mathbf{r}'|\Psi(t)\rangle$$

$$-\rho(\mathbf{r}t)\int \delta(\mathbf{r} - \mathbf{r}')d\mathbf{r}' \tag{2.32}$$

$$= N\rho(\mathbf{r}t) - \rho(\mathbf{r}t), \tag{2.33}$$

so that

$$\frac{1}{\rho(\mathbf{r}t)}\int P(\mathbf{r}\mathbf{r}'t)d\mathbf{r}' = N - 1. \tag{2.34}$$

The pair-correlation density is a *nonlocal* or *dynamic* charge distribution in that its structure changes as a function of electron position for nonuniform density systems. If there were no electron correlations, the density at $\mathbf{r}'$ would simply be $\rho(\mathbf{r}'t)$. However, due to electron correlations—the keeping apart of electrons—there is a reduction in the density at $\mathbf{r}'$. Hence, the pair-correlation density is the density $\rho(\mathbf{r}'t)$ at $\mathbf{r}'$ plus the reduction in this density at $\mathbf{r}'$ due to the electron correlations. The reduction in density about an electron which occurs as a result of the Pauli principle and Coulomb repulsion is the Fermi-Coulomb hole charge distribution $\rho_{xc}(\mathbf{r}\mathbf{r}'t)$. Thus, we may write the pair-correlation density as

$$g(\mathbf{r}\mathbf{r}'t) = \rho(\mathbf{r}'t) + \rho_{xc}(\mathbf{r}\mathbf{r}'t). \tag{2.35}$$

In this manner, the pair density is separated into its *local* and *nonlocal* components. Further, as a consequence, the total charge of the Fermi-Coulomb hole, for arbitrary electron position at **r**, is

$$\int \rho_{xc}(\mathbf{r}\mathbf{r}'t)d\mathbf{r}' = -1. \tag{2.36}$$

Note that there is no self-interaction in the pair-correlation density. This is evident from its definition (2.26). In its definition of (2.35), the self-interaction contribution to the Fermi-Coulomb hole charge is canceled by the corresponding term of the density.

An associated property is the pair-correlation function $h(\mathbf{r}\mathbf{r}'t)$ defined as

$$h(\mathbf{r}\mathbf{r}'t) = \frac{g(\mathbf{r}\mathbf{r}'t)}{\rho(\mathbf{r}'t)}, \tag{2.37}$$

which is symmetrical in an interchange of $\mathbf{r}$ and $\mathbf{r}'$:

$$h(\mathbf{r}\mathbf{r}'t) = h(\mathbf{r}'\mathbf{r}t). \tag{2.38}$$

This property of symmetry of the pair function is of value in various proofs to follow.

2.2.4 *Current Density* $\mathbf{j}(\mathbf{r}t)$

The current density $\mathbf{j}(\mathbf{r}t)$ at point $\mathbf{r}$ and at time t is defined as the expectation value

$$\mathbf{j}(\mathbf{r}t) = \langle \Psi(t) \mid \hat{\mathbf{j}}(\mathbf{r}) \mid \Psi(t) \rangle, \tag{2.39}$$

where $\hat{\mathbf{j}}(\mathbf{r})$ is the Hermitian current density operator:

$$\hat{\mathbf{j}}(\mathbf{r}) = \frac{1}{2i} \sum_j \left[\nabla_{\mathbf{r}_j} \delta\left(\mathbf{r}_j - \mathbf{r}\right) + \delta(\mathbf{r}_j - \mathbf{r}) \nabla_{\mathbf{r}_j} \right]. \tag{2.40}$$

The current density $\mathbf{j}(\mathbf{r})$ may also be written in terms of the canonical momentum operator $\hat{\mathbf{p}} = \frac{1}{i}\nabla$ as

$$\mathbf{j}(\mathbf{r}t) = N \mathrm{Re} \sum_\sigma \int \Psi^* \left(\mathbf{r}\sigma, \mathbf{X}^{N-1}, t\right) \frac{1}{i} \nabla \Psi \left(\mathbf{r}\sigma, \mathbf{X}^{N-1}, t\right) d\mathbf{X}^{N-1}. \tag{2.41}$$

It may also be expressed in terms of the single-particle density matrix $\gamma(\mathbf{r}\mathbf{r}'t)$ nonlocal source as

$$\mathbf{j}(\mathbf{r}t) = \frac{i}{2} \left[\nabla' - \nabla''\right] \gamma\left(\mathbf{r}'\mathbf{r}''t\right) |_{\mathbf{r}'=\mathbf{r}''=\mathbf{r}}. \tag{2.42}$$

The quantal sources defined above then give rise to 'classical' fields that pervade all space. These fields are defined below.

2.3 Definitions of 'Classical' Fields

The different fields associated with the quantum system defined by (2.1) are the electron-interaction $\mathcal{E}_{\rm ee}(\mathbf{r}t)$ field which is a sum of the Hartree $\mathcal{E}_{\rm H}(\mathbf{r}t)$ and Pauli-Coulomb $\mathcal{E}_{\rm xc}(\mathbf{r}t)$ fields, the differential density $\mathcal{D}(\mathbf{r}t)$, kinetic $\mathcal{Z}(\mathbf{r}t)$, and current-density $\mathcal{J}(\mathbf{r}t)$ fields.

2.3.1 Electron-Interaction Field $\mathcal{E}_{\rm ee}(\mathbf{r}t)$

The electron-interaction field $\mathcal{E}_{\rm ee}(\mathbf{r}t)$ is representative of electron correlations due to the Pauli principle and Coulomb repulsion. The quantal source of this field is the pair-correlation density $g(\mathbf{r}\mathbf{r}'t)$. It is obtained from this nonlocal charge distribution via Coulomb's law as

$$\mathcal{E}_{\rm ee}(\mathbf{r}t) = \int \frac{g(\mathbf{r}\mathbf{r}'t)(\mathbf{r}-\mathbf{r}')}{\mid \mathbf{r}-\mathbf{r}' \mid^3} d\mathbf{r}'. \tag{2.43}$$

The field $\mathcal{E}_{\rm ee}(\mathbf{r})$ may be rewritten in terms of an electron-interaction 'force' $\mathbf{e}_{ee}(\mathbf{r})$ and the density $\rho(\mathbf{r}t)$ as ('force'/charge)

$$\mathcal{E}_{\rm ee}(\mathbf{r}t) = \frac{\mathbf{e}_{ee}(\mathbf{r}t)}{\rho(\mathbf{r}t)}, \tag{2.44}$$

where $\mathbf{e}_{ee}(\mathbf{r}t)$ is obtained via Coulomb's law from the pair function $P(\mathbf{r}\mathbf{r}'t)$:

$$\mathbf{e}_{ee}(\mathbf{r}t) = \int \frac{P(\mathbf{r}\mathbf{r}'t)(\mathbf{r}-\mathbf{r}')}{|\mathbf{r}-\mathbf{r}'|^3} d\mathbf{r}'. \tag{2.45}$$

(The quantal source of the field $\mathcal{E}_{\rm ee}(\mathbf{r}t)$ can thus also be thought of as being the pair function $P(\mathbf{r}\mathbf{r}'t)$.) With the pair-correlation density expressed as in (2.35), the field $\mathcal{E}_{\rm ee}(\mathbf{r}t)$ may be written as a sum of its Hartree $\mathcal{E}_{\rm H}(\mathbf{r}t)$ and Pauli-Coulomb $\mathcal{E}_{\rm xc}(\mathbf{r}t)$ components as

$$\mathcal{E}_{\rm ee}(\mathbf{r}t) = \mathcal{E}_{\rm H}(\mathbf{r}t) + \mathcal{E}_{\rm xc}(\mathbf{r}t), \tag{2.46}$$

where

$$\mathcal{E}_{\rm H}(\mathbf{r}t) = \int \frac{\rho(\mathbf{r}'t)(\mathbf{r}-\mathbf{r}')}{\mid \mathbf{r}-\mathbf{r}' \mid^3} d\mathbf{r}', \tag{2.47}$$

and

$$\mathcal{E}_{\rm xc}(\mathbf{r}t) = \int \frac{\rho_{\rm xc}(\mathbf{r}\mathbf{r}'t)(\mathbf{r}-\mathbf{r}')}{\mid \mathbf{r}-\mathbf{r}' \mid^3} d\mathbf{r}'. \tag{2.48}$$

The Hartree field $\mathcal{E}_{\mathrm{H}}(\mathbf{r}t)$ is conservative as its source is a local charge distribution $\rho(\mathbf{r}t)$, so that $\nabla \times \mathcal{E}_{\mathrm{H}}(\mathbf{r}t) = 0$. In general, nonlocal sources such as the pair-correlation density and Fermi-Coulomb hole charge do not lead to conservative fields. Thus, the fields $\mathcal{E}_{\mathrm{ee}}(\mathbf{r}t)$ and $\mathcal{E}_{\mathrm{xc}}(\mathbf{r}t)$ are *in general* not conservative, *i.e.* $\nabla \times \mathcal{E}_{\mathrm{ee}}(\mathbf{r}t) \neq 0$ and $\nabla \times \mathcal{E}_{\mathrm{xc}}(\mathbf{r}t) \neq 0$.

2.3.2 *Differential Density Field* $\mathcal{D}(\mathbf{r}t)$

The differential density field $\mathcal{D}(\mathbf{r}t)$ is defined as ('force'/charge)

$$\mathcal{D}(\mathbf{r}t) = \frac{\mathbf{d}(\mathbf{r}t)}{\rho(\mathbf{r}t)}, \tag{2.49}$$

where the differential density 'force'

$$\mathbf{d}(\mathbf{r}t) = -\frac{1}{4}\nabla\nabla^2\rho(\mathbf{r}t). \tag{2.50}$$

This field also arises from a local source, the electronic density $\rho(\mathbf{r}t)$, so that it too is conservative, and $\nabla \times \mathcal{D}(\mathbf{r}t) = 0$. The vanishing of the curl of the 'force' $\mathbf{d}(\mathbf{r}t)$ is evident since the curl of the gradient of a scalar function vanishes.

2.3.3 *Kinetic Field* $\mathcal{Z}(\mathbf{r}t)$

The kinetic field $\mathcal{Z}(\mathbf{r}t)$ is so named because the kinetic energy density, and hence, the kinetic energy may be obtained from it. The field, whose source is the nonlocal single-particle density matrix $\gamma(\mathbf{r}\mathbf{r}'t)$, is defined as ('force'/charge)

$$\mathcal{Z}(\mathbf{r}t) = \frac{z(\mathbf{r}t; [\gamma])}{\rho(\mathbf{r}t)}, \tag{2.51}$$

where the kinetic 'force' $z(\mathbf{r}t)$ is defined in Cartesian coordinates by its component $z_\alpha(\mathbf{r}t)$ as

$$z_\alpha(\mathbf{r}t) = 2\sum_\beta \frac{\partial}{\partial r_\beta} t_{\alpha\beta}(\mathbf{r}t), \tag{2.52}$$

and where $t_{\alpha\beta}(\mathbf{r}t)$ is the second-rank kinetic-energy tensor defined in turn as

$$t_{\alpha\beta}(\mathbf{r}t) = \frac{1}{4}\left[\frac{\partial^2}{\partial r'_\alpha \partial r''_\beta} + \frac{\partial^2}{\partial r'_\beta \partial r''_\alpha}\right]\gamma(\mathbf{r}'\mathbf{r}''t)|_{\mathbf{r}'=\mathbf{r}''=\mathbf{r}}. \tag{2.53}$$

Note that the tensor $t_{\alpha\beta}(\mathbf{r}t)$, and hence the kinetic 'force' $z(\mathbf{r}t)$ and field $\mathcal{Z}(\mathbf{r}t)$, depend on both the diagonal and off-diagonal elements of the density matrix $\gamma(\mathbf{r}\mathbf{r}'t)$.

The quantal source of the field $\mathcal{Z}(\mathbf{r}t)$ which is $\gamma(\mathbf{r}\mathbf{r}'t)$ is nonlocal. Thus, in general the field $\mathcal{Z}(\mathbf{r}t)$ is not conservative, and $\nabla \times \mathcal{Z}(\mathbf{r}t) \neq 0$.

2.3.4 Current Density Field $\mathcal{J}(\mathbf{r}t)$

The current density field $\mathcal{J}(\mathbf{r}t)$, whose source is the nonlocal single particle density matrix $\gamma(\mathbf{r}\mathbf{r}'t)$, is defined as

$$\mathcal{J}(\mathbf{r}t) = \frac{1}{\rho(\mathbf{r}t)}\frac{\partial}{\partial t}\mathbf{j}(\mathbf{r}t), \tag{2.54}$$

where $\mathbf{j}(\mathbf{r}t)$ is the current density. As the source of this field is nonlocal, in general, this field too is nonconservative so that $\nabla \times \mathcal{J}(\mathbf{r}t) \neq 0$.

The fields $\mathcal{E}_{ee}(\mathbf{r}t)$, $\mathcal{E}_{xc}(\mathbf{r}t)$, $\mathcal{Z}(\mathbf{r}t)$, and $\mathcal{J}(\mathbf{r}t)$ are 'classical' in that they obey a classical equation of motion—the 'Quantal Newtonian' Second Law—as described in Sect. 2.4. They are also *in general* not conservative. However, their sum always is, so that

$$\nabla \times [\mathcal{E}_{ee}(\mathbf{r}t) + \mathcal{Z}(\mathbf{r}t) + \mathcal{J}(\mathbf{r}t)] = 0. \tag{2.55}$$

If the system in the presence of the time-dependent external field $\mathcal{F}^{ext}(\mathbf{r}t)$ has a symmetry which reduces these fields to being one dimensional, or when such a symmetry is imposed as by application of the central field approximation, the individual fields are then separately conservative. In such cases

$$\nabla \times \mathcal{E}_{ee}(\mathbf{r}t) = 0, \tag{2.56}$$

$$\nabla \times \mathcal{Z}(\mathbf{r}t) = 0, \tag{2.57}$$

$$\nabla \times \mathcal{J}(\mathbf{r}t) = 0. \tag{2.58}$$

The central field approximation can be achieved by spherically averaging the fields.

2.4 Energy Components in Terms of Quantal Sources and Fields

The kinetic $T(t)$, external $E_{ext}(t)$, and electron-interaction $E_{ee}(t)$ energies as defined by the expectations of (2.8)–(2.10), may be expressed directly in terms of the quantal sources, and also in integral virial form in terms of the respective fields described above.

2.4.1 Electron-Interaction Potential Energy $E_{ee}(t)$

The electron-interaction energy $E_{ee}(t)$ may be interpreted as the energy of interaction between the density $\rho(\mathbf{r}t)$ and the pair-correlation density $g(\mathbf{r}\mathbf{r}'t)$:

$$E_{ee}(t) = \frac{1}{2}\iint \frac{\rho(\mathbf{r}t)g(\mathbf{r}\mathbf{r}'t)}{|\mathbf{r}-\mathbf{r}'|} d\mathbf{r}d\mathbf{r}'. \tag{2.59}$$

Employing the decomposition of $g(\mathbf{r}\mathbf{r}'t)$ as in (2.35), we may write

$$E_{ee}(t) = E_{\mathrm{H}}(t) + E_{\mathrm{xc}}(t), \tag{2.60}$$

where $E_{\mathrm{H}}(t)$ is the Hartree or Coulomb self-energy:

$$E_H(t) = \frac{1}{2}\iint \frac{\rho(\mathbf{r}t)\rho(\mathbf{r}'t)}{|\mathbf{r}-\mathbf{r}'|} d\mathbf{r}d\mathbf{r}', \tag{2.61}$$

and $E_{\mathrm{xc}}(t)$ the *quantum-mechanical* exchange-correlation—Pauli-Coulomb—energy:

$$E_{\mathrm{xc}}(t) = \frac{1}{2}\iint \frac{\rho(\mathbf{r}t)\rho_{\mathrm{xc}}(\mathbf{r}\mathbf{r}'t)}{|\mathbf{r}-\mathbf{r}'|} d\mathbf{r}d\mathbf{r}'. \tag{2.62}$$

The energy $E_{\mathrm{xc}}(t)$ may in turn be interpreted as the energy of interaction between the density $\rho(\mathbf{r}t)$ and the Fermi-Coulomb hole charge distribution $\rho_{\mathrm{xc}}(\mathbf{r}\mathbf{r}'t)$.

These energy components may also be expressed in terms of the fields as follows. Since

$$\frac{1}{|\mathbf{r}-\mathbf{r}'|} = \frac{(\mathbf{r}-\mathbf{r}')\cdot(\mathbf{r}-\mathbf{r}')}{|\mathbf{r}-\mathbf{r}'|^3} = \frac{\mathbf{r}\cdot(\mathbf{r}-\mathbf{r}') - \mathbf{r}'\cdot(\mathbf{r}-\mathbf{r}')}{|\mathbf{r}-\mathbf{r}'|^3}, \tag{2.63}$$

we may write $E_{ee}(t)$ in terms of the pair-correlation function $h(\mathbf{r}\mathbf{r}'t)$ of (2.37) as

$$E_{ee}(t) = \frac{1}{2}\iint \frac{[\mathbf{r}\cdot(\mathbf{r}-\mathbf{r}') - \mathbf{r}'\cdot(\mathbf{r}-\mathbf{r}')]}{|\mathbf{r}-\mathbf{r}'|^3} \rho(\mathbf{r}t)\rho(\mathbf{r}'t)h(\mathbf{r}\mathbf{r}'t). \tag{2.64}$$

On interchanging $\mathbf{r}$ and $\mathbf{r}'$ in the second term of (2.64) and employing the symmetry property of $h(\mathbf{r}\mathbf{r}'t)$, we see that it is the same as the first, so that

$$\begin{aligned} E_{ee}(t) &= \iint \frac{\mathbf{r}\cdot(\mathbf{r}-\mathbf{r}')\rho(\mathbf{r}t)g(\mathbf{r}\mathbf{r}'t)}{|\mathbf{r}-\mathbf{r}'|^3} d\mathbf{r}d\mathbf{r}' \\ &= \int \rho(\mathbf{r}t)\mathbf{r}\cdot\left[\int \frac{g(\mathbf{r}\mathbf{r}'t)(\mathbf{r}-\mathbf{r}')}{|\mathbf{r}-\mathbf{r}'|^3} d\mathbf{r}'\right] d\mathbf{r} \\ &= \int \rho(\mathbf{r}t)\mathbf{r}\cdot\boldsymbol{\mathcal{E}}_{ee}(\mathbf{r}t)d\mathbf{r}. \end{aligned} \tag{2.65}$$

Employing the decomposition of $\mathcal{E}_{\text{ee}}(\mathbf{r}t)$ of (2.46), we then have

$$E_{\text{H}}(t) = \int \rho(\mathbf{r}t)\mathbf{r} \cdot \mathcal{E}_{\text{H}}(\mathbf{r}t)d\mathbf{r}, \tag{2.66}$$

and

$$E_{\text{xc}}(t) = \int \rho(\mathbf{r}t)\mathbf{r} \cdot \mathcal{E}_{\text{xc}}(\mathbf{r}t)d\mathbf{r}. \tag{2.67}$$

Note that the expressions for the energy components in terms of the fields is *independent* of whether or not the fields are conservative. These expressions are in integral virial form.

2.4.2 *Kinetic Energy $T(t)$*

The kinetic energy $T(t)$ may be written in terms of its quantal source, the single-particle density matrix $\gamma(\mathbf{r}\mathbf{r}'t)$ as

$$T(t) = \int t(\mathbf{r}t)d\mathbf{r}, \tag{2.68}$$

where $t(\mathbf{r}t)$ is the kinetic energy density. There are two traditional expressions for $t(\mathbf{r}t)$ as expressed in terms of the source $\gamma(\mathbf{r}\mathbf{r}'t)$. The first $t_A(\mathbf{r}t)$ corresponds to the integrand of the kinetic energy expectation value of (2.8) and is

$$t_A(\mathbf{r}t) = -\frac{1}{2}\nabla_{\mathbf{r}}^2\gamma(\mathbf{r}\mathbf{r}'t)\,|_{\mathbf{r}'=\mathbf{r}}\,. \tag{2.69}$$

The second $t_B(\mathbf{r}t)$ corresponds to the trace of the kinetic energy density tensor $t_{\alpha\beta}(\mathbf{r}t)$ of (2.53) and is

$$t_B(\mathbf{r}t) = \sum_{\alpha} t_{\alpha\alpha}(\mathbf{r}t) = \frac{1}{2}\boldsymbol{\nabla}_{\mathbf{r}'} \cdot \boldsymbol{\nabla}_{\mathbf{r}''}\gamma(\mathbf{r}'\mathbf{r}''t)\,|_{\mathbf{r}'=\mathbf{r}''=\mathbf{r}}\,. \tag{2.70}$$

Although both definitions of the kinetic energy density integrate to the same kinetic energy T, the functions differ:

$$t_B(\mathbf{r}t) - t_A(\mathbf{r}t) = \frac{1}{4}\nabla^2\rho(\mathbf{r}t). \tag{2.71}$$

The kinetic energy $T(t)$ may also be expressed in terms of the kinetic field $\mathcal{Z}(\mathbf{r}t)$ as

$$T(t) = -\frac{1}{2}\int \rho(\mathbf{r}t)\mathbf{r} \cdot \mathcal{Z}(\mathbf{r}t)d\mathbf{r}, \tag{2.72}$$

or in terms of the kinetic 'force' $z(\mathbf{r}t)$

$$T(t) = -\frac{1}{2}\int \mathbf{r} \cdot z(\mathbf{r}t) d\mathbf{r}. \tag{2.73}$$

Equation (2.72) can be shown to be equivalent to (2.68) by partial integration and by employing the fact that the wavefunction and hence the single-particle density matrix vanishes as $\mathbf{r}$, $\mathbf{r}'$ tend towards infinity. The corresponding kinetic energy density $t_C(\mathbf{r}t)$ as expressed in terms of the kinetic field $\mathcal{Z}(\mathbf{r}t)$ or 'force' $\mathbf{z}(\mathbf{r}t)$ is

$$t_C(\mathbf{r}t) = -\frac{1}{2}\rho(\mathbf{r})\mathbf{r} \cdot \mathcal{Z}(\mathbf{r}t) = -\frac{1}{2}\mathbf{r} \cdot \mathbf{z}(\mathbf{r}t). \tag{2.74}$$

The function $t_C(\mathbf{r}t)$ differs from those of $t_A(\mathbf{r}t)$ and $t_B(\mathbf{r}t)$ [21]. Once again, the expression for $T(t)$ in terms of the kinetic field $\mathcal{Z}(\mathbf{r}t)$ is *independent* of whether or not the field is conservative.

2.4.3 External Potential Energy $E_{\mathrm{ext}}(t)$

The external potential energy $E_{\mathrm{ext}}(t)$ may be expressed in terms of the electronic density $\rho(\mathbf{r}t)$ and the potential energy $v(\mathbf{r}t)$ of an electron in the external field $\mathcal{F}^{\mathrm{ext}}(\mathbf{r}t)$ as

$$E_{\mathrm{ext}}(t) = \int \rho(\mathbf{r}t) v(\mathbf{r}t) d\mathbf{r}. \tag{2.75}$$

Through the external potential energy $v(\mathbf{r}t)$, this component of the total energy depends on *all* the fields present in the quantal system. As the quantal sources of these fields are expectations taken with respect to the wave function $\Psi(t)$, the potential energy $v(\mathbf{r}t)$ is a functional of $\Psi(t)$, *i.e.* $v(\mathbf{r}t) = v[\Psi(t)](\mathbf{r}t)$. The explanation of this is arrived at via the 'Quantal Newtonian' Second Law to be enunciated next with further discussion provided in Chap. 3.

2.5 The 'Quantal Newtonian' Second Law

The Schrödinger theory description of a quantum system can alternatively be interpreted in terms of fields representative of the various electron correlations and properties. This description is based on the *pure* state 'Quantal Newtonian' Second Law or time-dependent differential virial theorem [5–7]. (A state is said to be *pure* if it is described by a wavefunction *i.e.* by the solution of (2.1). It is said to be *mixed* if it cannot be so described. A system in a mixed state can be characterized by a probability distribution over all accessible pure states).

As a prelude to the description of this quantal law, let us review the classical mechanics of a system of N particles that obey Newton's Third Law of action and reaction, and exert forces on each other that are equal and opposite, and lie along the line joining them. Then Newton's Second Law for the ith particle is

$$\mathbf{F}_i^{\text{ext}} + \sum_j{}' \mathbf{F}_{ji} = \frac{d}{dt}\mathbf{p}_i, \tag{2.76}$$

where $\mathbf{F}_i^{\text{ext}}$ is the external force, $\mathbf{F}_{ji}$ the internal force on the ith particle due to the jth particle, with the response of the particle to these forces being described by the rate of change of its linear momentum $\mathbf{p}_i$. Summing over all particles, (2.76) reduces to Newton's Second Law for the system of particles:

$$\mathbf{F}^{\text{ext}} = \frac{d^2}{dt^2}\sum_i \mathbf{r}_i, \tag{2.77}$$

where $\mathbf{F}^{\text{ext}} = \sum_i \mathbf{F}_i^{\text{ext}}$ is the total external force. The internal forces corresponding to the term $\sum_{i,j}' \mathbf{F}_{ji}$ vanish as a consequence of Newton's Third Law.

The 'Quantal Newtonian' Second Law is the quantum-mechanical counterpart of the classical equation of motion (2.76) for the individual particles. Its statement is the following:

$$\mathcal{F}^{\text{ext}}(\mathbf{r}t) + \mathcal{F}^{\text{int}}(\mathbf{r}t) = \mathcal{J}(\mathbf{r}t), \tag{2.78}$$

where *each* electron experiences the *external* field $\mathcal{F}^{\text{ext}}(\mathbf{r}t)$:

$$\mathcal{F}^{\text{ext}}(\mathbf{r}t) = -\boldsymbol{\nabla} v(\mathbf{r}t), \tag{2.79}$$

and a field *internal* to the system $\mathcal{F}^{\text{int}}(\mathbf{r}t)$ that is representative of the correlations between the electrons, the density, and the kinetic effects:

$$\mathcal{F}^{\text{int}}(\mathbf{r}t) = \mathcal{E}_{ee}(\mathbf{r}t) - \mathcal{D}(\mathbf{r}t) - \mathcal{Z}(\mathbf{r}t), \tag{2.80}$$

where the component fields $\mathcal{E}_{\text{ee}}(\mathbf{r}t)$, $\mathcal{D}(\mathbf{r}t)$, $\mathcal{Z}(\mathbf{r}t)$ are defined by (2.43, 2.44), (2.49), and (2.51). The response of each electron to the external and internal fields is the current density field $\mathcal{J}(\mathbf{r}t)$ defined by (2.54) which is the quantum analog of the time derivative of $\mathbf{p}_i$ of (2.76).

The law is valid for arbitrary gauge, and is derived so as to satisfy the continuity equation

$$\boldsymbol{\nabla} \cdot \mathbf{j}(\mathbf{r}t) + \frac{\partial \rho(\mathbf{r}t)}{\partial t} = 0, \tag{2.81}$$

which in turn is derived from the Schrödinger equation (2.1) [22].

The manner by which the 'Quantal Newtonian' Second Law is derived is by first writing the wave function in complex Cartesian form as $\Psi(\mathbf{X}t) = \Psi^R(\mathbf{X}t) +$

$i\Psi^I(\mathbf{X}t)$, where $\Psi^R(\mathbf{X}t)$ and $\Psi^I(\mathbf{X}t)$ are its real and imaginary parts. Substitution of this form of the complex wave function into the Schrödinger equation (2.1) then leads to the Law of (2.78). The details of the derivation are given in Appendix A. The proof is for arbitrary $\mathcal{F}^{\text{ext}}(\mathbf{r}t)$, and hence valid for both adiabatic and sudden switching on of the field [22].

The equation of motion for the individual electron, *viz.* the 'Quantal Newtonian' Second Law of (2.78) leads to many insights into the physical system and into the Schrödinger equation itself. Most significantly, one learns that in addition to the external field $\mathcal{F}^{\text{ext}}(\mathbf{r}t)$, each electron is also subject to an internal field $\mathcal{F}^{\text{int}}(\mathbf{r}t)$. Furthermore, this internal field is representative not only of Coulomb correlations between the electrons as one might expect but also of correlations due to the Pauli principle, with both these correlations being represented via the electron-interaction field $\mathcal{E}_{\text{ee}}(\mathbf{r}t)$. Additionally, there is a field due to the motion of the electrons, *viz.* the kinetic field $\mathcal{Z}(\mathbf{r}t)$, and a field representative of the electron density via the differential density field $\mathcal{D}(\mathbf{r}t)$. The response of the electron to the external and internal fields is then described by the current density field $\mathcal{J}(\mathbf{r}t)$. The 'Quantal Newtonian' Second Law thus provides a distinctly different way of describing a quantum system within the framework of Schrödinger theory. The perspective differs from the traditional one in that it is a description of the motion of the *individual electron*.

The other insights arrived at via the 'Quantal Newtonian' Second Law will be described in the sections to follow and in Chap. 3.

(An equation of motion similar [23] to the *pure* state expression (2.78) can be derived for nonequilibrium phenomena described by systems in a time-dependent external field $\mathcal{F}^{\text{ext}}(\mathbf{r}t)$ and finite temperature T. Such systems are described in terms of *a mixed* state, the expectation value of operators being *defined* in terms of the grand canonical ensemble of statistical mechanics. This grand canonical ensemble in turn is defined *at the initial time* in terms of the eigenfunctions and eigenvalues of the time-independent Hamiltonian. The physics underlying this similar equation of motion is intrinsically different since properties such as the density and current density are in terms of statistical averages. Furthermore, the expression in terms of the grand canonical ensemble is valid for sudden switching on of the external field at some initial time.)

2.6 The Internal Field and Ehrenfest's Theorem

The Schrödinger theory analogue of Newton's Second Law of motion is Ehrenhest's theorem [14, 22]. For a system of electrons in some arbitrary time-dependent external field $\mathcal{F}^{\text{ext}}(\mathbf{r}t)$, Ehrenhest's theorem states that the mean value of the field $\langle\mathcal{F}^{\text{ext}}(\mathbf{r})\rangle(t)$ is equal to the second temporal derivative of the average position $\langle\mathbf{r}\rangle(t)$ of the electrons. In order that the average position $\langle\mathbf{r}\rangle(t)$ actually follow Newton's classical equation, one must be able to replace the mean value of the external field $\langle\mathcal{F}^{\text{ext}}(\mathbf{r})\rangle(t)$ by its value $\mathcal{F}^{\text{ext}}(\langle\mathbf{r}\rangle)(t)$. This is the case when either the force vanishes or when it depends linearly on $\mathbf{r}$. The substitution is also justified if the wavefunction remains localized in a small region of space so that the force has a constant value over that region. Thus, Ehrenfest's theorem describes the evolution of the system in terms of its average position as governed by the averaged external field. What Ehrenfest's theorem does *not* describe is the evolution in time of *each* individual electron as the entire system evolves. As described by the 'Quantal Newtonian' Second Law (see Sect. 2.4 and 2.78), in addition to the external force field, *each* electron also experiences an *internal* field $\mathcal{F}^{\text{int}}(\mathbf{r}t)$. It is the sum of these fields that then describes the behavior of the electron and its evolution with time. Furthermore, for Ehrenfest's theorem to be satisfied, the averaged internal field $\langle\mathcal{F}^{\text{int}}(\mathbf{r})\rangle(t)$ must vanish. Similarly, the average torque of the internal field $\langle\mathbf{r}\times\mathcal{F}^{\text{int}}(\mathbf{r})\rangle(t)$ too must vanish. In this section, we draw a rigorous parallel with the equations of classical mechanics by proving that on summing over all electrons, the contribution of the internal field vanishes, thereby leading to Ehrenfest's theorem.

We first derive Ehrenfest's theorem in the traditional manner. The quantum-mechanical equation of motion for the expectation value of an operator $\mathbf{A}(t)$ is [22]

$$\frac{d\langle\mathbf{A}(t)\rangle}{dt}=-i\big\langle[\mathbf{A}(t),\hat{H}(t)]\big\rangle+\left\langle\frac{\partial\mathbf{A}(t)}{\partial t}\right\rangle. \tag{2.82}$$

Substituting the operator

$$\hat{\mathbf{r}}=\int\mathbf{r}\hat{\rho}(\mathbf{r})d\mathbf{r}, \tag{2.83}$$

into the equation of motion (2.82) leads to

$$\frac{d}{dt}\langle\hat{\mathbf{r}}\rangle=\frac{d}{dt}\int\mathbf{r}\rho(\mathbf{r}t)d\mathbf{r}=-i\langle[\hat{\mathbf{r}},\hat{H}(t)]\rangle. \tag{2.84}$$

On differentiating (2.84) again with respect to time and applying the equation of motion to the resulting right hand side, one obtains

$$\frac{d^2}{dt^2}\int\mathbf{r}\rho(\mathbf{r}t)d\mathbf{r}=-\langle[[\hat{\mathbf{r}},\hat{H}(t)],\hat{H}(t)]\rangle, \tag{2.85}$$

since $\partial[\hat{H}(t),\hat{\mathbf{r}}]/\partial t=0$. Evaluating the double commutator leads to Ehrenfest's theorem:

$$\int\rho(\mathbf{r}t)\mathcal{F}^{\text{ext}}(\mathbf{r}t)d\mathbf{r}=\frac{\partial^2}{\partial t^2}\int\mathbf{r}\rho(\mathbf{r}t)d\mathbf{r}. \tag{2.86}$$

This equation is the quantal analogue of Newton's Second Law of motion (2.77).

The quantal analog of Newton's equation of motion for the ith particle is the 'Quantal Newtonian' Second Law of (2.78). When summed over all the electrons, it must lead to Ehrenfest's theorem (2.86), with the contributions of the internal fields vanishing. Thus on operating with $\int d\mathbf{r}\rho(\mathbf{r}t)$ on (2.78) we have

$$\int \rho(\mathbf{r}t)\mathcal{F}^{\text{ext}}(\mathbf{r}t)d\mathbf{r} + \int \rho(\mathbf{r}t)\mathcal{F}^{\text{int}}(\mathbf{r}t)d\mathbf{r} = \int \rho(\mathbf{r}t)\mathcal{J}(\mathbf{r}t)d\mathbf{r}. \tag{2.87}$$

To simplify the right hand side of (2.87), consider the integral

$$\int j_x(\mathbf{r}t)d\mathbf{r} = -\int x\, dj_x\, dy\, dz = -\int x\frac{\partial j_x}{\partial x}dx\, dy\, dz, \tag{2.88}$$

where the second step is a consequence of the vanishing of the current density at the boundaries $x = +\infty, -\infty$. Now, for the same reason

$$\int x\frac{\partial j_y}{\partial y}dx\, dy\, dz = 0 \quad \text{and} \int x\frac{\partial j_z}{\partial z}dxdydz = 0, \tag{2.89}$$

so that

$$\int j_x(\mathbf{r}t)d\mathbf{r} = -\int x\nabla\cdot\mathbf{j}(\mathbf{r}t)d\mathbf{r}. \tag{2.90}$$

Thus,

$$\int \mathbf{j}(\mathbf{r}t)d\mathbf{r} = -\int \mathbf{r}\nabla\cdot\mathbf{j}(\mathbf{r}t)d\mathbf{r}, \tag{2.91}$$

and on employing the continuity equation (2.81) we have the right hand side of (2.87) to be

$$\begin{aligned}\int \rho(\mathbf{r}t)\mathcal{J}(\mathbf{r}t)d\mathbf{r} &= -\frac{\partial}{\partial t}\int \mathbf{r}\nabla\cdot\mathbf{j}(\mathbf{r}t)d\mathbf{r} \\ &= \frac{\partial^2}{\partial t^2}\int \mathbf{r}\rho(\mathbf{r}t)d\mathbf{r}.\end{aligned} \tag{2.92}$$

In order for Ehrenfest's theorem to be satisfied, what remains to be proved is that the average value of each component of $\mathcal{F}^{\text{int}}(\mathbf{r}t)$ of (2.80) vanish:

$$\int \rho(\mathbf{r}t)\mathcal{E}_{\text{ee}}(\mathbf{r}t)d\mathbf{r} = 0, \tag{2.93}$$

$$\int \rho(\mathbf{r}t)\mathcal{D}(\mathbf{r}t)d\mathbf{r} = 0, \tag{2.94}$$

and

$$\int \rho(\mathbf{r}t)\mathcal{Z}(\mathbf{r}t)d\mathbf{r} = 0. \tag{2.95}$$

In order to prove (2.93) we rewrite the left hand side in terms of the pair-correlation function $h(\mathbf{r}\mathbf{r}'t)$ of (2.37):

$$\int \rho(\mathbf{r}t)\mathcal{E}_{\text{ee}}(\mathbf{r}t)d\mathbf{r} = \int \rho(\mathbf{r}t)\rho(\mathbf{r}'t)h(\mathbf{r}\mathbf{r}'t)\frac{(\mathbf{r}-\mathbf{r}')}{|\mathbf{r}-\mathbf{r}'|^3}\,d\mathbf{r}d\mathbf{r}'. \tag{2.96}$$

On interchanging $\mathbf{r}$ and $\mathbf{r}'$, the right hand side of (2.96) is

$$\int \rho(\mathbf{r}t)\rho(\mathbf{r}'t)h(\mathbf{r}'\mathbf{r}t)\frac{(\mathbf{r}'-\mathbf{r})}{|\mathbf{r}-\mathbf{r}'|^3}\,d\mathbf{r}d\mathbf{r}'. \tag{2.97}$$

As $h(\mathbf{r}\mathbf{r}'t)$ is symmetric in an interchange of $\mathbf{r}$ and $\mathbf{r}'$ (see (2.38)), equation (2.96) is

$$\int \rho(\mathbf{r}t)\rho(\mathbf{r}'t)h(\mathbf{r}\mathbf{r}'t)\frac{(\mathbf{r}'-\mathbf{r})}{|\mathbf{r}-\mathbf{r}'|^3}\,d\mathbf{r}d\mathbf{r}' = -\int \rho(\mathbf{r}t)\mathcal{E}_{\text{ee}}(\mathbf{r}t)d\mathbf{r}, \tag{2.98}$$

which proves (2.93). Equation (2.94) follows from partial integration and the vanishing of the density at the boundary at infinity. To prove (2.95) we show that [7]

$$\int z(\mathbf{r}t)d\mathbf{r} = 0. \tag{2.99}$$

Consider the integral for the component

$$\int z_\alpha(\mathbf{r}t)d\mathbf{r} = 2\sum_\beta \int \frac{\partial}{\partial r_\beta}t_{\alpha\beta}(\mathbf{r}t)d\mathbf{r}. \tag{2.100}$$

The integral

$$\int \frac{\partial}{\partial x}t_{\alpha x}(\mathbf{r}t)dx \int dy\,dz = 0, \tag{2.101}$$

etc., since the tensor vanishes at the boundary $x = +\infty, -\infty$. Thus, (2.99) and hence (2.95) is proved.

As a consequence, the averaged internal force vanishes:

$$\int \rho(\mathbf{r}t)\mathcal{F}^{\text{int}}(\mathbf{r}t)d\mathbf{r} = 0, \tag{2.102}$$

and Ehrenfest's theorem is recovered. An alternate way of expressing Ehrenfest's theorem in terms of the response of the system to the external field as represented by the current density field $\mathcal{J}(\mathbf{r}t)$ is

$$\int \rho(\mathbf{r}t)\left[\mathcal{F}^{\text{ext}}(\mathbf{r}t) - \mathcal{J}(\mathbf{r}t)\right] = 0. \tag{2.103}$$

The vanishing of the average of the internal field $\langle\mathcal{F}^{\text{int}}\rangle$ may then be thought of as being a consequence of the quantal analog to Newton's Third Law. Note that although Coulomb's law, and hence the electron interaction field obeys Newton's

Third Law, the vanishing of the averaged differential density and kinetic fields is not a direct consequence of the Third Law.

Returning to Newton's Second Law for the ith particle (2.76), one obtains the total angular momentum $\mathbf{L}$ of the system by performing the cross product $\mathbf{r}_i \times$ on it and summing over all particles to obtain

$$\frac{d\mathbf{L}}{dt} = \mathbf{N}^{\text{ext}}, \tag{2.104}$$

where $\mathbf{L} = \sum_i (\mathbf{r} \times \mathbf{p}_i)$, and $\mathbf{N}^{\text{ext}} = \sum_i (\mathbf{r}_i \times \mathbf{F}_i^{\text{ext}})$ is the torque of the external force about a given point. The torque of the internal forces $\sum_{ij}' \mathbf{r}_i \times \mathbf{F}_{ji}$ once again vanishes as a consequence of Newton's Third Law.

For the quantal equivalent of (2.104), operate by $\int d\mathbf{r}\rho(\mathbf{r}t)\times$ on (2.78) to obtain

$$\int \rho(\mathbf{r}t)\mathbf{r} \times \mathcal{F}^{\text{ext}}(\mathbf{r}t)d\mathbf{r} = \frac{\partial}{\partial t}\int \mathbf{r} \times \mathbf{j}(\mathbf{r}t)d\mathbf{r}, \tag{2.105}$$

where once again it can be proved [7] along the lines described above, that the averaged torques of the individual components of the internal field vanish: $\langle \mathbf{r} \times \mathcal{F}^{\text{int}}(\mathbf{r}t)\rangle = 0$. *Defining* a velocity field $\boldsymbol{\nu}(\mathbf{r}t)$ of the electrons by the equation

$$\mathbf{j}(\mathbf{r}t) = \rho(\mathbf{r}t)\boldsymbol{\nu}(\mathbf{r}t), \tag{2.106}$$

and a momentum field $\mathbf{p}(\mathbf{r}t) = m\boldsymbol{\nu}(\mathbf{r}t)$, we have (with $m = 1$) the quantum analogue of the classical torque equation

$$\int \rho(\mathbf{r}t)\mathcal{N}^{\text{ext}}(\mathbf{r}t)d\mathbf{r} = \frac{\partial}{\partial t}\int \rho(\mathbf{r}t)\mathcal{L}(\mathbf{r}t)d\mathbf{r}, \tag{2.107}$$

where $\mathcal{L}(\mathbf{r}t) = \mathbf{r} \times \mathbf{p}(\mathbf{r}t)$ is the angular momentum field at each instant of time.

Thus, each electron in a sea of electrons, experiences in addition to the external field, an internal field. This internal field defined by (2.80) is representative of the density, the motion of the electrons, and the fact that they are kept apart as a result of the Pauli principle and Coulomb repulsion. As in classical physics, the average of this field and its averaged torque vanish at each instant of time. The 'Quantal Newtonian' Second Law thus provides a deeper insight into Ehrenfest's theorem.

2.7 Integral Virial Theorem

The time-dependent integral virial theorem can be obtained from the 'Quantal Newtonian' Second Law (2.78) by operating on it with $\int d\mathbf{r}\rho(\mathbf{r}t)\mathbf{r}\cdot$ to obtain

$$\int \rho(\mathbf{r}t)\mathbf{r}\cdot\boldsymbol{\mathcal{F}}^{\text{ext}}(\mathbf{r}t)d\mathbf{r} + E_{\text{ee}}(t) + 2T(t) = \int \rho(\mathbf{r}t)\mathbf{r}\cdot\boldsymbol{\mathcal{J}}(\mathbf{r}t)d\mathbf{r}. \tag{2.108}$$

where the energies $E_{\text{ee}}(t)$ and $T(t)$ are given in (2.65) and (2.72), respectively. (The term involving the differential density field $\boldsymbol{\mathcal{D}}(\mathbf{r}t)$ can be shown to vanish by transforming it into a surface integral and by employing the vanishing of the density there at infinity.)

The response component, (the right-hand side), of the virial theorem (2.108) can be written in a manner similar to that of Ehrenfest's theorem as follows. The integral

$$\int \rho(\mathbf{r}t)\mathbf{r}\cdot\boldsymbol{\mathcal{J}}(\mathbf{r}t)d\mathbf{r} = \frac{\partial}{\partial t}\int \mathbf{r}\cdot\mathbf{j}(\mathbf{r}t)d\mathbf{r}. \tag{2.109}$$

Thus, consider the integral

$$\begin{aligned}\int x j_x(\mathbf{r}t)d\mathbf{r} &= \frac{1}{2}\int j_x(\mathbf{r}t)dx^2dydz \\ &= -\frac{1}{2}\int x^2 dj_x(\mathbf{r}t)dydz \\ &= -\frac{1}{2}\int x^2\frac{\partial j_x(\mathbf{r}t)}{\partial x}d\mathbf{r},\end{aligned} \tag{2.110}$$

where we employ the vanishing of the current density $j_x(\mathbf{r}t)$ at the boundaries at $x = +\infty, -\infty$. Now, since for the same reason

$$\int x^2\frac{\partial j_y(\mathbf{r}t)}{\partial y}d\mathbf{r} = 0 \;\text{ and }\; \int x^2\frac{\partial j_z(\mathbf{r}t)}{\partial z}d\mathbf{r} = 0, \tag{2.111}$$

we have

$$\int x j_x(\mathbf{r}t)d\mathbf{r} = -\frac{1}{2}\int x^2\boldsymbol{\nabla}\cdot\mathbf{j}(\mathbf{r}t)d\mathbf{r}. \tag{2.112}$$

Therefore

$$\begin{aligned}\int \mathbf{r}\cdot\mathbf{j}(\mathbf{r}t)d\mathbf{r} &= -\frac{1}{2}\int r^2\boldsymbol{\nabla}\cdot\mathbf{j}(\mathbf{r}t)d\mathbf{r} \\ &= \frac{1}{2}\int r^2\frac{\partial\rho(\mathbf{r}t)}{\partial t}d\mathbf{r},\end{aligned} \tag{2.113}$$

where in the last step we have employed the continuity equation (2.81). Thus,

$$\int \rho(\mathbf{r}t)\mathbf{r}\cdot\boldsymbol{\mathcal{J}}(\mathbf{r}t)d\mathbf{r} = \frac{1}{2}\frac{\partial^2}{\partial t^2}\int r^2\rho(\mathbf{r}t)d\mathbf{r}, \tag{2.114}$$

and the integral virial theorem may alternatively be written as

$$\int \rho(\mathbf{r}t)\mathbf{r}\cdot\mathcal{F}^{\mathrm{ext}}(\mathbf{r}t)d\mathbf{r} + E_{ee}(t) + 2T(t) = \frac{1}{2}\frac{\partial^2}{\partial t^2}\int r^2\rho(\mathbf{r}t)d\mathbf{r}. \tag{2.115}$$

Note that the response term in Ehrenfest's theorem (2.86) and (2.103) depends on the expectation $\langle\mathbf{r}\rangle(t)$, that for the integral virial theorem (2.115) depends on the expectation $\langle r^2\rangle(t)$.

2.8 Time-Independent Schrödinger Theory: Ground and Bound Excited States

For a system of N electrons in a time-independent external field $\mathcal{F}^{\mathrm{ext}}(\mathbf{r})$ such that $\mathcal{F}^{\mathrm{ext}}(\mathbf{r}) = -\boldsymbol{\nabla}v(\mathbf{r})$, the Schrödinger equation (2.1) is

$$\hat{H}\Psi_{\mathrm{n}}(\mathbf{X}t) = E_{\mathrm{n}}\Psi_{\mathrm{n}}(\mathbf{X}t) = i\frac{\partial\Psi_n(\mathbf{X}t)}{\partial t}, \tag{2.116}$$

where now the Hamiltonian operator $\hat{H}$ is

$$\hat{H} = -\frac{1}{2}\sum_i \nabla_i^2 + \sum_i v(\mathbf{r}_i) + \frac{1}{2}\sum_{i,j}{}' \frac{1}{|\mathbf{r}_i - \mathbf{r}_j|}, \tag{2.117}$$

and where the wavefunction $\Psi_{\mathrm{n}}(\mathbf{X}t)$ are eigenfunctions of $\hat{H}$, and E_{n} the eigenvalues of the energy. The solutions of the equations (2.116) are of the form

$$\Psi_{\mathrm{n}}(\mathbf{X}t) = \psi_{\mathrm{n}}(\mathbf{X})e^{-iE_{\mathrm{n}}t}, \tag{2.118}$$

where the eigenfunctions $\psi_{\mathrm{n}}(\mathbf{X})$ and eigenvalues E_{n} of the energy are determined by the stationary-state Schrödinger equation

$$\hat{H}\psi_{\mathrm{n}}(\mathbf{X}) = E_{\mathrm{n}}\psi_{\mathrm{n}}(\mathbf{X}). \tag{2.119}$$

All observables O are then obtained as expectation values of Hermitian operators $\hat{O}$ taken with respect to the eigenfunctions $\psi_{\mathrm{n}}(\mathbf{X})$.

2.8.1 *The Quantal-Source and Field Perspective*

Stationary-state Schrödinger theory can also be described in terms of 'classical' fields and their quantal sources, and via the corresponding 'Quantal Newtonian' First Law. The description of the time-independent Schrödinger system for both the ground and bound excited states in terms of fields [3, 4, 8, 9, 15, 24] is the same as for the time-dependent case, but with the time-independent quantal sources and fields now determined by the functions $\psi_n(\mathbf{X})$. The definitions for the quantal sources: the density $\rho(\mathbf{r})$, the single-particle density matrix $\gamma(\mathbf{rr}')$, the pair-correlation density $g(\mathbf{rr}')$, the Fermi-Coulomb hole $\rho_{xc}(\mathbf{rr}')$, the current density $\mathbf{j}(\mathbf{r})$; and the fields: the electron-interaction $\mathcal{E}_{ee}(\mathbf{r})$, the Hartree $\mathcal{E}_{H}(\mathbf{r})$, the Pauli-Coulomb $\mathcal{E}_{xc}(\mathbf{r})$, the differential density $\mathcal{D}(\mathbf{r})$, and the kinetic $\mathcal{Z}(\mathbf{r})$, are all the same as defined for the time-dependent case but without the temporal parameter t. As such the definitions of these properties are not repeated. However, the expressions for the components of the total energy E, the 'Quantal-Newtonian' First Law, and the integral virial and Ehrenfest's theorem are given below.

2.8.2 Energy Components in Terms of Quantal Sources and Fields

The total energy E is the sum of the electron-interaction E_{ee}, the kinetic T, and external E_{ext} energy components. The electron-interaction energy E_{ee} can be further decomposed into its Hartree E_H and Pauli-Coulomb E_{xc} components. The expressions for these components of the total energy in terms of the respective quantal sources, and in integral-virial form in terms of the corresponding fields are given below. Thus,

$$E = T + E_{ee} + E_{ext} \tag{2.120}$$

$$= T + E_H + E_{xc} + E_{ext}. \tag{2.121}$$

The kinetic energy T in terms of its quantal source, the single-particle density matrix $\gamma(\mathbf{rr}')$, is

$$T = \int t(\mathbf{r}) d\mathbf{r}, \tag{2.122}$$

where the kinetic energy density $t(\mathbf{r})$ is either

$$t_A(\mathbf{r}) = -\frac{1}{2}\nabla_{\mathbf{r}}^2 \gamma(\mathbf{rr}')\bigg|_{\mathbf{r}'=r}, \tag{2.123}$$

or

$$t_B(\mathbf{r}) = \frac{1}{2}\nabla_{\mathbf{r}'} \cdot \nabla_{\mathbf{r}''}\gamma(\mathbf{r}'\mathbf{r}'')\bigg|_{\mathbf{r}'=\mathbf{r}''=r}. \tag{2.124}$$

In terms of the kinetic field $\mathcal{Z}(\mathbf{r})$, the energy T is

$$T = -\frac{1}{2}\int \rho(\mathbf{r})\mathbf{r} \cdot \mathcal{Z}(\mathbf{r}) d\mathbf{r}. \tag{2.125}$$

The electron-interaction energy E_{ee} in terms of its quantal source, the pair-correlation density $g(\mathbf{rr}')$, and in terms of the corresponding electron-interaction field $\mathcal{E}_{ee}(\mathbf{r})$, is

$$E_{ee} = \frac{1}{2}\int\int \frac{\rho(\mathbf{r})g(\mathbf{rr}')}{|\mathbf{r}-\mathbf{r}'|}d\mathbf{r}d\mathbf{r}' = \int \rho(\mathbf{r})\mathbf{r}\cdot\mathcal{E}_{ee}(\mathbf{r})d\mathbf{r}. \tag{2.126}$$

The Hartree energy E_H in terms of its quantal source $\rho(\mathbf{r})$, and in terms of the resulting field $\mathcal{E}_H(\mathbf{r})$, is

$$E_H = \frac{1}{2}\int\int \frac{\rho(\mathbf{r})\rho(\mathbf{r}')}{|\mathbf{r}-\mathbf{r}'|}d\mathbf{r}d\mathbf{r}' = \int \rho(\mathbf{r})\mathbf{r}\cdot\mathcal{E}_H(\mathbf{r})d\mathbf{r}. \tag{2.127}$$

The Pauli-Coulomb energy E_{xc} in terms of its quantal source, the Fermi-Coulomb hole charge $\rho_{xc}(\mathbf{rr}')$, and in terms of the corresponding Pauli-Coulomb field $\mathcal{E}_{xc}(\mathbf{r})$, is

$$E_{xc} = \frac{1}{2}\int\int \frac{\rho(\mathbf{r})\rho_{xc}(\mathbf{rr}')}{|\mathbf{r}-\mathbf{r}'|}d\mathbf{r}d\mathbf{r}' = \int \rho(\mathbf{r})\mathbf{r}\cdot\mathcal{E}_{xc}(\mathbf{r})d\mathbf{r}. \tag{2.128}$$

Finally, the external energy E_{ext} can be expressed in terms of the density $\rho(\mathbf{r})$ and the scalar potential energy $v(\mathbf{r})$ as

$$E_{ext}(\mathbf{r}) = \int \rho(\mathbf{r})v(\mathbf{r})d\mathbf{r}. \tag{2.129}$$

As will be explained in Chap. 3, the energy E_{ext} depends via the potential $v(\mathbf{r})$ on *all* the fields present in the quantum system. Since the quantal sources of these fields are expectation values taken with respect to the wave function $\psi_n(\mathbf{X})$, the potential $v(\mathbf{r})$ is then observed to be a functional of $\psi_n(\mathbf{X})$, *i.e.* $v(\mathbf{r}) = v[\psi_n(\mathbf{X})](\mathbf{r})$. This conclusion follows from the 'Quantal-Newtonian' First Law which is discussed next.

Note that the expressions for the various energy components in terms of the fields are independent of whether or not these individual fields are conservative.

2.8.3 The 'Quantal-Newtonian' First Law

The 'Quantal Newtonian' First Law is the time-independent version of the Second Law of (2.78). It states that the sum of the external $\mathcal{F}^{ext}(\mathbf{r})$ and internal $\mathcal{F}^{int}(\mathbf{r})$ fields experienced by each electron vanish (see Appendix A):

$$\mathcal{F}^{ext}(\mathbf{r}) + \mathcal{F}^{int}(\mathbf{r}) = 0, \tag{2.130}$$

where

$$\mathcal{F}^{\rm int}(\mathbf{r}) = \mathcal{E}_{\rm ee}(\mathbf{r}) - \mathcal{D}(\mathbf{r}) - \mathcal{Z}(\mathbf{r}) \tag{2.131}$$

$$= \mathcal{E}_H(\mathbf{r}) + \mathcal{E}_{xc}(\mathbf{r}) - \mathcal{D}(\mathbf{r}) - \mathcal{Z}(\mathbf{r}). \tag{2.132}$$

Once again we see that in addition to the external field, each electron in a given state experiences an electron-interaction field $\mathcal{E}_{\rm ee}(\mathbf{r})$ representative of electron correlations due to the Pauli principle and Coulomb repulsion, a differential density field $\mathcal{D}(\mathbf{r})$ representative of the density, and a kinetic field $\mathcal{Z}(\mathbf{r})$ representative of kinetic effects. These fields may not be individually conservative, but their sum defining the total internal field $\mathcal{F}^{\rm int}(\mathbf{r})$ always is *i.e.* $\nabla \times \mathcal{F}^{\rm int}(\mathbf{r}) = 0$. This follows from the 'Quantal Newtonian' First Law and the fact that the external field $\mathcal{F}^{\rm ext}(\mathbf{r})$ is conservative ($\nabla \times \nabla v(\mathbf{r}) = 0$). An important note is that the 'Quantal Newtonian' First Law is valid for *arbitrary* state, whether, nondegenerate or degenerate, ground or excited.

A study of the quantal sources and the corresponding fields then leads to a deeper and tangible physical understanding of a quantum mechanical system (see Chap. 5).

2.8.4 Integral Virial and Ehrenfest's Theorems

The integral virial theorem is the time-independent version of (2.108):

$$\int \rho(\mathbf{r})\mathbf{r} \cdot \mathcal{F}^{\rm ext}(\mathbf{r}) d\mathbf{r} + E_{ee} + 2T = 0. \tag{2.133}$$

where the energies $E_{\rm ee}$ and T are defined in integral virial form by (2.126) and (2.125), respectively.

In texts on quantum mechanics [25, 26], the theorem is stated as

$$\left\langle \sum_{i=1}^{N} \mathbf{r}_i \cdot \nabla_i \left[\hat{V}(\mathbf{r}_1, \ldots, \mathbf{r}_N) + \hat{U}(\mathbf{r}_1, \ldots, \mathbf{r}_N) \right] \right\rangle = 2T. \tag{2.134}$$

The equivalence of (2.133) and (2.134) follows from the fact that for Coulombic interactions the $\hat{U}(\mathbf{r}_1, \ldots, \mathbf{r}_N)$ is a homogenous function of degree -1. (A function $f(x_1, \ldots, x_j)$ is homogenous of degree n if it satisfies $f(sx_1, \ldots, sx_j) = s^n f(\mathbf{x}_1, \ldots, \mathbf{x}_N)$.) From Euler's theorem on homogenous functions, if $f(x_1, \ldots, x_j)$ is homogenous of degree n, then $\sum_{k=1}^{j} x_k(\partial f/\partial x_k) = nf$. Thus,

$$\left\langle \sum_{i=1}^{N} \mathbf{r}_i \cdot \nabla_i \hat{U}(\mathbf{r}_1, \ldots, \mathbf{r}_N) \right\rangle = -\langle \hat{U} \rangle = -E_{ee}. \tag{2.135}$$

Furthermore,

$$\left\langle \sum_{i=1}^{N} \mathbf{r}_i \cdot \nabla_i \hat{V}(\mathbf{r}_1, \ldots, \mathbf{r}_N) \right\rangle = -\int \rho(\mathbf{r}) \mathbf{r} \cdot \mathcal{F}^{\text{ext}}(\mathbf{r}) d\mathbf{r}, \tag{2.136}$$

since $\hat{V}(\mathbf{r}_1, \ldots, \mathbf{r}_N) = \sum_{i=1}^{N} v(\mathbf{r}_i)$ and $\mathcal{F}^{\text{ext}}(\mathbf{r}) = -\nabla v(\mathbf{r})$, and the equivalence follows.

Now if $\hat{V}(\mathbf{r}_1, \ldots, \mathbf{r}_N)$ is also Coulombic, then it too is a homogenous function of degree -1, so that

$$\left\langle \sum_{i=1}^{N} \mathbf{r}_i \cdot \nabla_i \left[\hat{V}(\mathbf{r}_1, \ldots, \mathbf{r}_N) \right] \right\rangle = -\langle \hat{V} \rangle = -E_{\text{ext}}, \tag{2.137}$$

where E_{ext} is given by (2.129).

Thus, for Coulombic systems, the integral virial theorem may be expressed as

$$E_{\text{ext}} + E_{\text{ee}} + 2T = 0. \tag{2.138}$$

The integral virial theorem for Columbic systems (2.138) may also be derived through scaling arguments. Thus, if $\psi(\mathbf{r}_1, \ldots, \mathbf{r}_N)$ is a normalized eigenstate of the Hamiltonian $\hat{H}$ of (2.117), then the scaled function defined as

$$\psi_\alpha(\mathbf{r}_1, \ldots, \mathbf{r}_N) = \alpha^{3N/2} \psi(\alpha \mathbf{r}_1, \ldots, \alpha \mathbf{r}_N), \tag{2.139}$$

is also normalized to unity. The kinetic, external, and electron-interaction energies scale as

$$< \psi_\alpha \mid \hat{T} \mid \psi_\alpha >= \alpha^2 < \psi \mid \hat{T} \mid \psi >, \tag{2.140}$$

$$< \psi_\alpha \mid \hat{V} \mid \psi_\alpha >= \alpha < \psi \mid \hat{V} \mid \psi >, \tag{2.141}$$

$$< \psi_\alpha \mid \hat{U} \mid \psi_\alpha >= \alpha < \psi \mid \hat{U} \mid \psi > . \tag{2.142}$$

Now the variational principle for the energy ensures that

$$\frac{d}{d\alpha} < \psi_\alpha \mid \hat{H} \mid \psi_\alpha > \Big|_{\alpha=1} = 0. \tag{2.143}$$

Application of (2.143) leads to the virial theorem (2.138). What this proof shows is that if an *approximate* wave function scales according to (2.139), then it will satisfy the integral virial theorem for Coulombic systems. This is an important fact since approximate wave functions that lead to poor results but which scale correctly still satisfy the virial theorem. The virial theorem is also satisfied for calculations that are fully self-consistent.

For an external potential energy that is harmonic: $v(\mathbf{r}) = \frac{1}{2}kr^2$, the function $\hat{V}(\mathbf{r}_1, \ldots, \mathbf{r}_N)$ is a homogeneous function of degree 2. Thus, on application of Euler's theorem,

$$\left\langle \sum_{i=1}^{N} \mathbf{r}_i \cdot \nabla_i \hat{V}(\mathbf{r}_1, \ldots, \mathbf{r}_N) \right\rangle = 2\langle \hat{V} \rangle = 2E_{\mathrm{ext}}. \tag{2.144}$$

For the harmonic external potential energy, the integral virial theorem is then

$$-2E_{\mathrm{ext}} + E_{\mathrm{ee}} + 2T = 0. \tag{2.145}$$

In classical physics, the total internal force between the N interacting particles and their torque vanish as a consequence of Newton's Third Law. In a similar manner, the quantal average of the internal field $\boldsymbol{\mathcal{F}}^{\mathrm{int}}(\mathbf{r})$ and the quantal average of the torque of this field also vanish. Thus,

$$\int \rho(\mathbf{r}) \boldsymbol{\mathcal{F}}^{\mathrm{int}}(\mathbf{r}) d\mathbf{r} = 0, \tag{2.146}$$

and

$$\int \rho(\mathbf{r}) \mathbf{r} \times \boldsymbol{\mathcal{F}}^{\mathrm{int}}(\mathbf{r}) d\mathbf{r} = 0. \tag{2.147}$$

This is the case because the quantal average of each component of the internal field $\boldsymbol{\mathcal{F}}^{\mathrm{int}}(\mathbf{r})$ as well as the quantal average of the torque of each component vanishes:

$$\int \rho(\mathbf{r}) \boldsymbol{\mathcal{E}}_{\mathrm{ee}}(\mathbf{r}) d\mathbf{r} = 0, \tag{2.148}$$

$$\int \rho(\mathbf{r}) \boldsymbol{\mathcal{D}}(\mathbf{r}) d\mathbf{r} = 0, \tag{2.149}$$

$$\int \rho(\mathbf{r}) \boldsymbol{\mathcal{Z}}(\mathbf{r}) d\mathbf{r} = 0, \tag{2.150}$$

$$\int \rho(\mathbf{r}) \mathbf{r} \times \boldsymbol{\mathcal{E}}_{\mathrm{ee}}(\mathbf{r}) d\mathbf{r} = 0, \tag{2.151}$$

$$\int \rho(\mathbf{r}) \mathbf{r} \times \boldsymbol{\mathcal{D}}(\mathbf{r}) d\mathbf{r} = 0, \tag{2.152}$$

$$\int \rho(\mathbf{r}) \mathbf{r} \times \boldsymbol{\mathcal{Z}}(\mathbf{r}) d\mathbf{r} = 0. \tag{2.153}$$

The vanishing of the averaged electron interaction field $\boldsymbol{\mathcal{E}}_{\mathrm{ee}}(\mathbf{r})$ (2.148) and that of its averaged torque (2.151) can be attributed to Newton's Third Law. This is because Coulomb's law and hence the field $\boldsymbol{\mathcal{E}}_{\mathrm{ee}}(\mathbf{r})$ obey the third law. The proof of these sum rules clearly demonstrates this. On the other hand, the vanishing of the average of the other components of the internal field and that of their averaged torque are not directly a consequence of the third law. The proof of the above sum rules is the same as in Sect. 2.5 for the time-dependent case.

The statement of Ehrenfest's theorem for the stationary state case is

$$\int \rho(\mathbf{r})\mathcal{F}^{\mathrm{ext}}(\mathbf{r})d\mathbf{r} = 0, \tag{2.154}$$

which is the quantal equivalent of Newton's First Law for classical particles.

Thus, *each electron* in the sea of electrons, experiences both the external field $\mathcal{F}^{\mathrm{ext}}(\mathbf{r})$ and the internal field $\mathcal{F}^{\mathrm{int}}(\mathbf{r})$, the sum of which vanish in order to satisfy the equation of motion or 'Quantal Newtonian' First Law. The internal field is a sum of fields each representative of a property of the system: electron correlations due to the Pauli principle and Coulomb repulsion, kinetic effects, and the electron density. And in a manner similar to that of classical physics, the quantal average of the internal field components and of their torques vanish, thereby leading to Ehrenfest's theorem.

The extension of the ideas of this chapter to include the electron spin moment as described by the Schrödinger-Pauli theory equation [27] is provided in Chap. 4. For an elucidation of the corresponding 'Quantal Newtonian' First Law by application to the spin-correlated first excited triplet 2^3S state of a 2-dimensional 2-electron semiconductor quantum dot in a magnetic field see Chap. 5. For an understanding of the time-evolution of this system when perturbed by a time-dependent electric field, and consequent satisfaction of the 'Quantal Newtonian' Second Law, see Chap. 9 on the Generalized Kohn Theorem [28].

2.9 Remarks on Quantum Fluid Dynamics and the 'Quantal Newtonian' Laws

In quantum fluid dynamics (QFD) [29–31], the electronic system is treated as a classical fluid moving under the action of an external field $\mathcal{F}(\mathbf{r}t)$. The equations of QFD—the continuity and Euler equations—are in terms of the collective properties of the electronic density $\rho(\mathbf{r}t)$ and physical current density $\mathbf{j}(\mathbf{r}t)$. These equations are derived from the time-dependent Schrödinger equation (2.1) by writing the wave function $\Psi(\mathbf{r}t)$ in polar form. The equations are thus a means of determining the amplitude and phase of the wave function. Consider the example of a *one-electron system* in an external field $\mathcal{F}^{ext}(\mathbf{r}t) = -\boldsymbol{\nabla} v(\mathbf{r}t)$. The corresponding Schrödinger equation is ($e = \hbar = m = 1$)

$$\left(-\frac{1}{2}\nabla^2 + v(\mathbf{r}t)\right)\Psi(\mathbf{r}t) = i\frac{\partial \Psi(\mathbf{r}t)}{\partial t}. \tag{2.155}$$

Writing the wave function in polar form as

$$\Psi(\mathbf{r}t) = R(\mathbf{r}t)\exp\left[iS(\mathbf{r}t)\right], \tag{2.156}$$

where $R(\mathbf{r}t)$ and $S(\mathbf{r}t)$ are real, and substituting it into the Schrödinger equation (2.155), one obtains the QFD continuity and Euler equations:

$$\frac{\partial\rho(\mathbf{r}t)}{\partial t} = -\nabla \cdot \mathbf{j}(\mathbf{r}t), \tag{2.157}$$

$$\frac{D\nu(\mathbf{r}t)}{Dt} = \mathcal{F}^{ext}(\mathbf{r}t) - \nabla f(\mathbf{r}t). \tag{2.158}$$

In these equations, the density $\rho(\mathbf{r}t) = R^2(\mathbf{r}t)$, the physical current density $\mathbf{j}(\mathbf{r}t) = \rho(\mathbf{r}t)\nabla S(\mathbf{r}t)$, the scalar function $f(\mathbf{r}t) = -\frac{1}{2}(\nabla^2 R/R)$, and the convective derivative

$$\frac{D\nu(\mathbf{r}t)}{Dt} = \frac{\partial\nu(\mathbf{r}t)}{\partial t} + [\nu(\mathbf{r}t) \cdot \nabla]\nu(\mathbf{r}t). \tag{2.159}$$

The above equations of QFD can be shown to be equivalent [8, 32] to the 'Quantal Newtonian' Second Law for the single electron:

$$\mathcal{F}^{ext}(\mathbf{r}t) + \mathcal{F}^{int}(\mathbf{r}t) = \mathcal{J}(\mathbf{r}t), \tag{2.160}$$

where the internal field

$$\mathcal{F}^{int}(\mathbf{r}t) = -\mathcal{D}(\mathbf{r}t) - \mathcal{Z}(\mathbf{r}t), \tag{2.161}$$

with the differential $\mathcal{D}(\mathbf{r}t)$ and kinetic $\mathcal{Z}(\mathbf{r}t)$ fields as defined in Sect. 2.2. The equivalence of the QFD continuity and Euler equations to that of the 'Quantal Newtonian' Second Law of (2.78) for the many-electron case can also be proved [8, 32].

The equivalence of the equations of QFD and the 'Quantal Newtonian' Laws is of course expected, since in the derivation of the former the wave function is expressed in polar form whereas for the latter it is expressed in Cartesian form. There is, however, a significant difference between the two descriptions of quantum mechanics. The equations of QFD provide no further physical insight into the many-electron system. On the other hand, from the 'Quantal Newtonian' Second and First Laws, we learn that in addition to any external binding fields, *each electron* also experiences an internal field. And that this field is comprised of components representative of the electron density and kinetic effects as well as a component representative of electron correlations due to the Pauli principle and Coulomb repulsion. In the following chapter it will be shown that in the added presence of a magnetic field, each electron not only experiences a Lorentz field as expected, but that there also exists an internal magnetic field component. Additionally, as will also be shown in the following chapter, a consequence of the 'Quantal Newtonian' Laws is that both the temporal and stationary-state Hamiltonians are *exactly known universal functionals* of the wave function. As such the time-dependent and stationary-state Schrödinger equations may be solved in a self-consistent manner. A new way of solution of the Schrödinger equations thus emerges.

References

1. E. Schrödinger, Ann. Physik **79**, 361, 489 (1926); **80**, 437 (1926); **81**, 109 (1926)
2. V. Sahni, J. Mol. Struct. (Theochem) **501**, 91 (2000)
3. V. Sahni, Top. Curr. Chem. **182**, 1 (1996)
4. V. Sahni, Phys. Rev. A **55**, 1846 (1997)
5. Z. Qian, V. Sahni, Phys. Lett. A **247**, 303 (1998)
6. Z. Qian, V. Sahni, Int. J. Quantum Chem. **78**, 341 (2000)
7. Z. Qian, V. Sahni, Phys. Rev. A **63**, 042508 (2001)
8. V. Sahni, *Quantal Density Functional Theory*, 2nd edn. (Springer, Berlin, Heidelberg, 2016)
9. V. Sahni, *Quantal Density Functional Theory: Approximation Methods and Applications* (Springer, Berlin, Heidelberg, 2010)
10. W. Pauli, Z. Physik **31**, 765 (1925)
11. P.A.M. Dirac, Proc. Roy. Soc. **A112**, 661 (1926); (The footnote on page 670 states: Prof. Born has informed me that Heisenberg has independently obtained results equivalent to these. (Added in proof)—See Heisenberg, 'Zeit fur Phys.,' vol. 38, p. 411 (1926))
12. W. Heisenberg, Z. Physik **38**, 411 (1926)
13. V. Sahni, Int. J. Quantum Chem. **97**, 953 (2004)
14. P. Ehrenfest, Z. Physik **45**, 455 (1927)
15. A. Holas, N.H. March, Phys. Rev. A **51**, 2040 (1995)
16. V. Sahni, X.-Y. Pan, Computation **5**, 15 (2017)
17. V. Sahni, J. Comp. Chem. **39**, 1083 (2018). https://doi.org/10.1002/jcc.24888(2017)
18. M. Born, J.R. Oppeheimer, Ann. Physik **84**, 457 (1927); V.F. Brattsev, Dokl. Acad. Nauk SSSR **160**, 570 (1965,English Transl: Soviet Physics-Doklady **10**, 44 (1965); S.T. Epstein, J. Chem. Phys. **44**, 836 (1966); Errata **44**, 4062 (1966)
19. V. Sahni, J.B. Krieger, Phys. Rev. A **11**, 409 (1975)
20. V. Sahni, J.B. Krieger, J. Gruenebaum, Phys. Rev. A **12**, 768 (1975)
21. M. Slamet, V. Sahni, Int. J. Quantum Chem. e25818 (2018). https://doi.org/10.1002/qua25818
22. A. Messiah, *Quantum Mechanics*, vol. I (North Holland, Amsterdam, 1966)
23. L.P. Kadanoff, G. Baym, *Statistical Mechanics* (W.A. Benjamin, New York, 1962), Sect. 10.3
24. V. Sahni, L. Massa, R. Singh, M. Slamet, Phys. Rev. Lett. **87**, 113002 (2001)
25. L.I. Schiff, *Quantum Mechanics*, 3rd edn. (McGraw Hill, New York, 1968)
26. D.J. Griffiths, D.F. Schroeter, *Introduction to Quantum Mechanics*, 3rd edn. (Cambridge University Press, 2017)
27. W. Pauli, Z. Physik **43**, 601 (1927)
28. H.-M. Zhu, J.-W. Chen, X.-Y. Pan, V. Sahni, J. Chem. Phys. **140**, 024318 (2014)
29. E. Madelung, Z. Physik **40**, 332 (1926)
30. H. Frölich, Physica (Amsterdam) **37**, 215 (1967)
31. B.M. Deb, S.K. Gosh, in *Single Particle Density in Physics and Chemistry*, ed. by N.H. March, B.M. Deb (Academic, New York, 1987)
32. M.K. Harbola, Phys. Rev. A **58**, 1779 (1998)

Chapter 3
Generalization of the Schrödinger Theory of Electrons

Abstract The Schrödinger theory of electrons in the presence of either a static or time-dependent external electromagnetic field is generalized such that the corresponding Hamiltonians $\{\hat{H}; \hat{H}(t)\}$ are shown to be *exactly known* functionals of the corresponding wave functions $\{\Psi; \Psi(t)\}$, that is, $\hat{H} = \hat{H}[\Psi]$ and $\hat{H}(t) = \hat{H}[\Psi(t)](t)$. The functionals are valid for *arbitrary* state, and are *universal* in that they are the *same* for *arbitrary* scalar binding potential of the electrons. Thus, the Schrödinger equations may be written as $\hat{H}[\Psi]\Psi = E[\Psi]\Psi$ and $\hat{H}[\Psi(t)](t)\Psi(t) = i\hbar\partial\Psi(t)/\partial t$. Written in this way, it becomes evident that the equations are intrinsically self-consistent. As such the Hamiltonians $\{\hat{H}[\Psi]; \hat{H}[\Psi(t)]\}$, the eigenfunctions and energy eigenvalues $\{\Psi, E\}$ of the stationary-state equation, and the wave function $\Psi(t)$ of the temporal equation, can *all* be determined self-consistently. The proofs are based on the 'Quantal Newtonian' First and Second Laws which are the equations of motion for the *individual* electron in the sea of electrons in the external fields. The generalization of the Schrödinger equation in this manner leads to additional new physics. The traditional description of the Schrödinger theory of electrons with *known* Hamiltonians $\{\hat{H}; \hat{H}(t)\}$ then constitutes a special case. Stationary-state Hartree-Fock and Hartree theories are described in terms of quantal sources and fields, and the derivations of the corresponding integro-differential equations provided. The Slater-Bardeen interpretation of Hartree-Fock theory, and theorems within the theory are explained. Although the Hartree-Fock and Hartree theory equations are intrinsically self-consistent, they can be further generalized via the corresponding 'Quantal Newtonian' First Laws. The traditional Hartree-Fock and Hartree theories then constitute a special case of the generalized forms.

Introduction

When Schrödinger, 'derived' the time-independent wave equation in his first paper [1], "He knew the equation before he devised the 'derivation', and his main purpose in this paper was to show that his equation gives the correct quantization of the energy levels of the hydrogen atom" [2]. In doing so he assumed knowledge of the Hamiltonian $\hat{H}$ of the hydrogen atom, i.e. the sum of the electron kinetic $\hat{T}$ and

V. Sahni, *Schrödinger Theory of Electrons: Complementary Perspectives*, Springer Tracts in Modern Physics 285,
https://doi.org/10.1007/978-3-030-97409-1_3

the external Coulombic binding potential $v(\mathbf{r})$ operators. Then solution of the wave equation led to the wave functions and energies of the atom.

Since the advent of the first paper, we continue to think of the Schrödinger eigenvalue equation in a similar manner, i.e. as one in which the Hamiltonian operator $\hat{H}$ is always *assumed known*, and it is only the *unknown* eigenfunctions Ψ and eigenenergies E that are to be determined. Thus, to obtain the properties of natural and artificial matter—atoms, molecules, solids, two-dimensional systems such as semiconductor heterostructures, quantum dots, wires, rings etc., one solves the Schrödinger partial differential equation

$$\hat{H}\Psi(\mathbf{X}) = E\Psi(\mathbf{X}), \tag{3.1}$$

for the $\{\Psi, E\}$ employing the known Hamiltonian $\hat{H}$ appropriate to the system being studied (see also 2.119). (Here, for an N electron system, $\mathbf{X} = \mathbf{x}_1, \mathbf{x}_2, \ldots, \mathbf{x}_N$; $\mathbf{x} = \mathbf{r}\sigma$, with $\{\mathbf{r}\sigma\}$ being the spatial and spin coordinates of each electron, respectively.)

In this chapter we generalize [3, 4] the Schrödinger equation by showing that the Hamiltonian $\hat{H}$ for a system of electrons in external electromagnetic fields is an *exactly known* and *universal* functional of the wave function Ψ, that is, $\hat{H} = \hat{H}[\Psi]$. By *universal* is meant that the analytical functional dependence of $\hat{H}$ on $\Psi(\mathbf{X})$ remains the *same* for *arbitrary* scalar binding potential function $v(\mathbf{r})$. The result is arrived at via the 'Quantal Newtonian' First Law of (2.130) for each electron. Hence, the Schrödinger equation can now be written in the more general form as

$$\hat{H}[\Psi]\Psi(\mathbf{X}) = E[\Psi]\Psi(\mathbf{X}). \tag{3.2}$$

Written as above, the known fact that the energy E is a functional of the wave function $\Psi(\mathbf{X})$ is explicitly enunciated. Since the 'Quantal Newtonian' First Law is valid for *arbitrary* state, so is the generalized form of the Schrödinger equation.

The Schrödinger equation expressed in its generalized form as in (3.2) introduces a new conceptual understanding of the equation. This generalized form shows that the equation is intrinsically *self-consistent*. This is a radical departure from the traditional understanding of how the wave functions $\Psi(\mathbf{X})$ and energies E of the Schrödinger equation as written in (3.1) are obtained. Solutions of the equations of the form $\hat{L}[\zeta]\zeta = \lambda[\zeta]\zeta$, where $\{\zeta, \lambda\}$ are the eigenfunctions and eigenvalues, respectively, of the operator $\hat{L}$ that depends on ζ, are obtained self-consistently. Additionally, so is the operator $\hat{L}[\zeta]$. What this means with regard to the Schrödinger equation is that the Hamiltonian $\hat{H}$, the interacting many-electron wave function $\Psi(\mathbf{X})$, and the energy E, are all determined simultaneously on achieving self-consistency. The self-consistency procedure is an *iterative* one. Beginning with an approximate input trial wave function $\Psi(\mathbf{X})$, one obtains the corresponding Hamiltonian $\hat{H}[\Psi]$. Then solution of the generalized Schrödinger equation (3.2) leads to a new output approximate $\Psi(\mathbf{X})$ and energy E. This new $\Psi(\mathbf{X})$ is then treated as the next input $\Psi(\mathbf{X})$, and the procedure is repeated until the input and output wave functions $\Psi(\mathbf{X})$ are the same. This is the solution to the eigenvalue equation. (There may exist other self-consistent solutions. It is only after self-consistency is achieved that one must judge and test with experiment whether or not the solution is physically meaningful.) Further explanations of the self-consistency procedure for the stationary-state case are described in the following Sect. 3.1. Note that for each iteration of the self-consistency procedure, the energy E explicitly depends on the $\Psi(\mathbf{X})$ for that iteration. It is to emphasize the explicit dependence of E on $\Psi(\mathbf{X})$ *at each iteration* that the eigenvalue E is written as $E[\Psi]$ in (3.2).

Finally, the non-self-consistent version of the Schrödinger equation of (3.1) with known Hamiltonian $\hat{H}$, then constitutes a special case of the more general self-consistent form of the equation as expressed by (3.2). The manner by which this is the case will also be explained in Sect. 3.1. There it will be shown, that with the Hamiltonian $\hat{H}$ known, the eigenfunctions $\Psi(\mathbf{X})$ and eigenvalues E can be determined in a self-consistent manner.

A considerable degree of new physics emerges from the generalized Schrödinger equation. These insights are to be described in Sect. 3.1.2. Here we make a few preliminary remarks with regard to the generalized equation for the stationary state.

(a) Let us consider a system of N electrons in some external conservative binding field $\boldsymbol{\mathcal{E}}(\mathbf{r}) = -\nabla v(\mathbf{r})/e$. When one speaks of a system of electrons, it is implicitly assumed that the electron kinetic $\hat{T}$ and electron-interaction $\hat{U}$ operators are known. Hence the functional dependence of the Hamiltonian $\hat{H}$ on Ψ arises from the dependence of the potential $v(\mathbf{r})$ on Ψ. That is $v(\mathbf{r}) = v[\Psi](\mathbf{r})$. Suppose one were to create an artificial electronic system, for example an 'artificial atom' or 'artificial molecule', for which the binding potential $v(\mathbf{r})$ is unknown. Then this potential could be determined by the self-consistent solution of the generalized Schrödinger equation of (3.2). Depending on the state of the system, the corresponding many-electron wave function Ψ and energy E too would be determined.

As a consequence of advances in semiconductor technology, it has been possible to create 'artificial atoms' or quantum dots and 'artificial molecules' or quantum dot molecules [5–7]. These are reduced dimensional nano systems with length scales ranging from 10 to 1000Å. The motion of electrons in a quantum dot is confined to two dimensions within a quantum well in a thin layer of semiconductor (GaAs) sandwiched between two layers of another semiconductor (AlGaAs). The two dimensional motion of the electrons is confined by an electrostatic field to create the quantum dot. The size of the quantum dot can be further reduced by application of a magnetic field perpendicular to the plane of motion. The length and energy scales of a quantum dot differ from that of natural atoms. As the quantum dot is housed in a semiconductor, its electron mass is the band effective mass $m^{\star}$. Additionally, the interaction between the electrons is modified by the dielectric function ϵ of the semiconductor. The issue of the binding potential $v(\mathbf{r})$ of the electrons in a quantum dot was addressed via both experiment [5–7] as well as by a Hartree level calculation [8]. It is now accepted that in contrast to that of natural atoms for which the binding potential is Coulombic, that for quantum dots is harmonic. The same conclusion could have been arrived at via the self-consistent solution of the generalized Schrödinger equation (3.2) as will be explained in a later chapter. Hence, if in the future, other electronic systems are created, their binding potential, energy, and wave function can then be determined via solution of the generalized Schrödinger equation.

(b) Writing the Schrödinger equation in its generalized form of (3.2) has the added advantage that the exact many-electron wave function can also be determined for systems for which the external potential function $v(\mathbf{r})$ is known. For example, this function is Coulombic for natural atoms and molecules, and harmonic for 'artificial atoms' and 'artificial molecules'. Recall that one knows the functional dependence of $v(\mathbf{r})$ on Ψ, i.e. $v(\mathbf{r}) = v[\Psi](\mathbf{r})$. For an approximate trial wave function Ψ, one then determines the corresponding $v[\Psi](\mathbf{r})$. This will be an approximation to the known function $v(\mathbf{r})$. On substituting this approximate function $v[\Psi](\mathbf{r})$ into the generalized Schrödinger equation (3.2), one then solves the latter till self-consistency is achieved at which point the exact known function $v(\mathbf{r})$ will be reproduced. The resulting solution for $\{\Psi, E\}$ is then exact.

(c) It is interesting to compare the self-consistent solution of the generalized Schrödinger equation (3.2) for the wave function $\Psi(\mathbf{X})$ for a system in a ground state to that of the variational [9] and constrained-search variational [10–12] methods. Both variational methods are associated principally with the property of the total energy. In the application of the variational principle, an approximate parametrized variational wave function correct to $O(\delta)$ leads to a rigorous upper bound for the energy that is correct to $O(\delta^2)$. Such a wave function is accurate in the region of space from where the principal contribution to the energy arises. However, all other observables obtained as the expectation value of Hermitian single- and two-particle operators are correct only to the same order as that of the wave function, *viz.* to $O(\delta)$. A better approximate variational wave function is one that leads to a lower value of the energy. There is no guarantee that other observables representative of different regions of configuration space are thereby more accurate. (For the exact wave function Ψ, one obtains the exact ground state energy E.) In the constrained-search variational method [10–12], the variational space of approximate parametrized wave functions is expanded by considering the wave function Ψ to be a functional of a function χ, i.e., $\Psi = \Psi[\chi]$. One searches over all functions χ such that the wave function $\Psi = \Psi[\chi]$ is normalized, gives the exact (theoretical or experimental) value of an observable, while leading to a rigorous upper bound to the energy. (It is thus possible to construct a wave function that is a functional of the density $\rho(\mathbf{r})$, *i.e.* $\Psi = \Psi[\rho]$.) In this manner, the wave function functional $\Psi[\chi]$ is accurate not only in the region contributing to the energy, but also that of the observable. Nonetheless, the wave function $\Psi[\chi]$ is still approximate. There also exist variational principles [13–19] whereby single particle properties such as the coherent atomic scattering factor, and hence the density; and the single-particle density matrix, and therefore, the momentum density, can be obtained correct to $O(\delta^2)$ while beginning with a trial wave function correct to $O(\delta)$. In contrast to the variational methods, the self-consistent procedure, achieved to a particular degree of numerical accuracy, *all* the properties are determined to the *same* degree of accuracy. An improved wave function is then one determined to a greater numerical accuracy. Note that the generalized Schrödinger equation is also valid for excited states. Hence, the above remarks are equally valid for excited states for which the variational results for upper bounds to the excited states may be obtained via the Ritz variational method for the Sturm-Liouville equation [9] or its application as the Hylleraas-Undheim method [9, 20] to the Schrödinger equation.

(*d*)The concept of obtaining the solution of an integro-differential equation self-consistently is a well-established one in electronic structure theory. For example, in the Hartree [21] and Hartree-Fock [22, 23] theories, the corresponding Hamiltonians $\hat{H}^H$ and $\hat{H}^{HF}$ are functionals of the single-particle orbitals $\phi_i(\mathbf{x})$ that constitute the solution of the respective equations. The corresponding equations are the Hartree and Hartree-Fock integro-differential equations $\hat{H}^H[\phi_i]\phi_i(\mathbf{x}) = \epsilon_i\phi_i(\mathbf{x})$ and $\hat{H}^{HF}[\phi_i]\phi_i(\mathbf{x}) = \epsilon_i\phi_i(\mathbf{x})$, where the ϵ_i are the single-particle eigenvalues. These equations are then solved self-consistently [24, 25] for the orbitals $\phi_i(\mathbf{x})$. The solutions in turn lead to a Hartree product and Slater determinant wave function, respectively. There are other formalisms, as for example within the context of local effective potential theory, in which the corresponding Hamiltonians too are functionals of the single-particle orbitals $\phi_i(\mathbf{x})$ which constitute the solution of an integro-differential equation. In these theories, the electrons are replaced by model non-interacting fermions with the many-body effects incorporated in the local effective potential. The resulting integro-differential equations are of the form $\hat{L}[\zeta]\zeta = \lambda[\zeta]\zeta$, where $\{\zeta, \lambda\}$ are the eigenfunctions and eigenvalues, respectively, and for which the solution is obtained in an iterative self-consistent manner. In these theories, the solutions $\phi_i(\mathbf{x})$ then reproduce in principle the same density $\rho(\mathbf{r})$ as that of the interacting system. Such theories are Kohn-Sham density functional theory [26, 27], the Optimized Potential Method [28–31], and Quantal density functional theory [32–35]. There is, however, an important distinction between all these theories and the generalized Schrödinger equation of (3.2). Whereas in Hartree, Hartree-Fock and local effective potential theories, the self-consistency is for the *single particle* orbitals $\phi_i(\mathbf{x})$ which then lead to either an approximate wave function or the exact density, the self-consistent solution of the generalized Schrödinger equation is the *exact many-electron* interacting system wave function $\Psi(\mathbf{X})$ and the total energy E.

In the Hartree and Hartree-Fock theory approximations, it is the electron-interaction operator $\hat{U}$ of the Hamiltonian (2.117) that is approximated. The kinetic $\hat{T}$ and external potential $\hat{V}$ operators remain the same. As noted above, the integro-differential equation of each theory is of the general form $\hat{L}[\zeta]\zeta = \lambda\zeta$, and must be solved self-consistently for the $\hat{L}[\zeta]$, ζ, and λ. The reason why the Hartree $\hat{H}^H[\phi_i]$ and Hartree Fock theory $\hat{H}^{HF}[\phi_i]$ Hamiltonians are functionals of the orbitals $\phi_i(\mathbf{x})$ is because the corresponding approximations to the electron-interaction operator $\hat{U}$ in these theories are functionals of $\phi_i(\mathbf{x})$. As in Schrödinger theory which can be described in terms of quantal sources and fields via the 'Quantal Newtonian' First Law (Chap. 2), both the Hartree and Hartree-Fock theories can also be similarly described via a corresponding 'Quantal Newtonian' First Law. Hence, as in Schrödinger theory, where as a consequence of the 'Quantal Newtonian' First Law the external potential is a functional of the wave function i.e. $v(\mathbf{r}) = v[\Psi](\mathbf{r})$, so also in Hartree and Hartree-Fock theory is the external potential $v(\mathbf{r})$ a functional of the respective wave functions, or equivalently of the orbitals $\phi_i(\mathbf{x})$ i.e. $v(\mathbf{r}) = v[\phi_i(\mathbf{x})](\mathbf{r})$. Thus, it is possible to generalize the Hartree and Hartree-Fock theory equations in a manner similar to that of (3.2) of Schrödinger theory.

(*e*)As noted above, it follows from the 'Quantal Newtonian' First Law that the external potential $v(\mathbf{r})$ is a functional of the wave function Ψ, *viz.* $v(\mathbf{r}) = v[\Psi](\mathbf{r})$, and that this functional dependence is valid for *arbitrary state*, whether nondegenerate or degenerate, ground or excited. Furthermore, the functional dependence is *unique* and *explicitly known*. That the external potential $v(\mathbf{r})$ is a unique functional of the *nondegenerate ground state* wave function is also arrived at via the proof of the first *Hohenberg-Kohn theorem* [26]. (A more explanatory discussion of the theorem and of its proof is to be provided in Chap. 7.) The proof of the theorem is also applicable to the lowest excited state of a system having a symmetry (spin and orbital angular momentum) different from that of the ground state, in which case it is referred to as the *Gunnarsson-Lundqvist theorem* [36, 37]. Hence, for such an excited state, one can conclude that the potential $v(\mathbf{r})$ is also a unique functional of that excited state wave function. The conclusion via these theorems of the functional dependence of $v(\mathbf{r})$ on Ψ is thus restrictive to these states. Further, neither theorem provides the explicit dependence of the potential $v(\mathbf{r})$ on Ψ. It is solely the knowledge of this functional dependence that is of significance in these theorems.

(*f*)In the extension of the stationary-state equation (3.1) to the case of a system of N electrons in a time-dependent external field $\mathcal{E}(\mathbf{y}) = -\nabla v(\mathbf{y})/e$; $\mathbf{y} = \mathbf{r}t$, Schrödinger once again assumed [2, 38], together with knowledge of the canonical kinetic $\hat{T}$ and electron-interaction $\hat{U}$ operators, that the Hamiltonian $\hat{H}(t)$ of the system was known. Thus, the time-dependent Schrödinger equation is (see (2.1))

$$\hat{H}(t)\Psi(\mathbf{Y}) = i\hbar\,\partial\Psi(\mathbf{Y})/\partial t, \tag{3.3}$$

where $\Psi(\mathbf{Y})$ is the system wave function; $\mathbf{Y} = \mathbf{X}t$. It is in this manner that the temporal equation is presently understood and applied.

However, as in the stationary-state case, the temporal equation too can be generalized [3, 4] as it is proved that the Hamiltonian $\hat{H}(t)$ is a functional of the wave function $\Psi(\mathbf{Y})$, i.e. $\hat{H}(t) = \hat{H}[\Psi(\mathbf{Y})](t)$. This follows from the 'Quantal Newtonian' Second Law of (2.78) whereby the external potential $v(\mathbf{y})$ is shown to be an *unique, exactly known and universal* functional of the wave function $\Psi(\mathbf{Y})$, i.e. $v(\mathbf{y}) = v[\Psi(\mathbf{Y})](\mathbf{y})$. With the electronic kinetic $\hat{T}$ and electron-interaction $\hat{U}$ operators assumed known, the temporal Schrödinger equation can then be written as [3, 4]

$$\hat{H}[\Psi(\mathbf{Y})](t)\Psi(\mathbf{Y}) = i\hbar\,\partial\Psi(\mathbf{Y})/\partial t. \tag{3.4}$$

This is the generalized form of the time-dependent Schrödinger equation. Written in this manner then elucidates its intrinsic self-consistent nature. Additionally, as in the stationary-state case, new physics emerges.

3.1 Generalization of the Stationary-State Schrödinger Equation

In order to explain the generalization of the stationary-state Schrödinger equation, let us consider the more general time-independent equation, *viz.* that for a system of N electrons in an external conservative electrostatic field $\mathcal{E}(\mathbf{r}) = -\nabla v(\mathbf{r})/e$ and a magnetostatic field $\mathcal{B}(\mathbf{r}) = \nabla \times \mathbf{A}(\mathbf{r})$, where $\{v(\mathbf{r}), \mathbf{A}(\mathbf{r})\}$ are the corresponding scalar and vector potentials, respectively. (To simplify the equation writing, we assume the charge of the electron to be $-e$ with $e = h = m = 1$ together with the further assumption of $c = 1$. To obtain the expressions in atomic units, replace $\mathbf{A}(\mathbf{r})$ by $\mathbf{A}(\mathbf{r})/c$.) The Hamiltonian $\hat{H}$ for the system is then

$$\hat{H} = \hat{T}_A + \hat{U} + \hat{V}, \tag{3.5}$$

where $\hat{T}_A$ is the *physical* kinetic energy operator:

$$\hat{T}_A = \frac{1}{2}\sum_{k=1}^{N}\left[\hat{\mathbf{p}}_k + \mathbf{A}(\mathbf{r}_k)\right]^2, \tag{3.6}$$

with $\hat{T}$ the 'canonical' kinetic energy operator

$$\hat{T} = \frac{1}{2}\sum_{k=1}^{N}\hat{\mathbf{p}}_k^2 = -\frac{1}{2}\sum_{k=1}^{N}\nabla_k^2. \tag{3.7}$$

The electron-interaction potential energy operator $\hat{U}$ is

$$\hat{U} = \sum_{k,\ell}^{N}{}' u(\mathbf{r}_k\mathbf{r}_\ell) = \sum_{k,\ell=1}^{N}{}' \frac{1}{|\mathbf{r}_k - \mathbf{r}_\ell|}, \tag{3.8}$$

and the external electrostatic potential energy operator $\hat{V}$ is

$$\hat{V} = \sum_{k=1}^{N} v(\mathbf{r}_k). \tag{3.9}$$

The time-independent Schrödinger equation is

$$\hat{H}(\mathbf{R}; \mathbf{A})\Psi(\mathbf{X}) = E\Psi(\mathbf{X}), \tag{3.10}$$

where $\{\Psi(\mathbf{X}), E\}$ are the eigenfunctions and eigenenergies of the system, with $\mathbf{R} = \mathbf{r}_1, \ldots, \mathbf{r}_N$; $\mathbf{X} = \mathbf{x}_1, \ldots, \mathbf{x}_N$; $\mathbf{x} = \mathbf{r}\sigma$, $\{\mathbf{r}\sigma\}$ being the spatial and spin coordinates.

We note the following salient features of the above Schrödinger equation:

(a) As a consequence of the correspondence principle [39], it is the vector potential $\mathbf{A}(\mathbf{r})$ and not the magnetic field $\mathcal{B}(\mathbf{r})$ that appears in the equation. This fact is significant because it constitutes a fundamental difference between quantum and classical physics. In classical electrodynamics, the vector potential $\mathbf{A}(\mathbf{r})$ is introduced as a construct to simplify the writing of certain equations such as the Ampere-Maxwell equation [40] [$\nabla \times \mathcal{B}(\mathbf{r}) = (4\pi/c)\mathbf{J}$ with $\mathbf{J}$ the current density] in terms of $\mathbf{A}(\mathbf{r})$.

The field $\mathcal{B}(\mathbf{r})$ then obtained from it as $\mathcal{B}(\mathbf{r}) = \nabla \times \mathbf{A}(\mathbf{r})$ is the physical quantity. In turn the expression for $\mathcal{B}(\mathbf{r})$ is obtained from Maxwell's equation (Gauss's law for magnetic fields) $\nabla \cdot \mathcal{B}(\mathbf{r}) = 0$ since $\nabla \cdot \nabla \times \mathbf{A}(\mathbf{r}) = 0$. In quantum mechanics, it is the vector potential $\mathbf{A}(\mathbf{r})$ that appears in the Schrödinger equation. This distinction between classical and quantum physics is important because a vector potential can exist in a region of space where there is no magnetic field. This fact is then employed to explain, for example, the Aharonov-Bohm [41] effect. In the Schrödinger equation, the magnetic field $\mathcal{B}(\mathbf{r})$ appears only following the choice of gauge, such as for example, the symmetric gauge for which $\mathbf{A}(\mathbf{r}) = \frac{1}{2}\mathcal{B}(\mathbf{r}) \times \mathbf{r}$, or the Landau gauge in which $\mathbf{A}(\mathbf{r}) = -\mathcal{B}x\hat{\mathbf{i}}_y$.

(b) The potential energy operator $v(\mathbf{r})$ has the following characteristics:

> ***(i)*** In speaking of a N electron system, it is assumed that the canonical kinetic $\hat{T}$ and electron-electron interaction potential $\hat{U}$ energy operators are *known*. As such the potential $v(\mathbf{r})$ is considered an *extrinsic* input to the Hamiltonian $\hat{H}$.
>
> ***(ii)*** The binding potential energy function $v(\mathbf{r})$ is also *assumed known*, *e.g.*. for natural atoms, molecules and solids, it is the Coulomb potential; for 'artificial atoms' or quantum dots [5–7], it is the harmonic potential; for screened-Coulomb interaction in an uniform electron gas, it is the Yukawa potential; at a jellium metal surface, it could be a finite linear potential [35, 42], etc.
>
> ***(iii)*** Since by assumption, the field $\mathcal{E}(\mathbf{r})$ is conservative, the potential $v(\mathbf{r})$ is *path-independent*.

With the Hamiltonian $\hat{H}$ operator known, the Schrödinger differential equation is then solved for the eigenfunctions $\Psi(\mathbf{X})$ and eigenenergies E. Physical observables are determined as expectations of Hermitian operators taken with respect to $\Psi(\mathbf{X})$. Further discussion to provide a more complete description of physical systems defined by the Hamiltonian of (3.5) is given in Chap. 4. Here we focus solely on the generalization of the corresponding Schrödinger equation.

To derive the generalization of the Schrödinger equation (3.10), we need to first derive the corresponding 'Quantal Newtonian' First Law.

3.1.1 'Quantal Newtonian' First Law in an Electrostatic and Magnetostatic Field

The 'Quantal Newtonian' First Law for the quantum system of electrons in both an electrostatic $\mathcal{E}(\mathbf{r})$ and magnetostatic $\mathcal{B}(\mathbf{r})$ field as defined by the Schrödinger equation of (3.10) is a generalization of the form derived (see (2.130)) for the system in the absence of the magnetic field. Hence, once again the law states that the sum of the external $\mathcal{F}^{\mathrm{ext}}(\mathbf{r})$ and internal $\mathcal{F}^{\mathrm{int}}(\mathbf{r})$ fields experienced by each electron vanish [34, 43, 44]:

$$\mathcal{F}^{\mathrm{ext}}(\mathbf{r}) + \mathcal{F}^{\mathrm{int}}(\mathbf{r}) = 0. \tag{3.11}$$

The law is gauge invariant, and derived employing the continuity condition $\nabla \cdot \mathbf{j}(\mathbf{r}) = 0$, where $\mathbf{j}(\mathbf{r})$ is the physical current density. (For the derivation of the law, see Appendix A.)

The external field $\mathcal{F}^{\mathrm{ext}}(\mathbf{r})$ in the 'Quantal Newtonian' First Law in this case is the sum of the electrostatic $\mathcal{E}(\mathbf{r})$ and Lorentz $\mathcal{L}(\mathbf{r})$ fields:

$$\mathcal{F}^{\mathrm{ext}}(\mathbf{r}) = \mathcal{E}(\mathbf{r}) - \mathcal{L}(\mathbf{r}) = -\nabla v(\mathbf{r}) - \mathcal{L}(\mathbf{r}), \tag{3.12}$$

where $\mathcal{L}(\mathbf{r})$ is defined in terms of the Lorentz 'force' $\boldsymbol{\ell}(\mathbf{r})$ and the density $\rho(\mathbf{r})$ as

$$\mathcal{L}(\mathbf{r}) = \frac{\boldsymbol{\ell}(\mathbf{r})}{\rho(\mathbf{r})}, \tag{3.13}$$

where

$$\boldsymbol{\ell}(\mathbf{r}) = \mathbf{j}(\mathbf{r}) \times \mathcal{B}(\mathbf{r}). \tag{3.14}$$

The density $\rho(\mathbf{r})$ is the expectation value

$$\rho(\mathbf{r}) = \langle \Psi(\mathbf{X}) | \hat{\rho}(\mathbf{r}) | \Psi(\mathbf{X}) \rangle, \tag{3.15}$$

with $\hat{\rho}(\mathbf{r})$ the density operator of (2.12). The physical current density $\mathbf{j}(\mathbf{r})$ is the expectation value

$$\mathbf{j}(\mathbf{r}) = \langle \Psi(\mathbf{X}) | \hat{\mathbf{j}}(\mathbf{r}) | \Psi(\mathbf{X}) \rangle, \tag{3.16}$$

where the current density operator $\hat{\mathbf{j}}(\mathbf{r})$ is defined as

$$\hat{\mathbf{j}}(\mathbf{r}) = \frac{1}{2i} \sum_k \left[\nabla_{\mathbf{r}_k} \delta(\mathbf{r}_k - \mathbf{r}) + \delta(\mathbf{r}_k - \mathbf{r}) \nabla_{\mathbf{r}_k} \right] + \hat{\rho}(\mathbf{r}) \mathbf{A}(\mathbf{r}). \tag{3.17}$$

The current density $\mathbf{j}(\mathbf{r})$ may also be written in terms of the *physical momentum operator* $\hat{\mathbf{p}}_{\mathrm{physical}} = (\hat{\mathbf{p}} + \mathbf{A}(\mathbf{r}))$, with $\hat{\mathbf{p}}$ the canonical momentum operator, as (see also (2.41))

$$\mathbf{j}(\mathbf{r}) = N \Re \sum_\sigma \int \Psi^\star(\mathbf{r}\sigma, \mathbf{X}^{N-1}) \big(\hat{\mathbf{p}} + \mathbf{A}(\mathbf{r}) \big) \Psi(\mathbf{r}\sigma, \mathbf{X}^{N-1}) d\mathbf{X}^{N-1}. \tag{3.18}$$

where $\mathbf{X}^{N-1} = \mathbf{x}_2, \ldots, \mathbf{x}_N$.

The internal field $\mathcal{F}^{\mathrm{int}}(\mathbf{r})$ is the sum of the electron-interaction $\mathcal{E}_{\mathrm{ee}}(\mathbf{r})$, kinetic $\mathcal{Z}(\mathbf{r})$, differential density $\mathcal{D}(\mathbf{r})$, and internal magnetic $\mathcal{I}_m(\mathbf{r})$ fields:

$$\mathcal{F}^{\mathrm{int}}(\mathbf{r}) = \mathcal{E}_{\mathrm{ee}}(\mathbf{r}) - \mathcal{Z}(\mathbf{r}) - \mathcal{D}(\mathbf{r}) - \mathcal{I}_m(\mathbf{r}). \tag{3.19}$$

These fields are defined in terms of the corresponding 'forces' $\mathbf{e}_{\mathrm{ee}}(\mathbf{r}), z(\mathbf{r}), \mathbf{d}(\mathbf{r})$, and $\boldsymbol{i}_m(\mathbf{r})$:

$$\mathcal{E}_{ee}(\mathbf{r}) = \frac{\mathbf{e}_{ee}(\mathbf{r})}{\rho(\mathbf{r})}\ ; \ \mathcal{Z}(\mathbf{r}) = \frac{\mathbf{z}(\mathbf{r})}{\rho(\mathbf{r})}\ ; \ \mathcal{D}(\mathbf{r}) = \frac{\mathbf{d}(\mathbf{r})}{\rho(\mathbf{r})}\ ; \ \mathcal{I}_m(\mathbf{r}) = \frac{\boldsymbol{i}_m(\mathbf{r})}{\rho(\mathbf{r})}. \tag{3.20}$$

The 'forces' $\mathbf{e}_{ee}(\mathbf{r})$, $z(\mathbf{r})$, and $\mathbf{d}(\mathbf{r})$ are the time-independent version of these forces as given by (2.45), (2.52), and (2.50), and obtained via them by removal of the time parameter t from the respective quantal sources. The internal magnetic 'force' $\boldsymbol{i}_m(\mathbf{r})$, whose quantal source is the physical current density $\mathbf{j}(\mathbf{r})$, is defined as

$$i_{m,\alpha}(\mathbf{r}) = \sum_{\beta} \nabla_\beta I_{\alpha\beta}(\mathbf{r}), \tag{3.21}$$

with the second rank internal magnetic tensor $I_{\alpha\beta}(\mathbf{r})$ being defined in turn as

$$I_{\alpha\beta}(\mathbf{r}) = \Big[j_\alpha(\mathbf{r}) A_\beta(\mathbf{r}) + j_\beta(\mathbf{r}) A_\alpha(\mathbf{r}) \Big] - \rho(\mathbf{r}) A_\alpha(\mathbf{r}) A_\beta(\mathbf{r}). \tag{3.22}$$

From the 'Quantal Newtonian' First Law of (3.11), we see that in addition to the external electrostatic field $\mathcal{E}(\mathbf{r})$, each electron also experiences a Lorentz field $\mathcal{L}(\mathbf{r})$, (one consistent in form with the force equation of classical physics). There is then an internal field $\mathcal{F}^{\text{int}}(\mathbf{r})$ experienced by each electron. This field via its electron-interaction $\mathcal{E}_{ee}(\mathbf{r})$ component is representative not only of Coulomb correlations as one might expect, but also those due to the Pauli principle as a consequence of the antisymmetric nature of the wave function $\Psi(\mathbf{X})$. Additionally, there is the kinetic field $\mathcal{Z}(\mathbf{r})$ component representative of the motion of the electrons; the differential density $\mathcal{D}(\mathbf{r})$ component representing the electron density, a fundamental property of the system; and an internal magnetic field $\mathcal{I}_m(\mathbf{r})$ contribution due to the presence of the magnetic field $\mathcal{B}(\mathbf{r})$. *Hence, each electron experiences an internal field that encapsulates all the basic properties of the system.* As in classical physics, in summing the 'Quantal Newtonian' First Law over all the electrons, the contribution to the internal field vanishes, leading thereby to Ehrenfest's First Law theorem: $\int \rho(\mathbf{r})\mathcal{F}^{\text{ext}}(\mathbf{r})d\mathbf{r} = 0$. (See also Sect. 2.5).

The Hamiltonian of a system of classical particles in an electrostatic and magnetostatic field contains both a scalar and vector potential representative respectively of these fields. From the correspondence principle, these same potentials appear in the quantum mechanical Hamiltonian (3.5). Hence, it is understood that each electron of the quantum system in such fields experiences a force due to the electrostatic field, and a Lorentz force due to the magnetic field. Whilst the electrostatic force is present via the scalar potential $v(\mathbf{r})$, the Lorentz force does not appear explicitly in the quantum-mechanical Hamiltonian as written in its traditional form of (3.5). (The Hamiltonian can be rewritten (see Chap. 4) such that the Lorentz force is also present.) The 'Quantal Newtonian' First Law on the other hand makes the existence of both forces acting on each electron *explicit* via the external field $\mathcal{F}^{\text{ext}}(\mathbf{r})$ which is the sum of the electrostatic $\mathcal{E}(\mathbf{r})$ and Lorentz $\mathcal{L}(\mathbf{r})$ fields, the latter involving the Lorentz 'force'.

3.1.2 New Insights to the Stationary-State Schrödinger Equation

We next discuss new insights into the stationary-state Schrödinger theory of electrons. These are arrived at via the single-particle perspective of the 'Quantal Newtonian' First Law. For this purpose we consider the Schrödinger equation of (3.10) in which the electrons experience both an external electrostatic and magnetostatic field. Amongst these new insights is the generalization of the Schrödinger equation in which the Hamiltonian is written in terms of the various fields defining the quantum system. And from this generalized form, it then becomes evident that the solution to the Schrödinger equation may be determined self-consistently. *The new understandings are valid for both ground and excited states.*

(i) The 'Quantal Newtonian' First Law of (3.11) affords a rigorous physical interpretation of the external electrostatic potential $v(\mathbf{r})$: *it is the work done to move an electron from some reference point at infinity to its position at* $\mathbf{r}$ *in the force of a conservative field* $\mathcal{F}(\mathbf{r})$*:*

$$v(\mathbf{r}) = \int_{\infty}^{\mathbf{r}} \mathcal{F}(\mathbf{r}') \cdot d\boldsymbol{\ell}', \tag{3.23}$$

where $\mathcal{F}(\mathbf{r}) = \mathcal{F}^{\text{int}}(\mathbf{r}) - \mathcal{L}(\mathbf{r}) = \mathcal{E}_{\text{ee}}(\mathbf{r}) - \mathcal{Z}(\mathbf{r}) - \mathcal{D}(\mathbf{r}) - \mathcal{I}_m(\mathbf{r}) - \mathcal{L}(\mathbf{r})$. Since $\nabla \times \mathcal{F}(\mathbf{r}) = 0$, this work done is *path-independent*. Thus, we now understand, in the rigorous classical sense of a potential being the work done in a conservative field, that $v(\mathbf{r})$ represents a potential energy, *viz.* that of an electron. Furthermore, that 'classical' field is now *explicitly* defined in terms of quantum-mechanical properties. It is reiterated that (3.23) (or equivalently the 'Quantal Newtonian' First Law) is valid for *arbitrary state*. What this means is that irrespective of whether the state is a ground, excited, or a degenerate state, the work done in the corresponding field $\mathcal{F}(\mathbf{r})$) for that state is always the same, *viz.* $v(\mathbf{r})$.

(ii) What the above physical interpretation of the potential $v(\mathbf{r})$ shows is that it can no longer be thought of as an *independent* entity. It is *inherently* dependent upon all the properties of the system via the various components of the internal field $\mathcal{F}^{\text{int}}(\mathbf{r})$, and the Lorentz field $\mathcal{L}(\mathbf{r})$ via the current density $\mathbf{j}(\mathbf{r})$.

(iii) The field $\mathcal{F}(\mathbf{r})$ in (3.23) is comprised of the constituent fields representative of the system. Hence, we now understand that the potential function $v(\mathbf{r})$ is a sum of constituent functions each of which represents an intrinsic property of the system. For a quantum system for which each of the individual component fields of $\mathcal{F}^{\text{int}}(\mathbf{r})$, and the Lorentz field $\mathcal{L}(\mathbf{r})$, are conservative, then the potential $v(\mathbf{r})$ may be written as

$$v(\mathbf{r}) = W_{\text{ee}}(\mathbf{r}) + W_z(\mathbf{r}) + W_d(\mathbf{r}) + W_i(\mathbf{r}) + W_\ell(\mathbf{r}), \tag{3.24}$$

where $W_{\text{ee}}(\mathbf{r})$, $W_z(\mathbf{r})$, $W_d(\mathbf{r})$, $W_i(\mathbf{r})$, and $W_\ell(\mathbf{r})$ are the work done in the fields $\mathcal{E}_{\text{ee}}(\mathbf{r})$, $\mathcal{Z}(\mathbf{r})$, $\mathcal{D}(\mathbf{r})$, $\mathcal{I}_m(\mathbf{r})$, and $\mathcal{L}(\mathbf{r})$, respectively. These works done are potentials in that they are each *path-independent*. Hence, the individual contributions to the potential $v(\mathbf{r})$ due to the electron correlations arising from the Pauli principle and

Coulomb repulsion, and those due to the kinetic effects, the density and the magnetic field can be delineated.

(iv) Define the field $\boldsymbol{\mathcal{M}}(\mathbf{r})$ as the sum of the Lorentz $\boldsymbol{\mathcal{L}}(\mathbf{r})$ and internal magnetic $\boldsymbol{\mathcal{I}}_m(\mathbf{r})$ field components:

$$\boldsymbol{\mathcal{M}}(\mathbf{r}) = -[\boldsymbol{\mathcal{L}}(\mathbf{r}) + \boldsymbol{\mathcal{I}}_m(\mathbf{r})]. \tag{3.25}$$

If the field $\boldsymbol{\mathcal{M}}(\mathbf{r})$ is conservative, i.e. $\boldsymbol{\nabla} \times \boldsymbol{\mathcal{M}}(\mathbf{r}) = 0$, then one can define a magnetic scalar potential $v_m(\mathbf{r})$ as

$$\boldsymbol{\mathcal{M}}(\mathbf{r}) = -\boldsymbol{\nabla} v_m(\mathbf{r}). \tag{3.26}$$

This implies that $v_m(\mathbf{r})$ is *path-independent*. The magnetic field contribution E_{mag} to the energy is then

$$E_{\text{mag}} = \int \rho(\mathbf{r}) v_m(\mathbf{r}) d\mathbf{r}. \tag{3.27}$$

This is similar to the electrostatic field contribution to the energy (see 2.129). The E_{mag} can also be written in integral virial form depending on the degree of the homogenous function $v_m(\mathbf{r})$. If $v_m(\mathbf{r})$ is of degree 2 as for the harmonic oscillator, then

$$E_{\text{mag}} = -\frac{1}{2} \int \rho(\mathbf{r}) \mathbf{r} \cdot \boldsymbol{\mathcal{M}}(\mathbf{r}) d\mathbf{r}. \tag{3.28}$$

In the general case when $\boldsymbol{\nabla} \times \boldsymbol{\mathcal{M}}(\mathbf{r}) \neq 0$, the expression is

$$E_{\text{mag}} = \int \rho(\mathbf{r}) \mathbf{r} \cdot \boldsymbol{\mathcal{M}}(\mathbf{r}) d\mathbf{r}. \tag{3.29}$$

The total energy is then

$$E = T + E_H + E_{xc} + E_{es} + E_{\text{mag}}, \tag{3.30}$$

where the definitions of the kinetic T, Hartree E_H, Pauli-Coulomb E_{xc}, and electrostatic E_{es} energies are given in Sect. 2.8.2. (Note that $E_{es} = E_{\text{ext}}$ of 2.129.)

(v) As each component of the internal field $\boldsymbol{\mathcal{F}}^{\text{int}}(\mathbf{r})$, and the Lorentz field $\boldsymbol{\mathcal{L}}(\mathbf{r})$, are obtained from quantal sources that are expectation values of Hermitian operators taken with respect to the wave function $\Psi(\mathbf{X})$, we see that the field $\boldsymbol{\mathcal{F}}(\mathbf{r})$ of (3.23) is a *functional* of $\Psi(\mathbf{X})$, i.e. $\boldsymbol{\mathcal{F}}(\mathbf{r}) = \boldsymbol{\mathcal{F}}[\Psi(\mathbf{X})](\mathbf{r})$. Thus, it follows from (3.23), that the potential $v(\mathbf{r})$ is a *functional* of $\Psi(\mathbf{X}) : v(\mathbf{r}) = v[\Psi(\mathbf{X})](\mathbf{r})$. Hence, (3.23) may be written as

$$v[\Psi(\mathbf{X})](\mathbf{r}) = \int_{\infty}^{\mathbf{r}} \boldsymbol{\mathcal{F}}[\Psi(\mathbf{X})](\mathbf{r}') \cdot d\boldsymbol{\ell}'. \tag{3.31}$$

The *functional* $v[\Psi(\mathbf{X})](\mathbf{r})$ is *exactly known* via (3.31).

(vi) The functional $v[\Psi(\mathbf{X})](\mathbf{r})$ is *universal*. By this meant that the definition of the potential $v(\mathbf{r})$, and of its explicit analytical functional dependence on $\Psi(\mathbf{X})$, is the *same* for *all* potentials $v(\mathbf{r})$, and thus the *same* for *all* electronic systems.

(vii) On substituting the functional $v[\Psi(\mathbf{X})](\mathbf{r})$ into the Schrödinger equation (3.10), we see that it can be written as

$$\left[\frac{1}{2}\sum_k\left(\hat{\mathbf{p}}_k+\mathbf{A}(\mathbf{r}_k)\right)^2+\frac{1}{2}\sum_{k,\ell}{}'\frac{1}{|\mathbf{r}_k-\mathbf{r}_\ell|}+\sum_k v\big[\Psi(\mathbf{X})\big](\mathbf{r}_k)\right]\Psi(\mathbf{X})$$
$$=E\big[\Psi(\mathbf{X})\big]\Psi(\mathbf{X}), \tag{3.32}$$

or equivalently as

$$\left[\frac{1}{2}\sum_k\left(\hat{\mathbf{p}}_k+\mathbf{A}(\mathbf{r}_k)\right)^2+\frac{1}{2}\sum_{k,\ell}{}'\frac{1}{|\mathbf{r}_k-\mathbf{r}_\ell|}+\sum_k\int_\infty^{\mathbf{r}_k}\mathcal{F}\big[\Psi(\mathbf{X})\big](\mathbf{r})\cdot d\boldsymbol{\ell}\right]\Psi(\mathbf{X})$$
$$=E\big[\Psi(\mathbf{X})\big]\Psi(\mathbf{X}). \tag{3.33}$$

The Hamiltonian $\hat{H}$ is thus a functional of $\Psi(\mathbf{X})$, i.e. $\hat{H}=\hat{H}[\Psi(\mathbf{X})]$, and the Schrödinger equation is then of the more general form $\hat{H}[\Psi(\mathbf{X})]\Psi(\mathbf{X})=E[\Psi(\mathbf{X})]\Psi(\mathbf{X})$. The equations (3.32) and (3.33) constitute the generalized form of the Schrödinger equation.

(viii) Written in this generalized form, *the intrinsic self-consistent nature of the Schrödinger equation becomes evident.* To understand this, recall what is meant by a functional such as $v[\Psi(\mathbf{X})](\mathbf{r})$. *The functional $v[\Psi(\mathbf{X})](\mathbf{r})$ means that for each different $\Psi(\mathbf{X})$, one obtains a different function $v(\mathbf{r})$.*

What then is the self-consistency procedure for the solution of the generalized Schrödinger equation of (3.32, 3.33)? To explain the self-consistency procedure, let us first consider how the wave functions $\Psi(\mathbf{X})$ and energies E are determined for the familiar case of when the Hamiltonian $\hat{H}$ is known. In other words, since the kinetic and electron-interaction potential operators are assumed known, the binding potential $v(\mathbf{r})$ is also assumed to be known.

Case A: Hamiltonian Known

Since for the electronic system, the kinetic and electron-interaction potential operators are assumed known, by a known Hamiltonian is meant that the binding potential $v(\mathbf{r})$ function too is assumed to be known.

One begins with an approximate input wave function $\Psi^{\rm in}(\mathbf{X})$. For this $\Psi^{\rm in}(\mathbf{X})$ one obtains the various quantal sources, and from these sources the various components of and hence the field $\mathcal{F}[\Psi^{\rm in}(\mathbf{X})](\mathbf{r})$ of (3.23), and from this field the potential $v[\Psi^{\rm in}(\mathbf{X})](\mathbf{r})$. (The approximate $v[\Psi^{\rm in}(\mathbf{X})](\mathbf{r})$ differs from the known $v(\mathbf{r})$.) This then determines the Hamiltonian $\hat{H}[\Psi^{\rm in}(\mathbf{X})]$. The Schrödinger equation $\hat{H}[\Psi^{\rm in}(\mathbf{X})]\Psi^{\rm out}=E[\Psi^{\rm out}]\Psi^{\rm out}$ is then solved to obtain a new output approximate solution $\Psi^{\rm out}(\mathbf{X})$ and eigenenergy $E[\Psi^{\rm out}(\mathbf{X})]$. The output solution $\Psi^{\rm out}(\mathbf{X})$ is then treated as the new approximate input wave function $\Psi^{\rm in}(\mathbf{X})$, and the entire process repeated until the $\Psi^{\rm in}(\mathbf{X})=\Psi^{\rm out}(\mathbf{X})$, or when self-consistency is achieved. At the

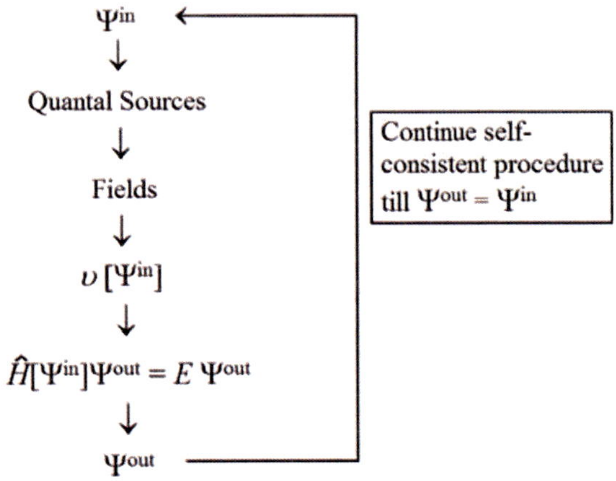

Fig. 3.1 Procedure for the self-consistent solution of the stationary state Schrödinger equation

final iteration, the known binding potential $v(\mathbf{r})$ will be recovered. (See also Fig. 3.1). In this manner, the *exact* wave function $\Psi(\mathbf{X})$, and eigenenergy E are obtained. As in any self-consistent procedure, the choice of the initial input wave function $\Psi^{\text{in}}(\mathbf{X})$ *apropos* to the physical system of interest is key.

Case B: Hamiltonian Unknown

By an unknown Hamiltonian $\hat{H}$ is meant that the binding potential $v(\mathbf{r})$ of the electrons is not known. As was the case with 'artificial atoms' or quantum dots for which the binding potential was initially unknown, there could be other electronic devices created in the future for which this may also be the case. The unknown potential $v(\mathbf{r})$ could then be determined via the self-consistent solution of (3.32, 3.33). The self-consistent procedure is the same as described above. Once again, the choice of the input wave function $\Psi^{\text{in}}(\mathbf{X})$ is key. One then determines the approximate $v[\Psi^{\text{in}}(\mathbf{X})](\mathbf{r})$, solves the corresponding Schrödinger equation to determine the $\Psi^{\text{out}}(\mathbf{X})$ and the $E[\Psi^{\text{out}}(\mathbf{X})]$, and one continues this iterative procedure till self-consistency is achieved. That is till the $\Psi^{\text{in}}(\mathbf{X}) = \Psi^{\text{out}}(\mathbf{X})$. One would then obtain the correct corresponding scalar potential $v(\mathbf{r})$. For natural atoms, molecules and solids, this potential would turn out to be Coulombic. For quantum dots or quantum dot molecules, the potential would be harmonic. There may exist other solutions for which the potential $v(\mathbf{r})$ is different. As in other self-consistent procedures, it is only after self-consistency is achieved that one must judge and test with experiment whether or not the solution is physically meaningful. Thus, the self-consistent solution of (3.2) (or (3.32, 3.33)) determines $\{v(\mathbf{r}), \Psi(\mathbf{X}), E\}$ or equivalently $\{\hat{H}[\Psi], \Psi(\mathbf{X}), E\}$ for the specific state.

It is evident that the self-consistent solution of the Schrödinger equation when the Hamiltonian is known constitutes a special case of the solution of the generalized Schrödinger equation. The solution $\{\Psi(\mathbf{X}), E\}$ of the Schrödinger equation (3.1) is equivalent to the final iteration of the self-consistent solution of (3.2) or ((3.32), (3.33)).

(ix) *Observe that in writing the Schrödinger equation in its generalized form as in (3.32) and (3.33), the magnetic field $\mathcal{B}(\mathbf{r})$ now appears in the Hamiltonian $\hat{H}[\Psi(\mathbf{X})]$ explicitly via the Lorentz field $\mathcal{L}(\mathbf{r})$* (see (3.31)). It is the intrinsic self-consistent nature of the equation that demands the presence of the field $\mathcal{B}(\mathbf{r})$ in the Hamiltonian. In other words, since the Hamiltonian $\hat{H}[\Psi(\mathbf{X})]$ is being determined self-consistently, all the information of the physical system, *viz.* that for the electrons and any external fields must be present in it. Thus, in addition to the vector potential $\mathbf{A}(\mathbf{r})$, the magnetic field $\mathcal{B}(\mathbf{r})$ too appears in the generalized Schrödinger equation. Of course, equivalently, the field $\mathcal{B}(\mathbf{r})$ could be expressed in terms of the vector potential $\mathbf{A}(\mathbf{r})$ by choice of a gauge. This then shows that when written in self-consistent form, there exists another component of the Hamiltonian involving the vector potential.

(x) The presence of solely an electrostatic field $\mathcal{E}(\mathbf{r}) = -\nabla v(\mathbf{r})$, i.e. with $\mathcal{B}(\mathbf{r}) = 0$, constitutes a special case of the stationary-state theory discussed above.

3.2 Hartree-Fock and Hartree Theories in Terms of Quantal Sources and Fields

The Hartree-Fock [22, 23] and Hartree [21] theories are approximations to the Schrödinger theory of electrons. These theories address the same physical system as defined by the Hamiltonian of Schrödinger theory. For purposes of our discussion, let us consider a system of N electrons in some binding external electrostatic field $\mathcal{E}(\mathbf{r}) = -\nabla v(\mathbf{r})/e$. The time-independent Hamiltonian $\hat{H}$ (in atomic units $e = \hbar = m = 1$) for the system is then the sum of the kinetic energy $\hat{T}$, electron interaction potential energy $\hat{U}$, and external potential energy $\hat{V}$ operators:

$$\hat{H} = \hat{T} + \hat{U} + \hat{V}, \tag{3.34}$$

where

$$\hat{T} = -\frac{1}{2}\sum_i \nabla_i^2, \tag{3.35}$$

$$\hat{U} = \frac{1}{2}\sum_{i,j}{}' \frac{1}{|\mathbf{r}_i - \mathbf{r}_j|}, \tag{3.36}$$

and

$$\hat{V} = \sum_i v(\mathbf{r}_i). \tag{3.37}$$

The next step within Schrödinger theory would be to solve the Schrödinger equation $\hat{H}\Psi(\mathbf{X}) = E\Psi(\mathbf{X})$ for the many-electron wave function $\Psi(\mathbf{X})$ and eigenenergy E for a particular state. The wave function $\Psi(\mathbf{X})$ has two principal attributes: *(i)* it is antisymmetric in an interchange of the coordinates of any two electrons so as to ensure the satisfaction of the Pauli exclusion principle, and *(ii)* it explicitly accounts for the Coulomb repulsion between the electrons. In other words, it accounts for electron correlations that arise as a result of the Pauli exclusion principle and those due to the Coulomb repulsion. Although Hartree-Fock and Hartree theories of electronic structure address the same physical system, they differ from Schrödinger theory in many fundamental ways. Listed below are salient features of these theories.

(a) The Hartree-Fock and Hartree theories are *single-particle* theories in which each electron is treated as being *quasi-independent* of the others. (These are thus *quasi-particles* although we will continue to refer to them as electrons.) Each is thus assigned a spin-orbital $\phi_i(\mathbf{x})$ which is a product of a spatial $\psi_i(\mathbf{r})$ and spin $\chi_i(\sigma)$ component: $\phi_i(\mathbf{x}) = \psi_i(\mathbf{r})\chi_i(\sigma)$.

(b) In Hartree-Fock theory, the wave function of the system $\Psi^{\mathrm{HF}}(\mathbf{X})$ is approximated to be a *single* Slater determinant of the orbitals $\phi_i(\mathbf{x})$. In Hartree theory, the wave function $\Psi^{\mathrm{H}}(\mathbf{X})$ is approximated to be a *product* of such orbitals. For the *ground* state, the N spin-orbitals $\phi_i(\mathbf{x})$ occupy the lowest energy levels consistent with the Pauli exclusion principle.

(c) In neither Hartree-Fock nor Hartree theory are the effects of Coulomb correlations explicitly incorporated in the wave function.

(d) The Hartree-Fock theory determinantal wave function $\Psi^{\mathrm{HF}}(\mathbf{X})$ is antisymmetric in an interchange of the coordinates, both spatial and spin, of any two electrons. As such $\Psi^{\mathrm{HF}}(\mathbf{X})$ explicitly accounts for electron correlations that arise due to the Pauli exclusion principle. (It is because the wave function $\Psi^{\mathrm{HF}}(\mathbf{X})$ accounts only for correlations due to the Pauli principle that the electrons are considered quasi-independent.)

(e) The Hartree theory wave function $\Psi^{\mathrm{H}}(\mathbf{X})$ is not antisymmetric in an interchange of the coordinates of any two electrons, and thus does not obey the Pauli exclusion principle. In the application of Hartree theory, it is the equivalent statement of the Pauli principle that no two electrons can occupy the same quantum state that is then employed in the occupation of states [24, 35] (see Sect 9.2.1 of [35]). In Hartree theory, each electron (quasi-particle) is considered to move in the external potential $v(\mathbf{r})$ and the potential due to the charge distribution of *all* the *other* electrons. Each electron thus has an *orbital-dependent potential energy*. Hence, in Hartree theory, the electrons are once again not independent, but rather quasi-independent.

(f) In both Hartree-Fock and Hartree theories, the N spin-orbitals $\phi_i(\mathbf{x})$ obtained are those that lead to the best value of the total energy. The respective total energies $\{E^{\mathrm{HF}}, E^{\mathrm{H}}\}$ are expectations of the Hamiltonian $\hat{H}$ of (3.34) taken with respect to the wave functions $\{\Psi^{\mathrm{HF}}(\mathbf{X}), \Psi^{\mathrm{H}}(\mathbf{X})\}$. As the wave functions $\{\Psi^{\mathrm{HF}}(\mathbf{X}), \Psi^{\mathrm{H}}(\mathbf{X})\}$ are a Slater determinant or product of the spin-orbitals $\phi_i(\mathbf{x})$, the total energies $\{E^{\mathrm{HF}}, E^{\mathrm{H}}\}$ are functionals of the corresponding orbitals $\phi_i(\mathbf{x})$, *i.e.* $E^{\mathrm{HF}} = E^{\mathrm{HF}}\{\phi_i(\mathbf{x})\}$ and $E^{\mathrm{H}} = E^{\mathrm{H}}[\phi_i(\mathbf{x})]$.

(g) The orbitals $\phi_i(\mathbf{x})$ that lead to the best value for the total energy $E^{\mathrm{HF}}\{\phi_i(\mathbf{x})\}$ and $E^{\mathrm{H}}\{\phi_i(\mathbf{x})\}$ are then obtained by application of the variational principle [9] for the energy for arbitrary variations $\delta\phi_i(\mathbf{x})$ of the orbitals in the functionals $E^{\mathrm{HF}}\{\phi_i(\mathbf{x})\}$ and $E^{\mathrm{H}}\{\phi_i(\mathbf{x})\}$. The condition of stationarity of the energy, i.e. the vanishing of the first order variation of the energy: $\delta E^{\mathrm{HF}} = 0$; $\delta E^{\mathrm{H}} = 0$, then leads to the Hartree-Fock and Hartree theory integro-differential eigenvalue equations for the orbitals $\phi_i(\mathbf{x})$ and the corresponding eigenvalues ϵ_i. The corresponding total energies $E^{\mathrm{HF}}\{\phi_i(\mathbf{x})\}$ and $E^{\mathrm{H}}[\phi_i(\mathbf{x})]$ as obtained for the energy optimized orbitals $\phi_i(\mathbf{x})$ then constitute *rigorous upper bounds* to the true total ground state energy E. As the Hartree-Fock theory determinantal wave function $\Psi^{\mathrm{HF}}(\mathbf{X})$ explicitly incorporates electron correlations due to the Pauli principle, and the Hartree theory wave function $\Psi^{\mathrm{H}}(\mathbf{X})$ does not, the energy E^{HF} is a superior upper bound to the energy E^{H} [24, 25, 35].

(h) In the Hartree-Fock and Hartree theory single-particle integro-differential equations, the kinetic $-\frac{1}{2}\nabla^2$ and external potential $v(\mathbf{r})$ energy operators remain unchanged. It is the electron interaction operator $\hat{U}$ of the Schrödinger theory Hamiltonian $\hat{H}$ of (3.34) that is modified and approximated. Furthermore, these terms approximating $\hat{U}$ are functionals of the orbitals $\phi_i(\mathbf{x})$. Thus, the Hartree-Fock and Hartree theory single-particle Hamiltonians are functionals of the orbitals $\phi_i(\mathbf{x})$, i.e. $\hat{H}^{\mathrm{HF}} = \hat{H}^{\mathrm{HF}}\{\phi_i(\mathbf{x})\}$ and $\hat{H}^{\mathrm{H}} = \hat{H}^{\mathrm{H}}\{\phi_i(\mathbf{x})\}$. The Hartree-Fock and Hartree theory differential equations can then be written, respectively, as

$$\hat{H}^{\mathrm{HF}}\{\phi_i(\mathbf{x})\}\phi_i(\mathbf{x}) = \epsilon_i\{\phi_i(\mathbf{x})\}\phi_i(\mathbf{x}), \tag{3.38}$$

and

$$\hat{H}^{\mathrm{H}}\{\phi_i(\mathbf{x})\}\phi_i(\mathbf{x}) = \epsilon_i\{\phi_i(\mathbf{x})\}\phi_i(\mathbf{x}). \tag{3.39}$$

The equations (3.38) and (3.39) are once again of the form $\hat{L}[\zeta]\zeta = \lambda[\zeta]\zeta$ (see the Introduction) and hence must be solved self-consistently for the determination of the Hamiltonians $\hat{H}^{\mathrm{HF}}\{\phi_i(\mathbf{x})\}$, $\hat{H}^{\mathrm{H}}\{\phi_i(\mathbf{x})\}$ and the corresponding $\{\phi_i(\mathbf{x}), \epsilon_i\}$. Note that in the above self-consistent procedure, the external potential operator $v(\mathbf{r})$ in each Hamiltonian is *assumed known*.

(i) In Hartree-Fock theory, the operator approximating the electron-interaction operator $\hat{U}$—the eponymous exchange operator $\hat{v}_{x,i}(\mathbf{x})$—is said to be a *nonlocal* operator. This is because operating with it on $\phi_i(\mathbf{x})$ depends upon the value of $\phi_i(\mathbf{x})$ throughout all space, not just at $\mathbf{x}$. (This is in contrast to *local* or *multiplicative* operators such as the external potential operator $v(\mathbf{r})$.)

(j) The exchange operator $\hat{v}_{x,i}(\mathbf{x})$ can be afforded a physical interpretation, due to Bardeen [45] and Slater [46], in which *each* electron is assigned an *orbital-dependent local or multiplicative* 'exchange potential'. Hence, within Hartree-Fock theory, *each* electron is associated with a potential energy. In the Slater-Bardeen interpretation, Hartree-Fock theory can then be considered an *orbital-dependent-potential* theory. This is akin to Hartree theory which too is an orbital-dependent-potential theory. Because in these theories, each electron has a *different* potential energy, we refer to them as being *nonlocal potential* theories. (This is in contrast to *local potential* theories such as Density Functional [26, 27] and Quantal Density Functional theory [32–35], in which each quasi-particle (noninteracting fermion) has the *same local* potential energy.)

(k) Just as in Schrödinger theory in which *each electron* satisfies the 'Quantal Newtonian' First Law equation of motion, each *quasi-particle* of Hartree-Fock and Hartree theories too satisfies a corresponding 'Quantal Newtonian' First Law. That is, each quasi-particle experiences an external $\mathcal{F}^{\mathrm{ext}}(\mathbf{r}) = \boldsymbol{\mathcal{E}}(\mathbf{r}) = -\boldsymbol{\nabla}v(\mathbf{r})/e$ and an internal $\mathcal{F}^{\mathrm{int}}(\mathbf{r})$ field, the sum of which vanishes. The internal field $\mathcal{F}^{\mathrm{int}}(\mathbf{r})$ is a sum of components representative of the electron correlations due to the Pauli principle, the density and the kinetic effects. These individual fields arise from quantal sources that are expectations of Hermitian operators taken with respect to the wave functions $\{\Psi^{\mathrm{HF}}(\mathbf{X}), \Psi^{\mathrm{H}}(\mathbf{X})\}$ which depend upon the orbitals $\phi_i(\mathbf{x})$. Thus, the internal field is a functional of these orbitals: $\mathcal{F}^{\mathrm{int}}(\mathbf{r}) = \mathcal{F}^{\mathrm{int}}\{\phi_i(\mathbf{x})\}(\mathbf{r})$. From the 'Quantal Newtonian' First Law, it follows that the *external potential* $v(\mathbf{r})$ *is the work done in the force of the internal field* $\mathcal{F}^{\mathrm{int}}(\mathbf{r})$. Hence, the external potential $v(\mathbf{r})$ is a functional of the orbitals $\phi_i(\mathbf{x})$, *i.e.* $v(\mathbf{r}) = v\{\phi_i(\mathbf{x})\}(\mathbf{r})$. Thus, all the terms in the Hamiltonians $\hat{H}^{\mathrm{HF}}$ and $\hat{H}^{\mathrm{H}}$, with the sole exception of the kinetic energy operator, are functionals of the orbitals $\phi_i(\mathbf{x})$. Hence, as was the case with the Schrödinger equation, it is possible to generalize both the Hartree-Fock and Hartree theory integro-differential equations. In this manner, even the potential $v(\mathbf{r})$ in the Hamiltonians $\hat{H}^{\mathrm{HF}}$ and $\hat{H}^{\mathrm{H}}$ can be determined in a self-consistent manner. Thus, one can perform Hartree-Fock and Hartree theory self-consistent calculations for physical systems for which the external potential $v(\mathbf{r})$ may not be known, and thereby determine it.

In the subsections to follow, Hartree-Fock and Hartree theories are described in terms of quantal sources and fields, and the derivation of the corresponding integro-differential equations provided. The interpretation of the Hartree-Fock theory equation by Slater and Bardeen as one in which each electron (quasi-particle) is assigned a separate potential is also described. Important theorems within Hartree-Fock theory are stated and explained. Finally, the 'Quantal Newtonian' First Law satisfied by each quasi-particle within Hartree-Fock and Hartree theory is described. The generalization of the corresponding integro-differential equations of these theories then follows from the respective laws.

3.2.1 Hartree-Fock Theory in Terms of Quantal Sources and Fields

The expressions for the quantal sources, and the fields to which they give rise to, within Hartree-Fock theory are defined in this section. The resulting expressions for the total energy and of its components as defined in terms of these quantal sources and fields are also provided. The fields are of significance in the description of the 'Quantal Newtonian' First Law within the theory, and thereby to the generalization of the Hartree-Fock theory integro-differential eigenvalue equation.

Wave Function

In Hartree-Fock theory, the wave function $\Psi(\mathbf{X})$ of the interacting system defined by the Hamiltonian $\hat{H}$ of (3.34) is approximated by $\Psi^{\mathrm{HF}}(\mathbf{X})$ which is a Slater determinant $\Phi\{\phi_i(\mathbf{x})\}$ of spin-orbitals $\phi_i(\mathbf{x}) = \psi_i(\mathbf{r})\chi_i(\sigma)$:

$$\Psi^{\mathrm{HF}}(\mathbf{X}) = \Phi\{\phi_i(\mathbf{x})\} = \frac{1}{\sqrt{N!}}\, det\ \phi_i(\mathbf{r}_j\sigma_j). \tag{3.40}$$

In this manner, the wave function satisfies the Pauli principle in that it is anti-symmetric in an interchange of the coordinates of two electrons including the spin coordinate.

Quantal Sources and Their Sum Rules

The quantal sources: the density $\rho(\mathbf{r})$, the Dirac spinless single-particle density matrix $\gamma^{\mathrm{HF}}(\mathbf{rr}')$, and the pair-correlation density $g^{\mathrm{HF}}(\mathbf{rr}')$, are obtained as the expectation values of the operators $\hat{\rho}(\mathbf{r})$, $\hat{\gamma}(\mathbf{rr}')$, and $\hat{P}(\mathbf{rr}')/\rho(\mathbf{r})$ of (2.12), (2.16) and (2.28) taken with respect to the wave function of (3.40). The expressions for these sources in terms of the spin orbitals $\phi_i(\mathbf{x})$, and the sum rules satisfied by them are given below:

Density

$$\rho(\mathbf{r}) = \sum_{\sigma}\sum_{i} |\phi_i(\mathbf{x})|^2, \tag{3.41}$$

$$\int \rho(\mathbf{r})d\mathbf{r} = N. \tag{3.42}$$

Dirac Density Matrix

$$\gamma^{\mathrm{HF}}(\mathbf{rr}') = \sum_{\sigma}\sum_{i} \phi_i^{\star}(\mathbf{r}\sigma)\phi_i(\mathbf{r}'\sigma), \tag{3.43}$$

$$\gamma^{\mathrm{HF}}(\mathbf{rr}) = \rho(\mathbf{r}), \tag{3.44}$$

$$\gamma^{\mathrm{HF}}(\mathbf{rr}') = \gamma^{\mathrm{HF}\star}(\mathbf{r}'\mathbf{r}), \tag{3.45}$$

$$\int \gamma^{\mathrm{HF}}(\mathbf{rr}'')\gamma^{\mathrm{HF}}(\mathbf{r}''\mathbf{r}')d\mathbf{r}'' = \gamma^{\mathrm{HF}}(\mathbf{rr}'). \tag{3.46}$$

Due to the equality in the sum rule of (3.46), the Dirac density matrix $\gamma^{\mathrm{HF}}(\mathbf{rr}')$ is said to be *idempotent*.

Pair-correlation Density

$$g^{\mathrm{HF}}(\mathbf{rr}') = \rho(\mathbf{r}') + \rho_x^{\mathrm{HF}}(\mathbf{rr}'). \tag{3.47}$$

$$\int g^{\mathrm{HF}}(\mathbf{rr}')d\mathbf{r}' = N - 1. \tag{3.48}$$

where $\rho_x^{\mathrm{HF}}(\mathbf{rr}')$ is the Fermi hole charge to be discussed below.

Note that, whereas the density $\rho(\mathbf{r})$ is a *local* (static) source, the Dirac single-particle density matrix $\gamma^{\mathrm{HF}}(\mathbf{rr}')$ and the pair-correlation density $g^{\mathrm{HF}}(\mathbf{rr}')$ are *nonlocal* (dynamic) quantal sources, i.e. their structure changes as a function of electron position at $\mathbf{r}$.

Fermi Hole

In (3.47), the term $\rho_x^{\mathrm{HF}}(\mathbf{rr}')$ is the Hartree-Fock theory *Fermi hole* charge distribution [47]. The Fermi hole is the reduction in density at $\mathbf{r}'$ due to the presence of an electron of *parallel* spin at $\mathbf{r}$. It represents the reduction in probability of two electrons of *parallel* spin approaching each other. (If both the spin and spatial coordinates of two orbitals are the same, then two columns of the Slater determinant of (3.40) are the same, and hence the determinant vanishes. Hence, the probability of two electrons of parallel spin being on top of each other vanishes as a consequence of the Pauli principle.) The expression for the Fermi hole in terms of the spin-orbitals $\phi_i(\mathbf{x})$ is thus

$$\rho_x^{\mathrm{HF}}(\mathbf{rr}') = -\frac{\sum\limits_{\substack{i,j \\ (\text{spin } j \| \text{spin } i)}} \phi_i^*(\mathbf{x})\phi_j^*(\mathbf{x}')\phi_i(\mathbf{x}')\phi_j(\mathbf{x})}{\rho(\mathbf{r})}. \tag{3.49}$$

It may be expressed in terms of the Dirac density matrix $\gamma^{\mathrm{HF}}(\mathbf{rr}')$ as

$$\rho_x^{\mathrm{HF}}(\mathbf{rr}') = -\frac{|\gamma^{\mathrm{HF}}(\mathbf{rr}')|^2}{2\rho(\mathbf{r})}. \tag{3.50}$$

The Fermi hole satisfies the sum rules

$$\rho_x^{\mathrm{HF}}(\mathbf{rr}') \leq 0, \tag{3.51}$$

$$\rho_x^{\mathrm{HF}}(\mathbf{rr}) = -\rho(\mathbf{r})/2, \tag{3.52}$$

$$\int \rho_x^{\mathrm{HF}}(\mathbf{rr}')d\mathbf{r}' = -1. \tag{3.53}$$

The Fermi hole charge thus constitutes a quantal source. Further, it is a *nonlocal* or *dynamic* charge distribution: it is the nonlocal component of the pair-correlation density $g^{\mathrm{HF}}(\mathbf{rr}')$. For nonuniform density systems, the structure of the Fermi hole changes as a function of electron position at $\mathbf{r}$. For uniform electron gas systems, its structure remains unchanged. Note also that the self-interaction term in the expression for the Fermi hole is canceled by a similar term in the density (see (3.47)), and as such there is no self-interaction in the pair-correlation density $g^{\mathrm{HF}}(\mathbf{rr}')$.

Fields and Forces

The fields descriptive of the system within the Hartree-Fock theory approximation as obtained from the respective quantal sources are the following. The electron-interaction field $\boldsymbol{\mathcal{E}}_{ee}^{\mathrm{HF}}(\mathbf{r})$ is obtained from its quantal source, the pair-correlation density $g^{\mathrm{HF}}(\mathbf{rr}')$, via Coulomb's Law as

$$\boldsymbol{\mathcal{E}}_{ee}^{\mathrm{HF}}(\mathbf{r}) = \int \frac{g^{\mathrm{HF}}(\mathbf{rr}')(\mathbf{r}-\mathbf{r}')}{|\mathbf{r}-\mathbf{r}'|^3}d\mathbf{r}'. \tag{3.54}$$

The field can also be expressed in terms of the corresponding electron-interaction 'force' $\mathbf{e}_{ee}^{HF}(\mathbf{r})$ and the density $\rho(\mathbf{r})$ as

$$\mathcal{E}_{ee}^{\mathrm{HF}}(\mathbf{r}) = \frac{\mathbf{e}_{ee}^{\mathrm{HF}}(\mathbf{r})}{\rho(\mathbf{r})}, \tag{3.55}$$

where $\mathbf{e}_{ee}^{HF}(\mathbf{r})$ is determined via Coulomb's Law from the pair-correlation function $P^{HF}(\mathbf{rr}') = \langle \Phi\{\phi_i(\mathbf{x})\}|\hat{P}(\mathbf{rr}')|\Phi\{\phi_i(\mathbf{x})\}\rangle$ obtained from the Slater determinant $\Phi\{\phi_i(\mathbf{x})\}$ with $\hat{P}(\mathbf{rr}')$ the pair-correlation operator of (2.28). The 'force' $\mathbf{e}_{ee}^{HF}(\mathbf{r})$ is thus

$$\mathbf{e}_{ee}^{\mathrm{HF}}(\mathbf{r}) = \int \frac{P^{\mathrm{HF}}(\mathbf{rr}')(\mathbf{r}-\mathbf{r}')}{|\mathbf{r}-\mathbf{r}'|^3} d\mathbf{r}'. \tag{3.56}$$

On employing the decomposition of $g^{\mathrm{HF}}(\mathbf{rr}')$ of (3.47) in (3.54), the electron-interaction field $\mathcal{E}_{ee}^{\mathrm{HF}}(\mathbf{r})$ may be written as a sum of its Hartree $\mathcal{E}_H(\mathbf{r})$ and Pauli $\mathcal{E}_x^{\mathrm{HF}}(\mathbf{r})$ components:

$$\mathcal{E}_{ee}^{\mathrm{HF}}(\mathbf{r}) = \mathcal{E}_H(\mathbf{r}) + \mathcal{E}_x^{\mathrm{HF}}(\mathbf{r}), \tag{3.57}$$

where

$$\mathcal{E}_H(\mathbf{r}) = \int \frac{\rho(\mathbf{r}')(\mathbf{r}-\mathbf{r}')}{|\mathbf{r}-\mathbf{r}'|^3} d\mathbf{r}', \tag{3.58}$$

and

$$\mathcal{E}_x^{\mathrm{HF}}(\mathbf{r}) = \int \frac{\rho_x^{\mathrm{HF}}(\mathbf{rr}')(\mathbf{r}-\mathbf{r}')}{|\mathbf{r}-\mathbf{r}'|^3} d\mathbf{r}'. \tag{3.59}$$

Thus, whereas the local (static) density $\rho(\mathbf{r})$ is the quantal source of the Hartree field $\mathcal{E}_H(\mathbf{r})$, the nonlocal (dynamic) Fermi hole $\rho_x^{\mathrm{HF}}(\mathbf{rr}')$ constitutes the source for the Pauli field $\mathcal{E}_x^{\mathrm{HF}}(\mathbf{r})$.

The kinetic field $\mathcal{Z}^{\mathrm{HF}}(\mathbf{r})$ is written in terms of the kinetic 'force' $z^{\mathrm{HF}}(\mathbf{r}; [\gamma^{\mathrm{HF}}])$ and the density $\rho(\mathbf{r})$ as

$$\mathcal{Z}^{\mathrm{HF}}(\mathbf{r}) = \frac{z(\mathbf{r}; [\gamma^{\mathrm{HF}}])}{\rho(\mathbf{r})}, \tag{3.60}$$

where the kinetic 'force' is defined by its component $z_\alpha^{\mathrm{HF}}(\mathbf{r})$ as

$$z_\alpha^{\mathrm{HF}}(\mathbf{r}) = 2\sum_\beta \frac{\partial}{\partial_{r_\beta}} t_{\alpha\beta}^{\mathrm{HF}}(\mathbf{r}), \tag{3.61}$$

and where $t_{\alpha\beta}^{\mathrm{HF}}(\mathbf{r})$ is the second rank kinetic energy tensor defined in turn as

$$t_{\alpha\beta}^{\mathrm{HF}}(\mathbf{r}) = \frac{1}{4}\left[\frac{\partial^2}{\partial r'_\alpha \partial''_\beta} + \frac{\partial^2}{\partial r'_\beta \partial r''_\alpha}\right]\gamma^{\mathrm{HF}}(\mathbf{r}'\mathbf{r}'')\bigg|_{\mathbf{r}'=\mathbf{r}''=\mathbf{r}}. \tag{3.62}$$

Total Energy in terms of Quantal Sources and Fields

The total energy E^{HF} is the expectation of the interacting system Hamiltonian $\hat{H}$ of (3.34) taken with respect to the Slater determinant wave function $\Phi\{\phi_i(\mathbf{x})\}$:

$$E^{\mathrm{HF}} = \Big\langle \Phi\{\phi_i(\mathbf{x})\}|\hat{H}|\Phi\{\phi_i(\mathbf{x})\}\Big\rangle \tag{3.63}$$

$$= T^{\mathrm{HF}} + E_{ee}^{\mathrm{HF}} + E_{ext}^{\mathrm{HF}}, \tag{3.64}$$

where T_{HF}, E_{ee}^{HF}, and E_{ext}^{HF} are respectively the kinetic, and the electron-interaction and external potential energy components. The energy E_{ee}^{HF} can be further subdivided into a Hartree component E_H, and a term referred to as the *exchange energy* E_x^{HF} that arises due to the use of a determinantal type wave function in the expectation value. Thus, the total energy E^{HF} may be written as

$$E^{\mathrm{HF}} = T^{\mathrm{HF}} + E_H + E_x^{\mathrm{HF}} + E_{ext}^{\mathrm{HF}}. \tag{3.65}$$

These total energy components are defined below in terms of their quantal sources, and in terms of the corresponding fields.

The kinetic energy T^{HF} in terms of its quantal source, the Dirac density matrix $\gamma^{\mathrm{HF}}(\mathbf{rr}')$ of (3.43) is

$$T^{\mathrm{HF}} = \int t^{\mathrm{HF}}(\mathbf{r})d\mathbf{r}, \tag{3.66}$$

where the kinetic energy density $t^{\mathrm{HF}}(\mathbf{r})$ is either

$$t_A^{\mathrm{HF}}(\mathbf{r}) = -\frac{1}{2}\nabla_{\mathbf{r}}^2\gamma^{\mathrm{HF}}(\mathbf{rr}')\Big|_{\mathbf{r}'=\mathbf{r}}, \tag{3.67}$$

or

$$t_B^{\mathrm{HF}}(\mathbf{r}) = \frac{1}{2}\nabla_{\mathbf{r}'}\cdot\nabla_{\mathbf{r}''}\gamma^{\mathrm{HF}}(\mathbf{r}'\mathbf{r}'')\Big|_{\mathbf{r}'=\mathbf{r}''=\mathbf{r}}. \tag{3.68}$$

In terms of the kinetic field $\mathcal{Z}^{\mathrm{HF}}(\mathbf{r})$, the kinetic energy is

$$T^{\mathrm{HF}} = -\frac{1}{2}\int \rho(\mathbf{r})\mathbf{r}\cdot\mathcal{Z}^{\mathrm{HF}}(\mathbf{r})d\mathbf{r}. \tag{3.69}$$

The electron interaction energy E_{ee}^{HF} in terms of its quantal source, the pair correlation density $g^{\mathrm{HF}}(\mathbf{rr}')$, and in terms of the electron-interaction field $\mathcal{E}_{ee}^{\mathrm{HF}}(\mathbf{r})$, is

$$E_{ee}^{\mathrm{HF}} = \frac{1}{2}\iint \frac{\rho(\mathbf{r})g^{\mathrm{HF}}(\mathbf{rr}')}{|\mathbf{r}-\mathbf{r}'|}d\mathbf{r}d\mathbf{r}' = \int \rho(\mathbf{r})\mathbf{r}\cdot\mathcal{E}_{ee}^{\mathrm{HF}}(\mathbf{r})d\mathbf{r}. \tag{3.70}$$

The Hartree energy E_H in terms of its quantal source, the density $\rho(\mathbf{r})$, and in terms of the Hartree field $\mathcal{E}_{\mathrm{H}}(\mathbf{r})$, is

$$E_H = \frac{1}{2}\iint \frac{\rho(\mathbf{r})\rho(\mathbf{r}')}{|\mathbf{r}-\mathbf{r}'|} d\mathbf{r}d\mathbf{r}' = \int \rho(\mathbf{r})\mathbf{r}\cdot\boldsymbol{\mathcal{E}}_{\mathrm{H}}(\mathbf{r})d\mathbf{r}. \tag{3.71}$$

The exchange energy E_x^{HF} in terms of its quantal source, the Fermi hole $\rho_x^{\mathrm{HF}}(\mathbf{rr}')$, and in terms of the Pauli field $\boldsymbol{\mathcal{E}}_x^{\mathrm{HF}}(\mathbf{r})$, is

$$E_x^{\mathrm{HF}} = \frac{1}{2}\iint \frac{\rho(\mathbf{r})\rho_x^{\mathrm{HF}}(\mathbf{rr}')}{|\mathbf{r}-\mathbf{r}'|} d\mathbf{r}d\mathbf{r}' = \int \rho(\mathbf{r})\mathbf{r}\cdot\boldsymbol{\mathcal{E}}_x^{\mathrm{HF}}(\mathbf{r})d\mathbf{r}. \tag{3.72}$$

Finally, the external potential energy E_{ext}^{HF} may be expressed in terms of the density $\rho(\mathbf{r})$ as

$$E_{ext}^{\mathrm{HF}} = \int \rho(\mathbf{r})v(\mathbf{r})d\mathbf{r}. \tag{3.73}$$

The external energy may also be expressed in terms of all the fields descriptive of the system via the 'Quantal Newtonian' First Law. The law within Hartree-Fock theory, and the subsequent expression for this energy component in terms of fields, are discussed in Sect. 3.2.5.

Note that the expressions for the individual components of the total energy in terms of fields are independent of whether or not the fields are conservative.

3.2.2 *Derivation of the Hartree-Fock Theory Integro-Differential Equation*

We next derive the Hartree-Fock theory integro-differential equations that generate the single particle orbitals $\phi_i(\mathbf{x})$ to be employed in the Slater determinant $\Phi\{\phi_i(\mathbf{x})\}$ of (3.40). As the quantal sources, and hence the corresponding fields, are functionals of the orbitals $\phi_i(\mathbf{x})$, so is the expression for the total energy and its components as written in terms of these sources and fields: $E^{\mathrm{HF}} = E^{\mathrm{HF}}\{\phi_i(\mathbf{x})\}$. The best single-particle orbitals $\phi_i(\mathbf{x})$ from the perspective of the total energy are obtained by application of the variational principle for the energy [9]. This requires the first order variation of the energy, for arbitrary variations of the wavefunction, to vanish. In HF theory, the orbital $\phi_i(\mathbf{x})$ is varied by an arbitrarily small amount $\delta\phi_i(\mathbf{x})$ such that $\phi_i(\mathbf{x}) \rightarrow \phi_i(\mathbf{x}) + \delta\phi_i(\mathbf{x})$, and the stationary condition written as

$$\delta\left[E^{\mathrm{HF}}[\Phi] - \sum_{i,j=1}^{\mathrm{N}} \lambda_{ij}\langle\phi_i|\phi_j\rangle\right] = 0, \tag{3.74}$$

where the $\lambda_{ij} = \lambda_{ji}^*$ are the Langrange multipliers introduced to satisfy the $N(N+1)/2$ orthonormality conditions $\langle\phi_i|\phi_j\rangle = \delta_{ij}$. This leads to the Hartree-Fock theory equations:

$$\left[-\frac{1}{2}\nabla^2 + v(\mathbf{r}) + \sum_{\substack{j=1 \\ j\neq i}}^{N} \langle \phi_j | \hat{U} | \phi_j \rangle \right] \phi_i(\mathbf{x}) - \sum_{\substack{j=1 \\ j\neq i}}^{N} \langle \phi_j | \hat{U} | \phi_i \rangle \phi_j(\mathbf{x}) = \sum_{j=1}^{N} \lambda_{ij}\phi_j(\mathbf{x}), \tag{3.75}$$

where

$$\langle \phi_j | \hat{U} | \phi_i \rangle = \sum_{\sigma'} \int \frac{\phi_j^*(\mathbf{x}')\phi_i(\mathbf{x}')}{|\mathbf{r}-\mathbf{r}'|} d\mathbf{r}' \tag{3.76}$$

Including the self-interaction term in both the third (Hartree) and fourth (exchange) components of the left hand side of (3.75) leads to the definition of the Hermitian exchange operator $\hat{v}_{x,i}(\mathbf{x})$:

$$\hat{v}_{x,i}(\mathbf{x})\phi_i(\mathbf{x}) = -\sum_{j=1}^{N} \langle \phi_j | \hat{U} | \phi_i \rangle \phi_j(\mathbf{x}). \tag{3.77}$$

The exchange operator is said to be *nonlocal* because operating with it on $\phi_i(\mathbf{x})$ depends upon the value of $\phi_i(\mathbf{x})$ throughout all space, not just at $\mathbf{x}$, as is evident from (3.77). With the inclusion of the self-interaction term, the resulting Hamiltonian on the left hand side of (3.75) can be readily shown to be Hermitian. We prove below the hermiticity of the exchange operator as defined by (3.77). Writing out (3.77) we have

$$\hat{v}_{x,i}(\mathbf{x})\phi_i(\mathbf{x}) = -\sum_{j=1}^{N} \int \frac{\phi_j^\star(\mathbf{x}')\phi_j(\mathbf{x})\phi_i(\mathbf{x}')}{|\mathbf{r}-\mathbf{r}'|} d\mathbf{x}' \tag{3.78}$$

$$= -\int \frac{\gamma^{\mathrm{HF}}(\mathbf{x}'\mathbf{x})\phi_i(\mathbf{x}')}{|\mathbf{r}-\mathbf{r}'|} d\mathbf{x}' \tag{3.79}$$

where $\gamma^{\mathrm{HF}}(\mathbf{x}\mathbf{x}') = \sum_j \phi_j^\star(\mathbf{x})\phi_j(\mathbf{x}')$. Thus, for any function $\phi(\mathbf{x})$ we need to prove that

$$\langle \phi | \hat{v}_{x,i} | \phi_i \rangle = \langle \phi_i | \hat{v}_{x,i} | \phi \rangle^\star. \tag{3.80}$$

Proof

$$\langle \phi | \hat{v}_{x,i} | \phi_i \rangle^\star = \left[-\iint \frac{\phi^\star(\mathbf{x})\gamma^{\mathrm{HF}}(\mathbf{x}'\mathbf{x})\phi_i(\mathbf{x}')}{|\mathbf{r}-\mathbf{r}'|} d\mathbf{x}d\mathbf{x}' \right]^* \tag{3.81}$$

$$= -\iint \frac{\phi(\mathbf{x})\gamma^{\mathrm{HF}\star}(\mathbf{x}'\mathbf{x})\phi_i^\star(\mathbf{x}')}{|\mathbf{r}-\mathbf{r}'|} d\mathbf{x}d\mathbf{x}' \tag{3.82}$$

But since $\gamma^{\mathrm{HF}\star}(\mathbf{x}'\mathbf{x}) = \gamma^{\mathrm{HF}}(\mathbf{x}\mathbf{x}')$, we have

$$\langle \phi | \hat{v}_{x,i} | \phi_i \rangle^\star = - \iint \frac{\phi(\mathbf{x}) \gamma^{\text{HF}}(\mathbf{x}\mathbf{x}') \phi_i^\star(\mathbf{x}')}{|\mathbf{r} - \mathbf{r}'|} d\mathbf{x} d\mathbf{x}'. \tag{3.83}$$

Now since $\mathbf{x}$ and $\mathbf{x}'$ are dummy variables, we can interchange them. Of course, we also have to interchange the vectors $\mathbf{r}$ and $\mathbf{r}'$, but since only the magnitude of their difference is required, the denominator on the right hand side remains the same.

On performing the interchange $\mathbf{x} \leftrightarrow \mathbf{x}'$, equation (3.83) becomes

$$\langle \phi | \hat{v}_{x,i} | \phi_i \rangle^\star = - \iint \frac{\phi_i^\star(\mathbf{x}) \gamma^{\text{HF}}(\mathbf{x}'\mathbf{x}) \phi(\mathbf{x}')}{|\mathbf{r} - \mathbf{r}'|} d\mathbf{x} d\mathbf{x}' \tag{3.84}$$

$$= \langle \phi_i | \hat{v}_{x,i} | \phi \rangle, \tag{3.85}$$

and the hermiticity of the exchange operator is proved.

As the Hamiltonian on the left hand side of (3.75) is Hermitian, the Lagrange multipliers may be chosen as $\lambda_{ij} = \epsilon_i \delta_{ij}$. This then leads to the Hartree-Fock theory eigenvalue equation, which in terms of the spatial component $\psi_i(\mathbf{r})$ is

$$\left[-\frac{1}{2}\nabla^2 + v(\mathbf{r}) + W_{\text{H}}(\mathbf{r}) \right] \psi_i(\mathbf{r}) - \sum_{\substack{j=1 \\ \text{spin } j \| \text{spin } i}}^{\text{N}} \left[\int \frac{\psi_j^*(\mathbf{r}')\psi_i(\mathbf{r}')}{|\mathbf{r} - \mathbf{r}'|} d\mathbf{r}' \right] \psi_j(\mathbf{r}) = \epsilon_i \psi_i(\mathbf{r}), \tag{3.86}$$

where $W_{\text{H}}(\mathbf{r})$ is the Hartree potential energy

$$W_{\text{H}}(\mathbf{r}) = \int \frac{\rho(\mathbf{r}')}{|\mathbf{r} - \mathbf{r}'|} d\mathbf{r}'. \tag{3.87}$$

Equation (3.86) constitutes the Hartree-Fock theory integro-differential eigenvalue equation. It is evident from the integro-differential equation (3.87) that the Hartree-Fock theory effective single particle Hamiltonian is identical for each orbital. (By identical is not meant the same, i.e. the integral exchange operator term is not multiplicative or local.)

In terms of the HF theory eigenvalues ϵ_i, the total energy may then be written as

$$E^{\text{HF}} = \sum_i \epsilon_i - E_{\text{H}} - E_{\text{x}}^{\text{HF}} = \sum_i \epsilon_i - E_{\text{ee}}^{\text{HF}}, \tag{3.88}$$

with E_{H}, E_{x}^{HF}, and $E_{\text{ee}}^{\text{HF}}$ as defined above.

It is also evident from both the Hartree $W_H(\mathbf{r})$ and the exchange potential terms of (3.86) that the Hartree-Fock theory Hamiltonian is a functional of the spatial component $\psi_i(\mathbf{r})$ of the spin-orbitals $\phi_i(\mathbf{x})$, i.e. $\hat{H}^{\text{HF}} = \hat{H}^{\text{HF}}[\psi_i(\mathbf{r})]$, so that the eigenvalue equation can be written as

$$\hat{H}^{\mathrm{HF}}[\psi_i(\mathbf{r})]\psi_i(\mathbf{r}) = \epsilon_i \psi_i(\mathbf{r}). \tag{3.89}$$

This equation must be solved *self-consistently* for the eigenfunctions $\psi_i(\mathbf{r})$ and eigenvalues ϵ_i. To do so, one would begin with an approximate $\psi_i(\mathbf{r})$, and then determine the resulting approximate Hamiltonian $\hat{H}^{\mathrm{HF}}[\psi_i(\mathbf{r})]$. The solution of the Hartree-Fock equation (3.86) would then lead to new set of $\{\psi_i(\mathbf{r}), \epsilon_i\}$ which are then employed to determine the new Hamiltonian which in turn would generate the next set of orbitals and eigenvalues. And this procedure is continued till the orbitals $\psi_i(\mathbf{r})$ that are the input to the Hamiltonian $\hat{H}^{\mathrm{HF}}[\psi_i(\mathbf{r})]$ are the same as those generated by that Hamiltonian. Such solutions are entirely numerical, and said to be *fully-self-consistent*.

For an example of such calculations as applied to the atoms of the Periodic Table, see [25]. The results of these calculations for the total energy, highest occupied eigenvalue, atomic shell structure, the typical structure of the dynamic Fermi hole as a function of electron position, and the expectation value of single-particle operators $\hat{O} = \sum_i r_i^n, n = 1, 2, -1, -2$, and $\hat{O} = \sum_i \delta(\mathbf{r}_i)$ are also quoted in Chap. 10 of [35].

For the significantly different structure of the Fermi hole as a function of electron position at a jellium-metal - vacuum interface as an electron is removed from a position within the metal bulk to well into the vacuum region, see [48–50] or Chap. 17 of [35].

3.2.3 The Slater-Bardeen Interpretation of Hartree-Fock Theory

The Hartree-Fock theory equation of (3.86) has been re-described by Bardeen [45] and Slater [46] in a manner that provides a 'physical' interpretation of the integral exchange operator term. (The reason for the quotation marks about the adjective "physical" will be explained following the new description.)

By multiplying and dividing the exchange term of (3.86) by $\psi_i(\mathbf{r})$, it may be rewritten as

$$\left[\int \frac{\left(- \sum\limits_{\substack{j=1 \\ \text{spin } j \parallel \text{spin } i}}^{\mathrm{N}} \psi_j^*(\mathbf{r}')\psi_i(\mathbf{r}')\psi_j(\mathbf{r})/\psi_i(\mathbf{r}) \right)}{|\mathbf{r} - \mathbf{r}'|} d\mathbf{r}' \right] \psi_i(\mathbf{r}). \tag{3.90}$$

The integral in the square parentheses may thus be interpreted as an *orbital-dependent multiplicative* exchange 'potential' energy

$$v_{x,i}(\mathbf{r}) = \int \frac{\rho_{x,i}(\mathbf{r}\mathbf{r}')}{|\mathbf{r}-\mathbf{r}'|} d\mathbf{r}', \tag{3.91}$$

due to the *orbital-dependent Fermi hole* charge distribution $\rho_{x,i}(\mathbf{r}\mathbf{r}')$ at $\mathbf{r}'$ for an electron at $\mathbf{r}$ defined as

$$\rho_{x,i}(\mathbf{r}\mathbf{r}') = - \sum_{\substack{j=1 \\ \text{spin } j \| \text{spin } i}}^{N} \frac{\psi_j^*(\mathbf{r}')\psi_i(\mathbf{r}')\psi_j(\mathbf{r})}{\psi_i(\mathbf{r})}. \tag{3.92}$$

The orbital-dependent Fermi hole satisfies the same sum rules as those of the Fermi hole $\rho_x^{\text{HF}}(\mathbf{r}\mathbf{r}')$. (See 3.51–3.53). Thus

$$\rho_{x,i}(\mathbf{r}\mathbf{r}') \leq 0 \tag{3.93}$$

$$\rho_{x,i}(\mathbf{r}\mathbf{r}) = -\rho(\mathbf{r})/2, \tag{3.94}$$

$$\int \rho_{x,i}(\mathbf{r}\mathbf{r}')d\mathbf{r}' = -1, \text{ (for each electron position } \mathbf{r}). \tag{3.95}$$

The Hartree-Fock theory eigenvalue equation (3.86) may then be written as

$$\left[-\frac{1}{2}\nabla^2 + v(\mathbf{r}) + W_{\text{H}}(\mathbf{r}) + v_{x,i}(\mathbf{r}) \right] \psi_i(\mathbf{r}) = \epsilon_i \psi_i(\mathbf{r}), \tag{3.96}$$

and the theory interpreted as *each* electron having a 'potential' energy that is the sum of the external $v(\mathbf{r})$ and Hartree $W_{\text{H}}(\mathbf{r})$ potential energies, which are the same for all the electrons, and an exchange 'potential' energy $v_{x,i}(\mathbf{r})$ that depends on the orbital the electron is in. Thus, Hartree-Fock theory may be thought of as being an *orbital-dependent-potential* theory, with each electron having a *different* 'potential' energy.

In a rigorous physical sense, the expression for $v_{x,i}(\mathbf{r})$ of (3.91) does not represent a potential energy for nonuniform electron density systems. This is because the orbital-dependent Fermi hole $\rho_{x,i}(\mathbf{r}\mathbf{r}')$ is a *nonlocal* or *dynamic* charge distribution whose structure changes as a function of the electron position at $\mathbf{r}$. The expression for $v_{x,i}(\mathbf{r})$ would represent a potential energy provided the charge distribution $\rho_{x,i}(\mathbf{r}\mathbf{r}')$ were a *local* or *static* charge whose structure remained the same *independent* of the electron position as is the case for the uniform electron gas. (For a more detailed discussion, the reader is referred to Sect. 10.2 of [34].)

To account for the dynamic nature of the orbital-dependent Fermi hole $\rho_{x,i}(\mathbf{r}\mathbf{r}')$, one must first obtain the orbital dependent field $\mathcal{E}_{x,i}(\mathbf{r})$ which is

$$\mathcal{E}_{x,i}(\mathbf{r}) = \int \frac{\rho_{x,i}(\mathbf{r}\mathbf{r}')(\mathbf{r}-\mathbf{r}')}{|\mathbf{r}-\mathbf{r}'|^3} d\mathbf{r}', \tag{3.97}$$

and then obtain the orbital-dependent work done to bring an electron from some reference point at infinity to its position at $\mathbf{r}$:

$$W_{x,i}(\mathbf{r}) = -\int_{\infty}^{\mathbf{r}} \boldsymbol{\mathcal{E}}_{x,i}(\mathbf{r}') \cdot d\boldsymbol{\ell}'. \tag{3.98}$$

This work done is *path-independent* provided the fields $\boldsymbol{\mathcal{E}}_{x,i}(\mathbf{r})$ are conservative or curl free. The work done would then constitute an *exchange potential energy of each electron*. Hence, the Hartree-Fock theory differential equation of (3.96) would be replaced by the equation

$$\left[-\frac{1}{2}\nabla^2 + v(\mathbf{r}) + W_H(\mathbf{r}) + W_{x,i}(\mathbf{r})\right]\psi_i(\mathbf{r}) = \epsilon_i\psi_i(\mathbf{r}), \tag{3.99}$$

whereby the terms $v(\mathbf{r})$, $W_H(\mathbf{r})$, and $W_{x,i}(\mathbf{r})$ each constitute a potential energy. As the orbitals $\psi_i(\mathbf{r})$ of (3.99) differ from those of (3.96), the corresponding total energy E would be an upper bound to the Hartree-Fock theory value E^{HF}.

It is interesting to note that in Hartree-Fock theory the best single particle orbitals from the perspective of the total energy are obtained from functions $v_{x,i}(\mathbf{r})$ that do not represent the potential energy of an electron in the rigorous classical sense.

3.2.4 'Quantal Newtonian' First Law in Hartree-Fock Theory

As was the case for the stationary-state Schrödinger theory (see Sect. 2.7), there is a 'Quantal Newtonian' First Law for the Hartree-Fock theory description of an electronic system. The general form of the law remains unchanged, but the fields are now described in terms of the Hartree-Fock theory quantal sources. Thus, the law states that the sum of the external $\boldsymbol{\mathcal{F}}^{\mathrm{ext}}(\mathbf{r})$ and internal $\boldsymbol{\mathcal{F}}^{\mathrm{int,HF}}(\mathbf{r})$ fields experienced by each electron vanishes:

$$\boldsymbol{\mathcal{F}}^{\mathrm{ext}}(\mathbf{r}) + \boldsymbol{\mathcal{F}}^{\mathrm{int,HF}}(\mathbf{r}) = 0. \tag{3.100}$$

The external field is

$$\boldsymbol{\mathcal{F}}^{\mathrm{ext}}(\mathbf{r}) = -\nabla v(\mathbf{r}), \tag{3.101}$$

and the internal field is

$$\boldsymbol{\mathcal{F}}^{\mathrm{int,HF}}(\mathbf{r}) = \boldsymbol{\mathcal{E}}_{ee}^{\mathrm{HF}} - \boldsymbol{\mathcal{Z}}^{\mathrm{HF}}(\mathbf{r}) - \boldsymbol{\mathcal{D}}(\mathbf{r}) \tag{3.102}$$

$$= \boldsymbol{\mathcal{E}}_H(\mathbf{r}) + \boldsymbol{\mathcal{E}}_x^{\mathrm{HF}}(\mathbf{r}) - \boldsymbol{\mathcal{Z}}^{\mathrm{HF}}(\mathbf{r}) - \boldsymbol{\mathcal{D}}(\mathbf{r}), \tag{3.103}$$

where the electron-interaction $\boldsymbol{\mathcal{E}}_{ee}^{\mathrm{HF}}$, Hartree $\boldsymbol{\mathcal{E}}_H(\mathbf{r})$, Pauli $\boldsymbol{\mathcal{E}}_x^{\mathrm{HF}}(\mathbf{r})$, and kinetic $\boldsymbol{\mathcal{Z}}^{\mathrm{HF}}(\mathbf{r})$ fields are defined in Sect. 3.2.1. The form of the differential density field $\boldsymbol{\mathcal{D}}(\mathbf{r})$ is the same as in (2.49) for the fully interacting system but with the Hartree-Fock theory density $\rho(\mathbf{r})$ of (3.41) employed instead. (The law is derived [32, 33, 51] in a manner similar to that for the Schrödinger equation (see Appendix A) by writing the spatial part $\psi_i(\mathbf{r})$ of the spin-orbital $\phi_i(\mathbf{x})$ in complex Cartesian form as $\psi_i(\mathbf{r}) =$

$\psi_i^R(\mathbf{r}) + i\psi_i^I(\mathbf{r})$, where $\psi_i^R(\mathbf{r})$ and $\psi_i^I(\mathbf{r})$ are its real and imaginary parts. Substitution of this form of the spatial component of the spin-orbital into the Hartree-Fock theory eigenvalue equation (3.86) then leads to the law of (3.100).)

The 'Quantal Newtonian' First Law of (3.100) then provides a rigorous physical interpretation of the electrostatic potential $v(\mathbf{r})$ in terms of the internal field $\mathcal{F}^{\text{int},\text{HF}}(\mathbf{r})$ experienced by each electron. *It is the work done to move an electron from some reference point at infinity to its position at* $\mathbf{r}$ *in the force of the conservative internal field* $\mathcal{F}^{\text{int},\text{HF}}(\mathbf{r})$ of (3.102):

$$v(\mathbf{r}) = \int_{\infty}^{\mathbf{r}} \mathcal{F}^{\text{int},\text{HF}}(\mathbf{r}') \cdot d\boldsymbol{\ell}'. \tag{3.104}$$

As the $\nabla \times \mathcal{F}^{\text{int},\text{HF}}(\mathbf{r}) = 0$ because $\nabla \times \mathcal{F}^{\text{ext}}(\mathbf{r}) = 0$, this work done is *path-independent*. Thus, this is another way to define the external potential $v(\mathbf{r})$.

The expression (3.104) for the electrostatic potential $v(\mathbf{r})$ is valid for *arbitrary* state of the system.

Furthermore, in writing the scalar potential $v(\mathbf{r})$ as in (3.104), it becomes clear that within Hartree-Fock theory, this function is a sum of components due to electron correlations arising from the Pauli principle, the electron density and kinetic effects. In particular, if the component fields of $\mathcal{F}^{\text{int},\text{HF}}(\mathbf{r})$ are individually conservative, then $v(\mathbf{r})$ may be written as

$$v(\mathbf{r}) = W_H(\mathbf{r}) + W_x^{\text{HF}}(\mathbf{r}) + W_z^{\text{HF}}(\mathbf{r}) + W_d(\mathbf{r}), \tag{3.105}$$

where $W_H(\mathbf{r})$, $W_x^{\text{HF}}(\mathbf{r})$, $W_z^{\text{HF}}(\mathbf{r})$ and $W_d(\mathbf{r})$ are, respectively, the work done in the Hartree $\mathcal{E}_H(\mathbf{r})$, Pauli $\mathcal{E}_x^{\text{HF}}(\mathbf{r})$, kinetic $\mathcal{Z}^{\text{HF}}(\mathbf{r})$, and differential density $\mathcal{D}(\mathbf{r})$ fields. Each work done is separately *path-independent*.

3.2.5 *Generalization of the Hartree-Fock Theory Equations*

In the expression for the scalar potential $v(\mathbf{r})$ of (3.104), the components of the internal field $\mathcal{F}^{\text{int},\text{HF}}(\mathbf{r})$ are obtained from quantal sources that are expectation values of Hermitian operators taken with respect to the Slater determinant $\Phi\{\phi_i(\mathbf{x})\}$. Hence, $\mathcal{F}^{\text{int},\text{HF}}(\mathbf{r})$ is a functional of the Slater determinant $\Phi\{\phi_i(\mathbf{x})\}$, *i.e.* $\mathcal{F}^{\text{int},\text{HF}}(\mathbf{r}) = \mathcal{F}^{\text{int},\text{HF}}[\Phi](\mathbf{r})$. As a consequence, the potential $v(\mathbf{r})$ is a functional of $\Phi\{\phi_i(\mathbf{x})\} : v(\mathbf{r}) = v[\Phi](\mathbf{r})$. Since in the solution of the Hartree-Fock theory equations (see (3.86) or (3.96)), one is concerned with the spatial part $\psi_i(\mathbf{r})$ of the spin-orbital $\phi_i(\mathbf{x})$, the potential $v(\mathbf{r})$ is equivalently a functional of the orbitals $\psi_i(\mathbf{r})$, i.e. $v(\mathbf{r}) = v[\psi_i(\mathbf{r})](\mathbf{r})$. Thus, one may write the expression for $v(\mathbf{r})$ as

$$v(\mathbf{r}) = v[\psi_i](\mathbf{r}) = \int_{\infty}^{\mathbf{r}} \mathcal{F}^{\text{int},\text{HF}}[\psi_i](\mathbf{r}') \cdot d\boldsymbol{\ell}'. \tag{3.106}$$

This is an *exactly known* functional. Each potential energy term of the Hartree-Fock theory Hamiltonian (see either (3.86) or (3.96)) is then a *known functional* of the spatial orbitals $\psi_i(\mathbf{r})$. The Hartree-Fock theory differential equation (3.96) may thus be written as

$$\Big[-\frac{1}{2}\nabla^2 + v[\psi_i](\mathbf{r}) + W_H[\psi_i](\mathbf{r}) + v_{x,i}[\psi_i](\mathbf{r})\Big]\psi_i(\mathbf{r}) = \epsilon_i\psi_i(\mathbf{r}), \tag{3.107}$$

or as

$$\Big[-\frac{1}{2}\nabla^2 + \int_\infty^{\mathbf{r}} \boldsymbol{\mathcal{F}}^{\mathrm{int,HF}}[\psi_i](\mathbf{r}')\cdot d\boldsymbol{\ell}' + W_H[\psi_i](\mathbf{r}) + v_{x,i}[\psi_i](\mathbf{r})\Big]\psi_i(\mathbf{r}) = \epsilon_i\psi_i(\mathbf{r}), \tag{3.108}$$

with the functionals $W_H[\psi_i](\mathbf{r})$ and $v_{x,i}[\psi_i](\mathbf{r})$ defined by (3.87) and (3.91), respectively. The equations (3.107) and (3.108) then constitute the generalization of the Hartree-Fock theory equations in which the Hamiltonian $\hat{H}^{\mathrm{HF}}$ (see (3.96)), with the exception of the kinetic energy operator, is a known functional of the orbitals $\psi_i(\mathbf{r})$. The eigenvalue equation may thus again be written as in (3.89), but now with the understanding that even the external potential operator $v(\mathbf{r})$ is a functional of $\psi_i(\mathbf{r})$. Once again, the solution to the equation must be determined self-consistently, but in this case not only are the Hartree potential $W_H[\psi_i](\mathbf{r})$ and exchange 'potential' $v_{x,i}[\psi_i](\mathbf{r})$ determined self-consistently, but so also is the external potential $v[\psi_i(\mathbf{r})](\mathbf{r})$. This is of value when one does not know the structure of the external potential $v(\mathbf{r})$ as for example when a physical system is artificially constructed. And to determine the potential $v(\mathbf{r})$ by performing a self-consistent Hartree-Fock theory calculation is simpler than performing such a calculation for the fully-interacting Schrödinger equation.

Finally, employing (3.104), the external energy $E_{\mathrm{ext}}^{\mathrm{HF}}$ can be defined in terms of the internal field $\boldsymbol{\mathcal{F}}^{\mathrm{int,HF}}(\mathbf{r})$ (and its components) as

$$E_{\mathrm{ext}}^{\mathrm{HF}} = \int \rho(\mathbf{r})\Big[\int_\infty^{\mathbf{r}} \boldsymbol{\mathcal{F}}^{\mathrm{int,HF}}(\mathbf{r}')\cdot d\boldsymbol{\ell}'\Big]d\mathbf{r}. \tag{3.109}$$

Hence, *each* term in the expression (3.65) for the total energy E^{HF} can be expressed in terms of a field descriptive of the system: the kinetic $\boldsymbol{\mathcal{Z}}^{\mathrm{HF}}(\mathbf{r})$, Hartree $\boldsymbol{\mathcal{E}}_H(\mathbf{r})$, Pauli $\boldsymbol{\mathcal{E}}_x^{\mathrm{HF}}(\mathbf{r})$, and differential density $\boldsymbol{\mathcal{D}}(\mathbf{r})$ (see (3.69)–(3.72)).

3.2.6 Theorems in Hartree-Fock Theory

There are four theorems of significance within Hartree-Fock theory which are described next with explanatory remarks. The reader is referred to the original literature or to a standard text on quantum chemistry [52] for their proofs.

1. According to Koopmans' theorem [53], the eigenvalues ϵ_i of the Hartree-Fock theory equation (3.86) or (3.96) may be interpreted as *removal* energies. The proof *assumes* that the orbitals $\psi_i(\mathbf{r})$ of the neutral system and those of the resulting ionized system with an electron removed are the same, and that there is no relaxation of the orbitals of the latter. This is rigorously the case for a many-electron system with extended orbitals as in a simple metal with s-p band character. Thus, within Hartree-Fock theory, the work function of a metal—the removal energy of an electron from the highest occupied energy or Fermi level to infinity—is the difference in energy between its barrier height and the Fermi energy [35, 54] (see Chap. 17 of 35). However, for finite systems such as atoms or molecules, there is a relaxation of the orbitals on electron removal. *Hence, the interpretation of the eigenvalues ϵ_i as removal energies is not quite rigorous.* Consequently, the highest occupied eigenvalue ϵ_m of the Hartree-Fock theory of atoms is not as good an approximation to the experimental ionization potential as that of local-effective-potential energy theories such as the Pauli-correlated approximation of Quantal Density Functional theory (Q-DFT) [35, 55, 56] (see Chap. 10 of 35). (As is the case in Hartree-Fock theory, the Pauli-correlated approximation of Q-DFT incorporates electron correlations solely due to the Pauli principle. However, it does so within the context of a *local* or multiplicative exchange potential. It is the lowest-order approximation in a many-body theory of Q-DFT [35] (See Chap. 18 of 35).) The theorem and above remarks are equally valid for the case of the addition of an electron to the neutral system. As such the highest occupied eigenvalue ϵ_m of Hartree-Fock theory of negative ions is again not as accurate [35, 57] (See Chap. 10 of 35) as the Pauli-correlated approximation of Q-DFT when compared to experimental electron affinities.

2. For external potentials that vanish at infinity as in atoms, the orbitals [58] of Hartree-Fock theory *all* have the *same* asymptotic structure in the classically forbidden region: $\psi_i(\mathbf{r}) \underset{r\to\infty}{\sim} f_{mi}(\mathbf{r}) \exp(-\sqrt{-2\epsilon_m} r)$, where ϵ_m is the corresponding highest occupied eigenvalue. (Here $i = 1, \ldots, N$; $N = m$.) Thus, within Hartree-Fock theory, *all* the orbitals contribute to the asymptotic structure of the density $\rho(\mathbf{r})$ in the classically forbidden region. This structure is not solely proportional to the exponential term $\exp(-2\sqrt{-2\epsilon_m} r)$, but also involves the sum of the squares of the functions $f_{mi}(\mathbf{r})$.

In contrast, within local effective potential theory, the asymptotic structure of each orbital depends exponentially on the corresponding eigenvalue $\epsilon_i : \psi_i(\mathbf{r}) \underset{r\to\infty}{\sim} \exp(-\sqrt{-2\epsilon_i}\, r)$. Hence, the asymptotic structure of the density is *solely* due to the highest occupied orbital $\epsilon_m : \lim_{r\to\infty} \rho(\mathbf{r}) = |\psi_m(\mathbf{r})|^2 \sim \exp(-2\sqrt{-2\epsilon_m}\, r)$. By solution of the Schrödinger equation (2.119) for atoms, it can be shown that the asymptotic structure of the density in this limit is $\lim_{r\to\infty} (\mathbf{r}) \sim \exp(-2\sqrt{2 I_{k,n}} r)$, where $I_{k,n} = E_k^{N-1} - E_n$, with E_n, E_k^{N-1} are the total energies of the interacting N- and $(N-1)-$ electron systems in states n and k, respectively. Thus, the highest occupied eigenvalue ϵ_m within local effective potential theory is the negative of the first ionization potential.

From the perspective of the discussion in the previous paragraph, it is evident that the highest occupied eigenvalue ϵ_m of Hartree-Fock theory cannot be directly related to the first ionization potential. Consequently, the relationship between ϵ_m and the experimental ionization potential has meaning only within the context of Koopmans' theorem.

3. According to the Brillouin-Møller-Plesset theorem [59, 60], the matrix element of the Hamiltonian $\hat{H}$ of (3.34) taken with respect to the ground-state Hartree-Fock theory determinant and any determinant with a singly excited electron vanishes.

4. As a consequence of Brillouin-Møller-Plesset theorem [59, 60], the expectation values of single particle operators of the form $W = \sum_i W(r_i)$, where $W(r_i) = r_i^n$, $n = -3, -2, -1, 1, 2$ and $\delta(\mathbf{r}_i)$, taken with respect to the Hartree-Fock theory ground state Slater determinant wave function, are correct to second-order [60, 61] as is the energy. That is, the Hartree-Fock theory wave function correct to $O(\delta)$, which leads to rigorous upper bounds to the total ground state energy correct to $O(\delta^2)$, also leads to these expectation values being correct to $O(\delta^2)$. Hence these expectations are accurate.

To prove this [61] consider the exact wave function Ψ of the Hamiltonian $\hat{H}$ of (3.34) to be approximated as

$$\Psi = \left(1 - \frac{1}{2}\delta^2\right)\Phi + \delta\chi \tag{3.110}$$

where the functions Ψ, Φ, χ are normalized, the functions Φ and χ are orthogonal, and δ is the smallness parameter such that as Φ becomes a better approximation, $\delta \to 0$. The wave function Ψ written as (3.110) is normalized to $O(\delta^4)$. From (3.110) it follows that to $O(\delta^2)$

$$\Phi = \Psi - \delta\chi + \frac{1}{2}\delta^2\Psi. \tag{3.111}$$

Thus, to $O(\delta^2)$, the expectation value of an operator $\hat{O}$ taken with respect to Φ is

$$\begin{aligned}\langle\hat{O}\rangle_{\text{approx}} = \langle\hat{O}\rangle_{\text{exact}} &- \delta\Big[\langle\chi|\hat{O}|\Psi\rangle + \langle\Psi|\hat{O}|\chi\rangle\Big] \\ &+ \delta^2\Big[\langle\hat{O}\rangle_{\text{exact}} + \langle\chi|\hat{O}|\Psi\rangle\Big].\end{aligned} \tag{3.112}$$

Now according to Brillouin-Møller-Plesset, if the function Φ is the ground state Hartree-Fock theory determinant, then the first-order correction χ is orthogonal to Φ in two electronic coordinates. If the operator $\hat{O}$ is the sum of the single-particle operators W, then the integral $\langle\chi|W|\Psi\rangle$ vanishes since it reduces to

$$\langle\chi|W|\Psi\rangle = \sum_i \langle\chi|W(\mathbf{r}_i)|\Psi\rangle = 0 \ \text{to} \ O(\delta^2), \tag{3.113}$$

because each matrix element vanishes due to the orthogonality of χ to Φ in at least one electronic coordinate other than $\mathbf{r}_i$. Thus,

$$\langle W\rangle_{\text{approx}} = \langle W\rangle_{\text{exact}} + O(\delta^2). \tag{3.114}$$

(As noted in the Introduction there also exist variational principles [13–19] whereby beginning with a trial wave function correct to $O(\delta)$, one can obtain the expectation values of the single-particle operators of the form $W = \sum_i W(\mathbf{r}_i)$ and that of the complex single-particle density matrix operator $\hat{\gamma}(\mathbf{rr}')$ of (2.16) to $O(\delta^2)$. In this manner it is possible to accurately obtain the electronic density $\rho(\mathbf{r})$, the coherent atomic scattering factor $F(\mathbf{k})$ which is the Fourier transform of the density $\rho(\mathbf{r})$, and the momentum density $\rho(\mathbf{k})$ which is the double Fourier transform of the single-particle matrix $\gamma(\mathbf{rr}')$.)

3.2.7 *Hartree Theory in Terms of Quantal Sources and Fields*

The quantal sources, the sum rules satisfied by these sources, and the corresponding fields applicable to Hartree theory are defined in this section. The expressions for the energy in terms of both the sources and fields are provided. The 'Quantal Newtonian' First Law within Hartree theory is in terms of these fields. The law in turn leads to the generalization of the Hartree theory integro-differential equation.

Wave Function

In Hartree theory, the wave function $\Psi(\mathbf{X})$ of the interacting system defined by the Hamiltonian $\hat{H}$ of (3.34) is approximated by $\Psi^H(\mathbf{X})$ which is a product of spin-orbitals $\phi_i(\mathbf{x}) = \psi_i(\mathbf{r})\chi_i(\sigma)$:

$$\Psi^H(\mathbf{X}) = \prod_{i=1}^N \phi_i(\mathbf{x}) = \prod_{i=1}^N \psi_i(\mathbf{r})\chi_i(\sigma). \tag{3.115}$$

Note that this wave function is not anti-symmetric in an interchange of the coordinates of two electrons, and hence does not satisfy the Pauli principle. (A consequence of the requirement of anti-symmetry of the wave function is that no two electrons can occupy the same state. As noted in Sect. 3.2 (e), in the application of Hartree theory, it is this statement of the Pauli principle that is imposed.)

Quantal Sources and Their Sum Rules

The quantal sources: the density $\rho(\mathbf{r})$, the Hartree spinless single-particle density matrix $\gamma^H(\mathbf{rr}')$, and the pair-correlation density $g^H(\mathbf{rr}')$, are obtained as the expectation values of the operators $\hat{\rho}(\mathbf{r})$, $\hat{\gamma}(\mathbf{rr}')$, and $\hat{P}(\mathbf{rr}')/\rho(\mathbf{r})$ of (2.12), (2.16) and

(2.28) taken with respect to the wave function of (3.115). The expressions for these sources in terms of the spin orbitals $\phi_i(\mathbf{x})$, and the sum rules satisfied by them are given below:

Density

$$\rho(\mathbf{r}) = \sum_{\sigma} \sum_{i} \phi_i^{\star}(x)\phi_i(\mathbf{x}), \tag{3.116}$$

$$\int \rho(\mathbf{r})d\mathbf{r} = N. \tag{3.117}$$

Hartree Theory Density Matrix

$$\gamma^H(\mathbf{rr}') = \sum_{\sigma} \sum_{i} \phi_i^{\star}(\mathbf{r}\sigma)\phi_i(\mathbf{r}'\sigma), \tag{3.118}$$

$$\gamma^H(\mathbf{rr}) = \rho(\mathbf{r}), \tag{3.119}$$

$$\gamma^H(\mathbf{rr}') = \gamma^{H\star}(\mathbf{r}'\mathbf{r}), \tag{3.120}$$

$$\int \gamma^H(\mathbf{rr}'')\gamma^H(\mathbf{r}''\mathbf{r}')d\mathbf{r}'' = \gamma^H(\mathbf{rr}'). \tag{3.121}$$

Thus, the Hartree density matrix $\gamma^H(\mathbf{rr}')$ is *idempotent*.

Pair-correlation Density

$$g^H(\mathbf{rr}') = \rho(\mathbf{r}') + \rho^{\mathrm{SIC}}(\mathbf{rr}'), \tag{3.122}$$

$$\int g^H(\mathbf{rr}')d\mathbf{r}' = N - 1, \tag{3.123}$$

where $\rho^{\mathrm{SIC}}(\mathbf{rr}')$ is the self-interaction-correction (SIC) density to be discussed below.

Self-interaction-correction Density

The self-interaction-correction density $\rho^{\mathrm{SIC}}(\mathbf{rr}')$ is the *nonlocal* component of the Hartree pair-correlation density $g^H(\mathbf{rr}')$, and is defined as

$$\rho^{\mathrm{SIC}}(\mathbf{rr}') = -\frac{\sum_{\sigma}\sum_{i} q_i(\mathbf{r}\sigma)q_i(\mathbf{r}'\sigma)}{\rho(\mathbf{r})}, \tag{3.124}$$

where $q_i(\mathbf{r}\sigma) = -\phi_i^{\star}(\mathbf{r}\sigma)\phi_i(\mathbf{r}\sigma)$. The sum rules satisfied by $\rho^{\mathrm{SIC}}(\mathbf{rr}')$ are

$$\rho^{\mathrm{SIC}}(\mathbf{rr}') \le 0, \tag{3.125}$$

$$\int \rho^{\mathrm{SIC}}(\mathbf{rr}')d\mathbf{r}' = -1. \tag{3.126}$$

Fields

The fields descriptive of the system within the Hartree approximation are obtained from their respective quantal sources. The electron-interaction field $\boldsymbol{\mathcal{E}}_{ee}^{H}(\mathbf{r})$ is obtained via Coulomb's law from the pair-correlation density $g^{H}(\mathbf{rr}')$:

$$\boldsymbol{\mathcal{E}}_{ee}^{H}(\mathbf{r}) = \int \frac{g^{H}(\mathbf{rr}')(\mathbf{r}-\mathbf{r}')}{|\mathbf{r}-\mathbf{r}'|^{3}} d\mathbf{r}', \tag{3.127}$$

which on employing (3.122) may be written as

$$\boldsymbol{\mathcal{E}}_{ee}^{H}(\mathbf{r}) = \boldsymbol{\mathcal{E}}_{H}(\mathbf{r}) + \boldsymbol{\mathcal{E}}_{ee}^{\mathrm{SIC}}(\mathbf{r}), \tag{3.128}$$

with $\boldsymbol{\mathcal{E}}_{H}(\mathbf{r})$ the Hartree field due to the density $\rho(\mathbf{r})$ being

$$\boldsymbol{\mathcal{E}}_{H}(\mathbf{r}) = \int \frac{\rho(\mathbf{r}')(\mathbf{r}-\mathbf{r}')}{|\mathbf{r}-\mathbf{r}'|^{3}} d\mathbf{r}', \tag{3.129}$$

and $\boldsymbol{\mathcal{E}}_{ee}^{\mathrm{SIC}}(\mathbf{r})$ the SIC field arising from the nonlocal SIC density $\rho^{\mathrm{SIC}}(\mathbf{rr}')$ as

$$\boldsymbol{\mathcal{E}}_{ee}^{\mathrm{SIC}}(\mathbf{r}) = \int \frac{\rho^{\mathrm{SIC}}(\mathbf{rr}')(\mathbf{r}-\mathbf{r}')}{|\mathbf{r}-\mathbf{r}'|^{3}} d\mathbf{r}'. \tag{3.130}$$

The electron-interaction field $\boldsymbol{\mathcal{E}}_{ee}^{H}(\mathbf{r})$ may also be written in terms of the electron-interaction 'force' $\mathbf{e}_{ee}^{H}(\mathbf{r})$ and the density $\rho(\mathbf{r})$ as

$$\boldsymbol{\mathcal{E}}_{ee}^{H}(\mathbf{r}) = \mathbf{e}_{ee}^{H}(\mathbf{r})/\rho(\mathbf{r}), \tag{3.131}$$

where $\mathbf{e}_{ee}^{H}(\mathbf{r})$ is determined via Coulomb's law from the Hartree theory pair-correlation function $P^{H}(\mathbf{rr}') = \langle \psi^{H}(\mathbf{X})|\hat{P}(\mathbf{rr}')|\psi^{H}(\mathbf{X})\rangle$ obtained from the Hartree wave function $\psi^{H}(\mathbf{X})$ with $\hat{P}(\mathbf{rr}')$ the pair-correlation operator of (2.28).

The kinetic field $\boldsymbol{\mathcal{Z}}^{H}(\mathbf{r})$ is written in terms of the kinetic 'force' $\mathbf{z}^{H}(\mathbf{r}; [\gamma^{H}])$ and the density $\rho(\mathbf{r})$ as

$$\boldsymbol{\mathcal{Z}}^{H}(\mathbf{r}) = \frac{\mathbf{z}(\mathbf{r}; [\gamma^{H}])}{\rho(\mathbf{r})}, \tag{3.132}$$

where the kinetic 'force' is defined by its component $z_{\alpha}^{H}(\mathbf{r})$ as

$$z_{\alpha}^{H}(\mathbf{r}) = 2\sum_{\beta} \frac{\partial}{\partial r_{\beta}} t_{\alpha\beta}^{H}(\mathbf{r}), \tag{3.133}$$

and where the second rank kinetic energy tensor $t_{\alpha\beta}^{H}(\mathbf{r})$ is defined as

$$t_{\alpha\beta}^{H}(\mathbf{r}) = \frac{1}{4}\left[\frac{\partial^{2}}{\partial r'_{\alpha}\partial r''_{\beta}} + \frac{\partial^{2}}{\partial r'_{\beta}\partial r''_{\alpha}}\right]\gamma^{H}(\mathbf{r}'\mathbf{r}'')\bigg|_{\mathbf{r}'=\mathbf{r}''=\mathbf{r}}. \tag{3.134}$$

Total Energy in Terms of Quantal Sources and Fields

The total energy E^H is the expectation value of the Hamiltonian $\hat{H}$ of (3.34) taken with respect to the Hartree theory wave function $\Psi^H(X)$:

$$E^H = \langle \Psi^H(\mathbf{X}) | \hat{H} | \Psi^H(\mathbf{X}) \rangle \tag{3.135}$$

$$= T^H + E^H_{ee} + E^H_{\text{ext}}, \tag{3.136}$$

where T^H, E^H_{ee}, and E^H_{ext} are respectively the kinetic, electron-interaction, and external potential energy components being the expectation values of the corresponding operators. The energy E^H_{ee} can be further subdivided into a Hartree component E^H, and a SIC contribution E^{SIC}_H. Thus, the total energy E^H may be written as

$$E^H = T^H + E_H + E^{\text{SIC}}_H + E^H_{\text{ext}}. \tag{3.137}$$

The energy components may also be defined in terms of both their quantal sources and corresponding fields. Thus, the kinetic energy T^H in terms of its quantal source, the Hartree theory density matrix $\gamma^H(\mathbf{rr}')$ of (3.118) is

$$T^H = \int t^H(\mathbf{r}) d\mathbf{r}, \tag{3.138}$$

where the kinetic energy density $t^H(\mathbf{r})$ is either

$$t^H_A(\mathbf{r}) = -\frac{1}{2} \nabla^2_{\mathbf{r}} \gamma^H(\mathbf{rr}') \Big|_{\mathbf{r}'=r}, \tag{3.139}$$

or

$$t^H_B(\mathbf{r}) = \frac{1}{2} \nabla_{\mathbf{r}'} \cdot \nabla_{\mathbf{r}''} \gamma^H(\mathbf{r}'\mathbf{r}'') \Big|_{\mathbf{r}'=\mathbf{r}''=\mathbf{r}} \tag{3.140}$$

In terms of the Hartree theory kinetic field $\mathcal{Z}^H(\mathbf{r})$, the kinetic energy is

$$T^H = -\frac{1}{2} \int \rho(\mathbf{r}) \mathbf{r} \cdot \mathcal{Z}^H(\mathbf{r}) d\mathbf{r}. \tag{3.141}$$

The electron interaction energy E^H_{ee} in terms of its quantal source, the pair correlation density $g^H(\mathbf{rr}')$, and in terms of the electron-interaction field $\mathcal{E}^H_{ee}(\mathbf{r})$, is

$$E^H_{ee} = \frac{1}{2} \iint \frac{\rho(\mathbf{r}) g^H(\mathbf{rr}')}{|\mathbf{r} - \mathbf{r}'|} d\mathbf{r} d\mathbf{r}' = \int \rho(\mathbf{r}) \mathbf{r} \cdot \mathcal{E}^H_{ee}(\mathbf{r}) d\mathbf{r}. \tag{3.142}$$

The Hartree energy E_H in terms of its quantal source, the density $\rho(\mathbf{r})$, and in terms of the Hartree field $\mathcal{E}_H(\mathbf{r})$, is

$$E_H = \frac{1}{2} \iint \frac{\rho(\mathbf{r})\rho(\mathbf{r}')}{|\mathbf{r}-\mathbf{r}'|} d\mathbf{r}d\mathbf{r}' = \int \rho(\mathbf{r})\mathbf{r} \cdot \mathcal{E}_H(\mathbf{r}) d\mathbf{r}. \tag{3.143}$$

The SIC energy E_H^{SIC} in terms of its quantal source, the SIC density $\rho^{\mathrm{SIC}}(\mathbf{r}\mathbf{r}')$, and in terms of the SIC field $\mathcal{E}_H^{\mathrm{SIC}}(\mathbf{r})$, is

$$E_H^{\mathrm{SIC}} = \frac{1}{2} \iint \frac{\rho(\mathbf{r})\rho^{\mathrm{SIC}}(\mathbf{r}\mathbf{r}')}{|\mathbf{r}-\mathbf{r}'|} d\mathbf{r}d\mathbf{r}' = \int \rho(\mathbf{r})\mathbf{r} \cdot \mathcal{E}_H^{\mathrm{SIC}}(\mathbf{r}) d\mathbf{r}. \tag{3.144}$$

Finally, the external potential energy E_{ext}^{H} may be expressed in terms of the density $\rho(\mathbf{r})$ as

$$E_{\mathrm{ext}}^{H} = \int \rho(\mathbf{r}) v(\mathbf{r}) d\mathbf{r}. \tag{3.145}$$

The external energy may also be expressed in terms of all the fields descriptive of the system via the 'Quantal Newtonian' First Law as will be shown later. Note that the expressions for the individual components of the total energy in terms of fields are independent of whether or not the fields are conservative.

3.2.8 Derivation of the Hartree Theory Integro-Differential Equation

In a manner similar to the derivation of the Hartree-Fock theory integro-differential equations (Sect. 3.2.2), the corresponding Hartree theory equations leading to the best single-particle spin-orbitals $\phi_i(\mathbf{x})$ of the Hartree theory product wave function $\Psi^{H}(\mathbf{X})$ of (3.115) from the perspective of the energy, are obtained by application of the variational principle [9] to the energy expectation value E^H of (3.135). The arbitrary variations of the spin-orbitals $\phi_i(\mathbf{x})$ subject to the normalization constraint $\langle \phi_i(\mathbf{x})|\phi_i(\mathbf{x})\rangle = 1$, leads to the Hartree theory differential equations

$$\left[-\frac{1}{2}\nabla^2 + v(\mathbf{r}) + \sum_{j \neq i} \int \frac{\phi_j^*(\mathbf{x}')\phi_j(\mathbf{x}')}{|\mathbf{r}-\mathbf{r}'|} d\mathbf{x}' \right] \phi_i(\mathbf{x}) = \epsilon_i \phi_i(\mathbf{x}) \;\; ; \;\; i = 1, \ldots, N. \tag{3.146}$$

The Hartree theory single particle Hamiltonian is Hermitian, and as such the orbitals are orthogonal: $\langle \phi_i(\mathbf{x})|\phi_j(\mathbf{x})\rangle = \delta_{ij}$.

Hartree theory approximates the interacting system as defined by the Hamiltonian of (3.34). The approximation reduces the complexity of that Hamiltonian to one in which *each* electron experiences the external electrostatic field, and a field due to the charge distribution of *all the other electrons*. With this description of the physical system, the Hartree theory differential equation could be written directly without the application of the variational principle.

The Hartree theory differential equation can also be written in terms of quantal sources and the spatial part $\psi_i(\mathbf{x})$ of the spin orbitals $\phi_i(\mathbf{x})$ as,

$$\left[-\frac{1}{2}\nabla^2 + v(\mathbf{r}) + W_H(\mathbf{r}) + v_i^{\text{SIC}}(\mathbf{r})\right]\psi_i(\mathbf{r}) = \epsilon_i\psi_i(\mathbf{r}) \;\; ; \;\; i = 1, \ldots, N, \quad (3.147)$$

where $W_H(\mathbf{r})$ is the Hartree potential energy (see (3.87)), and $v_i^{\text{SIC}}(\mathbf{r})$ the orbital-dependent SIC potential energy due to the orbital charge density $q_i(\mathbf{r}\sigma)$:

$$v_i^{\text{SIC}}(\mathbf{r}) = \int \frac{q_i(\mathbf{r}'\sigma)}{|\mathbf{r}-\mathbf{r}'|}d\mathbf{x}'. \quad (3.148)$$

The Hartree energy E^H, which is the expectation value of the Hamiltonian $\hat{H}$ of (3.34) taken with respect to the wave function $\Psi^H(\mathbf{X})$ (see (3.135, 3.136)), may be written in terms of the eigenvalues ϵ_i of the Hartree differential equation (3.147). Multiplying (3.146) by $\phi_i^\star(\mathbf{x})$, summing over all the electrons, and integrating over the spatial and spin coordinates, leads to an expression for the kinetic energy T^H in terms of the eigenvalues ϵ_i. Substituting this expression for T^H into (3.136) for E^H then leads to

$$E^H = \sum_I \epsilon_i - E_{ee}^H, \quad (3.149)$$

where E_{ee}^H is the expectation value of the electron-interaction operator $\hat{U}$ of (3.36), or as given in (3.142). In a similar manner, employing the Hartree differential equation as in (3.147), the energy may also be written as

$$E^H = \sum_I \epsilon_i - E_H - E_H^{\text{SIC}}, \quad (3.150)$$

with E_H and E_H^{SIC} defined in (3.143) and (3.144).

It is evident from either form of the differential equation ((3.146) or (3.147)) that in Hartree theory each electron has a unique potential energy. Therefore, Hartree theory is an *orbital-dependent-potential* theory. This is akin to the Slater-Bardeen interpretation (Sect. 3.2.3) which showed that Hartree-Fock theory too was an *orbital-dependent-potential* theory. It is also evident that the Hartree theory Hamiltonian is an exactly known functional of the spin-orbitals $\phi_i(\mathbf{x})$ or equivalently of the spatial part $\psi_i(\mathbf{r})$, i.e. $\hat{H}^H = \hat{H}^H[\psi_i(\mathbf{r})]$. Thus, the Hartree differential equation may be written as

$$\hat{H}^H[\psi_i(\mathbf{r})]\psi_i(\mathbf{r}) = \epsilon_i\psi_i(\mathbf{r}), \quad (3.151)$$

and must be solved in a fully-self-consistent manner.

The results of such fully-self-consistent calculations as applied to the atoms of the Periodic Table are given in [24, 62]. (The results for the total energy, highest occupied eigenvalue, atomic shell structure, a typical structure of the SIC density, and the satisfaction of the *aufbau* principle, are given in Chap. 9 of [35].)

3.2.9 'Quantal Newtonian' First Law in Hartree Theory

As in Schrödinger and Hartree-Fock theories, the equation of motion or 'Quantal Newtonian' First Law for the individual electron can be derived [32] for Hartree theory. The law is unchanged, except for the fact that the fields are obtained from quantal sources that are determined from the Hartree theory wave function $\Psi^H(\mathbf{X})$. Thus, the sum of the external $\mathcal{F}^{\text{ext}}(\mathbf{r})$ and internal $\mathcal{F}^{\text{int,H}}(\mathbf{r})$ fields experienced by each electron vanishes:

$$\mathcal{F}^{\text{ext}}(\mathbf{r}) + \mathcal{F}^{\text{int,H}}(\mathbf{r}) = 0, \tag{3.152}$$

where the external field is

$$\mathcal{F}^{\text{ext}}(\mathbf{r}) = -\nabla v(\mathbf{r}), \tag{3.153}$$

and the internal field is

$$\mathcal{F}^{\text{int,H}}(\mathbf{r}) = \mathcal{E}_{ee}^{H}(\mathbf{r}) - \mathcal{Z}^{H}(\mathbf{r}) - \mathcal{D}(\mathbf{r}) \tag{3.154}$$

$$= \mathcal{E}_{H}(\mathbf{r}) + \mathcal{E}_{H}^{\text{SIC}}(\mathbf{r}) - \mathcal{Z}^{H}(\mathbf{r}) - \mathcal{D}(\mathbf{r}), \tag{3.155}$$

where the electron-interaction $\mathcal{E}_{ee}^{H}(\mathbf{r})$, Hartree $\mathcal{E}_{H}(\mathbf{r})$, SIC $\mathcal{E}_{H}^{\text{SIC}}(\mathbf{r})$, and kinetic $\mathcal{Z}^{H}(\mathbf{r})$ fields are defined in Sect. 3.2.7. The differential density field $\mathcal{D}(\mathbf{r})$ is of the same form as in (2.49) for the fully interacting system but with the Hartree theory density $\rho(\mathbf{r})$ of (3.116) employed instead.

Once again, the 'Quantal Newtonian' First Law leads to a rigorous interpretation of the external electrostatic potential potential $v(\mathbf{r})$ in terms of the internal field $\mathcal{F}^{\text{int,H}}(\mathbf{r})$ experienced by each electron. *It is the work done to move an electron from some reference point at infinity to its position at* $\mathbf{r}$ *in the force of the conservative internal field* $\mathcal{F}^{\text{int,H}}(\mathbf{r})$ of (3.154):

$$v(\mathbf{r}) = \int_{\infty}^{\mathbf{r}} \mathcal{F}^{\text{int,H}}(\mathbf{r}') \cdot d\boldsymbol{\ell}'. \tag{3.156}$$

This work done is *path-independent* because $\nabla \times \mathcal{F}^{\text{int,H}}(\mathbf{r}) = 0$. Note that this is still yet another way by which the potential $v(\mathbf{r})$ is defined.

If the symmetry of the system is such that the individual field components of $\mathcal{F}^{\text{int,H}}(\mathbf{r})$ are separately conservative, then the potential $v(\mathbf{r})$ may be written as

$$v(\mathbf{r}) = W_H(\mathbf{r}) + W_H^{\text{SIC}}(\mathbf{r}) + W_z^H(\mathbf{r}) + W_d(\mathbf{r}), \tag{3.157}$$

where $W_H(\mathbf{r})$, $W_H^{\mathrm{SIC}}(\mathbf{r})$, $W_z^H(\mathbf{r})$ and $W_d(\mathbf{r})$ are, respectively, the work done in the Hartree $\boldsymbol{\mathcal{E}}_H(\mathbf{r})$, SIC $\boldsymbol{\mathcal{E}}_H^{\mathrm{SIC}}(\mathbf{r})$, kinetic $\boldsymbol{\mathcal{Z}}^H(\mathbf{r})$ and differential density $\boldsymbol{\mathcal{D}}(\mathbf{r})$ fields. Each work done is separately *path-independent*.

3.2.10 Generalization of the Hartree Theory Equations

In the expression (3.156) for the external potential $v(\mathbf{r})$ in terms of $\boldsymbol{\mathcal{F}}^{\mathrm{int,H}}(\mathbf{r})$, the component fields of the latter are obtained from quantal sources that are expectation values of Hermitian operators taken with respect to the Hartree wave function $\Psi^H(\mathbf{X})$. Hence, $\boldsymbol{\mathcal{F}}^{\mathrm{int,H}}(\mathbf{r})$ is an *exactly known* functional of the spatial part $\psi_i(\mathbf{r})$ of the spin-orbital $\phi_i(\mathbf{x})$, i.e. $\boldsymbol{\mathcal{F}}^{\mathrm{int,H}}(\mathbf{r}) = \boldsymbol{\mathcal{F}}^{\mathrm{int,H}}[\psi_i(\mathbf{r})](\mathbf{r})$. As a consequence, the potential $v(\mathbf{r})$ is an *exactly known* functional of $\psi_i(\mathbf{r})$, *i.e.* $v(\mathbf{r}) = v[\psi_i(\mathbf{r})](\mathbf{r})$. Thus, the potential $v(\mathbf{r})$ may be written as

$$v(\mathbf{r}) = v[\psi_i](\mathbf{r}) = \int_{\infty}^{\mathbf{r}} \boldsymbol{\mathcal{F}}^{\mathrm{int,H}}[\psi_i](\mathbf{r}') \cdot d\boldsymbol{\ell}'. \tag{3.158}$$

Consequently, each potential energy term of the Hartree theory Hamiltonian (see (3.147)) is a known functional of the orbitals $\psi_i(\mathbf{r})$. As such the Hartree theory differential equation may be written as

$$\Big[-\frac{1}{2}\nabla^2 + v[\psi_i](\mathbf{r}) + W_H[\psi_i](\mathbf{r}) + v_i^{\mathrm{SIC}}[\psi_i](\mathbf{r}) \Big] \psi_i(\mathbf{r}) = \epsilon_i \psi_i(\mathbf{r}) \ ; \ i = 1, \ldots, N, \tag{3.159}$$

or as

$$\Big[-\frac{1}{2}\nabla^2 + \int_{\infty}^{\mathbf{r}} \boldsymbol{\mathcal{F}}^{\mathrm{int},H}[\psi_i](\mathbf{r}') \cdot d\boldsymbol{\ell}' + W_H[\psi_i](\mathbf{r}) + v_i^{\mathrm{SIC}}[\psi_i](\mathbf{r}) \Big] \psi_i(\mathbf{r}) = \epsilon_i \psi_i(\mathbf{r}) \ ; \ i = 1, \ldots, N. \tag{3.160}$$

This then is the *generalized* form of the Hartree theory equations. It is of particular value when the external potential $v(\mathbf{r})$ is unknown. When solved self-consistently, one obtains the orbitals $\psi_i(\mathbf{r})$, the eigen values ϵ_i, and the potential $v(\mathbf{r})$, and hence the Hamiltonian $\hat{H}^H$. Simultaneously, the total energy E^H may be obtained from any of the expressions (3.135), (3.149), (3.150).

Employing the expression (3.156) for the potential $v(\mathbf{r})$ in (3.145) then allows the external energy E_{ext}^H to be written in terms of the internal field $\boldsymbol{\mathcal{F}}^{\mathrm{int},H}(\mathbf{r})$:

$$E_{\mathrm{ext}}^H = \int \rho(\mathbf{r}) \Big[\int_{\infty}^{\mathbf{r}} \boldsymbol{\mathcal{F}}^{\mathrm{int},H}(\mathbf{r}') \cdot d\boldsymbol{\ell}' \Big] d\mathbf{r}. \tag{3.161}$$

Hence, the total energy E^H in Hartree theory (see (3.137)) may be written entirely in terms of the fields descriptive of the system, *viz.* the kinetic $\mathcal{Z}^H(\mathbf{r})$, Hartree $E_H(\mathbf{r})$, SIC $\mathcal{E}_H^{\mathrm{SIC}}(\mathbf{r})$, and differential density $\mathcal{D}(\mathbf{r})$ fields.

3.3 Generalization of the Time-Dependent Schrödinger Equation

In a manner similar to the generalizations of the Schrödinger, Hartree-Fock, and Hartree theory equations, the time-dependent Schrödinger equation too can be written as in (3.4) so as to prove that it is an intrinsically self-consistent equation. Thus, consider a system of N electrons in a time-dependent electric field $\mathcal{E}(\mathbf{y}) = -\nabla v(\mathbf{y})/e$. (We employ the notation $\mathbf{y} = \mathbf{r}t$; $\mathbf{y}_k = \mathbf{r}_k t$; $\mathbf{y}' = \mathbf{r}'t$; $\mathbf{Y} = \mathbf{X}t$; $\mathbf{X} = \mathbf{x}_1, \ldots, \mathbf{x}_N$; $\mathbf{x} = \mathbf{r}\sigma$; $\mathbf{r}$ and σ the spatial and spin coordinates; and assume $e = \hbar = m = 1$.) The Hamiltonian $\hat{H}(t)$ for this system is

$$\hat{H}(t) = \hat{T} + \hat{U} + \hat{V}(t), \tag{3.162}$$

where the kinetic energy operator $\hat{T}$ is

$$\hat{T} = \frac{1}{2}\sum_k p_k^2, \tag{3.163}$$

the electron-interaction operator $\hat{U}$ is

$$\hat{U} = \frac{1}{2}\sum_{k,\ell}{}' \frac{1}{|\mathbf{r}_k - \mathbf{r}_\ell|}, \tag{3.164}$$

and the external potential operator $\hat{V}(t)$ is

$$\hat{V} = \sum_k v(\mathbf{y}_k), \tag{3.165}$$

with the canonical momentum operator $\mathbf{p}_k = -i\nabla_{\mathbf{r}_k}$. The Schrödinger equation for the time-dependent wave function $\Psi(\mathbf{Y})$ is then

$$\hat{H}(t)\Psi(\mathbf{Y}) = i\partial\Psi(\mathbf{Y})/\partial t. \tag{3.166}$$

The corresponding equation of motion for each electron or the 'Quantal Newtonian' Second Law (see Appendix A for the derivation) is

$$\mathcal{F}^{\mathrm{ext}}(\mathbf{y}) + \mathcal{F}^{\mathrm{int}}(\mathbf{y}) = \mathcal{J}(\mathbf{y}), \tag{3.167}$$

where

$$\mathcal{F}^{\text{ext}}(\mathbf{y}) = \mathcal{E}(\mathbf{y}), \tag{3.168}$$

$$\mathcal{F}^{\text{int}}(\mathbf{y}) = \mathcal{E}_{ee}(\mathbf{y}) - \mathcal{Z}(\mathbf{y}) - \mathcal{D}(\mathbf{y}), \tag{3.169}$$

$$\mathcal{J}(\mathbf{y}) = \frac{1}{\rho(\mathbf{y})}\frac{\partial \mathbf{j}(\mathbf{y})}{\partial t}, \tag{3.170}$$

The definitions of the fields $\mathcal{E}_{ee}(\mathbf{y})$, $\mathcal{Z}(\mathbf{y})$, $\mathcal{D}(\mathbf{y})$ and the density $\rho(\mathbf{y})$ and current density $\mathbf{j}(\mathbf{y})$ are given in (2.43), (2.51), (2.49), (2.11) and (2.39), respectively. These fields and densities are expectation values of Hermitian operators taken with respect to the wave function $\Psi(\mathbf{Y})$. Thus, each is a known functional of $\Psi(\mathbf{Y})$. From the 'Quantal Newtonian' Second Law, from which the scalar potential $v(\mathbf{y})$ can be written in terms of these fields, it follows that $v(\mathbf{y})$ is a known functional of $\Psi(\mathbf{Y})$. From the Second Law of (3.167), the potential $v(\mathbf{y})$ has the *rigorous physical interpretation of being the work done at each instant of time t to bring an electron from some reference point at infinity to its position at* $\mathbf{r}$ *in a conservative field* $\mathcal{F}[\Psi(\mathbf{Y})](\mathbf{y})$ *which is a known functional of* $\Psi(\mathbf{Y})$. Thus,

$$v(\mathbf{y}) = v[\Psi(\mathbf{Y})](\mathbf{y}) = \int_{\infty}^{\mathbf{r}} \mathcal{F}[\Psi(\mathbf{Y})](\mathbf{y}') \cdot d\boldsymbol{\ell}', \tag{3.171}$$

where $\mathcal{F}[\Psi(\mathbf{Y})](\mathbf{y}) = \mathcal{F}^{\text{int}}(\mathbf{y}) - \mathcal{J}(\mathbf{y}) = \mathcal{E}_{ee}(\mathbf{y}) - \mathcal{Z}(\mathbf{y}) - \mathcal{D}(\mathbf{y}) - \mathcal{J}(\mathbf{y})$. Furthermore, as $\nabla \times \mathcal{F}[\Psi(\mathbf{Y})](\mathbf{y}) = 0$, this work done is *path-independent.*

On substitution of the functional of $v[\Psi(\mathbf{Y})](\mathbf{y})$ of (3.171) into the Hamiltonian $\hat{H}(t)$ of (3.162), the Schrödinger equation may be written in the generalized form of (3.4) which for the particular case is

$$\left[\frac{1}{2}\sum_k \hat{p}_k^2 + \frac{1}{2}\sum_{k,\ell}{}' \frac{1}{|\mathbf{r}_k - \mathbf{r}_\ell|} + \sum_k v[\Psi(\mathbf{Y})](\mathbf{y}_k)\right]\Psi(\mathbf{Y}) = i\frac{\partial \Psi(\mathbf{Y})}{\partial t}, \tag{3.172}$$

or

$$\left[\frac{1}{2}\sum_k \hat{p}_k^2 + \frac{1}{2}\sum_{k,\ell}{}' \frac{1}{|\mathbf{r}_k - \mathbf{r}_\ell|} + \sum_k \int_{\infty}^{\mathbf{r}_k} \mathcal{F}[\Psi(\mathbf{Y})](\mathbf{y}) \cdot d\boldsymbol{\ell}\right]\Psi(\mathbf{Y}) = i\frac{\partial \Psi(\mathbf{Y})}{\partial t}. \tag{3.173}$$

Consider the case when the external binding potential $v(\mathbf{r})$ is known so that $v(\mathbf{y}) = v(\mathbf{r}) + v_1(\mathbf{r}t)$ The time evolution of the wave function $\Psi(\mathbf{Y})$ is then obtained by self-consistent solution of (3.173) at each instant of time t given an initial condition $\Psi(\mathbf{Y}) = \Psi(\mathbf{Y}_0)$ at $t = t_0$. This initial state could for example be the solution of the corresponding stationary-state Schrödinger equation.

The Schrödinger equation of (3.166) constitutes a special case of the generalized form of (3.172).

References

1. E. Schrödinger, Ann. Physik **79**, 361 (1926)
2. W. Moore, *Schrödinger, Life and Thought*, Cambridge University Press, New York, (1989)
3. V. Sahni, X.-Y. Pan, Computation **5**, 15 (2017). https://doi.org/10.3390/computation5010015
4. V. Sahni, J. Comp. Chem. **39**, 1083 (2018). https://doi.org/10.1002/jcc.24888 (2017)
5. R.C. Ashoori, Nature **379**, 423 (1996)
6. S.M. Reimann, M. Manninen, Rev. Mod. Phys. **74**, 1283 (2002)
7. H. Saarikovski, S.M. Reimann, A. Harju, M. Manninen, Rev. Mod. Phys. **82**, 2785 (2010)
8. A. Kumar, S.E. Laux, F. Stern, Phys. Rev. B. **42**, 5166 (1990)
9. B.L. Moiseiwitsch, *Variational Principles* (John Wiley, London, UK, 1966)
10. X.-Y. Pan, V. Sahni, L. Massa, Phys. Rev. Lett. **93**, 130401 (2004)
11. X.-Y. Pan, M. Slamet, V. Sahni, Phys. Rev. A **81**, 042524 (2010)
12. M. Slamet, X.-Y. Pan, V. Sahni, Phys. Rev. A **84**, 052504 (2011)
13. V. Sahni, J.B. Krieger, Int. J. Quantum Chem. **5**, 47 (1971)
14. V. Sahni, J.B. Krieger, Int. J. Quantum Chem. **6**, 103 (1972)
15. J.B. Krieger, V. Sahni, Phys. Rev. A **6**, 919 (1972)
16. V. Sahni, J.B. Krieger, Phys. Rev. A **6**, 928 (1972)
17. V. Sahni, J.B. Krieger, Phys. Rev. A **8**, 65 (1973)
18. V. Sahni, J.B. Krieger, Phys. Rev. A **11**, 409 (1975)
19. V. Sahni, J.B. Krieger, J. Gruenebaum, Phys. Rev. A **12**, 768 (1975)
20. E.A. Hylleraas, B. Undheim, Z. Physik **65**, 759 (1930)
21. D.R. Hartree, Proc. Cambridge Philos. Soc. **24**, 39 (1928); **24**, 111 (1928); **24**, 426 (1928)
22. V. Fock, Z. Phys. **61**, 126 (1930)
23. J.C. Slater, Phys. Rev. **35**, 210 (1930); ibid **34**, 1293 (1929)
24. D.R. Hartree, *The Calculation of Atomic Structures* (Wiley, Inc., New York, 1957)
25. C.F. Fischer, *The Hartree-Fock Theory for Atoms* (Wiley, Inc., New York, 1977)
26. P. Hohenberg, W. Kohn, Phys. Rev. **136**, B864 (1964)
27. W. Kohn, L.J. Sham, Phys. Rev. **140**, A1133 (1965)
28. R.T. Sharp, G.K. Horton, Phys. Rev. **30**, 317 (1953)
29. J.D. Talman, W.F. Shadwick, Phys. Rev. A **14**, 36 (1976)
30. J.B. Krieger, Y. Li, G.J. Iafrate, Phys. Lett. A **146**, 256 (1990); Phys. Rev. A **45**, 5453 (1992); in *Density Functional Theory*, vol. 337, ed. by E.K.U. Gross, R.M. Dreizler, NATO ASI Series, Series B: Physics (Plenum, New York, 1995)
31. E. Engel, S.H. Vosko, Phys. Rev. A **47**, 2800 (1993)
32. V. Sahni, Phys. Rev. A **55**, 1846 (1997)
33. V. Sahni, Top. Curr. Chem. **182**, 1 (1996)
34. V. Sahni, *Quantal Density Functional Theory*, 2nd edn. (Springer, Berlin, Heidelberg, 2016)
35. V. Sahni, *Quantal Density Functional Theory II: Approximation Methods and Applications* (Springer, Berlin, Heidelberg, 2010)
36. O. Gunnarsson, B. Lundqvist, Phys. Rev. B **13**, 4274 (1976)
37. Y.-Q. Li, X.-Y. Pan, B. Li, V. Sahni, Phys. Rev. A **85**, 032517 (2012)
38. E. Schrödinger, Ann. Phys. **81**, 109 (1926)
39. A. Messiah, *Quantum Mechanics*, vol. I (North-Holland Publishing Co., Amsterdam, 1966)
40. J.D. Jackson, *Classical Electrodynamics* (Wiley, Inc., New York, 1963)
41. Y. Aharonov, D. Bohm, Phys. Rev. **115**, 485 (1959)
42. V. Sahni, C.Q. Ma, J.S. Flamholz, Phys. Rev. B **18**, 3931 (1978)
43. T. Yang, X.-Y. Pan, V. Sahni, Phys. Rev. A **83**, 042518 (2011)

44. A. Holas, N.H. March, Phys. Rev. A **56**, 4595 (1997)
45. J. Bardeen, Phys. Rev. **49**, 653 (1936) (In footnote 18, Bardeen notes: 'It should be emphasized that this interpretation cannot be made in all cases. The [orbital-dependent] exchange potential defined in this manner may be complex or may have infinite values. In our case [the metal-vacuum interface], the potential is real and finite.' See also V. Sahni, Surf. Sci. **213**, 226 (1989).)
46. J.C. Slater, Phys. Rev. **81**, 385 (1951)
47. M.K. Harbola, V. Sahni, J. Chem. Educ. **70**, 920 (1993)
48. V. Sahni, K.-P. Bohnen, Phys. Rev. B **29**, 1045 (1984)
49. V. Sahni, K.-P. Bohnen, Phys. Rev. B **31**, 7651 (1985)
50. M.K. Harbola, V. Sahni, Phys. Rev. B **37**, 745 (1988)
51. A. Holas, N.H. March, Top. Curr. Chem. **180**, 57 (1996)
52. A. Szabo, N.S. Ostlund, *Modern Quantum Chemistry* (McGraw-Hill, New York, 1989)
53. T. Koopmans, Physica **1**, 104 (1933)
54. V. Sahni, C.Q. Ma, Phys. Rev. B **22**, 5987 (1980)
55. V. Sahni, Y. Li, M.K. Harbola, Phys. Rev. A **45**, 1434 (1992)
56. Y. Li, M.K. Harbola, J.B. Krieger, V. Sahni, Phys. Rev. A **40**, 6084 (1989)
57. V. Sahni, Int. J. Quantum Chem. **56**, 265 (1995)
58. N.C. Handy, M.T. Marron, H.J. Silverstone, Phys. Rev. **180**, 45 (1969)
59. L. Brillouin, Actualités sci.et.ind. **71** (1933); 159 (1934); 160 (1934)
60. C. Møller, M.S. Plesset, Phys. Rev. **46**, 618 (1934)
61. J. Goodisman, W. Klemperer, J. Chem. Phys. **38**, 721 (1963)
62. V. Sahni, Z. Qian, K.D. Sen, J. Chem. Phys. **114**, 8784 (2001)

Chapter 4
Schrödinger-Pauli Theory of Electrons: A Complementary Perspective

Abstract The Schrödinger-Pauli theory of electrons explicitly considers the spin moment of the electrons, and therefore goes beyond the Schrödinger theory description of *spinless* electrons. As a consequence of the electrons possessing a spin moment, the Schrödinger-Pauli theory Hamiltonian is derived non-relativistically via the Feynman kinetic energy operator. In this chapter, the Schrödinger-Pauli theory of electrons in the presence of static and time-dependent electromagnetic fields is described from the new perspective of the *individual* electron via the corresponding 'Quantal Newtonian' First and Second Laws. These laws are a description in terms of 'classical' fields experienced by *each* electron, the fields arising from sources that are quantum-mechanical expectation values of Hermitian operators taken with respect to the system wave function. In the temporal case–the Second Law–each electron experiences an external field comprised of the Coulomb and Lorentz fields, and an internal field whose components are representative of electron correlations due to the Pauli principle and Coulomb repulsion, kinetic effects, the electron density, and an internal magnetic field. The response of the electron is described by a field representative of the physical current density which is a sum of its paramagnetic, diamagnetic and magnetization components. The First Law, descriptive of the stationary-state theory, constitutes a special case. The Schrödinger-Pauli theory is *generalized* such that the Hamiltonian operator is proved to be an *exactly known universal* functional of the wave function. This then shows the stationary-state and time-dependent Schrödinger-Pauli equations to be intrinsically self-consistent. To facilitate the understanding of this new description and of proofs within it, further relevant aspects of the stationary-state Schrödinger theory of spinless electrons in an electromagnetic field are discussed. The Hamiltonian operator, as obtained by the correspondence principle, is expressed in terms of operators representative of the gauge invariant properties of the electronic density and physical current density. It is also written so as to explicitly show the existence of the Lorentz force via the corresponding operator. Thus, with any scalar potential representative of external electrostatic forces, the Hamiltonian can now be seen to explicitly encompass both the external Coulomb and Lorentz forces. Finally, it is proved that the stationary state wave function is a functional of a gauge function. (As will be proved in a future chapter, for a uniform magnetic field, the wave function is also a functional of the

V. Sahni, *Schrödinger Theory of Electrons: Complementary Perspectives*,
Springer Tracts in Modern Physics 285,
https://doi.org/10.1007/978-3-030-97409-1_4

gauge invariant ground state density and physical current density). The wave function is thus ensured to be gauge variant.

Introduction

In the previous two chapters, the Schrödinger theory [1] of *spinless* electrons in the presence of static and time-dependent electromagnetic fields was described from the new perspective of the *individual* electron via the 'Quantal Newtonian' Second and First Laws. This was a description in terms of 'classical' fields experienced by each electron, the fields arising from sources that are quantum-mechanical expectation values of Hermitian operators taken with respect to the system wave function. (By spinless electrons is meant that the spin moment of the electron is absent in the Hamiltonian.)

This description of Schrödinger theory was then generalized by proving that the corresponding stationary-state and time-dependent Hamiltonian operators $\{\hat{H}_{\text{spinless}}, \hat{H}_{\text{spinless}}(t)\}$ were *exactly known universal* functionals of the wave functions $\{\psi, \Psi(t)\}$, *i.e.* $\{\hat{H}_{\text{spinless}}[\psi], \hat{H}_{\text{spinless}}[\Psi](t)\}$. This then reveals that the corresponding stationary-state and time-dependent Schrödinger equations are in fact *intrinsically* self-consistent. In turn this means that the Hamiltonian operators $\{\hat{H}_{\text{spinless}}[\psi], \hat{H}_{\text{spinless}}[\Psi](t)\}$, the wave functions $\{\psi, \Psi(t)\}$, and the stationary-state eigen energy $E[\psi]$ are all determined on achieving self-consistency. This generalization allows for situations in which the external binding potential of the electrons, and therefore the Hamiltonian operators, are unknown and thereby determined through self-consistency. The Schrödinger equations with the known Hamiltonian operators $\{\hat{H}_{\text{spinless}}, \hat{H}_{\text{spinless}}(t)\}$, then constitutes a special case of the generalized form.

In this chapter, we extend these ideas to the case of electrons possessing a spin moment via Schrödinger-Pauli theory [2]. Whereas in the Schrödinger theory of spinless electrons, the interaction of the magnetic field is solely with the orbital angular momentum, in Schrödinger-Pauli theory, the magnetic field interacts with both the orbital and spin moment of the electrons. Hence, there is an additional term present in the stationary-state and time-dependent Hamiltonian operators $\{\hat{H}_{\text{spin}}, \hat{H}_{\text{spin}}(t)\}$. A nonrelativistic derivation of the Schrödinger-Pauli theory Hamiltonian operators is provided via the use of the Feynman kinetic energy operator [3, 4]. (The Schrödinger-Pauli equation can also be derived as the nonrelativistic limit of the Dirac equation [3, 5].) Once again the theory can be described in terms of 'classical' fields and quantal sources [6, 7] via the corresponding 'Quantal Newtonian' First and Second Laws. Finally, the Schrödinger-Pauli theory is *generalized* by proving the Hamiltonians are *exactly known universal* functionals of the wave functions $\{\psi, \Psi(t)\}$, *i.e.* $\{\hat{H}_{\text{spin}}[\psi], \hat{H}_{\text{spin}}[\Psi](t)\}$. As a consequence, it is once again revealed that the stationary-state and time-dependent Schrödinger-Pauli equations are *intrinsically* self-consistent. The case when the Hamiltonians $\{\hat{H}_{\text{spin}}, \hat{H}_{\text{spin}}(t)\}$ are known constitutes a special case of the generalized form.

We begin the chapter by continuing the discussion of stationary-state Schrödinger theory of spinless electrons in an electromagnetic field. The Hamiltonian operator $\hat{H}_{\text{spinless}}$ is obtained via the correspondence principle from the classical Hamiltonian which too is derived. That this Hamiltonian operator accounts for the interaction of the magnetic field with the orbital angular moment is shown for the case of a uniform field. The Hamiltonian is also expressed in terms of the operators representing the gauge invariant properties of the electronic density $\rho(\mathbf{r})$ and physical current density $\mathbf{j}(\mathbf{r})$ thereby emphasizing the importance of these properties in electronic structure theory. This form of the Hamiltonian will prove of value when deriving the 'Quantal Newtonian' First Law for the case of electrons with spin. Employing this form of the Hamiltonian, it is then rewritten in a third way in which the presence of the Lorentz force is shown explicitly. Thus, with any scalar potential representative of external electrostatic forces, the Hamiltonian can now be seen to *explicitly* encompass both the external Coulomb and Lorentz forces. In the traditional way of writing this Hamiltonian, the effects of the Lorentz force are understood to be implicit. Finally, it is shown [8] that the wave function $\psi(\mathbf{X})$ is a functional of a gauge function $\alpha(\mathbf{R})$, *i.e.* $\psi(\mathbf{X}) = \psi[\alpha(\mathbf{R})](\mathbf{X})$. (Here $\mathbf{X} = \mathbf{x}_1, \ldots, \mathbf{x}_N$; $\mathbf{x} = r\sigma$; $\mathbf{r}, \sigma$ the spatial and spin coordinates; $\mathbf{R} = \mathbf{r}_1, \ldots, \mathbf{r}_N$) This is of significance because in a later chapter on Hohenberg-Kohn [9] density functional theory it will be proved [10] that for electrons in a uniform magnetic field, the wave function is also a functional of the *ground state* density and *physical* current density, *i.e.* $\psi(\mathbf{X}) = \psi[\rho(\mathbf{r}), \mathbf{j}(\mathbf{r})](\mathbf{X})$. But these are gauge invariant properties. The wave function, on the other hand, is gauge variant. The fact that the wave function is a functional of a gauge function $\alpha(\mathbf{R})$ *i.e.* $\psi(\mathbf{X}) = \psi[\alpha(\mathbf{R}), \rho(\mathbf{r}), \mathbf{j}(\mathbf{r})](\mathbf{X})$, then allows for it to be gauge variant.

4.1 The Classical Hamiltonian in an Electromagnetic Field

Consider a classical particle of charge Q, mass M, velocity $\mathbf{v}(t) = \dot{\mathbf{r}}(t)$ in a time-dependent electric $\mathcal{E}(\mathbf{y})$ and magnetic $\mathcal{B}(\mathbf{y})$ field (with $\mathbf{y} = \mathbf{r}, t$). These fields may be expressed in terms of a scalar $\phi(\mathbf{y})$ and vector $\mathbf{A}(\mathbf{y})$ potential as

$$\mathcal{E}(\mathbf{y}) = -\nabla\phi(\mathbf{y}) - \frac{1}{c}\frac{\partial A(\mathbf{y})}{\partial t} \; ; \; \mathcal{B}(\mathbf{y}) = \nabla \times \mathbf{A}(\mathbf{y}). \tag{4.1}$$

To determine the Hamiltonian H of the particle, we first determine its Lagrangian L. We derive the Lagrangian from the Coulomb-Lorentz equation of motion:

$$M\frac{d\mathbf{v}}{dt} = Q\left[\mathcal{E}(\mathbf{y}) + \frac{1}{c}\mathbf{v} \times \mathcal{B}(\mathbf{y})\right]. \tag{4.2}$$

This equation is gauge invariant because the fields $\mathcal{E}(\mathbf{y})$ and $\mathcal{B}(\mathbf{y})$ are gauge invariant. Substituting the field expressions into (4.2) leads to

$$M\frac{d\mathbf{v}}{dt} = Q\bigg[-\nabla\phi - \frac{1}{c}\frac{\partial\mathbf{A}}{\partial t} + \frac{1}{c}\{\mathbf{v}\times\nabla\times\mathbf{A}\}\bigg]. \tag{4.3}$$

The vector relation

$$\begin{aligned}\mathbf{v}\times\nabla\times\mathbf{A} &= \nabla(\mathbf{v}\cdot\mathbf{A}) - \mathbf{A}\times(\nabla\times\mathbf{v}) - (\mathbf{v}\cdot\nabla)\mathbf{A} + \mathbf{A}\cdot\nabla\mathbf{v} \qquad (4.4)\\ &= \nabla(\mathbf{v}\cdot\mathbf{A}) - (\mathbf{v}\cdot\nabla)\mathbf{A}, \qquad (4.5)\end{aligned}$$

as $\mathbf{v}(t)$ is not a function of position, and the terms $\mathbf{A}\times(\nabla\times\mathbf{v})$ and $(\mathbf{A}\cdot\nabla)\mathbf{v}$ vanish. Thus, (4.3) reduces to

$$M\frac{d\mathbf{v}}{dt} = Q\bigg[-\nabla\phi - \frac{1}{c}\bigg\{\frac{\partial\mathbf{A}}{\partial t} + (\mathbf{v}\cdot\nabla)\mathbf{A}\bigg\} + \frac{1}{c}\nabla(\mathbf{v}\cdot\mathbf{A})\bigg]. \tag{4.6}$$

Now the convective time derivative

$$\frac{d\mathbf{A}}{dt} = \frac{\partial\mathbf{A}}{\partial t} + (\mathbf{v}\cdot\nabla)\mathbf{A}, \tag{4.7}$$

so that the Coulomb-Lorentz equation is

$$M\frac{d\mathbf{v}}{dt} = -\nabla\bigg[Q\phi - \frac{Q}{c}\mathbf{v}\cdot\mathbf{A}\bigg] - \frac{Q}{c}\frac{d\mathbf{A}}{dt} \tag{4.8}$$

or equivalently

$$\frac{d}{dt}\bigg(M\mathbf{v} + \frac{Q}{c}\mathbf{A}\bigg) = -\nabla U, \tag{4.9}$$

where U is the generalized or velocity-dependent potential:

$$U = Q\phi - \frac{Q}{c}\mathbf{v}\cdot\mathbf{A}. \tag{4.10}$$

A comparison of (4.9) with the Lagrange's equation

$$\frac{d}{dt}\bigg(\frac{\partial L}{\partial\mathbf{v}}\bigg) = \nabla L, \tag{4.11}$$

produces the Coulomb-Lorentz equation if the Lagrangian L for the charged particle with kinetic energy $T = \frac{1}{2}Mv^2$ is

$$L(\mathbf{r},\mathbf{v}) = T - U = \frac{1}{2}Mv^2 - Q\phi(\mathbf{y}) + \frac{Q}{c}\mathbf{v}\cdot\mathbf{A}(\mathbf{y}). \tag{4.12}$$

Canonical, Physical, and Field Momentum

The total or canonical momentum $\mathbf{p}$ is defined as

$$\mathbf{p} = \frac{\partial L}{\partial \mathbf{v}} = M\mathbf{v} + \frac{Q}{c}\mathbf{A}. \tag{4.13}$$

This is a sum of its kinetic or physical momentum

$$\mathbf{p}_{\text{physical}} = M\mathbf{v}, \tag{4.14}$$

and its field momentum

$$\mathbf{p}_{\text{field}} = \frac{Q}{c}\mathbf{A}. \tag{4.15}$$

Thus, another expression for the physical momentum, usually designated as $\boldsymbol{\Pi}$ is

$$\mathbf{p}_{\text{physical}} \equiv \boldsymbol{\Pi} = \mathbf{p} - \frac{Q}{c}\mathbf{A}. \tag{4.16}$$

(It is the canonical momentum **p** on which we impose the canonical commutation relations when we write the quantum-mechanical Hamiltonian. Whereas the kinetic momentum is gauge invariant, the canonical momentum is gauge variant.)

Canonical and Physical Angular Momentum

The canonical angular momentum is defined in terms of the canonical momentum as

$$\mathbf{L} = \mathbf{r} \times \mathbf{p}. \tag{4.17}$$

The kinetic or physical angular momentum is then

$$\boldsymbol{\Lambda} = \mathbf{r} \times \boldsymbol{\Pi} = \mathbf{r} \times \left(\mathbf{p} - \frac{Q}{c}\mathbf{A}\right). \tag{4.18}$$

whereas the canonical angular momentum is gauge variant, the physical angular momentum is gauge invariant.

The Classical Hamiltonian

From the definition of the canonical momentum **p** of (4.13), the velocity **v** of the particle is

$$\mathbf{v} = \frac{1}{M}\left(\mathbf{p} - \frac{Q}{c}\mathbf{A}\right). \tag{4.19}$$

Thus, the classical Hamiltonian $H(\mathbf{p}, \mathbf{r})$ is

$$H(\mathbf{p}, \mathbf{r}) = \mathbf{p} \cdot \mathbf{v} - L \tag{4.20}$$

$$= Mv^2 + \frac{Q}{c}\mathbf{v} \cdot \mathbf{A} - \frac{1}{2}Mv^2 + Q\phi(\mathbf{y}) - \frac{Q}{c}\mathbf{v} \cdot \mathbf{A}(\mathbf{y}) \tag{4.21}$$

$$= \frac{1}{2M}\left(\mathbf{p} - \frac{Q}{c}\mathbf{A}\right)^2 + Q\phi. \tag{4.22}$$

To draw an analogy between the Coulomb-Lorentz Law of (4.2) and the 'Quantal Newtonian' Second and First Laws (see Appendix A, equations (A.39) and (A.54), we rewrite the former as

$$M\frac{d\mathbf{v}}{dt} = Q\big(\boldsymbol{\mathcal{E}}(\mathbf{y}) + \boldsymbol{\mathcal{L}}(\mathbf{y})\big), \tag{4.23}$$

where the Lorentz field $\boldsymbol{\mathcal{L}}(\mathbf{y})$ is defined in terms of the Lorentz force $\boldsymbol{\ell}(\mathbf{y})$ as

$$\boldsymbol{\mathcal{L}}(\mathbf{y}) = \frac{\boldsymbol{\ell}(\mathbf{y})}{Q}, \tag{4.24}$$

with $\boldsymbol{\ell}(\mathbf{y}) = \frac{1}{c}\mathbf{v} \times \boldsymbol{\mathcal{B}}(\mathbf{y})$. (The Lagrangian L of (4.12) and the Hamiltonian H of (4.22) are in terms of the scalar $\phi(\mathbf{y})$ and vector $\mathbf{A}(\mathbf{y})$ potentials. From both Lagrange's equation (4.11) and Hamilton's equations ($\mathbf{v} = \partial H/\partial \mathbf{p}$; $d\mathbf{p}/dt = -\frac{\partial H}{\partial \mathbf{r}}$), the Coulomb-Lorentz Law of (4.2) can be derived. Hence, the effects of the electric and Lorentz fields are implicit in both the Lagrangian L and Hamiltonian H.) The 'Quantal Newtonian' Laws of quantum mechanics which explicitly include the electric $\boldsymbol{\mathcal{E}}(\mathbf{y})$ and Lorentz $\boldsymbol{\mathcal{L}}(\mathbf{y})$ fields are thus analogous to the Coulomb-Lorentz law of classical mechanics. However, the 'Quantal Newtonian' Laws additionally include internal fields experienced by each electron representative of properties of the quantum system. Finally, as in the case for the Coulomb-Lorentz Law, the 'Quantal Newtonian' Laws are gauge invariant.

4.2 Stationary-State Schrödinger Theory in an Electromagnetic Field

In this section we consider a system of N electrons in an electrostatic $\boldsymbol{\mathcal{E}}(\mathbf{r}) = -\boldsymbol{\nabla} v(\mathbf{r})/e$ and magnetostatic $\boldsymbol{\mathcal{B}}(\mathbf{r}) = \boldsymbol{\nabla} \times \mathbf{A}(\mathbf{r})$ field, where $[v(\mathbf{r}), \mathbf{A}(\mathbf{r})]$ are scalar and vector potentials, respectively. The quantum-mechanical Hamiltonian $\hat{H}$ is obtained by application of the correspondence principle to the classical Hamiltonian H of (4.22). This Hamiltonian is for *spinless* electrons. The Hamiltonian $\hat{H}$ leads to the understanding that in the presence of a *uniform* magnetic field $\boldsymbol{\mathcal{B}}(\mathbf{r}) = \boldsymbol{\mathcal{B}}$, there exists a term that corresponds to the interaction of the field with the orbital angular momentum $\mathbf{L}$. It is also meaningful to rewrite the quantum-mechanical Hamiltonian $\hat{H}$ in terms of the density $\hat{\rho}(\mathbf{r})$ and physical current density $\hat{\mathbf{j}}(\mathbf{r})$ operators. This form of the Hamiltonian $\hat{H}$ will be of importance when we derive the 'Quantal-Newtonian' First Law within Schrödinger-Pauli theory in Sect. 4.2. The energy E can then also be expressed in terms of the density $\rho(\mathbf{r})$ and the physical current density $\mathbf{j}(\mathbf{r})$. These properties have added significance in quantum mechanics, and are referred to as *basic variables*. It can be proved [10] that knowledge of the *ground state* $\{\rho(\mathbf{r}), \mathbf{j}(\mathbf{r})\}$ of electrons in a *uniform* magnetic field $\boldsymbol{\mathcal{B}}$ determines the Hamiltonian $\hat{H}$ to within a constant and the gradient of a scalar function. What this means is that the wave function

$\psi(\mathbf{X})$ is a functional of the $\{\rho(\mathbf{r}), \mathbf{j}(\mathbf{r})\}$, *i.e.* $\psi(\mathbf{X}) = \psi[\rho(\mathbf{r}), \mathbf{j}(\mathbf{r})]$. (See Chap. [6] on Density Functional Theory for the proof and further discussion.) Finally, it is possible to determine additional properties of the quantum system by mapping the interacting system of electrons in an electromagnetic field to one of noninteracting fermions possessing the same $\{\rho(\mathbf{r}), \mathbf{j}(\mathbf{r})\}$ for arbitrary state. This mapping will be described in Chap. 6 on Quantal Density Functional Theory.

4.2.1 Schrödinger Theory Hamiltonian

In atomic units where we assume the charge of the electron $Q = -e$, with $|e| = \hbar = m = 1$, the Hamiltonian operator, on application of the correspondence principle to the classical Hamiltonian of (4.22), is

$$\hat{H} = \hat{T}_A + \hat{W} + \hat{V}, \tag{4.25}$$

where $\hat{T}_A$ is the *physical* kinetic energy operator:

$$\hat{T}_A = \frac{1}{2}\sum_k \left(\hat{\mathbf{p}}_k + \frac{1}{c}\mathbf{A}(\mathbf{r}_k)\right)^2 \tag{4.26}$$

$$= \hat{T} + \frac{1}{2c}\sum_k \left[\hat{\mathbf{p}}_k \cdot \mathbf{A}(\mathbf{r}_k) + \mathbf{A}(\mathbf{r}_k) \cdot \hat{\mathbf{p}}_k\right] + \frac{1}{2c^2}\sum_k A^2(\mathbf{r}_k), \tag{4.27}$$

with $\hat{T}$ the *canonical* kinetic energy operator:

$$\hat{T} = \frac{1}{2}\sum_k p_k^2; \quad \hat{\mathbf{p}}_k = -i\nabla_{\mathbf{r}_k}, \tag{4.28}$$

with $\hat{\mathbf{p}}_k$ the canonical momentum operator.

The electron-interaction potential energy operator is

$$\hat{W} = \frac{1}{2}\sum_{k,\ell}{}' \frac{1}{|\mathbf{r}_k - \mathbf{r}_l|}, \tag{4.29}$$

and scalar potential energy operator is

$$\hat{V} = \sum_k v(\mathbf{r}_k). \tag{4.30}$$

The time-independent Schrödinger equation is then

$$\hat{H}(\mathbf{R})\psi(\mathbf{X}) = E\psi(\mathbf{X}), \tag{4.31}$$

where $\{\psi(\mathbf{X}), E\}$ are the eigenfunctions and eigenergies of the system with $\mathbf{R} = \mathbf{r}_1, \ldots, \mathbf{r}_N$; $\mathbf{X} = \mathbf{x}_1, \ldots, \mathbf{x}_N$; $\mathbf{x} = \mathbf{r}\sigma$, $\{\mathbf{r}, \sigma\}$ being the spatial and spin coordinates of the electron. (The generalization of this equation to demonstrate its intrinsic self-consistent nature and other properties are discussed in Chap. 3 Sect. 3.1.)

Employing the commutator relationship between the momentum operator and any function of the coordinates, we have

$$\hat{\mathbf{p}} \cdot \mathbf{A} - \mathbf{A} \cdot \hat{\mathbf{p}} = -i\nabla \cdot \mathbf{A}. \tag{4.32}$$

Thus, in the Coulomb gauge $\nabla \cdot \mathbf{A} = 0$, we see that $\hat{\mathbf{p}}$ and $\mathbf{A}(\mathbf{r})$ commute. Using this fact, the physical kinetic energy operator may be written as

$$\hat{T}_A = \hat{T} + \frac{1}{c}\sum_k \mathbf{A}(\mathbf{r}_k) \cdot \hat{\mathbf{p}}_k + \frac{1}{2c^2}\sum_k A^2(\mathbf{r}_k). \tag{4.33}$$

4.2.2 *Magnetic Field—Orbital Angular Momentum Interaction*

In the symmetric gauge $\mathbf{A}(\mathbf{r}) = \frac{1}{2}\boldsymbol{\mathcal{B}}(\mathbf{r}) \times \mathbf{r}$ with a uniform magnetic field $\boldsymbol{\mathcal{B}}$, the term in $\hat{T}_A$ of (4.33) that is linear in the magnetic field via the vector potential $\mathbf{A}(\mathbf{r})$ may be written as

$$\begin{aligned} \frac{1}{c}\sum_k \mathbf{A}(\mathbf{r}_k) \cdot \hat{\mathbf{p}}_k = \frac{1}{2c}\sum \boldsymbol{\mathcal{B}} \times \mathbf{r}_k \cdot \hat{\mathbf{p}}_k &= \frac{1}{2c}\sum_k \boldsymbol{\mathcal{B}} \cdot \mathbf{r}_k \times \hat{\mathbf{p}}_k \\ &= \frac{1}{2c}\boldsymbol{\mathcal{B}} \cdot \hat{\mathbf{L}}, \end{aligned} \tag{4.34}$$

where $\hat{\mathbf{L}}$ is the total canonical electronic angular momentum operator with

$$\hat{\mathbf{L}} = \sum_k \hat{\boldsymbol{\ell}}_k\,; \quad \hat{\boldsymbol{\ell}}_k = \mathbf{r}_k \times \hat{\mathbf{p}}_k, \tag{4.35}$$

with $\hat{\boldsymbol{\ell}}_k$ the electronic angular momentum operator. Thus, the Hamiltonian $\hat{H}$ of (4.25) intrinsically incorporates the interaction of the magnetic field with the orbital angular momentum. It is interesting to note that this interaction emerges from the physical kinetic energy operator $\hat{T}_A$. Next, the term in $\hat{T}_A$ of second-order in the magnetic field $\boldsymbol{\mathcal{B}}$ (via A^2) can be expressed simply by employing the symmetric gauge and the vector identity $(\mathbf{a} \times \mathbf{b}) \cdot (\mathbf{a} \times \mathbf{b}) = a^2b^2 - (\mathbf{a} \cdot \mathbf{b})^2$ as

$$\frac{1}{2c^2}\sum_k A^2(\mathbf{r}_k) = \frac{1}{8c^2}\sum_k (\boldsymbol{\mathcal{B}} \times \mathbf{r}_k) \cdot (\boldsymbol{\mathcal{B}} \times \mathbf{r}_k) \tag{4.36}$$

$$= \frac{1}{8c^2} \sum_k \big(\mathcal{B}^2 r_k^2 - (\mathcal{B} \cdot \mathbf{r}_k)^2 \big). \tag{4.37}$$

Thus, for a uniform magnetic field $\mathcal{B}$, the physical kinetic energy operator of (4.33) may be written as

$$\hat{T}_A = \hat{T} + \frac{1}{2c} \mathcal{B} \cdot \hat{\mathbf{L}} + \frac{1}{8c^2} \sum_k \big(\mathcal{B}^2 r_k^2 - (\mathcal{B} \cdot \mathbf{r}_k)^2 \big). \tag{4.38}$$

4.2.3 Schrödinger Theory in Terms of the Density and Physical Current Density

We next rewrite the Hamiltonian $\hat{H}$ of (4.25) in terms of the density $\hat{\rho}(\mathbf{r})$ and physical current density $\hat{\mathbf{j}}(\mathbf{r})$ operators. To define $\hat{\mathbf{j}}(\mathbf{r})$ for the present physical case, we revert to the definition of the current density $\mathbf{j}(\mathbf{r})$ of (2.41). In terms of the *physical momentum operator* $\hat{\mathbf{p}}_{\text{physical}} = (\hat{\mathbf{p}} + \frac{1}{c}\mathbf{A})$, the physical current density $\mathbf{j}(\mathbf{r})$ is defined as

$$\mathbf{j}(\mathbf{r}) = N \Re \sum_\sigma \int \psi^\star(\mathbf{r}\sigma, \mathbf{X}^{N-1}) \left(\hat{\mathbf{p}} + \frac{1}{c}\mathbf{A}(\mathbf{r}) \right) \psi(\mathbf{r}\sigma, \mathbf{X}^{N-1}) d\mathbf{X}^{N-1}, \tag{4.39}$$

with $\mathbf{X}^{N-1} = \mathbf{x}_2, \ldots, \mathbf{x}_N$ and $d\mathbf{X}^{N-1} = d\mathbf{x}_2, \ldots, d\mathbf{x}_N$. Separating the terms we have

$$\mathbf{j}(\mathbf{r}) = \mathbf{j}_p(\mathbf{r}) + \frac{\mathbf{A}(\mathbf{r})}{c} N \Re \sum_\sigma \int \psi^\star(\mathbf{r}\sigma, \mathbf{X}^{N-1}) \psi(\mathbf{r}\sigma, \mathbf{X}^{N-1}) d\mathbf{X}^{N-1} \tag{4.40}$$

$$= \mathbf{j}_p(\mathbf{r}) + \frac{1}{c} \rho(\mathbf{r}) \mathbf{A}(\mathbf{r}) \tag{4.41}$$

$$= \mathbf{j}_p(\mathbf{r}) + \mathbf{j}_d(\mathbf{r}), \tag{4.42}$$

where $\mathbf{j}_p(\mathbf{r})$ is the paramagnetic component:

$$\mathbf{j}_p(\mathbf{r}) = N \Re \sum_\sigma \int \psi^\star(\mathbf{r}\sigma, \mathbf{X}^{N-1}) \hat{\mathbf{p}} \psi(\mathbf{r}\sigma, \mathbf{X}^{N-1}) d\mathbf{X}^{N-1}, \tag{4.43}$$

(which is the same as that of (2.41)), and $\mathbf{j}_d(\mathbf{r})$ the diamagnetic component:

$$\mathbf{j}_d(\mathbf{r}) = \frac{1}{c} \rho(\mathbf{r}) \mathbf{A}(\mathbf{r}). \tag{4.44}$$

Thus, the physical current density operator is (see Sect. 2.2.4)

$$\hat{\mathbf{j}}(\mathbf{r}) = \hat{\mathbf{j}}_p(\mathbf{r}) + \hat{\mathbf{j}}_d(\mathbf{r}), \tag{4.45}$$

with the paramagnetic and diamagnetic component operators defined as

$$\hat{\mathbf{j}}_p(\mathbf{r}) = \frac{1}{2}\sum_k \left[\hat{\mathbf{p}}_k \delta(\mathbf{r}_k - \mathbf{r}) + \delta(\mathbf{r}_k - \mathbf{r})\hat{\mathbf{p}}_k\right], \tag{4.46}$$

$$\hat{\mathbf{j}}_d(\mathbf{r}) = \frac{1}{c}\hat{\rho}(\mathbf{r})\mathbf{A}(\mathbf{r}), \tag{4.47}$$

and where $\hat{\rho}(\mathbf{r})$ is the density operator of (2.12).

With the operator $\hat{T}_A$ expressed as in (4.33), the Hamiltonian $\hat{H}$ of (4.25) may be written in terms of the density $\hat{\rho}(\mathbf{r})$ and the paramagnetic current density $\hat{\mathbf{j}}_p(\mathbf{r})$ operators as

$$\hat{H} = \hat{T} + \hat{W} + \hat{V} + \frac{1}{c}\int \hat{\mathbf{j}}_p(\mathbf{r}) \cdot \mathbf{A}(\mathbf{r})d\mathbf{r} + \frac{1}{2c^2}\int \hat{\rho}(\mathbf{r})A^2(\mathbf{r})d\mathbf{r}. \tag{4.48}$$

From this expression it is evident that in the presence of a magnetic field, one can define the physical current density operator $\hat{\mathbf{j}}(\mathbf{r})$ of (4.45) as

$$\hat{\mathbf{j}}(\mathbf{r}) = c\,\frac{\partial \hat{H}}{\partial \mathbf{A}(\mathbf{r})} = \hat{\mathbf{j}}_p(\mathbf{r}) + \hat{\mathbf{j}}_d(\mathbf{r}). \tag{4.49}$$

In terms of the density $\hat{\rho}(\mathbf{r})$ and physical current density $\hat{\mathbf{j}}(\mathbf{r})$ operators, the Hamiltonian (4.48) is

$$\hat{H} = \hat{T} + \hat{W} + \hat{V} + \frac{1}{c}\int \hat{\mathbf{j}}(\mathbf{r}) \cdot \mathbf{A}(\mathbf{r})d\mathbf{r} - \frac{1}{2c^2}\int \hat{\rho}(\mathbf{r})A^2(\mathbf{r})d\mathbf{r}. \tag{4.50}$$

Hence, the system energy E which is the expectation valve

$$E = \langle \psi | \hat{H} | \psi \rangle, \tag{4.51}$$

may be written in terms of the density $\rho(\mathbf{r})$, and either the paramagnetic $\mathbf{j}_p(\mathbf{r})$ or physical $\mathbf{j}(\mathbf{r})$ current densities as

$$E = T + E_{ee} + E_{es} + \frac{1}{c}\int \mathbf{j}_p(\mathbf{r}) \cdot \mathbf{A}(\mathbf{r})d\mathbf{r} + \frac{1}{2c^2}\int \rho(\mathbf{r})A^2(\mathbf{r})d\mathbf{r}, \tag{4.52}$$

or as

$$E = T + E_{ee} + E_{es} + \frac{1}{c}\int \mathbf{j}(\mathbf{r}) \cdot \mathbf{A}(\mathbf{r})d\mathbf{r} - \frac{1}{2c^2}\int \rho(\mathbf{r})A^2(\mathbf{r})d\mathbf{r}. \tag{4.53}$$

Here the kinetic T, electron-interaction potential E_{ee} and external electrostatic E_{es} energies are the expectations (see Sect. 2.3)

$$T = \langle\psi|\hat{T}|\psi\rangle, \tag{4.54}$$
$$E_{ee} = \langle\psi|\hat{W}|\psi\rangle, \tag{4.55}$$
$$E_{es} = \langle\psi|\hat{V}|\psi\rangle, \tag{4.56}$$

and the paramagnetic $\mathbf{j}_p(\mathbf{r})$ and diamagnetic $\mathbf{j}(\mathbf{r})$ current densities the expectations

$$\mathbf{j}_p(\mathbf{r}) = \langle\psi|\hat{\mathbf{j}}_p(\mathbf{r})|\psi\rangle, \tag{4.57}$$

and

$$\mathbf{j}(\mathbf{r}) = \langle\psi|\hat{\mathbf{j}}(\mathbf{r})|\psi\rangle, \tag{4.58}$$

respectively.

Finally, for this stationary-state system, the continuity equation for the physical current density $\mathbf{j}(\mathbf{r})$ is

$$\nabla \cdot \mathbf{j}(\mathbf{r}) = \nabla \cdot \mathbf{j}_p + \nabla \cdot \mathbf{j}_d(\mathbf{r}) = 0. \tag{4.59}$$

4.2.4 Schrödinger Theory Hamiltonian in Terms of the Lorentz 'Force' Operator

In the presence of both an electrostatic $\boldsymbol{\mathcal{E}}(\mathbf{r}) = -\boldsymbol{\nabla} v(\mathbf{r})$ and magnetostatic $\boldsymbol{\mathcal{B}}(\mathbf{r}) = \boldsymbol{\nabla} \times \mathbf{A}(\mathbf{r})$ field each electron experiences both a force due to the electric field as well as the Lorentz force. In the corresponding Hamiltonian $\hat{H}$ of (3.5) or (4.25) it is the scalar $v(\mathbf{r})$ and vector $\mathbf{A}(\mathbf{r})$ potentials that appear. Whilst the force due to the electric field $\boldsymbol{\mathcal{E}}(\mathbf{r})$ is explicit via the scalar potential $v(\mathbf{r})$, the Lorentz force is implicit and does not appear explicitly in the Hamiltonian $\hat{H}$. However, with the Hamiltonian $\hat{H}$ as written in terms of the physical current density operator $\hat{\mathbf{j}}(\mathbf{r})$ as in (4.50), it is possible to elucidate the presence of the Lorentz force via a Lorentz 'force' operator $\hat{\boldsymbol{\ell}}(\mathbf{r})$ defined below.

Consider the term involving the operator $\hat{\mathbf{j}}(\mathbf{r})$ in the Hamiltonian of (4.50). In the symmetric gauge with $\mathbf{A}(\mathbf{r}) = \frac{1}{2}\boldsymbol{\mathcal{B}}(\mathbf{r}) \times \mathbf{r}$, this term is

$$\frac{1}{c}\int \hat{\mathbf{j}}(\mathbf{r}) \cdot \mathbf{A}(\mathbf{r}) d\mathbf{r} = \frac{1}{2c}\int \hat{\mathbf{j}}(\mathbf{r}) \cdot \big(\boldsymbol{\mathcal{B}}(\mathbf{r}) \times \mathbf{r}\big) d\mathbf{r}, \tag{4.60}$$

which on employing the vector relation $\mathbf{a} \cdot (\mathbf{b} \times \mathbf{c}) = \mathbf{c} \cdot (\mathbf{a} \times \mathbf{b})$ becomes

$$= \frac{1}{2c}\int \mathbf{r} \cdot \Big(\hat{\mathbf{j}}(\mathbf{r}) \times \boldsymbol{\mathcal{B}}(\mathbf{r})\Big) d\mathbf{r} \tag{4.61}$$
$$= \frac{1}{2}\int \mathbf{r} \cdot \hat{\boldsymbol{\ell}}(\mathbf{r}) d\mathbf{r}, \tag{4.62}$$

with the Lorentz 'force' operator defined as

$$\hat{\boldsymbol{\ell}}(\mathbf{r}) = \frac{\hat{\mathbf{j}}(\mathbf{r}) \times \boldsymbol{\mathcal{B}}(\mathbf{r})}{c}. \tag{4.63}$$

Hence, the Hamiltonian can now be written as

$$\hat{H} = \hat{T} + \hat{W} + \hat{V} + \frac{1}{2} \int \mathbf{r} \cdot \hat{\boldsymbol{\ell}}(\mathbf{r}) d\mathbf{r} - \frac{1}{2c^2} \int \hat{\rho}(\mathbf{r}) A^2(\mathbf{r}) d\mathbf{r}, \tag{4.64}$$

which makes clear that the electrons experience both an electrostatic force as well as the Lorentz force.

The total energy E can then be written as

$$E = T + E_{ee} + E_{es} + E_L - \frac{1}{2c^2} \int \rho(\mathbf{r}) A^2(\mathbf{r}) d\mathbf{r}, \tag{4.65}$$

where the expectations T, E_{ee}, E_{es} are defined in (4.54)–(4.56), and the Lorentz energy E_L is

$$E_L = \frac{1}{2} \int \mathbf{r} \cdot \boldsymbol{\ell}(\mathbf{r}) d\mathbf{r}, \tag{4.66}$$

with the Lorentz 'force' being

$$\boldsymbol{\ell}(\mathbf{r}) = \frac{\mathbf{j}(\mathbf{r}) \times \boldsymbol{\mathcal{B}}(\mathbf{r})}{c}, \tag{4.67}$$

and $\mathbf{j}(\mathbf{r})$ the physical current density the expectation of (4.58).

4.2.5 *The Wave Function, a Functional of the Gauge Function*

By performing a unitary or gauge transformation we next show that the wave function $\Psi(\mathbf{X})$ of the Schrödinger equation (4.31) is a functional of the gauge function [8].

Consider a density and physical current density *preserving* unitary transformation for which the unitary operator U is

$$U = e^{i\alpha(\mathbf{R})} \quad ; \quad \alpha(\mathbf{R}) = \sum_j \alpha(\mathbf{r}_j), \tag{4.68}$$

where $\alpha(\mathbf{r})$ is an arbitrary smooth function of position. The transformed wave function $\psi'(\mathbf{X})$ and Hamiltonian $\hat{H}'(\mathbf{R})$ of (4.31) are, respectively,

$$\psi'(\mathbf{X}) = U^{\dagger} \psi(\mathbf{X}), \tag{4.69}$$

and

$$\hat{H}' = U^{\dagger}\hat{H}(\mathbf{R})U \tag{4.70}$$

$$= \frac{1}{2}\sum_{k}\left(\hat{\mathbf{p}}_k + \mathbf{A}(\mathbf{r}_k) + \nabla\alpha(\mathbf{r}_k)\right)^2 + \hat{W} + \hat{V}. \tag{4.71}$$

The transformed Schrödinger equation is then

$$\hat{H}'(\mathbf{R})\psi'(\mathbf{X}) = E'\psi'(\mathbf{X}) \tag{4.72}$$

with

$$E' = E. \tag{4.73}$$

Equivalently, if one performs a gauge transformation of the vector potential $\mathbf{A}(\mathbf{r})$ such that

$$\mathbf{A}'(\mathbf{r}) = \mathbf{A}(\mathbf{r}) + \nabla\alpha(\mathbf{r}) \tag{4.74}$$

but let $v'(\mathbf{r}) = v(\mathbf{r})$, the Hamiltonian of (4.25) changes to that of (4.71). Thus, the Hamiltonian is *gauge variant*. Because the physical system remains the same, the wave function $\psi(\mathbf{X})$ must be multiplied by a phase factor $exp[-i\alpha(\mathbf{R})]$, which is (4.69). The system wave function is therefore also *gauge variant*. However, all the physical properties of the system such as the energy E and its individual components T_A, E_{ee}, V, the density $\rho(\mathbf{r})$ and physical current density $\mathbf{j}(\mathbf{r})$ which are all expectations of Hermitian operators remain the same and are *gauge invariant*. The component paramagnetic $\mathbf{j}_p(\mathbf{r})$ and diamagnetic $\mathbf{j}_d(\mathbf{r})$ current densities, on the other hand, are *gauge variant*.

The unitary or gauge transformation is summarized in Fig. 4.1. When the wave function is $\psi(\mathbf{X})$, the Hamiltonian $\hat{H}(\mathbf{R}) = \hat{H}(v, \mathbf{A}, \alpha = \text{constant})$ of (4.25). When the wave function is $\psi_1(\mathbf{X})$ corresponding to a phase factor of $\alpha_1(\mathbf{R})$, the Hamiltonian

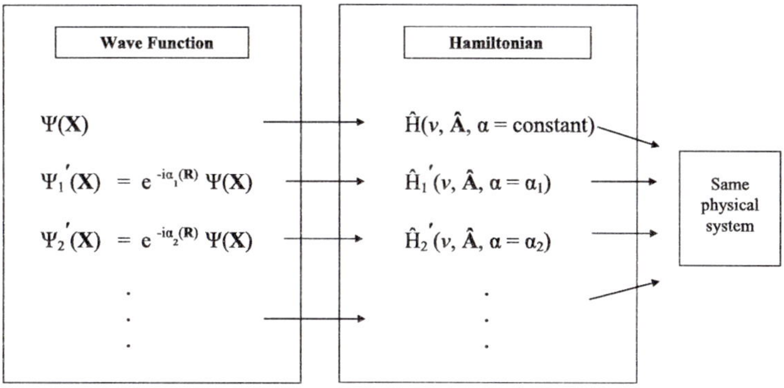

Fig. 4.1 Pictorial representation of the unitary or gauge transformation of (4.69) and (4.71)

is $H_1'(\mathbf{R})$ of (4.71) with $\alpha(\mathbf{R}) = \alpha_1(\mathbf{R})$, and so on. But all these infinite number of Hamiltonians correspond to the same physical system. Thus, one can conclude that the wave function $\psi(\mathbf{X})$ is a functional of the gauge function $\alpha(\mathbf{R})$, *i.e.* $\psi(\mathbf{X}) = \psi[\alpha(\mathbf{R})](\mathbf{X})$.

As noted in the introductory paragraph of Sect. 4.1, the wave function $\psi(\mathbf{X})$ is also a functional of the ground state density $\rho(\mathbf{r})$ and physical current density $\mathbf{j}(\mathbf{r})$. Thus, the wave function is a functional of $\{\rho, \mathbf{j}, \alpha\} : \psi(\mathbf{X}) = \psi[\rho, \mathbf{j}, \alpha]$. As the wave function $\psi(\mathbf{X})$ is gauge variant and $\rho(\mathbf{r})$ and $\mathbf{j}(\mathbf{r})$ gauge invariant, it is the presence of the gauge function $\alpha(\mathbf{R})$ in the wave function written as a functional that ensures it is gauge variant. Note that the case of $\alpha(\mathbf{R}) =$ constant is equally valid as it corresponds to the same physical system (see Fig. 4.1).

4.3 Stationary-State Schrödinger-Pauli Theory in an Electromagnetic Field

As we have seen in the previous section, the Schrödinger theory of a system of N electrons in an electrostatic $\mathcal{E}(\mathbf{r}) = -\nabla v(\mathbf{r})/e$ and magnetostatic $\mathcal{B}(\mathbf{r}) = \nabla \times \mathbf{A}(\mathbf{r})$ field, where $[v(\mathbf{r}), \mathbf{A}(\mathbf{r})]$ are scalar and vector potentials, respectively, explicitly accounts for the interaction of the magnetic field with the total electronic orbital angular momentum. Schrödinger-Pauli theory goes beyond Schrödinger theory in that it also includes the interaction of the magnetic field with the electron spin moment. Originally, this interaction was added on *ad hoc* by Pauli [2] to the Schrödinger theory Hamiltonian $\hat{H}$ of (4.25). The Schrödinger-Pauli theory equation is obtained as the non-relativistic limit of the Dirac equation [5]. However, as emphasized by Feynman [3, 4], the Schrödinger-Pauli Hamiltonian $\hat{H}_{\mathrm{spin}}$ can also be derived directly within non-relativistic quantum mechanics by employing (what we now refer to as) the Feynman kinetic energy operator which assumes knowledge of the fact that electrons possess a spin angular momentum.

In this section, we therefore begin by deriving the stationary-state Schrödinger-Pauli theory Hamiltonian $\hat{H}_{\mathrm{spin}}$ via the Feynman kinetic energy operator. The Hamiltonian is then rewritten in the alternate representation in terms of the density $\hat{\rho}(\mathbf{r})$ and the physical current density $\hat{\mathbf{j}}(\mathbf{r})$ operators. The eigenenergies E are therefore expressed in terms of the gauge invariant properties of the density $\rho(\mathbf{r})$ and physical current density $\mathbf{j}(\mathbf{r})$. In Schrödinger-Pauli theory, the physical current density $\mathbf{j}(\mathbf{r})$ is comprised, in addition to the paramagnetic $\mathbf{j}_p(\mathbf{r})$ and diamagnetic $\mathbf{j}_d(\mathbf{r})$ components, of a magnetization current density $\mathbf{j}_m(\mathbf{r})$ component which is a consequence of the electron spin moment.

4.3.1 Schrödinger-Pauli Theory Hamiltonian and Equation

The stationary-state Schrödinger-Pauli theory Hamiltonian for spin $\frac{1}{2}$ particles is the sum of the Feynman [3, 4] kinetic $\hat{T}_F$, electron-interaction $\hat{W}$, and external electrostatic potential $\hat{V}$ operators. In atomic units (charge of electron $-e$, $|e| = \hbar = m = 1$), the Hamiltonian is

$$\hat{H}_{\text{spin}} = \hat{T}_F + \hat{W} + \hat{V}, \tag{4.75}$$

where

$$\hat{T}_F = \frac{1}{2}\sum_k \big(\boldsymbol{\sigma}_k \cdot \hat{\mathbf{p}}_{k,\text{phys}}\big)\big(\boldsymbol{\sigma}_k \cdot \hat{\mathbf{p}}_{k,\text{phys}}\big), \tag{4.76}$$

$$\hat{W} = \frac{1}{2}\sum_{k,\ell}{}' \frac{1}{|\mathbf{r}_k - \mathbf{r}_\ell|}, \tag{4.77}$$

$$\hat{V} = \sum_k v(\mathbf{r}_k). \tag{4.78}$$

Here the physical momentum operator $\hat{\mathbf{p}}_{\text{phys}} = (\hat{\mathbf{p}} + \frac{1}{c}\mathbf{A})$, with $\hat{\mathbf{p}} = -i\boldsymbol{\nabla}$ the canonical momentum operator. The $\boldsymbol{\sigma}$ is the Pauli spin matrix: $\mathbf{s} = \frac{1}{2}\boldsymbol{\sigma}$, with $\mathbf{s}$ the electron spin angular momentum vector operator. Substituting for $\hat{\mathbf{p}}_{\text{phys}}$, the kinetic energy operator for the individual electron is

$$\hat{T}_F = \frac{1}{2}\boldsymbol{\sigma} \cdot \left(\hat{\mathbf{p}} + \frac{1}{c}\mathbf{A}\right)\boldsymbol{\sigma} \cdot \left(\hat{\mathbf{p}} + \frac{1}{c}\mathbf{A}\right). \tag{4.79}$$

On employing the vector relation

$$(\boldsymbol{\sigma} \cdot \mathbf{A})(\boldsymbol{\sigma} \cdot \mathbf{B}) = \mathbf{A} \cdot \mathbf{B} + i\boldsymbol{\sigma} \cdot (\mathbf{A} \times \mathbf{B}) \tag{4.80}$$

which holds even with $\mathbf{A}$ and $\mathbf{B}$ being operators, the kinetic energy operator is

$$\hat{T}_F = \frac{1}{2}(\hat{\mathbf{p}} + \frac{1}{c}\mathbf{A})^2 + \frac{i}{2}\boldsymbol{\sigma} \cdot \big[(\hat{\mathbf{p}} + \frac{1}{c}\mathbf{A}) \times (\hat{\mathbf{p}} + \frac{1}{c}\mathbf{A})\big] \tag{4.81}$$

$$= \frac{1}{2}(\hat{\mathbf{p}} + \frac{1}{c}\mathbf{A})^2 + \frac{i}{2}\boldsymbol{\sigma} \cdot \big[\frac{1}{c}\mathbf{A} \times \hat{\mathbf{p}} + \frac{1}{c}\hat{\mathbf{p}} \times \mathbf{A}\big]. \tag{4.82}$$

Using the operator relation

$$\hat{\mathbf{p}} \times \mathbf{A} = -i\boldsymbol{\nabla} \times \mathbf{A} - \mathbf{A} \times \hat{\mathbf{p}}, \tag{4.83}$$

we then arrive at

$$\hat{T}_F = \frac{1}{2}(\hat{\mathbf{p}} + \frac{1}{c}\mathbf{A})^2 + \frac{1}{2c}\boldsymbol{\sigma} \cdot \boldsymbol{\mathcal{B}} \tag{4.84}$$

$$= \frac{1}{2}(\hat{\mathbf{p}} + \frac{1}{c}\mathbf{A})^2 + \frac{1}{c}\mathbf{s} \cdot \boldsymbol{\mathcal{B}}, \tag{4.85}$$

where we have employed $\boldsymbol{\mathcal{B}} = \boldsymbol{\nabla} \times \mathbf{A}$ and $\mathbf{s} = \frac{1}{2}\boldsymbol{\sigma}$. The spin magnetic moment generated in this way has the correct gyromagnetic ratio $g = 2$. (Note that the operator $\hat{T}_F$ of (4.76) reduces to $p^2/2$ in the absence of the vector potential.)

The N-electron Hamiltonian is then

$$\hat{H}_{\text{spin}} = \frac{1}{2}\sum_k \left(\hat{\mathbf{p}}_k + \frac{1}{c}\mathbf{A}(\mathbf{r}_k)\right)^2 + \frac{1}{c}\sum_k \boldsymbol{\mathcal{B}}(\mathbf{r}_k) \cdot \mathbf{s}_k + \hat{W} + \hat{V}, \tag{4.86}$$

which now incorporates the interaction of the magnetic field with both the orbital and spin moment of the electron. Note that both interactions arise from the Feynman kinetic energy operator $\hat{T}_F$. (The second term of (4.86) not in atomic units is $(g\mu_B/\hbar)\sum_k \boldsymbol{\mathcal{B}}(\mathbf{r}_k) \cdot \mathbf{s}_k$ with $g = 2$, $\mu_B = e\hbar/2mc$ is the Bohr magneton, and $\mathbf{s} = (\hbar/2)\boldsymbol{\sigma}$.) The corresponding Schrödinger-Pauli equation is then

$$\hat{H}_{\text{spin}}\psi(\mathbf{X}) = E\psi(\mathbf{X}), \tag{4.87}$$

with $\{\psi(\mathbf{X}), E\}$ the eigenfunctions and eigenvalues, $\mathbf{X} = \mathbf{x}_1, \ldots, \mathbf{x}_N$; $\mathbf{x} = \mathbf{r}\sigma$; and $\mathbf{r}\sigma$ the spatial and spin coordinates.

4.3.2 Schrödinger-Pauli Theory in Terms of the Density and Physical Current Density

The Schrödinger-Pauli Hamiltonian of (4.86) may be written in terms of the physical current density $\hat{\mathbf{j}}(\mathbf{r})$ and density $\hat{\rho}(\mathbf{r})$ operators (or equivalently in terms of the paramagnetic $\hat{\mathbf{j}}_p(\mathbf{r})$ and diamagnetic $\hat{\mathbf{j}}_d(\mathbf{r})$ current density components (see (4.45)–(4.47)), and the local magnetization density $\hat{\mathbf{m}}(\mathbf{r})$ operators as

$$\hat{H}_{\text{spin}} = \hat{T} + \hat{W} + \hat{V}_A - \int \hat{\mathbf{m}}(\mathbf{r}) \cdot \boldsymbol{\mathcal{B}}(\mathbf{r})d\mathbf{r}, \tag{4.88}$$

where the total external potential operator $\hat{V}_A$ is

$$\hat{V}_A = \hat{V} + \frac{1}{c}\int \hat{\mathbf{j}}(\mathbf{r}) \cdot \mathbf{A}(\mathbf{r})d\mathbf{r} - \frac{1}{2c^2}\int \hat{\rho}(\mathbf{r})A^2(\mathbf{r})d\mathbf{r} \tag{4.89}$$

$$= \hat{V} + \frac{1}{c}\int \hat{\mathbf{j}}_p(\mathbf{r}) \cdot \mathbf{A}(\mathbf{r})d\mathbf{r} + \frac{1}{2c}\int \hat{\mathbf{j}}_d[\mathbf{A}](\mathbf{r}) \cdot \mathbf{A}(\mathbf{r})d\mathbf{r}, \tag{4.90}$$

and $\hat{\mathbf{m}}(\mathbf{r})$ is defined as

$$\hat{\mathbf{m}}(\mathbf{r}) = -\frac{1}{c}\sum_k \mathbf{s}_k \delta(\mathbf{r}_k - \mathbf{r}). \tag{4.91}$$

The magnetization density $\mathbf{m}(\mathbf{r})$ is then the expectation

$$\mathbf{m}(\mathbf{r}) = \langle \psi | \hat{\mathbf{m}}(\mathbf{r}) | \psi \rangle. \tag{4.92}$$

In the expression for the Hamiltonian $\hat{H}_{\text{spin}}$ of (4.88), the definition of the current density operator $\hat{\mathbf{j}}(\mathbf{r})$ employed is that of (4.39) or equivalently (4.42), *viz.* in terms of the paramagnetic and diamagnetic components.

However, for finite systems, yet another component—the magnetization current density—due to the electron spin can be introduced [11]. Consider the last term of the Hamiltonian of (4.88):

$$\int \hat{\mathbf{m}}(\mathbf{r}) \cdot \boldsymbol{\mathcal{B}}(\mathbf{r}) d\mathbf{r} = \int \hat{\mathbf{m}}(\mathbf{r}) \cdot (\boldsymbol{\nabla} \times \mathbf{A}(\mathbf{r})) d\mathbf{r} \tag{4.93}$$

$$= \int \mathbf{A}(\mathbf{r}) \cdot (\boldsymbol{\nabla} \times \hat{\mathbf{m}}(\mathbf{r})) d\mathbf{r} + \int \boldsymbol{\nabla} \cdot (\mathbf{A}(\mathbf{r}) \times \hat{\mathbf{m}}(\mathbf{r})) d\mathbf{r}, \tag{4.94}$$

where the vector identity

$$\boldsymbol{\nabla} \cdot (\mathbf{C} \times \mathbf{D}) = \mathbf{D} \cdot (\boldsymbol{\nabla} \times \mathbf{C}) - \mathbf{C} \cdot (\boldsymbol{\nabla} \times \mathbf{D}) \tag{4.95}$$

is employed. The last term of (4.94) may be converted to an integral over a surface: $\int \boldsymbol{\nabla} \cdot (\mathbf{A} \times \hat{\mathbf{m}}) d\mathbf{r} = \int (\mathbf{A} \times \hat{\mathbf{m}}) \cdot d\mathbf{S}$, which vanishes in the usual way for an infinitely distant surface. Thus, the Hamiltonian of (4.88) can be written as

$$\hat{H}_{\text{spin}} = \hat{T} + \hat{W} + \hat{V} + \frac{1}{c}\int \hat{\mathbf{j}}_p(\mathbf{r}) \cdot \mathbf{A}(\mathbf{r}) d\mathbf{r} + \frac{1}{2c^2}\int \hat{\rho}(\mathbf{r}) A^2(\mathbf{r}) d\mathbf{r} + \frac{1}{c}\int \hat{\mathbf{j}}_m(\mathbf{r}) \cdot \mathbf{A}(\mathbf{r}) d\mathbf{r}, \tag{4.96}$$

where the magnetization current density operator $\hat{\mathbf{j}}_m(\mathbf{r})$ is defined as

$$\hat{\mathbf{j}}_m(\mathbf{r}) = -c\boldsymbol{\nabla} \times \hat{\mathbf{m}}(\mathbf{r}). \tag{4.97}$$

Hence the *physical current density* operator $\hat{\mathbf{j}}_{\text{spin}}(\mathbf{r})$ may also be defined as

$$\hat{\mathbf{j}}_{\text{spin}}(\mathbf{r}) = c\frac{\partial \hat{H}_{\text{spin}}}{\partial \mathbf{A}(\mathbf{r})} = \hat{\mathbf{j}}_p(\mathbf{r}) + \hat{\mathbf{j}}_d(\mathbf{r}) + \hat{\mathbf{j}}_m(\mathbf{r}), \tag{4.98}$$

the sum of the paramagnetic $\hat{\mathbf{j}}_p(\mathbf{r})$, diamagnetic $\hat{\mathbf{j}}_d(\mathbf{r})$, and magnetization density $\hat{\mathbf{j}}_m(\mathbf{r})$ operators, with $\hat{\mathbf{j}}_p(\mathbf{r})$ and $\hat{\mathbf{j}}_d(\mathbf{r})$ defined by (4.46) and (4.47), respectively. Thus in terms of the density $\hat{\rho}(\mathbf{r})$ and physical current density $\hat{\mathbf{j}}_{\text{spin}}(\mathbf{r})$ operators, the Schrödinger-Pauli Hamiltonian $\hat{H}_{\text{spin}}$ of (4.88) is

$$\hat{H}_{\text{spin}} = \hat{T} + \hat{W} + \hat{V} + \frac{1}{c}\int \hat{\mathbf{j}}_{\text{spin}}(\mathbf{r}) \cdot \mathbf{A}(\mathbf{r}) - \frac{1}{2c^2}\int \hat{\rho} A^2(\mathbf{r}) d\mathbf{r}. \tag{4.99}$$

Note that in this case, the physical current density operator $\hat{\mathbf{j}}_{\text{spin}}(\mathbf{r})$ is defined by (4.98). The system eigenenergy E, which is the expectation

$$E = \langle \psi | \hat{H}_{\text{spin}} | \psi \rangle, \tag{4.100}$$

may then be expressed in terms of the density $\rho(\mathbf{r})$ and physical current density $\mathbf{j}(\mathbf{r})$ as

$$E = T + E_{ee} + V + \frac{1}{c}\int \mathbf{j}_{\text{spin}}(\mathbf{r}) \cdot \mathbf{A}(\mathbf{r}) d\mathbf{r} - \frac{1}{2c^2}\int \rho(\mathbf{r}) A^2(\mathbf{r}) d\mathbf{r}. \tag{4.101}$$

where $\rho(\mathbf{r})$ and $\mathbf{j}(\mathbf{r})$ are the expectations

$$\rho(\mathbf{r}) = \langle \psi | \hat{\rho}(\mathbf{r}) | \psi \rangle, \tag{4.102}$$

$$\mathbf{j}_{\text{spin}}(\mathbf{r}) = \langle \psi | \hat{\mathbf{j}}_{\text{spin}}(\mathbf{r}) | \psi \rangle, \tag{4.103}$$

with $\hat{\rho}(\mathbf{r})$ defined by (2.12) and $\hat{\mathbf{j}}_{\text{spin}}(\mathbf{r})$ by (4.98). The kinetic T, electron-interaction E_{ee}, and external electrostatic E_{es} energy components are the expectations

$$T = \langle \psi | \hat{T} | \psi \rangle, \tag{4.104}$$

$$E_{ee} = \langle \psi | \hat{W} | \psi \rangle, \tag{4.105}$$

$$E_{es} = \langle \psi | \hat{V} | \psi \rangle. \tag{4.106}$$

Finally, as a consequence of the vector relation $\boldsymbol{\nabla} \cdot \boldsymbol{\nabla} \times \mathbf{a} = 0$, we have $\boldsymbol{\nabla} \cdot \mathbf{j}_m(\mathbf{r}) = 0$. Thus, together with (4.59), the continuity condition

$$\boldsymbol{\nabla} \cdot \mathbf{j}_{\text{spin}}(\mathbf{r}) = 0, \tag{4.107}$$

is satisfied.

4.4 Stationary-State Schrödinger-Pauli Theory in Terms of Quantal Sources and Fields

As noted in the previous section, Schrödinger-Pauli theory of electrons in a static electromagnetic field: $\boldsymbol{\mathcal{E}}(\mathbf{r}) = -\nabla v(\mathbf{r})/e, \boldsymbol{\mathcal{B}}(\mathbf{r}) = \nabla \times \mathbf{A}(\mathbf{r})$, where $[v(\mathbf{r}), \mathbf{A}(\mathbf{r})]$ are scalar and vector potentials, differs from that of Schrödinger theory in that electron spin is now explicitly accounted for. The Hamiltonian for the spinless electrons $\hat{H}_{\text{spinless}}$ (Schrödinger theory) and that for electrons with spin $\hat{H}_{\text{spin}}$ (Schrödinger-Pauli theory) are, respectively,

$$\hat{H}_{\text{spinless}} = \frac{1}{2}\sum_k \left(\hat{\mathbf{p}}_k + \frac{1}{c}\mathbf{A}(\mathbf{r}_k)\right)^2 + \hat{W} + \hat{V}, \tag{4.108}$$

and

$$\hat{H}_{\text{spin}} = \frac{1}{2}\sum_k \left(\hat{\mathbf{p}}_k + \frac{1}{c}\mathbf{A}(\mathbf{r}_k)\right)^2 + \frac{1}{c}\sum_k \boldsymbol{\mathcal{B}}(\mathbf{r}_k)\cdot\mathbf{s}_k + \hat{W} + \hat{V}, \tag{4.109}$$

where $\hat{W}$ and $\hat{V}$ are the electron-interaction and electrostatic potential operators defined by (4.29) and (4.30), and $\mathbf{s}$ the spin angular momentum vector. The corresponding Schrödinger and Schrödinger-Pauli theory equations, respectively, are

$$\hat{H}_{\text{spinless}}\psi = E\psi, \tag{4.110}$$

and

$$\hat{H}_{\text{spin}}\psi = E\psi \tag{4.111}$$

Although a priori not self-evident, it turns out that as was the case for Schrödinger theory, it is also possible to describe Schrödinger-Pauli theory in terms of quantal sources and 'classical' fields as experienced by each electron but now possessing spin.

In order to understand that this is the case, it is best to first derive the equation of motion or 'Quantal Newtonian' First Law for the case of an electron with spin because the law incorporates all the fields that describe the physical system. One approach to deriving the law is the traditional one as described in the Appendix A. One writes the wave function as $\psi = \psi^R + i\psi^I$, where ψ^R and ψ^I are the real and imaginary parts, substitute it into the corresponding differential equation, perform the various derivatives, manipulate the magnetic field-spin interaction term to identify the magnetization current density, employ the equation of continuity, and arrive at the law. However, the law for electrons with spin can also be obtained by analogy with the case for spinless electrons. This is so because the Hamiltonian $\hat{H}_{\text{spinless}}$ of Schrödinger theory and that of $\hat{H}_{\text{spin}}$ of Schrödinger-Pauli theory when written in terms of the density $\hat{\rho}(\mathbf{r})$ and the physical current density $\hat{\mathbf{j}}(\mathbf{r})$ operators are similar. Thus,

$$\hat{H}_{\text{spinless}} = \hat{T} + \frac{1}{c}\int \hat{\mathbf{j}}_{\text{spinless}}(\mathbf{r}) \cdot \mathbf{A}(\mathbf{r})d\mathbf{r} - \frac{1}{2c^2}\int \hat{\rho}(\mathbf{r})A^2(\mathbf{r})d\mathbf{r} + \hat{W} + \hat{V}, \tag{4.112}$$

with $\hat{T}$ the canonical kinetic energy operator, and with the physical current density $\hat{\mathbf{j}}_{\text{spinless}}(\mathbf{r})$ operator being the sum of the paramagnetic $\hat{\mathbf{j}}_p(\mathbf{r})$ and diamagnetic $\hat{\mathbf{j}}_d(\mathbf{r})$ components defined by (4.46) and (4.47):

$$\hat{\mathbf{j}}_{\text{spinless}}(\mathbf{r}) = \hat{\mathbf{j}}_p(\mathbf{r}) + \hat{\mathbf{j}}_d(\mathbf{r}), \tag{4.113}$$

whereas

$$\hat{H}_{\text{spin}} = \hat{T} + \frac{1}{c}\int \hat{\mathbf{j}}_{\text{spin}}(\mathbf{r}) \cdot \mathbf{A}(\mathbf{r})d\mathbf{r} - \frac{1}{2c^2}\int \hat{\rho}(\mathbf{r})A^2(\mathbf{r})d\mathbf{r} + \hat{W} + \hat{V}, \tag{4.114}$$

with the physical current density $\hat{\mathbf{j}}_{\text{spin}}(\mathbf{r})$ operator being the sum of the paramagnetic $\hat{\mathbf{j}}_p(\mathbf{r})$, diamagnetic $\hat{\mathbf{j}}_d(\mathbf{r})$, and magnetization $\hat{\mathbf{j}}_m(\mathbf{r})$ components, the last component being defined by (4.97):

$$\hat{\mathbf{j}}_{\text{spin}}(\mathbf{r}) = \hat{\mathbf{j}}_p(\mathbf{r}) + \hat{\mathbf{j}}_d(\mathbf{r}) + \hat{\mathbf{j}}_m(\mathbf{r}). \tag{4.115}$$

Employing the similarity of the Hamiltonian of electrons with spin to that of spinless electrons, it is then possible to directly write down the corresponding 'Quantal Newtonian' First Law for the former.

4.4.1 *The 'Quantal Newtonian' First Law for an Electron with Spin*

The equation of motion or 'Quantal Newtonian' First Law for an electron with spin of the physical system defined by the Schrödinger-Pauli equation of (4.111) states that the sum of the external $\mathcal{F}^{\text{ext}}(\mathbf{r})$ and internal $\mathcal{F}^{\text{int}}(\mathbf{r})$ fields experienced by each electron vanishes:

$$\mathcal{F}^{\text{ext}}(\mathbf{r}) + \mathcal{F}^{\text{int}}(\mathbf{r}) = 0. \tag{4.116}$$

The law satisfies the continuity condition

$$\nabla \cdot \mathbf{j}_{\text{spin}}(\mathbf{r}) = 0, \tag{4.117}$$

with the physical current density $\mathbf{j}_{\text{spin}}(\mathbf{r}) = \langle \psi | \hat{\mathbf{j}}_{\text{spin}}(\mathbf{r}) | \psi \rangle$, the operator $\hat{\mathbf{j}}_{\text{spin}}(\mathbf{r})$ defined by (4.115). Thus, the quantal source-field perspective of Schrödinger-Pauli theory is consistent with Schrödinger's insight [1] that satisfaction of this condition is the explanation of the lack of radiation in a stationary state. The 'Quantal Newtonian' first law is valid for *arbitrary* state. It is also gauge invariant.

The definitions of the external and internal fields given next are similar to those of spinless electrons.

External field $\mathcal{F}^{\mathrm{ext}}(\mathbf{r})$

The external field $\mathcal{F}^{\mathrm{ext}}(\mathbf{r})$ experienced by each electron with spin is the sum of the binding electrostatic $\mathcal{E}(\mathbf{r})$ and Lorentz $\mathcal{L}(\mathbf{r})$ fields:

$$\mathcal{F}^{\mathrm{ext}}(\mathbf{r}) = \mathcal{E}(\mathbf{r}) - \mathcal{L}(\mathbf{r}) = -\nabla v(\mathbf{r}) - \mathcal{L}(\mathbf{r}), \tag{4.118}$$

where the Lorentz field $\mathcal{L}(\mathbf{r})$ is defined in terms of the Lorentz 'force' $\boldsymbol{\ell}(\mathbf{r})$ and electronic density $\rho(\mathbf{r})$ as

$$\mathcal{L}(\mathbf{r}) = \frac{\boldsymbol{\ell}(\mathbf{r})}{\rho(\mathbf{r})}, \tag{4.119}$$

with

$$\boldsymbol{\ell}(\mathbf{r}) = \mathbf{j}_{\mathrm{spin}}(\mathbf{r}) \times \mathcal{B}(\mathbf{r}). \tag{4.120}$$

The electronic $\rho(\mathbf{r})$ and physical current $\mathbf{j}_{\mathrm{spin}}(\mathbf{r})$ densities are, respectively, the expectations defined by (4.102) and (4.103).

Internal Field $\mathcal{F}^{\mathrm{int}}(\mathbf{r})$

The internal field $\mathcal{F}^{\mathrm{int}}(\mathbf{r})$ is a sum of components each descriptive of a property of the system: an electron-interaction field $\mathcal{E}_{\mathrm{ee}}(\mathbf{r})$ representative of electron correlations due to the Pauli principle and Coulomb repulsion; a kinetic field $\mathcal{Z}(\mathbf{r})$ from which the kinetic energy density and kinetic energy can be obtained; the differential density field $\mathcal{D}(\mathbf{r})$ representative of the electron density; and finally an internal magnetic field component $\mathcal{I}_m(\mathbf{r})$. Thus,

$$\mathcal{F}^{\mathrm{int}}(\mathbf{r}) = \mathcal{E}_{\mathrm{ee}}(\mathbf{r}) - \mathcal{Z}(\mathbf{r}) - \mathcal{D}(\mathbf{r}) - \mathcal{I}_m(\mathbf{r}). \tag{4.121}$$

The electron-interaction field $\mathcal{E}_{\mathrm{ee}}(\mathbf{r})$ is defined in terms of the electron-interaction 'force' $\mathbf{e}_{\mathrm{ee}}(\mathbf{r})$ and density $\rho(\mathbf{r})$ as

$$\mathcal{E}_{\mathrm{ee}}(\mathbf{r}) - \frac{\mathbf{e}_{\mathrm{ee}}(\mathbf{r})}{\rho(\mathbf{r})}, \tag{4.122}$$

where $\mathbf{e}_{\mathrm{ee}}(\mathbf{r})$ is obtained via Coulomb's Law from its *nonlocal* (dynamic) quantal source, the pair-correlation function $P(\mathbf{r}\mathbf{r}')$:

$$\mathbf{e}_{\mathrm{ee}}(\mathbf{r}) = \int \frac{P(\mathbf{r}\mathbf{r}')(\mathbf{r} - \mathbf{r}')}{|\mathbf{r} - \mathbf{r}'|^3} d\mathbf{r}', \tag{4.123}$$

with $P(\mathbf{r}\mathbf{r}')$ the expectation value

$$P(\mathbf{r}\mathbf{r}') = \langle \Psi(\mathbf{X}) | \hat{P}(\mathbf{r}\mathbf{r}') | \Psi(\mathbf{X}) \rangle, \tag{4.124}$$

of the pair-correlation operator $\hat{P}(\mathbf{r}\mathbf{r}')$ of (2.28).

The electron-interaction field $\mathcal{E}_{\text{ee}}(\mathbf{r})$ may equivalently be thought of as arising via Coulomb's Law from the quantal source of the pair-correlation density $g(\mathbf{rr}') = P(\mathbf{rr}')/\rho(\mathbf{r})$. The pair-correlation density can be separated into its *local* $\rho(\mathbf{r}')$ and *nonlocal* $\rho_{xc}(\mathbf{rr}')$ components: $g(\mathbf{rr}') = \rho(\mathbf{r}') + \rho_{xc}(\mathbf{rr}')$, where $\rho_{xc}(\mathbf{rr}')$ is the quantum-mechanical Fermi-Coulomb hole charge distribution. Thus, the field $\mathcal{E}_{\text{ee}}(\mathbf{r})$ may be written as a sum of its Hartree $\mathcal{E}_H(\mathbf{r})$ and Pauli-Coulomb $\mathcal{E}_{\text{xc}}(\mathbf{r})$ components:

$$\mathcal{E}_{\text{ee}}(\mathbf{r}) = \mathcal{E}_H(\mathbf{r}) + \mathcal{E}_{\text{xc}}(\mathbf{r}), \tag{4.125}$$

where

$$\mathcal{E}_H(\mathbf{r}) = \int \frac{\rho(\mathbf{r}')(\mathbf{r}-\mathbf{r}')}{|\mathbf{r}-\mathbf{r}'|^3} d\mathbf{r}', \tag{4.126}$$

and

$$\mathcal{E}_{\text{xc}}(\mathbf{r}) = \int \frac{\rho_{\text{xc}}(\mathbf{rr}')(\mathbf{r}-\mathbf{r}')}{|\mathbf{r}-\mathbf{r}'|^3} d\mathbf{r}'. \tag{4.127}$$

Note that in traditional quantum mechanics, it is not possible to further split the Fermi-Coulomb hole into its Fermi $\rho_x(\mathbf{rr}')$ and Coulomb $\rho_c(\mathbf{rr}')$ components. In other words, it is not possible to separate the correlations due to the Pauli principle and Coulomb repulsion. This separation will be accomplished via Quantal Density Functional theory as shown in Chap. 6.

The kinetic field $\mathcal{Z}(\mathbf{r})$ is defined in terms of the kinetic 'force' $z(\mathbf{r})$ and the density $\rho(\mathbf{r})$ as

$$\mathcal{Z}(\mathbf{r}) = \frac{z(\mathbf{r})}{\rho(\mathbf{r})}. \tag{4.128}$$

The kinetic 'force' $z(\mathbf{r})$ in Cartesian coordinates is obtained from its *nonlocal* (dynamic) quantal source, the single-particle density matrix $\gamma(\mathbf{rr}')$ as follows:

$$z_\alpha(\mathbf{r}) = 2\sum_\beta \nabla_\beta t_{\alpha\beta}(\mathbf{r};\gamma), \tag{4.129}$$

where the second-rank kinetic energy tensor $t_{\alpha\beta}(\mathbf{r};\gamma)$ is

$$t_{\alpha\beta}(\mathbf{r};\gamma) = \frac{1}{4}\left[\frac{\partial^2}{\partial r'_\alpha \partial r''_\beta} + \frac{\partial^2}{\partial r'_\beta \partial r''_\alpha}\right]\gamma(\mathbf{r}'\mathbf{r}'')\bigg|_{\mathbf{r}'=\mathbf{r}''=\mathbf{r}}. \tag{4.130}$$

The quantal source $\gamma(\mathbf{rr}')$is the expectation value

$$\gamma(\mathbf{rr}') = \langle\psi(\mathbf{X})|\hat{\gamma}(\mathbf{rr}')|\psi(\mathbf{X})\rangle, \tag{4.131}$$

with the complex density matrix operator $\hat{\gamma}(\mathbf{rr}')$ given by (2.16).

The differential density field $\mathcal{D}(\mathbf{r})$ whose quantal source is the *local* electron density $\rho(\mathbf{r})$, is defined in terms of the corresponding 'force' $d(\mathbf{r})$ and density $\rho(\mathbf{r})$

as

$$\mathcal{D}(\mathbf{r}) = \frac{d(\mathbf{r})}{\rho(\mathbf{r})}, \tag{4.132}$$

where

$$d(\mathbf{r}) = -\frac{1}{4}\nabla\nabla^2\rho(\mathbf{r}). \tag{4.133}$$

The magnetic field contribution $\mathcal{I}_m(\mathbf{r})$ to the internal field is defined in terms of the 'force' $\boldsymbol{i}_m(\mathbf{r})$ and the density $\rho(\mathbf{r})$ as

$$\mathcal{I}_m(\mathbf{r}) = \frac{\boldsymbol{i}_m(\mathbf{r})}{\rho(\mathbf{r})}, \tag{4.134}$$

where in Cartesian coordinates

$$i_{m,\alpha}(\mathbf{r}) = \sum_{\beta} \nabla_\beta I_{\alpha\beta}(\mathbf{r}), \tag{4.135}$$

and the second-rank tensor $I_{\alpha\beta}(\mathbf{r})$ is

$$I_{\alpha\beta}(\mathbf{r}) = \left[j_\alpha(\mathbf{r})A_\beta(\mathbf{r}) + j_\beta(\mathbf{r})A_\alpha(\mathbf{r})\right] - \rho(\mathbf{r})A_\alpha(\mathbf{r})A_\beta(\mathbf{r}), \tag{4.136}$$

with $\mathbf{j}_{\text{spin}}(\mathbf{r})$ the quantal source of the field.

The individual components of the internal field $\mathcal{F}^{\text{int}}(\mathbf{r})$ of (4.121) are in general not conservative. However, as will be shown, their sum taken together with the Lorentz field is conservative. Under conditions of certain symmetry, the individual components can each be separately conservative.

4.4.2 *Energy Components in Terms of Fields*

The canonical kinetic T, electron-interaction E_{ee}, and its Hartree E_H and Pauli-Coulomb E_{xc} energy components of the total energy E can each be expressed in integral virial form in terms of the corresponding fields $\mathcal{Z}(\mathbf{r}), \mathcal{E}_{\text{ee}}(\mathbf{r}), \mathcal{E}_H(\mathbf{r}), \mathcal{E}_{xc}(\mathbf{r})$. With the exception of $\mathcal{E}_H(\mathbf{r})$ which is conservative, these expressions are valid irrespective of whether the fields are conservative. Thus,

$$T = -\frac{1}{2}\int \rho(\mathbf{r})\mathbf{r}\cdot\mathcal{Z}(\mathbf{r})d\mathbf{r}, \tag{4.137}$$

$$E_{\text{ee}} = \int \rho(\mathbf{r})\mathbf{r}\cdot\mathcal{E}_{\text{ee}}(\mathbf{r})d\mathbf{r}, \tag{4.138}$$

$$E_H = \int \rho(\mathbf{r})\mathbf{r}\cdot\mathcal{E}_H(\mathbf{r})d\mathbf{r}, \tag{4.139}$$

$$E_{xc} = \int \rho(\mathbf{r})\mathbf{r} \cdot \boldsymbol{\mathcal{E}}_{xc}(\mathbf{r})d\mathbf{r}. \tag{4.140}$$

The contribution of the conservative external electrostatic field $\boldsymbol{\mathcal{E}}(\mathbf{r}) = -\boldsymbol{\nabla} v(\mathbf{r})$ to the energy E_{es} can be written directly in terms of the potential $v(\mathbf{r})$ as

$$E_{es} = \int \rho(\mathbf{r})v(\mathbf{r})d\mathbf{r}. \tag{4.141}$$

Note that $v(\mathbf{r})$ is path-independent. The energy E_{es} can also be written in integral virial form, but the coefficient of the expression depends upon the degree of the homogeneous function $v(\mathbf{r})$. Hence, for the Coulombic potential for which the degree is -1, the expression is

$$E_{es} = \int \rho(\mathbf{r})\mathbf{r} \cdot \boldsymbol{\mathcal{E}}(\mathbf{r})d\mathbf{r}. \tag{4.142}$$

For the magnetic field contribution to the energy, *i.e.*. the contribution of the Lorentz $\boldsymbol{\mathcal{L}}(\mathbf{r})$ and internal magnetic $\boldsymbol{\mathcal{I}}_m(\mathbf{r})$ field components, define the field

$$\boldsymbol{\mathcal{M}}(\mathbf{r}) = -\big[\boldsymbol{\mathcal{L}}(\mathbf{r}) + \boldsymbol{\mathcal{I}}_m(\mathbf{r})\big]. \tag{4.143}$$

If the field $\boldsymbol{\mathcal{M}}(\mathbf{r})$ is conservative, *i.e.*. $\boldsymbol{\nabla} \times \boldsymbol{\mathcal{M}}(\mathbf{r}) = 0$, then one can define a magnetic scalar potential $v_m(\mathbf{r})$ as

$$\boldsymbol{\mathcal{M}}(\mathbf{r}) = -\boldsymbol{\nabla} v_m(\mathbf{r}). \tag{4.144}$$

This implies that $v_m(\mathbf{r})$ is path-independent. The magnetic contribution E_{mag} to the energy is then

$$E_{\text{mag}} = \int \rho(\mathbf{r})v_m(\mathbf{r})d\mathbf{r}. \tag{4.145}$$

The E_{mag} can also be written in integral virial form depending on the degree of the homogeneous function $v_m(\mathbf{r})$. If $v_m(\mathbf{r})$ is of degree 2 as for the harmonic oscillator, then

$$E_{\text{mag}} = -\frac{1}{2}\int \rho(\mathbf{r})\mathbf{r} \cdot \boldsymbol{\mathcal{M}}(\mathbf{r})d\mathbf{r}. \tag{4.146}$$

In the general case when $\boldsymbol{\nabla} \times \boldsymbol{\mathcal{M}}(\mathbf{r}) \neq 0$, the expression is

$$E_{\text{mag}} = \int \rho(\mathbf{r})\mathbf{r} \cdot \boldsymbol{\mathcal{M}}(\mathbf{r})d\mathbf{r}. \tag{4.147}$$

The total energy E with its components expressed in terms of fields is

$$E = T + E_{ee} + E_{es} + E_{mag} \quad (4.148)$$

$$= T + E_H + E_{xc} + E_{es} + E_{mag}. \quad (4.149)$$

It is evident from the above that the quantum-mechanical system defined via the Schrödinger-Pauli equation can be alternatively described from the perspective of the individual electron. This description is in terms of 'classical' fields experienced by each electron, with the fields arising from quantal sources. The fields satisfy the 'Quantal Newtonian' First Law or equation of motion for each electron. The total energy E and its components can also be expressed in terms of these fields.

4.4.3 *Physical and Mathematical Insights*

The 'Quantal Newtonian' First Law perspective of Schrödinger-Pauli theory in terms of fields leads to further understandings, both physical and mathematical, of the system of electrons with spin. These are described below:

(*i*) The manner of writing the Schrödinger-Pauli theory Hamiltonian of (4.86) is in terms of a scalar potential $v(\mathbf{r})$ and a vector potential $\mathbf{A}(\mathbf{r})$, these potentials being representative, respectively, of the electrostatic $\mathcal{E}(\mathbf{r})$ and magnetostatic $\mathcal{B}(\mathbf{r})$ fields experienced by each electron. (The field $\mathcal{B}(\mathbf{r})$ also appears explicitly in its interaction with the electron's spin moment.) Thus, each electron experiences a force due to the electrostatic field, and a Lorentz force due to the magnetic field. Whilst the electrostatic force is explicit via the scalar potential $v(\mathbf{r})$, the Lorentz force does not appear explicitly in the Hamiltonian. However, with the Hamiltonian written in terms of the physical current density $\mathbf{j}_{\text{spin}}(\mathbf{r})$, which is the sum of the paramagnetic $\mathbf{j}_p(\mathbf{r})$, diamagnetic $\mathbf{j}_d(\mathbf{r})$, and magnetization $\mathbf{j}_m(\mathbf{r})$ components, as in (4.99), it is possible to rewrite the term involving the current density in terms of the Lorentz 'force' operator $\hat{\boldsymbol{\ell}}(\mathbf{r})$ of (4.63). Thus, as was shown for spinless electrons in Sect. 4.2.4, the term $\int \hat{\mathbf{j}}_{\text{spin}}(\mathbf{r}) \cdot \mathbf{A}(\mathbf{r}) d\mathbf{r} = \frac{1}{2} \int \mathbf{r} \cdot \hat{\boldsymbol{\ell}}(\mathbf{r}) d\mathbf{r}$. Thus, the presence of both the electrostatic and Lorentz force is now evident via the many-electron Hamiltonian.

The 'Quantal Newtonian' First Law (4.116), on the other hand, makes the existence of both forces acting on each electron *explicit* via the external field $\mathcal{F}^{\text{ext}}(\mathbf{r})$ which is the sum of the electrostatic $\mathcal{E}(\mathbf{r})$ and Lorentz $\mathcal{L}(\mathbf{r})$ fields, the latter involving the quantum-mechanical Lorentz 'force'.

(*ii*) The 'Quantal Newtonian' First Law, however, provides a deeper understanding of the physical system. In addition to the external field $\mathcal{F}^{\text{ext}}(\mathbf{r})$, each electron is now seen to also experience an internal field $\mathcal{F}^{\text{int}}(\mathbf{r})$. This field is comprised of a sum of fields each representative of a property of the system: An electron-interaction field $\mathcal{E}_{ee}(\mathbf{r})$ that accounts for the electron correlations due to the Pauli principle and Coulomb repulsion; a kinetic field $\mathcal{Z}(\mathbf{r})$ that accounts for the kinetic effects; a differential density field $\mathcal{D}(\mathbf{r})$ representative of the electron density $\rho(\mathbf{r})$; and an internal magnetic field $\mathcal{I}_m(\mathbf{r})$ component. The existence of the internal field $\mathcal{F}^{\text{int}}(\mathbf{r})$ and its property related components comes to light *solely* as a consequence of the 'Quantal Newtonian' First Law.

(*iii*) In summing the 'Quantal Newtonian' First Law (4.116) over all the electrons, the contribution of the internal field vanishes (see Sect. 2.5):

$$\int \rho(\mathbf{r})\mathcal{F}^{\mathrm{int}}(\mathbf{r})d\mathbf{r} = 0. \tag{4.150}$$

This then leads to Ehrenfest's theorem [12] for a stationary state:

$$\int \rho(\mathbf{r})\mathcal{F}^{\mathrm{ext}}(\mathbf{r})d\mathbf{r} = 0. \tag{4.151}$$

(*iv*) The external scalar potential $v(\mathbf{r})$ which appears in the quantum-mechanical Hamiltonian (4.86) represents the potential energy of each electron in the presence of the field of the positively charged nucleus in atoms, molecules, and solids. It could represent the potential due to the field of the positive jellium background model of solids (metals) employed to study the uniform electron gas or the study of the metal-vacuum interface [13–23], or the fractional quantum Hall effect [24–26]. The potential, furthermore, is *path-independent*. The 'Quantal Newtonian' First Law, however, provides a deeper physical understanding of this potential in terms of the properties of the system. Further, it affords an interpretation of the potential in the rigorous classical sense. It follows from the 'Quantal Newtonian' First Law of (4.116) that *the potential* $v(\mathbf{r})$ *is the work done to move an electron from some reference point at infinity to its position at* $\mathbf{r}$ *in the force of a conservative field* $\mathcal{F}(\mathbf{r})$:

$$v(\mathbf{r}) = \int_{\infty}^{\mathbf{r}} \mathcal{F}(\mathbf{r}') \cdot d\boldsymbol{\ell}', \tag{4.152}$$

where $\mathcal{F}(\mathbf{r}) = \mathcal{F}^{\mathrm{int}}(\mathbf{r}) - \mathcal{L}(\mathbf{r}) = \mathcal{E}_{\mathrm{ee}}(\mathbf{r}) - \mathcal{Z}(\mathbf{r}) - \mathcal{D}(\mathbf{r}) - \mathcal{I}_m(\mathbf{r}) - \mathcal{L}(\mathbf{r})$. As the field $\mathcal{F}(\mathbf{r})$ is conservative, the $\nabla \times \mathcal{F}(\mathbf{r}) = 0$. Hence, the work done is *path-independent*, and therefore $v(\mathbf{r})$ constitutes a potential energy. It is reiterated that the 'Quantal Newtonian' First Law is valid for *arbitrary state*. Hence, the potential function $v(\mathbf{r})$ as expressed in (4.152) remains the same irrespective of the state of the system.

(*v*) In the Hamiltonian $\hat{H}$ of (4.75), the potential energy function $v(\mathbf{r})$ binding the electrons is usually assumed known analytically. Thus, it could be Coulombic $(-Ze^2/r)$, harmonic $(\frac{1}{2}kr^2)$, screened-Coulomb Yukawa $(-Ze^2\exp(-\lambda r)/r)$, etc. It could also be a function that needs to be determined as for example at the surface of the jellium model metal-vacuum interface. The 'Quantal Newtonian' First Law as written in (4.116) then shows that this function $v(\mathbf{r})$ depends on all the components of the internal field $\mathcal{F}^{\mathrm{int}}(\mathbf{r})$ of the system and the Lorentz field $\mathcal{L}(\mathbf{r})$. As such the potential $v(\mathbf{r})$ is inherently related to and constructed via the properties of the system. In particular, if the various components of the internal field $\mathcal{F}^{\mathrm{int}}(\mathbf{r})$ are separately conservative, then the function $v(\mathbf{r})$ is comprised of a sum of constituent functions, each representative of a property of the system. Thus, the potential $v(\mathbf{r})$ may be expressed as

$$v(\mathbf{r}) = W_{ee}(\mathbf{r}) + W_z(\mathbf{r}) + W_d(\mathbf{r}) + W_i(\mathbf{r}) + W_\ell(\mathbf{r}), \tag{4.153}$$

where $W_{ee}(\mathbf{r})$, $W_z(\mathbf{r})$, $W_d(\mathbf{r})$, $W_i(\mathbf{r})$ and $W_\ell(\mathbf{r})$ are the work done in the fields $\mathcal{E}_{ee}(\mathbf{r})$, $\mathcal{Z}(\mathbf{r})$, $\mathcal{D}(\mathbf{r})$, $\mathcal{I}_m(\mathbf{r})$, and $\mathcal{L}(\mathbf{r})$, respectively. Each work done is *path-independent*, and hence each constitutes a potential energy.

(*vi*) Provided the sum of the Lorentz $\mathcal{L}(\mathbf{r})$ and internal magnetic $\mathcal{I}_m(\mathbf{r})$ fields is conservative, it is then possible to define a scalar potential $v_m(\mathbf{r})$ representative of all the magnetic effects of the system. This potential is the work done in the sum of the fields $\mathcal{L}(\mathbf{r})$ and $\mathcal{I}_m(\mathbf{r})$. This work done is *path-independent*.

(*vii*) The 'Quantal Newtonian' First Law also provides a deeper mathematical understanding of the potential $v(\mathbf{r})$. As the components of the conservative field $\mathcal{F}(\mathbf{r})$ of (4.152) are obtained from quantal sources that are expectation values of Hermitian operators taken with respect to the wave function ψ, the field $\mathcal{F}(\mathbf{r})$ is a functional of ψ, *i.e.* $\mathcal{F}(\mathbf{r}) = \mathcal{F}[\psi](\mathbf{r})$. This functional is *exactly known* since the individual component fields are explicitly defined. This in turn means that the scalar potential energy $v(\mathbf{r})$ as defined by (4.152) is an *exactly known* functional of the wave function $\psi : v(\mathbf{r}) = v[\psi](\mathbf{r})$. It is emphasized that this functional dependence is valid for *arbitrary* state.

(*viii*) Finally, the functional $v[\psi](\mathbf{r})$ is *universal*. This means that the functional is the same for all types of electronic systems whether they be natural atoms or molecules, or artificial atoms such as quantum dots, or jellium metal clusters, etc.

(*ix*) The expression for the functional $v[\psi](\mathbf{r})$ of (4.152) may then be thought of as being the quantum-mechanical definition of the binding potential $v(\mathbf{r})$.

4.4.4 Generalization of the Schrödinger-Pauli Theory Equation

As noted, the 'Quantal Newtonian' First Law of (4.116, 4.152) leads to the understanding that the scalar potential $v(\mathbf{r})$ of the Hamiltonian (4.75) is a known and universal functional of the wave function ψ, *i.e.* $v(\mathbf{r}) = v[\psi](\mathbf{r})$. This allows for the generalization of the Schrödinger-Pauli equation which in turn then exhibits its intrinsic self-consistent nature. On substituting the functional $v[\psi](\mathbf{r})$ for $v(\mathbf{r})$, the Schrödinger-Pauli equation (4.86) may be written as

$$\Big[\frac{1}{2}\sum_k \big(\hat{\mathbf{p}}_k + \frac{1}{c}\mathbf{A}(\mathbf{r}_k)\big)^2 + \frac{1}{c}\sum_k \mathcal{B}(\mathbf{r}_k)\cdot \mathbf{s}_k + \frac{1}{2}\sum_{k,\ell}{}' + \frac{1}{|\mathbf{r}_k - \mathbf{r}_\ell|} + \sum_k v[\psi](\mathbf{r}_k)\Big]\psi(\mathbf{X}) = E[\psi]\psi(\mathbf{X}), \tag{4.154}$$

or, on employing (4.152), as

$$\left[\frac{1}{2}\sum_k\left(\hat{\mathbf{p}}_k+\frac{1}{c}\mathbf{A}(\mathbf{r}_k)\right)^2+\frac{1}{c}\sum_k\boldsymbol{\mathcal{B}}(\mathbf{r}_k)\cdot\mathbf{s}_k+\frac{1}{2}{\sum_{k,\ell}}'+\frac{1}{|\mathbf{r}_k-\mathbf{r}_\ell|}\right.$$
$$\left.+\sum_k\int_\infty^{\mathbf{r}_k}\boldsymbol{\mathcal{F}}[\psi](\mathbf{r})\cdot d\boldsymbol{\ell}\right]\psi(\mathbf{X})=E[\psi]\psi(\mathbf{X}). \tag{4.155}$$

The Hamiltonian is thus a functional of the wave function ψ : $\hat{H}_{\mathrm{spin}}=\hat{H}_{\mathrm{spin}}[\psi]$, and hence the Schrödinger-Pauli equation can be written in *generalized* form as

$$\hat{H}_{\mathrm{spin}}[\psi]\psi(\mathbf{X})=E[\psi]\psi(\mathbf{X}). \tag{4.156}$$

The Hamiltonian functional $\hat{H}_{\mathrm{spin}}[\psi]$ is *exactly known*. It is valid for *arbitrary* state. It is *universal* in that it is applicable to any electronic system defined by this Hamiltonian.

The generalized form of the Schrödinger-Pauli equation makes evident that its solution may be obtained self-consistently. One begins with an appropriate approximate wave function $\psi(\mathbf{X})$ to first determine the quantal sources and fields constituting the field $\boldsymbol{\mathcal{F}}[\psi](\mathbf{r})$ of (4.152), and from it the potential $v(\mathbf{r})$, and thereby the approximate Hamiltonian $\hat{H}_{\mathrm{spin}}[\psi]$. The Schrödinger-Pauli equation of (4.156) is then solved to obtain the next approximation to the wave function ψ and energy E, from which the corresponding $\boldsymbol{\mathcal{F}}[\psi](\mathbf{r})$ and potential $v(\mathbf{r})$ lead to the next approximate $\hat{H}_{\mathrm{spin}}[\psi]$. For this new approximate $\hat{H}_{\mathrm{spin}}[\psi]$, the Schrödinger-Pauli equation is again solved for the next approximate wave function ψ and energy E. And this procedure is continued till the input wave function ψ to $\hat{H}_{\mathrm{spin}}[\psi]$ is the same ψ as that generated by this $\hat{H}_{\mathrm{spin}}[\psi]$ via solution of the Schrödinger-Pauli equation. Note that the meaning of the functional $v[\psi](\mathbf{r})$ is that for each new ψ, one obtains a new $v[\psi](\mathbf{r})$, and therefore the Hamiltonian functional $\hat{H}_{\mathrm{spin}}[\psi]$ changes with each new iterative ψ. This then allows for the self-consistent procedure.

Other attributes of the generalized Schrödinger-Pauli equation of (4.154) are similar to those for the spinless electrons in a magnetic field described in Sect. 3.1.2. These conclusions are summarized below.

(a) In the traditional way of writing the Schrödinger-Pauli equation as in (4.86), the potential $v(\mathbf{r})$ is considered as being *extrinsic* to the system of N electrons, and as such assumed to be a known but independent input to the Hamiltonian $\hat{H}_{\mathrm{spin}}$. In other words, it does not depend on any of the other terms of the Hamiltonian $\hat{H}_{\mathrm{spin}}$. From the generalized form of the equation (4.154), it becomes evident that the potential $v[\psi](\mathbf{r})$ is in fact *intrinsic* to the physical system being related to it via the components of the internal field (see (4.152)). It is thereby (self-consistently) *dependent* on all the properties of the system via the other operators of the Hamiltonian $\hat{H}_{\mathrm{spin}}$.

(b) On achieving self-consistency, the wave function $\psi(\mathbf{X})$, the eigen energy E, and the potential $v[\psi](\mathbf{r})$ or equivalently the Hamiltonian $\hat{H}_{\mathrm{spin}}$ are determined. This is of particular significance in those cases for which the potential $v(\mathbf{r})$ may be initially *unknown*.

(c) The self-consistency procedure could also be employed to determine the wave function $\psi(\mathbf{X})$ and energy E for the case when the potential $v(\mathbf{r})$ is *known*. Starting with an accurate approximate wave function ψ, the corresponding approximate $v[\psi](\mathbf{r})$ could be determined. Of course, this would not correspond to the known $v(\mathbf{r})$ function. But the solution of the resulting Schrödinger-Pauli equation with this approximate $v[\psi](\mathbf{r})$ would be an improvement to the original wave function. Continuing with the self-consistency procedure would lead to the exact $\{\psi(\mathbf{X}), E\}$. On achieving self-consistency, the known function $v(\mathbf{r})$ would be reproduced.

(d) The generalized Schrödinger-Pauli equation (4.154, 4.155) is valid for both ground and excited states. The excited state wave functions obtained self-consistently will automatically be orthogonal to the ground and other states of the system.

(e) In the Schrödinger theory [1] of electrons in the presence of electrostatic $\mathcal{E}(\mathbf{r})$ and magnetostatic $\mathcal{B}(\mathbf{r})$ fields, one hews to the philosophy that electromagnetic interactions occur by the substitution $\hat{\mathbf{p}} \rightarrow \hat{\mathbf{p}} + (e/c)\mathbf{A}$. Thus, it is the vector potential $\mathbf{A}(\mathbf{r})$ and not the magnetic field $\mathcal{B}(\mathbf{r})$ that appears in the corresponding Schrödinger equation. This is a fundamental difference between classical and quantum physics. The magnetic field $\mathcal{B}(\mathbf{r})$ appears in the Schrödinger equation only after a choice of gauge for the vector potential $\mathbf{A}(\mathbf{r})$. In the Schrödinger-Pauli Hamiltonian (4.86), the magnetic field $\mathcal{B}(\mathbf{r})$ appears explicitly as a consequence of the use of the Feynman kinetic energy operator $\hat{T}_F$: It is the term corresponding to the interaction between the magnetic field and the spin angular momentum operator. The generalized form of the Schrödinger-Pauli equation further shows that the magnetic field $\mathcal{B}(\mathbf{r})$ also appears in the term involving $v[\Psi](\mathbf{r})$ via the conservative field $\mathcal{F}(\mathbf{r})$ which includes the Lorentz field $\mathcal{L}(\mathbf{r})$ (see (4.152)). (Of course, whenever a term involving the magnetic field $\mathcal{B}(\mathbf{r})$ appears it can always be written in terms of the vector potential $\mathbf{A}(\mathbf{r})$ by a choice of gauge.)

4.5 Time-Dependent Schrödinger-Pauli Theory and Its Generalization: The 'Quantal Newtonian' Second Law for an Electron with Spin

In this final section, the quantal-source–field individual electron perspective is extended to the temporal Schrödinger-Pauli theory of electrons. Hence, consider a system of N electrons each possessing spin in the presence of a time-dependent binding electric field $\mathcal{E}(\mathbf{y}) = -\nabla v(\mathbf{y})/e$ and an electromagnetic field $E(\mathbf{y}) = -\nabla\phi(\mathbf{y}) - \partial\mathbf{A}(\mathbf{y})/\partial t$; $\mathcal{B}(\mathbf{y}) = \nabla \times \mathbf{A}(\mathbf{y})$, where $v(\mathbf{y}), \phi(\mathbf{y})$ are scalar potentials and $\mathbf{A}(\mathbf{y})$ the vector potential. (The notation is $\mathbf{y} = \mathbf{r}t$; $\mathbf{y}_k = \mathbf{r}_k t$; $\mathbf{y}' = \mathbf{r}'t$; $\mathbf{Y} = \mathbf{X}t$; $\mathbf{X} = \mathbf{x}_1, \ldots, \mathbf{x}_N$; $\mathbf{x} = \mathbf{r}\sigma$, with $\mathbf{r}$ and σ the spatial and spin coordinates. Atomic units are employed: charge of electrons $-e$; $|e| = \hbar = m = 1$.)

The Hamiltonian $\hat{H}_{\text{spin}}(t)$ is the sum of the Feynman kinetic $\hat{T}_F(t)$, electron-interaction potential $\hat{W}$ and total scalar potential $\hat{U}(t)$ operators:

$$\hat{H}_{\text{spin}}(t) = \hat{T}_F(t) + \hat{W} + \hat{U}(t), \tag{4.157}$$

where

$$\hat{T}_F(t) = \frac{1}{2}\sum_k \left(\hat{\mathbf{p}}_k + \frac{1}{c}\mathbf{A}(\mathbf{y}_k)\right)^2 + \frac{1}{c}\sum_k \boldsymbol{\mathcal{B}}(\mathbf{y}_k)\cdot\mathbf{s}_k, \tag{4.158}$$

$$\hat{W} = \frac{1}{2}\sum_{k,\ell}{}' \frac{1}{|\mathbf{r}_k - \mathbf{r}_\ell|}, \tag{4.159}$$

$$\hat{U}(t) = \sum_k u(\mathbf{y}_k)\ ;\ u(\mathbf{y}) = v(\mathbf{y}) - \phi(\mathbf{y}), \tag{4.160}$$

with $\hat{\mathbf{p}} = -i\boldsymbol{\nabla}$ the canonical momentum operator, and $\mathbf{s}$ the electron spin angular momentum operator.

The Hamiltonian $\hat{H}_{\text{spin}}(t)$ may also be expressed in terms of the density $\hat{\rho}(\mathbf{r})$ and physical current density $\hat{\mathbf{j}}(\mathbf{r})$ operators as

$$\hat{H}_{\text{spin}}(t) = \hat{T} + \frac{1}{c}\int \hat{\mathbf{j}}_{\text{spin}}(\mathbf{r})\cdot\mathbf{A}(\mathbf{y})d\mathbf{r} - \frac{1}{2c^2}\int \hat{\rho}(\mathbf{r})A^2(\mathbf{y})d\mathbf{r} + \hat{W} + \hat{U}(t), \tag{4.161}$$

where $\hat{T} = \frac{1}{2}\sum_k p_k^2$ is the canonical kinetic energy operator. The operator $\hat{\mathbf{j}}_{\text{spin}}(\mathbf{r})$ is the sum of the paramagnetic $\hat{\mathbf{j}}_p(\mathbf{r})$, diamagnetic $\hat{\mathbf{j}}_d(\mathbf{r})$, and magnetization current density $\hat{\mathbf{j}}_m(\mathbf{r})$ components as defined in (4.46), (4.47) and (4.97), respectively.

The time-dependent Schrödinger-Pauli equation is then

$$\hat{H}_{\text{spin}}(t)\Psi(\mathbf{Y}) = i\frac{\partial\Psi(\mathbf{Y})}{\partial t}, \tag{4.162}$$

whose solution $\Psi(\mathbf{Y})$ is obtained given a known initial condition $\Psi(\mathbf{X}t_0)$ at t_0.

The system of electrons as defined by the Schrödinger-Pauli equation (4.162) can be equivalently described from the perspective of the individual electron via the 'Quantal Newtonian' Second Law equation of motion satisfied by each electron. As was the case for spinless electrons in a time-dependent electric field $\boldsymbol{\mathcal{E}}(\mathbf{y})$ as described in Chap. 2, the 'Quantal Newtonian' Second Law for the electron with spin leads to a deeper and fundamental understanding of this more general representation of the physical system.

The derivation of the 'Quantal Newtonian' Second Law for *spinless* electrons in the above temporal external electric binding $\boldsymbol{\mathcal{E}}(\mathbf{y})$ and electromagnetic $\{\mathbf{E}(\mathbf{y}), \boldsymbol{\mathcal{B}}(\mathbf{y})\}$ fields is given in Appendix A. The law for electrons with *spin* can be obtained directly on the recognition that the Hamiltonian $\hat{H}_{\text{spin}}(t)$ as written in terms of the density $\hat{\rho}(\mathbf{r})$ and physical current density $\hat{\mathbf{j}}_{\text{spin}}(\mathbf{r})$ operators as in (4.161) is of the same form as the Hamiltonian $\hat{H}_{\text{spinless}}(t)$ for spinless electrons. The only difference is that the current density operator $\hat{\mathbf{j}}_{\text{spin}}(\mathbf{r})$ for electrons with spin now additionally incorporates the magnetization current density operator $\hat{\mathbf{j}}_m(\mathbf{r})$ of (4.97). Thus, the

'Quantal Newtonian' Second Law for each electron in the physical system described by the Schrödinger-Pauli equation (4.162) is

$$\mathcal{F}^{\text{ext}}(\mathbf{y}) + \mathcal{F}^{\text{int}}(\mathbf{y}) = \mathcal{J}(\mathbf{y}), \tag{4.163}$$

where the external field $\mathcal{F}^{\text{ext}}(\mathbf{y})$ is

$$\mathcal{F}^{\text{ext}}(\mathbf{y}) = \mathcal{E}(\mathbf{y}) - \mathcal{L}(\mathbf{y}) - E(\mathbf{y}), \tag{4.164}$$

the internal field $\mathcal{F}^{\text{int}}(\mathbf{y})$ is

$$\mathcal{F}^{\text{int}}(\mathbf{y}) = \mathcal{E}_{ee}(\mathbf{y}) - \mathcal{Z}(\mathbf{y}) - \mathcal{D}(\mathbf{y}) - \mathcal{I}_m(\mathbf{y}), \tag{4.165}$$

and the electron response field $\mathcal{J}(\mathbf{y})$ to these fields is

$$\mathcal{J}(\mathbf{y}) = \frac{1}{\rho(\mathbf{y})} \frac{\partial \mathbf{j}_{\text{spin}}(\mathbf{y})}{\partial t}. \tag{4.166}$$

Here the density $\rho(\mathbf{y}) = \langle \Psi(\mathbf{Y}) | \hat{\rho}(\mathbf{r}) | \Psi(\mathbf{Y}) \rangle$, and the physical current density $\mathbf{j}_{\text{spin}}(\mathbf{y}) = \langle \Psi(\mathbf{Y}) | \hat{\mathbf{j}}_{\text{spin}}(\mathbf{r}) | \Psi(\mathbf{Y}) \rangle$, with $\hat{\rho}(\mathbf{r})$ defined in (2.12) and $\hat{\mathbf{j}}(\mathbf{r})$ by (4.98). The Lorentz $\mathcal{L}(\mathbf{y})$, electron-interaction $\mathcal{E}_{ee}(\mathbf{y})$, kinetic $\mathcal{Z}(\mathbf{y})$, differential density $\mathcal{D}(\mathbf{y})$, and internal magnetic $\mathcal{I}_m(\mathbf{y})$ fields are as defined in Sect. 4.3.1 but from time-dependent quantal sources obtained via the wave function $\Psi(\mathbf{Y})$. The 'Quantal Newtonian' Second Law is *gauge invariant* and satisfies the equation of continuity:

$$\nabla \cdot \mathbf{j}_{\text{spin}}(\mathbf{y}) + \frac{\partial \rho(\mathbf{y})}{\partial t} = 0. \tag{4.167}$$

From the Second Law we once again see that each electron in the presence of an external time-dependent electromagnetic field also experiences a temporal internal field representative of the properties of the system: correlations due to the Pauli principle and Coulomb repulsion $\mathcal{E}_{ee}(\mathbf{y})$; kinetic effect $\mathcal{Z}(\mathbf{y})$; the electronic density $\mathcal{D}(\mathbf{y})$. There also exists an internal magnetic field $\mathcal{I}_m(\mathbf{y})$ component. The knowledge of existence of these internal fields experienced by each electron is a direct consequence of the Second Law.

On summing over all the electrons, the contribution of the internal field vanishes: $\int \rho(\mathbf{y})\mathcal{F}^{\text{int}}(\mathbf{y})d\mathbf{r} = 0$. Thus, Ehrenfest's theorem follows: $\int \rho(\mathbf{y})\mathcal{F}^{\text{ext}}(\mathbf{y})d\mathbf{r} = \int \rho(\mathbf{y})\mathcal{J}(\mathbf{y})d\mathbf{r}$ as written in terms of the electron response field $\mathcal{J}(\mathbf{y})$. The 'Quantal Newtonian' Second Law also provides a rigorous physical interpretation of the scalar function $u(\mathbf{y})$ of (4.160). It is *the work done, at each instant of time, to bring an electron from some reference point at infinity to its position at* $\mathbf{r}$ *in the force of a conservative field* $\mathcal{F}$:

$$u(\mathbf{y}) = \int_{\infty}^{\mathbf{r}} \mathcal{F}(\mathbf{y}') \cdot d\boldsymbol{\ell}', \tag{4.168}$$

where $\mathcal{F}(\mathbf{y}) = \mathcal{E}_{ee}(\mathbf{y}) - \mathcal{Z}(\mathbf{y}) - \mathcal{D}(\mathbf{y}) - \mathcal{I}_m(\mathbf{y}) - \mathcal{L}(\mathbf{y}) - \mathcal{Q}(\mathbf{y})$, and where

$$\mathcal{Q}(\mathbf{y}) = \frac{1}{\rho(\mathbf{y})}\left[\frac{\partial \mathbf{j}_{\text{spin}}(t)}{\partial t} - \rho(\mathbf{y})\frac{\partial \mathbf{A}(\mathbf{y})}{\partial t}\right]. \tag{4.169}$$

As $\nabla \times \mathcal{F}(\mathbf{y}) = 0$, the work done $\mathbf{u}(\mathbf{y})$ is path independent. Further, the potential $u(\mathbf{y})$ is comprised of components representative of the properties of the system, and as such is *intrinsically* related to the physical system.

It is also evident from (4.168) that the potential $u(\mathbf{y})$ is a *known functional* of the wave function $\Psi(\mathbf{Y})$ as each component of $\mathcal{F}(\mathbf{y})$ is such a functional, *i.e.* $u(\mathbf{y}) = u[\Psi](\mathbf{y})$, where

$$u[\Psi](\mathbf{y}) = \int_{\infty}^{\mathbf{r}} \mathcal{F}[\Psi](\mathbf{y}') \cdot d\boldsymbol{\ell}'. \tag{4.170}$$

Thus, the Hamiltonian $\hat{H}_{\text{spin}}(t)$ is a known functional of the wave function $\Psi(\mathbf{Y})$, *i.e.* $\hat{H}_{\text{spin}}(t) = \hat{H}_{\text{spin}}[\Psi](t)$, where

$$\hat{H}_{\text{spin}}[\Psi](t) = \hat{T}_F(t) + \hat{W} + \hat{U}[\Psi](t), \tag{4.171}$$

and where

$$\hat{U}[\Psi](t) = \sum_k u_k[\Psi](\mathbf{y}_k). \tag{4.172}$$

The Schrödinger-Pauli equation (4.162) can then be expressed in *generalized* form as

$$\hat{H}_{\text{spin}}[\Psi](t)\Psi(\mathbf{Y}) = i\frac{\partial \Psi(\mathbf{Y})}{\partial t}. \tag{4.173}$$

Written in this form, the intrinsic self-consistent nature of the Schrödinger-Pauli equation becomes evident. The time evolution of the wave function $\Psi(\mathbf{Y})$, and the Hamiltonian $\hat{H}_{\text{spin}}[\Psi](t)$ are obtained by self-consistent solution of (4.173) at each instant of time t given an initial condition $\Psi(\mathbf{Y}_0)$ at $t = t_0$. The self-consistency procedure is the same as described in the previous chapter for the cases when the external binding potential $v(\mathbf{y})$ is known or unknown. In either case, it is the potential $v(\mathbf{y})$ that is obtained via the self-consistent procedure. In the former case, on achieving self-consistency, $\hat{H}_{\text{spin}}[\Psi](t) = \hat{H}_{\text{spin}}(t)$ of (4.162). Hence, the time-dependent Schrödinger-Pauli equation written in traditional form as in (4.162) constitutes a special case of the generalized form as given by (4.173).

References

1. E. Schrödinger, Ann. Physik **79**, 361 (1926); **80**, 437 (1926); **81**, 109 (1926)
2. W. Pauli, Z. Physik **43**, 601 (1927)
3. J.J. Sakurai, *Advanced Quantum Mechanics* (Addison-Wesley, Reading, MA, 1967)

4. In the footnote of [3] on page 79, it is noted that within nonrelativistic quantum mechanics, the correct gyromagnetic ratio of $g = 2$ is obtained from the kinetic energy operator of (4.76). This was emphasized particularly by Feynman. It is for this reason that this operator is referred to as the Feynman kinetic energy operator $\hat{T}_F$
5. P.A.M. Dirac, Proc. Roy. Soc. (London) **A117**, 610 (1927)
6. V. Sahni, Int. J. Quantum Chem. **121**, e26556 (2021)
7. M. Slamet, V. Sahni, Chem. Phys. **546**, 111073 (2021)
8. X.-Y. Pan, V. Sahni, Int. J. Quantum Chem. **110**, 2833 (2010)
9. P. Hohenberg, W. Kohn, Phys. Rev. **136**, B864 (1964)
10. X.-Y. Pan, V. Sahni, J. Chem. Phys. **143**, 174105 (2015)
11. L.D. Landau, E.M. Lifshitz, *Quantum Mechanics* (Pergamon Press, 1965)
12. P. Ehrenfest, Z. Physik **45**, 455 (1927)
13. V. Sahni, *Quantal Density Functional Theory II: Approximation Methods and Applications*, (Springer, Berlin, Heidelberg, 2010) (see Chapter 17)
14. J. Bardeen, Phys. Rev. **49**, 653 (1936)
15. N.D. Lang, in *Solid State Physics*, vol. 28, ed. by H. Ehrenreich, F. Seitz, D. Turnbull (Academic, New York, 1973)
16. V. Sahni, C.Q. Ma, Phys. Rev. B **22**, 5987 (1980)
17. V. Sahni, Prog. Surf. Sci. **54**, 115 (1997)
18. A. Solomatin, V. Sahni, Ann. Phys. **259**, 97 (1997)
19. A. Solomatin, V. Sahni, Ann. Phys. **268**, 149 (1998)
20. Z. Qian, V. Sahni, Phys. Rev. B **66**, 205103 (2002)
21. Z. Qian, V. Sahni, Int. J. Quantum Chem. **104**, 929 (2005)
22. Z. Qian, Phys. Rev. B **85**, 115124 (2012)
23. H. Luo, W. Hackbusch, H.-J. Flad, D. Kolb, Phys. Rev. B **78**, 035136 (2008)
24. H.L. Stormer et al., Phys. Rev. Lett. **50**, 1953 (1983)
25. R.B. Laughlin, Phys. Rev. Lett. **50**, 1395 (1983)
26. *The Quantum Hall Effect*, ed. by R.E. Prange, S.M. Girvin (Springer, New York, (1990)

Chapter 5
Elucidation of Complimentary Perspective to Schrödinger-Pauli Theory: Application to the 2^3S State of a Quantum Dot in a Magnetic Field

Abstract The Schrödinger-Pauli theory of electrons in the presence of a static electromagnetic field can be described via a perspective that is 'Newtonian'. The perspective is that of the individual electron via its quantum-mechanical equation of motion or 'Quantal Newtonian' First Law. The law is in terms of 'classical' fields whose sources are quantal expectations of Hermitian operators taken with respect to the wave function. The law states that the sum of the external and internal fields experienced by each electron vanishes. The external field is the sum of the binding electrostatic and Lorentz fields. The internal field is a sum of fields representative of properties of the system: electron correlations due to the Pauli principle and Coulomb repulsion; the electron density; kinetic effects; and physical current density. Thus, the internal field is a sum of the electron-interaction, differential density, kinetic and internal magnetic fields. The energy can be expressed in integral virial form in terms of these fields. This new perspective is explicated by application to the triplet 2^3S state of a quantum dot in a magnetic field. The quantal sources of the electron density; the paramagnetic, diamagnetic, and magnetization (spin) current densities and their circulation; the pair-correlation density; the Fermi-Coulomb hole; and the single-particle density matrix are obtained. From these sources, the corresponding fields are determined. The fields are shown to satisfy the 'Quantal Newtonian' First Law. The energy and its components too are obtained from these fields. The first law further proves the Hamiltonian to be an exactly known universal functional of the wave function. This generalizes the Schrödinger-Pauli equation, and shows it to be self-consistent. This intrinsic self-consistency of the equation is explained via the example of the quantum dot.

Introduction

The quantal-source – field 'Newtonian' perspective of a quantum-mechanical system of electrons in an electromagnetic field as developed in the prior three chapters is elucidated here by application. In this manner new properties of the physical system will be determined. Further, the application will demonstrate the tangibility of this description of a quantum system. We consider the case of stationary-state

V. Sahni, *Schrödinger Theory of Electrons: Complementary Perspectives*,
Springer Tracts in Modern Physics 285,
https://doi.org/10.1007/978-3-030-97409-1_5

Schrödinger-Pauli theory [1, 2] in which the electrons are subject to an electrostatic binding field and a uniform magnetostatic field. The application of the new perspective [3–5] is to a 2-electron 2-dimensional quantum dot in a magnetic field in its excited triplet 2^3S state [6]. In this state the electrons have the same spin orientation with one electron in the ground and the other in the first excited state. With the total spin moment being finite, the effects of the corresponding magnetization (spin) current density can be studied. Additionally, the solution to the Schrödinger-Pauli equation for this system of interacting electrons is known in closed analytical form. Hence, all the properties obtained are *exact*, with the majority of them being expressed in analytical or semi-analytical form.

Quantum dots [7–10] are 'artificial atoms' whose size and state can be manipulated by electric and magnetic fields, and thus they constitute fertile ground for the study of reduced dimensionality electronic structure [11–25]. In quantum dots, the motion of the electrons is confined to 2 dimensions in a quantum well in a thin layer of semiconductor such as GaAs which is sandwiched between two layers of another semiconductor AlGaAs. The 2-dimensional motion is restricted by an electrostatic field that can be varied. This motion can be further constrained by a magnetic field perpendicular to the plane of motion. As the 'artificial atom' is in a semiconductor, the free electron mass m must be replaced by the band effective mass m^*, and the electron interaction modified by the dielectric constant ϵ. For GaAs the effective mass is $m^* = 0.067m$, and $\epsilon = 12.4$. Finally, the binding potential $v(\mathbf{r})$ of the electrons has been established by both theory and experiment to be harmonic [9–11] so that $v(\mathbf{r}) = \frac{1}{2}m^*\omega_0^2 r$, with ω_0 the binding frequency so that $\omega_0^2 = k_0$, where k_0 is the binding force constant. As a consequence of these facts, the size of a quantum dot is an order of magnitude greater than that of a natural atom. Thus, there is a lowering of the electron density.

The stationary-state Schrödinger-Pauli theory eigenvalue equation for an N-electron quantum dot in an external electrostatic binding field $\boldsymbol{\mathcal{E}}(\mathbf{r}) = -\boldsymbol{\nabla} v(\mathbf{r})/e$ and a magnetostatic field $\boldsymbol{\mathcal{B}}(\mathbf{r}) = \boldsymbol{\nabla} \times \mathbf{A}(\mathbf{r})$ is thus (charge of electron is $-e$)

$$\left[\frac{1}{2m^\star}\sum_k \left(\hat{\mathbf{p}}_k + \frac{e}{c}\mathbf{A}(\mathbf{r}_k)\right)^2 + (g^\star\mu_B/\hbar)\sum_k \boldsymbol{\mathcal{B}}(\mathbf{r}_k)\cdot\mathbf{s}_k + \frac{1}{2\epsilon}\sum_{k,\ell}{}' \frac{e^2}{|\mathbf{r}_k - \mathbf{r}_\ell|} + \frac{1}{2}m^\star\sum_k \omega_0^2 r_k^2\right]\Psi(\mathbf{X}) = E\Psi(\mathbf{X}), \tag{5.1}$$

where the canonical momentum operator $\hat{p} = -i\hbar\boldsymbol{\nabla}$, the corresponding gyromagnetic ratio is g^*, the Bohr magneton $\mu_B = e\hbar/2mc$, the velocity of light is c, the spin angular momentum vector is $\mathbf{s}$, the wave function is $\Psi(\mathbf{X})$, the eigenenergy is E, and $\mathbf{X} = \mathbf{x}_1, \ldots, \mathbf{x}_N$, with $\mathbf{x} = \mathbf{r}\sigma$, the spatial and spin coordinates, respectively.

In the sections to follow, we discuss the various properties of the 2-electron quantum dot in a uniform magnetic field in its triplet 2^3S state. These properties are the following:

(a) The structure and properties of the closed-form analytical complex wave function $\Psi(\mathbf{x}_1\mathbf{x}_2)$. In particular, the nodal structure of the wave function, and the satisfaction by the wave function of the integral *nodal* electron-electron coalescence condition [26–29]. Further, the *parity* of the wave function about the various nodes, particularly about the origin and points of electron-electron coalescence is described. (For the derivation of the wave function, see Chap. 9. For the derivation of the coalescence condition, see Chap. 8.)

(b) The *local* quantal sources of the electronic density $\rho(\mathbf{r})$, and physical current density $\mathbf{j}(\mathbf{r})$ together with its paramagnetic $\mathbf{j}_p(\mathbf{r})$, diamagnetic $\mathbf{j}_d(\mathbf{r})$, and magnetization $\mathbf{j}_m(\mathbf{r})$ density components, and the *nonlocal* sources of the pair-correlation density $g(\mathbf{r}\mathbf{r}')$, the Fermi-Coulomb hole charge $\rho_{xc}(\mathbf{r}\mathbf{r}')$, and the single-particle density matrix $\gamma(\mathbf{r}\mathbf{r}')$.

(c) The various 'forces' and fields that arise from these quantal sources. These fields comprise the electron-interaction $\mathcal{E}_{ee}(\mathbf{r})$ and its Hartree $\mathcal{E}_H(\mathbf{r})$ and Pauli-Coulomb $\mathcal{E}_{xc}(\mathbf{r})$ components; the kinetic $\mathcal{Z}(\mathbf{r})$; the differential density $\mathcal{D}(\mathbf{r})$; internal magnetic $\mathcal{I}_m(\mathbf{r})$; and the Lorentz $\mathcal{L}(\mathbf{r})$ field.

(d) The total energy E, and the corresponding components as obtained from these fields are also given.

(e) The expectation values of the single particle operators $r^2, r, 1/r, \delta(\mathbf{r})$.

(f) The satisfaction of the 'Quantal Newtonian' First Law by these fields.

(g) The results of this application are employed to explain the self-consistent nature of the Schrödinger-Pauli equation.

The definition of each property is given as it is discussed. The analytical and semi-analytical expressions for these properties are given in Appendix B.

5.1 Triplet 2^3S State Wave Function

The derivation of the triplet 2^3S state wave function $\Psi(\mathbf{x}_1\mathbf{x}_2)$ is given in Chap. 9. The derivation follows the method of Taut [30–32]. Here we discuss its salient properties.

In the symmetric gauge $\mathbf{A}(\mathbf{r}) = \frac{1}{2}\mathcal{B}(\mathbf{r}) \times \mathbf{r}$, with the magnetic field in the z-direction $\mathcal{B}(\mathbf{r}) = \mathcal{B}\hat{\mathbf{i}}_z$, the Schrödinger-Pauli equation (5.1) can be solved for the triplet 2^3S state of the 2D 2 electron quantum dot in closed analytical form for a denumerably infinite set of frequencies ω_0 and ω_L such that the effective force constant $k_{\text{eff}} = \Omega^2 = \omega_0^2 + \omega_L^2$, where the binding force constant $k_o = \omega_0^2$ and $\omega_L = B/2c$ is the Larmor frequency. Effective atomic units are employed: $e^2/\epsilon = \hbar = m^\star = c = 1$. The effective Bohr radius is $a_0^\star = a_0(m\epsilon/m^\star)$, where m is the free electron mass. The effective energy unit is $(a.u.)^\star = (a.u.)(m^\star/m\epsilon^2)$. As there is no spin-orbital angular momentum coupling term in the Hamiltonian of (5.1), the spatial and spin coordinates are separable. Thus, the wave function $\Psi(\mathbf{x}_1\mathbf{x}_2)$ for this excited state is a product of a spatial $\psi(\mathbf{r}_1\mathbf{r}_2)$ and spin $\chi(\sigma_1\sigma_2)$ component:

$$\Psi(\mathbf{x}_1\mathbf{x}_2) = \psi(\mathbf{r}_1\mathbf{r}_2)\chi(\sigma_1\sigma_2). \tag{5.2}$$

The spatial component $\psi(\mathbf{r}_1\mathbf{r}_2)$ is

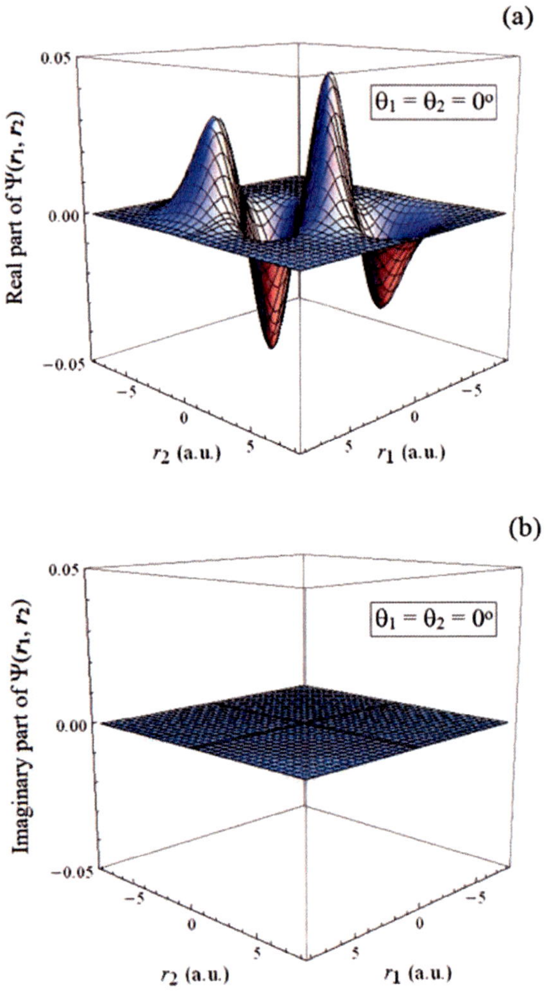

Fig. 5.1 (**a**) Structure of the Real component of the spatial part $\Psi(\mathbf{r}_1\mathbf{r}_2)$ of the triplet 2^3S wave function of the quantum dot in a magnetic field. The angles θ_1, θ_2 of the vectors $\mathbf{r}_1$ and $\mathbf{r}_2$ are measured from the $+x$-axis (not shown in the figure). In this Fig. 5.1, these angles are $\theta_1 = \theta_2 = 0°$, which means vectors $\mathbf{r}_1$ and $\mathbf{r}_2$ are oriented along the x axis. (**b**) The corresponding structure of the Imaginary part of $\Psi(\mathbf{r}_1\mathbf{r}_2)$

$$\psi(\mathbf{r}_1\mathbf{r}_2) = Ne^{im\theta}e^{-\Omega(r_1^2+r_2^2)/2}\big[|\mathbf{r}_2 - \mathbf{r}_1| + c_2|\mathbf{r}_2 - \mathbf{r}_1|^2 + c_3|\mathbf{r}_2 - \mathbf{r}_1|^3 + c_4|\mathbf{r}_2 - \mathbf{r}_1|^4\big], \tag{5.3}$$

where the normalization constant $N = 0.022466$; the angular quantum number $m = 0, \pm 1, \pm 2, \ldots$ is chosen to be $m = +1$, the coefficients $c_2 = \frac{1}{3}$; $c_3 = -0.059108$; $c_4 = -0.015884$; the effective force constant $k_{\text{eff}} = 0.072217$; the angle θ is that of the relative coordinate vector $\mathbf{u} = \mathbf{r}_2 - \mathbf{r}_1$; and $\mathbf{r}_1 = (r_1\theta_1)$, $\mathbf{r}_2 = (r_2\theta_2)$.

The wave function $\Psi(\mathbf{x}_1\mathbf{x}_2)$ is of course antisymmetric in an interchange of the coordinates $\mathbf{x}_1$ and $\mathbf{x}_2$, i.e. $\Psi(\mathbf{x}_1\mathbf{x}_2) = -\Psi(\mathbf{x}_2\mathbf{x}_1)$. Since the spin component $\chi(\sigma_1\sigma_2)$ for the triplet state is symmetric in an interchange of the coordinates σ_1 and σ_2, the

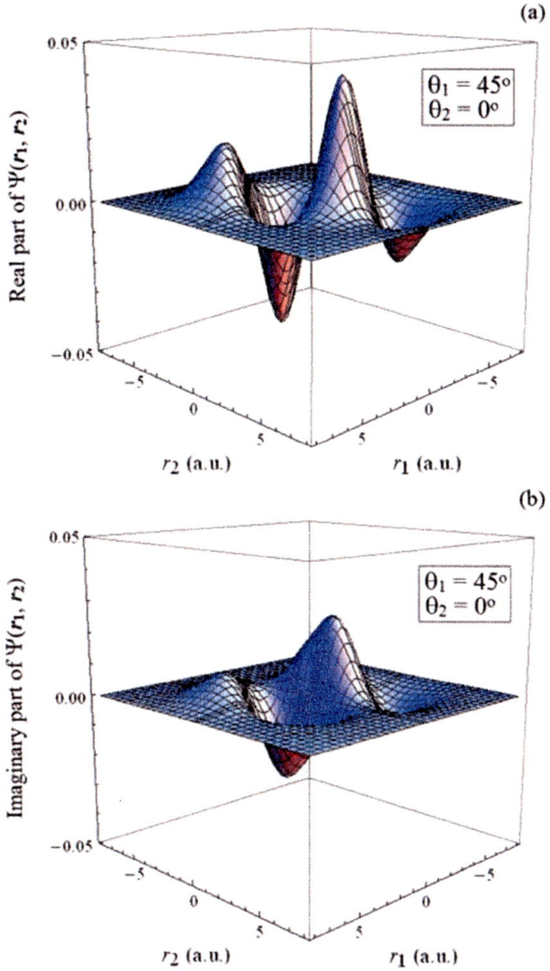

Fig. 5.2 Same as in Fig. 5.1 except that $\theta_1 = 45°, \theta_2 = 0°$. In this figure, the vector $\mathbf{r}_2$ is along the x axis

spatial component $\psi(\mathbf{r}_1\mathbf{r}_2)$ is antisymmetric in an interchange of $\mathbf{r}_1$ and $\mathbf{r}_2$, i.e. $\psi(\mathbf{r}_1\mathbf{r}_2) = -\psi(\mathbf{r}_2\mathbf{r}_1)$. Note that it is the phase factor $e^{im\theta}$ that ensures this antisymmetry. (When $\mathbf{r}_1$ and $\mathbf{r}_2$ are interchanged, the magnitude of the relative coordinate vector $\mathbf{u}$ does not change, but its angle θ (which points from the tip of $\mathbf{r}_1$ to the tip of $\mathbf{r}_2$), changes to $\theta + \pi$. This changes the sign of the phase factor $e^{im\theta}$.)

The spatial part of the wave function $\Psi(\mathbf{r}_1\mathbf{r}_2)$ is complex, and has many interesting properties [33], some of which are exhibited here. Other properties are simply stated.

1. In Figs. 5.1, 5.2, 5.3 and 5.4 we plot the function $\psi(\mathbf{r}_1\mathbf{r}_2)$ as a function of r_1 and r_2 for different θ_1 and θ_2. In each figure, panel (a) corresponds to the real part of $\psi(\mathbf{r}_1\mathbf{r}_2)$, and panel (b) to its imaginary part. Observe that in Fig. 5.1 for $\theta_1 = \theta_2 = 0°$, the function $\psi(\mathbf{r}_1\mathbf{r}_2)$ is real. As θ_1 increases to $\theta_1 = 45°$ in Fig. 5.2,

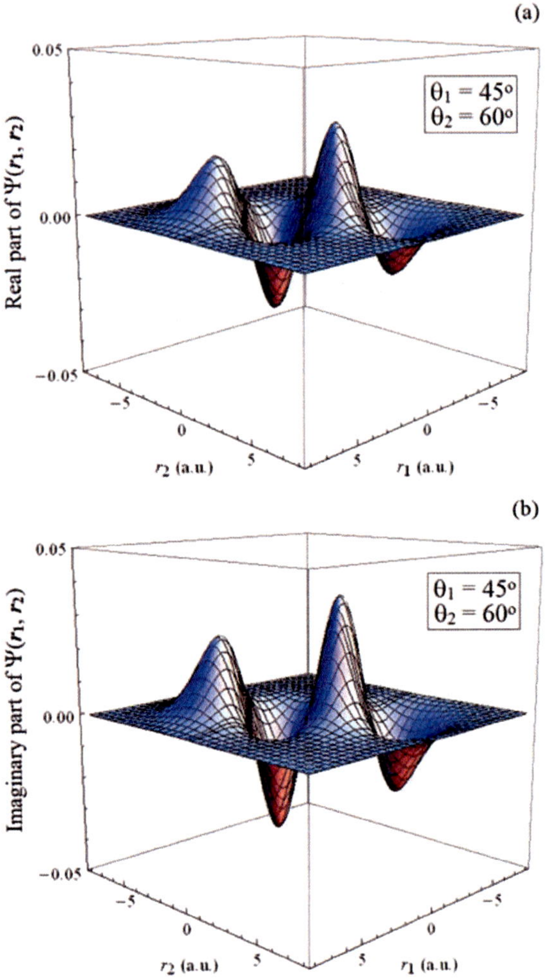

Fig. 5.3 Same as in Fig. 5.1 except that $\theta_1 = 45°, \theta_2 = 60°$

the real part shrinks and the imaginary part becomes finite. For fixed $\theta_1 = 45°$ and increasing $\theta_2 = 60°, 90°$ as in Figs. 5.3 and 5.4, respectively, the real part continues to diminish whilst the imaginary part increases in magnitude.

2. As a consequence of the electron-interaction term, the Hamiltonian of a many-electron system is singular at the coalescence of any two-electrons. For the wave functions $\Psi(\mathbf{X})$ to satisfy the corresponding Schrödinger or Schrödinger-Pauli equations, and remain bounded, they must satisfy a coalescence constraint. Accordingly, as derived in Chap. 8, for an N particle system, the coalescence condition [26] in D dimensions of 2 particles of masses m_1 and m_2, and charges Z_1 and Z_2, (with the spin index suppressed) is

Fig. 5.4 Same as in Fig. 1 except that $\theta_1 = 45°, \theta_2 = 90°$. In this figure, the vector $\mathbf{r}_2$ is along the y-axis

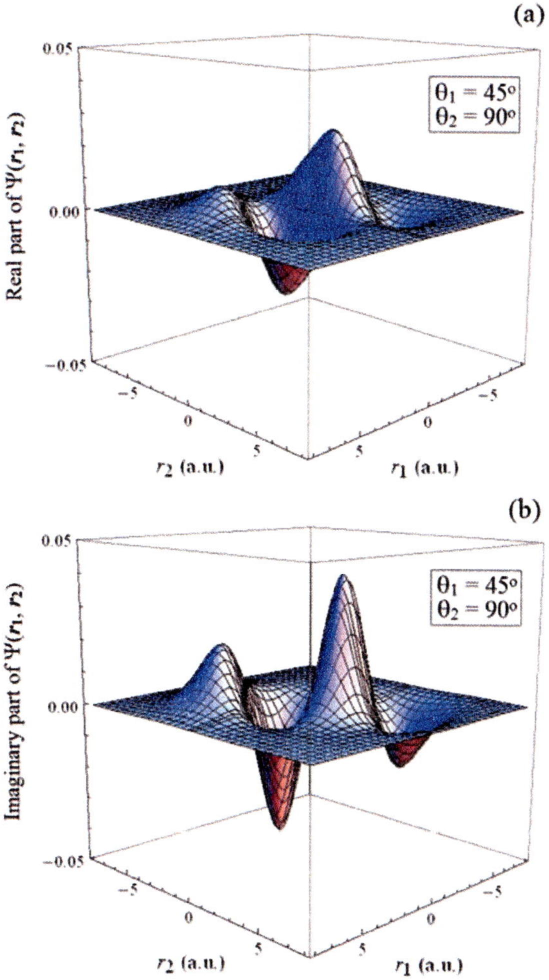

$$\psi(\mathbf{r}_1, \mathbf{r}_2, \ldots, \mathbf{r}_N) = \psi(\mathbf{r}_2, \mathbf{r}_2, \mathbf{r}_3, \ldots, \mathbf{r}_N)\left(1 + \frac{2Z_1Z_2\mu_{12}}{D-1}u\right) + \mathbf{u} \cdot \mathbf{C}(\mathbf{r}_2, \mathbf{r}_3, \ldots, \mathbf{r}_N), \quad (5.4)$$

where $\mu_{12} = m_1m_2/m_1 + m_2$ is the reduced mass, and $\mathbf{C}(\mathbf{r}_2, \mathbf{r}_3, \ldots, \mathbf{r}_N)$ is an undetermined vector. This is the *cusp coalescence* condition. It is equally valid when the wave function vanishes at the point of coalescence, i.e. when $\psi(\mathbf{r}_2, \mathbf{r}_2, \mathbf{r}_3, \ldots, \mathbf{r}_N) = 0$, and is then referred to as the *node coalescence* condition.

The wave function $\Psi(\mathbf{x}_1\mathbf{x}_2)$ for the triplet state via its spatial component $\psi(\mathbf{r}_1\mathbf{r}_2)$ satisfies the *node* electron-electron *coalescence* condition.

3. The function $\psi(\mathbf{r}_1\mathbf{r}_2)$ exhibits the following nodes.

(a) There is a node at the origin. This is evident in Figs. 5.1, 5.2, 5.3 and 5.4 for both the cases of $\theta_1 = \theta_2$ and $\theta_1 \neq \theta_2$. This is because the probability of 2 electrons of the same spin being at the same position in space at $\mathbf{r}_1 = \mathbf{r}_2 = 0$ is zero as a result of the Pauli principle. Observe also that the *parity* of the function $\psi(\mathbf{r}_1\mathbf{r}_2)$ about the origin is *odd*, i.e. $\psi(\mathbf{r}_1, \mathbf{r}_2) = -\psi(-\mathbf{r}_1, -\mathbf{r}_2)$.

(b) There is a node [33] at all points of electron-electron coalescence, again as a consequence of the Pauli principle. The function $\psi(\mathbf{r}_1\mathbf{r}_2)$ has *odd parity* about *all* these points of coalescence.

(c) The real part of $\psi(\mathbf{r}_1\mathbf{r}_2)$ has a node [33] when the projections of the vectors $\mathbf{r}_1$ and $\mathbf{r}_2$ on the x-axis are the same. The function $\psi(\mathbf{r}_1\mathbf{r}_2)$ is then purely imaginary. The *parity* of the $\psi(\mathbf{r}_1\mathbf{r}_2)$ is *odd* about the line $\mathbf{r}_2 = (cos\theta_1/cos\theta_2)\mathbf{r}_1$.

(d) The imaginary part of $\psi(\mathbf{r}_1\mathbf{r}_2)$ has a node [33] when the projections of the vectors $\mathbf{r}_1$ and $\mathbf{r}_2$ on the y-axis are the same. The wave function is then real. The *parity* of $\psi(\mathbf{r}_1\mathbf{r}_2)$ is *odd* about the line $\mathbf{r}_2 = (sin\theta_1/sin\theta_2)\mathbf{r}_1$.

(e) There is a node of $\psi(\mathbf{r}_1\mathbf{r}_2)$ as a result of it being a first excited state. These nodes are located where $\psi(\mathbf{r}_1\mathbf{r}_2)$ is zero along the lines at non-zero values of r_1 and r_2 as shown in Figs. 5.1, 5.2, 5.3 and 5.4. There is *no parity* of $\psi(\mathbf{r}_1\mathbf{r}_2)$ about this node.

5.2 Quantal Sources

In this section the various quantal sources for the fields that satisfy the 'Quantal Newtonian' First Law are described. As the spin and spatial coordinates are separable, and the corresponding spin $\chi(\sigma_1\sigma_2)$ and spatial $\psi(\mathbf{r}_1\mathbf{r}_2)$ components are separately normalized, the quantal sources are simply expectation values taken with respect to the spatial component. The analytical and semi-analytical expressions for the sources are given in Appendix B.

5.2.1 *Electron Density* $\rho(\mathbf{r})$

The electron density $\rho(\mathbf{r})$ is the expectation value

$$\rho(\mathbf{r}) = \langle \Psi | \hat{\rho}(\mathbf{r}) | \Psi \rangle, \tag{5.5}$$

where the density operator $\hat{\rho}(\mathbf{r})$ is defined as

$$\hat{\rho}(\mathbf{r}) = \sum_k \delta(\mathbf{r}_k - \mathbf{r}). \tag{5.6}$$

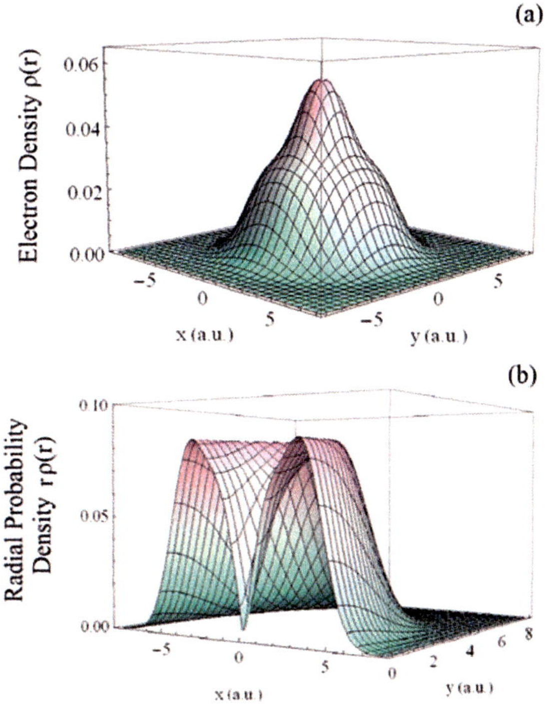

Fig. 5.5 (**a**) Electron density $\rho(\mathbf{r})$ of the triplet 2^3S state of the quantum dot in a magnetic field. (**b**) The radial probability density $r\rho(r)$

In Fig. 5.5(a) the electron density $\rho(\mathbf{r})$ is plotted. It is spherically symmetric about the origin, and exhibits shell structure. As a consequence of the binding potential being harmonic, the density is finite at the origin approaching it with zero slope. Thus, there is no cusp in the density there. (For natural atoms with a Coulombic binding potential, there is a singularity in the Hamiltonian at the nucleus. This is a consequence of the coalescence of the electron with the nucleus. Thus, the wave function exhibits a cusp at the nucleus, as does the density.)

Also observe that while the density $\rho(\mathbf{r})$ has different values for each electron position, the overall structure remains unchanged as the electron position is varied. Thus, the density $\rho(\mathbf{r})$ is a *local* or *static* property. The radial probability density $r\rho(\mathbf{r})$ is plotted in Fig. 5.5(b). Observe that shell structure is evident in the inner shoulder of the figure.

5.2.2 *Physical Current Density* $\mathbf{j}(\mathbf{r})$ *and Its Paramagnetic* $\mathbf{j}_p(\mathbf{r})$*, Diamagnetic* $\mathbf{j}_d(\mathbf{r})$ *and Magnetization* $\mathbf{j}_m(\mathbf{r})$ *Components*

The physical current density $\mathbf{j}(\mathbf{r})$, a *local* quantal source, is the expectation value

$$\mathbf{j}(\mathbf{r}) = \langle \psi | \hat{\mathbf{j}}(\mathbf{r}) | \psi \rangle, \tag{5.7}$$

where the current density operator $\hat{\mathbf{j}}(\mathbf{r})$ is the sum of its paramagnetic $\hat{\mathbf{j}}_p(\mathbf{r})$, diamagnetic $\hat{\mathbf{j}}_d(\mathbf{r})$, and magnetization $\hat{\mathbf{j}}_m(\mathbf{r})$ current density components:

$$\hat{\mathbf{j}}(\mathbf{r}) = \hat{\mathbf{j}}_p(\mathbf{r}) + \hat{\mathbf{j}}_d(\mathbf{r}) + \hat{\mathbf{j}}_m(\mathbf{r}), \tag{5.8}$$

with

$$\hat{\mathbf{j}}_p(\mathbf{r}) = \frac{1}{2} \sum_k \Big[\hat{\mathbf{p}}_k \delta(\mathbf{r}_k - \mathbf{r}) + \delta(\mathbf{r}_k - \mathbf{r}) \hat{\mathbf{p}}_k \Big], \tag{5.9}$$

$$\hat{\mathbf{j}}_d(\mathbf{r}) = \hat{\rho}(\mathbf{r}) \mathbf{A}(\mathbf{r}), \tag{5.10}$$

$$\hat{\mathbf{j}}_m(\mathbf{r}) = -\boldsymbol{\nabla} \times \hat{\mathbf{m}}(\mathbf{r}) \tag{5.11}$$

and the magnetization density $\hat{\mathbf{m}}(\mathbf{r})$ operator is

$$\hat{\mathbf{m}}(\mathbf{r}) = \sum_k \mathbf{s}_k \delta(\mathbf{r}_k - \mathbf{r}). \tag{5.12}$$

The paramagnetic current density $\mathbf{j}_p(\mathbf{r})$ may also be defined via the non local quantal source of the single-particle density matrix $\gamma(\mathbf{r}\mathbf{r}')$ as

$$\mathbf{j}_p(\mathbf{r}) = \frac{i}{2} \Big[\boldsymbol{\nabla}' - \boldsymbol{\nabla}'' \Big] \gamma(\mathbf{r}'\mathbf{r}'') \Big|_{\mathbf{r}'=\mathbf{r}''=r}, \tag{5.13}$$

where $\gamma(\mathbf{r}\mathbf{r}')$ is defined below in Sect. 5.2.4. (The expression for the current density $\mathbf{j}_p(\mathbf{r})$ for this triplet state given in Appendix B is derived independently through the definitions of (5.9) and (5.13).)

In Figs. 5.6, 5.7, 5.8 and 5.9 panels (a), the physical current density $\mathbf{j}(\mathbf{r})$, and its paramagnetic $\mathbf{j}_p(\mathbf{r})$, diamagnetic $\mathbf{j}_d(\mathbf{r})$, and magnetization $\mathbf{j}_m(\mathbf{r})$ components, respectively, are plotted. The diamagnetic component $\mathbf{j}_d(\mathbf{r})$ which is the only component that depends explicitly on the magnetic field is plotted for a value of the Larmor frequency of $\omega_L = 0.1$. Hence, the plot of the total current density $\mathbf{j}(\mathbf{r})$ is for $\omega_L = 0.1$. Each density component is a function solely of the radial component r, but points in the $\hat{\mathbf{i}}_\theta$ direction. Hence, the divergence of each component vanishes, and therefore $\boldsymbol{\nabla} \cdot \mathbf{j}(\mathbf{r}) = 0$.

Observe that shell structure is clearly evident in the plot of the current density $\mathbf{j}(\mathbf{r})$ (see Fig. 5.6(a)). This structure is also evident in the individual components

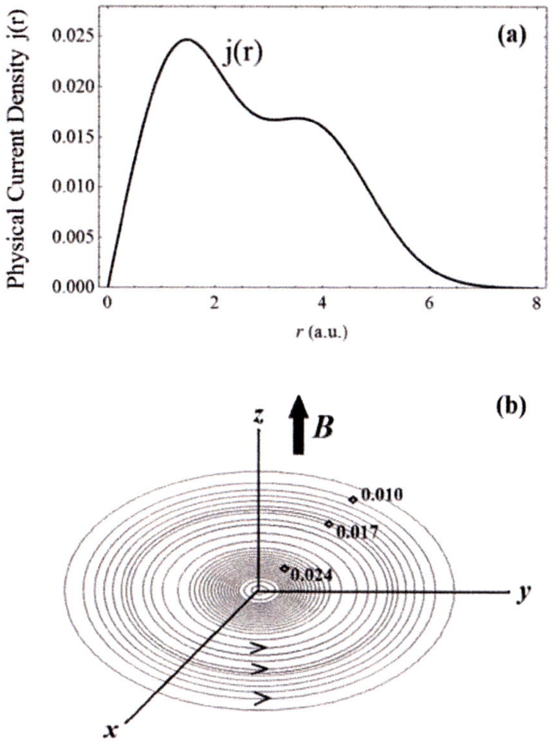

Fig. 5.6 (**a**) The physical current density $\mathbf{j}(\mathbf{r})$ of the triplet 2^3S state of the quantum dot in a magnetic field for a value of the Larmor frequency $\omega_L = 0.1$. (**b**) The flow line contours of $\mathbf{j}(\mathbf{r})$

(see Figs. 5.7(a), 5.8(a) and 5.9(a)), although their individual structures are different. For the choice of $\omega_L = 0.1$, the magnitude of the paramagnetic $\mathbf{j}_p(\mathbf{r})$, diamagnetic $\mathbf{j}_d(\mathbf{r})$, and magnetization $\mathbf{j}_m(\mathbf{r})$ components is essentially the same. (Depending on the value of ω_L, the diamagnetic component $\mathbf{j}_d(\mathbf{r})$ and thus $\mathbf{j}(\mathbf{r})$ can be made larger or smaller.)

In Figs. 5.6, 5.7, 5.8 and 5.9, panels (b), the flow line contours of each current density component are plotted. These contour lines are closest in the regions of greater density. Observe the difference in the contours for each density component. The circulation direction of the component $\mathbf{j}_p(\mathbf{r})$ depends explicitly on the choice of angular momentum quantum number m. This is also the case for $\mathbf{j}_m(\mathbf{r})$ whose dependency on m is via the electronic density $\rho_{\uparrow\uparrow}$ (corresponding to m = 1) or $\rho_{\downarrow\downarrow}$ (corresponding to $m = -1$). The circulation direction of these two current densities $\mathbf{j}_p(\mathbf{r})$ and $\mathbf{j}_m(\mathbf{r})$ is always the same, but the direction depends upon whether $m = 1$ or $m = -1$. On the other hand, the diamagnetic current density $\mathbf{j}_d(\mathbf{r})$ does not depend on m. Thus, its circulation can be either in the same or opposite direction to that of $\mathbf{j}_p(\mathbf{r})$ depending on the value of m. For our choice of $m = 1$, the circulation direction for $\mathbf{j}_p(\mathbf{r})$, $\mathbf{j}_d(\mathbf{r})$, and $\mathbf{j}_m(\mathbf{r})$ are all the same (counterclockwise). (The fact that the circulation directions of $\mathbf{j}_p(\mathbf{r})$ and $\mathbf{j}_d(\mathbf{r})$ are the same, for the chosen value of m, is confirmed by an independent derivation related to the contribution of the Lorentz $\mathcal{L}(\mathbf{r})$ and internal magnetic $\mathcal{I}_m(\mathbf{r})$ fields to the total energy.)

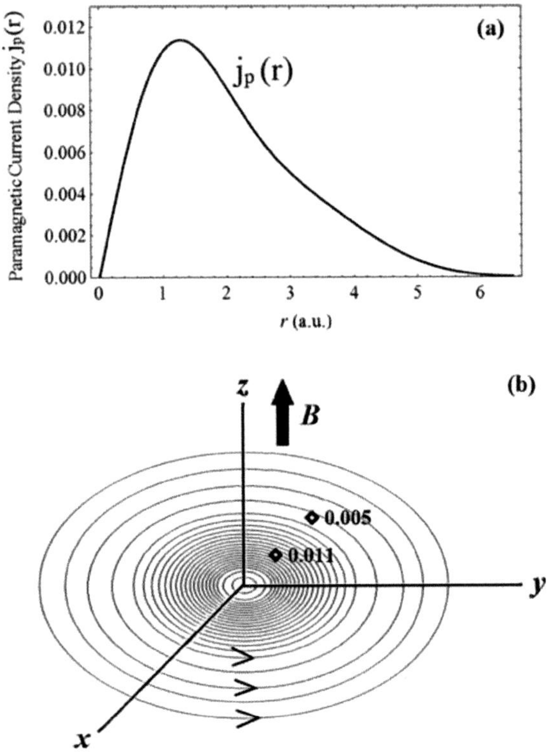

Fig. 5.7 (**a**) The paramagnetic current density $\mathbf{j}_p(\mathbf{r})$ of the triplet 2^3S state of the quantum dot in a magnetic field. (**b**) The flow line contours of $\mathbf{j}_p(\mathbf{r})$

5.2.3 *Pair-Correlation Density* $g(\mathbf{rr}')$ *and the Fermi-Coulomb Hole* $\rho_{xc}(\mathbf{rr}')$

The pair-correlation density $g(\mathbf{rr}')$ is defined as the ratio of the pair-correlation function $P(\mathbf{rr}')$ to the density $\rho(\mathbf{r})$:

$$g(\mathbf{rr}') = P(\mathbf{rr}')/\rho(\mathbf{r}), \tag{5.14}$$

where $P(\mathbf{rr}')$ is the expectation value

$$P(\mathbf{rr}') = \langle\psi|\hat{P}(\mathbf{rr}')|\psi\rangle, \tag{5.15}$$

with the pair operator defined as

$$\hat{P}(\mathbf{rr}') = \sum_{k,\ell}{}' \delta(\mathbf{r}_k - \mathbf{r})\delta(\mathbf{r}_\ell - \mathbf{r}'). \tag{5.16}$$

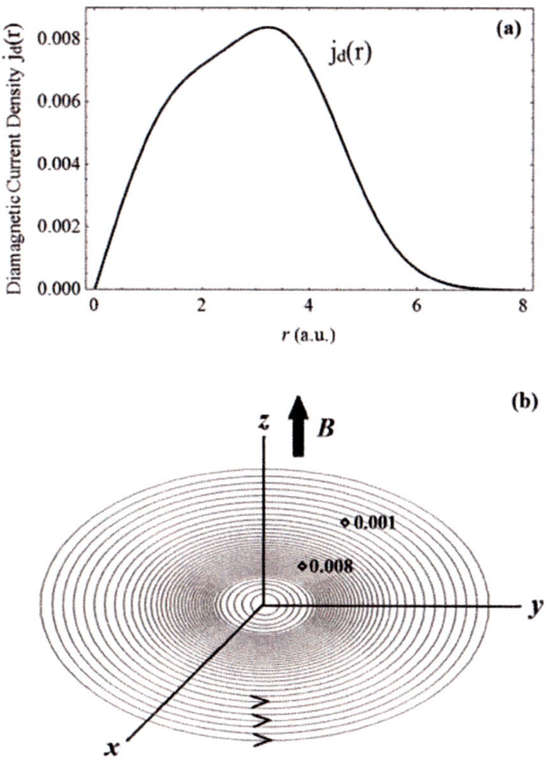

Fig. 5.8 (**a**) The diamagnetic current density $\mathbf{j}_d(\mathbf{r})$ of the triplet 2^3S state of the quantum dot in a magnetic field for a value of the Larmor frequency $\omega_L = 0.1$. (**b**) The flow line contours of $\mathbf{j}_d(\mathbf{r})$

The pair-correlation density $g(\mathbf{rr}')$ may also be written in terms of its local and nonlocal components as

$$g(\mathbf{rr}') = \rho(\mathbf{r}') + \rho_{xc}(\mathbf{rr}'), \tag{5.17}$$

where $\rho_{xc}(\mathbf{rr}')$ is the Fermi-Coulomb hole charge.

The pair-correlation density $g(\mathbf{rr}')$ and Fermi-Coulomb hole $\rho_{xc}(\mathbf{rr}')$ are *nonlocal* quantal sources in that their structure changes as a function of the electron position. This is demonstrated in Fig. 5.10 where $g(\mathbf{rr}')$ is plotted for the following different electron positions: (a) the center of the quantum dot at $r = 0$; (b) at $r = 0.5\ a.u.$; (c) at $r = 1.0\ a.u.$;(d) at $r = 1.5\ a.u.$ Observe that in each figure, the pair-correlation density vanishes at the electron position. This is a consequence of the node coalescence condition satisfied by the wave function. Also note that except for the electron position at the center of the quantum dot, $g(\mathbf{rr}')$ is not spherically symmetric about the electron position. In Fig. 5.11, the $g(\mathbf{rr}')$ is plotted for asymptotic positions of the electron: (a) at $r = 8.0\ a.u.$; (b) at $r = 12.0\ a.u.$ For these asymptotic positions, observe that the figures are very similar. This is a reflection of the fact that for such asymptotic positions of the electron, the nonlocal charge is becoming essentially static. Since the total charge of the pair-correlation density $g(\mathbf{rr}')$ is 1 (obtained from

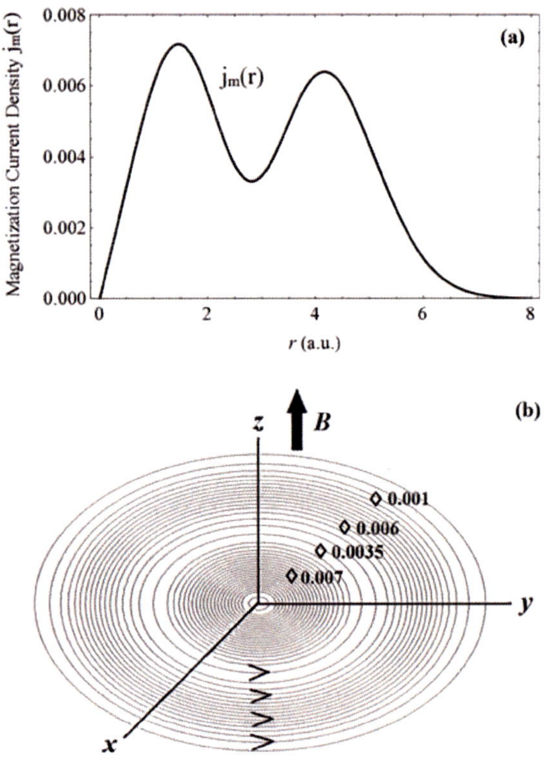

Fig. 5.9 (**a**) The magnetization current density $\mathbf{j}_m(\mathbf{r})$ of the triplet 2^3S state of the quantum dot in a magnetic field. (**b**) The flow line contours of $\mathbf{j}_m(\mathbf{r})$

$N-1$), the asymptotic structure of the electron-interaction field $\mathcal{E}_{ee}(\mathbf{r})$ derived from it via Coulomb's Law is analytically known (see Sect. 5.3.1). Since for an electron at the center of the quantum dot, the density $g(\mathbf{rr}')$ is spherically symmetric about this position (Fig. 5.10a), the field vanishes there.

The nonlocal structure of the Fermi-Coulomb hole $\rho_{xc}(\mathbf{rr}')$ (see Fig. 5.12) can be obtained from (5.17). The hole represents the reduction in density at $\mathbf{r}'$ for an electron at $\mathbf{r}$ due to the Pauli exclusion principle and Coulomb repulsion. Although this structure differs significantly from that of $g(\mathbf{rr}')$, its properties are similar. Thus, at the electron position, the hole is finite and continuous and has the lowest value. (There is no cusp at this point as is the case for a singlet excited state. (See Chap. 6)) For an electron position at the center of the quantum dot, the hole is spherically symmetric about it, and thus the Pauli-Coulomb field $\mathcal{E}_{xc}(\mathbf{r})$ vanishes at the origin. The hole is not spherically symmetric about the other electron positions. As the hole becomes an essentially static charge for asymptotic positions of the electron, and since the total hole charge is -1, the asymptotic structure of $\mathcal{E}_{xc}(\mathbf{r})$ is also analytically known (see Sect. 5.3.1).

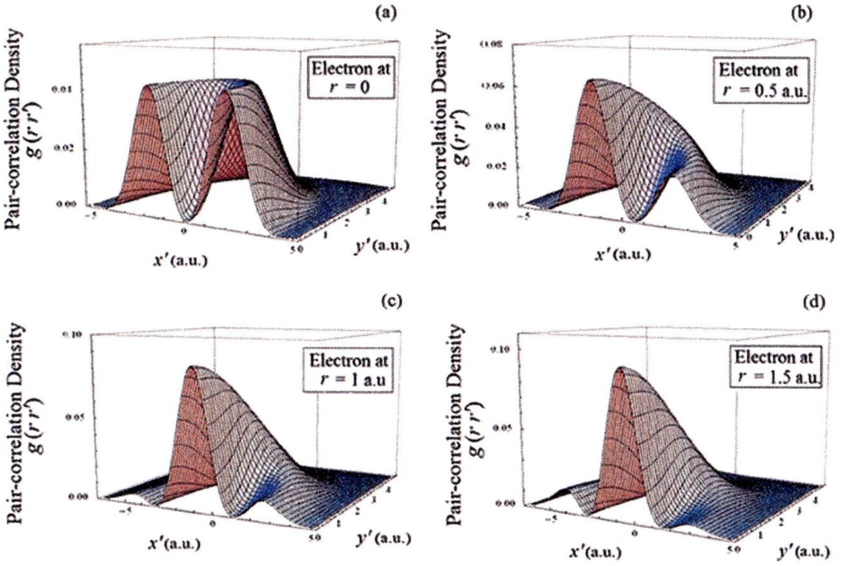

Fig. 5.10 Surface plot of the pair-correlation density $g(\mathbf{rr}')$ of the triplet 2^3S state of the quantum dot in a magnetic field for different electron positions located on the x-axis: (**a**) the center of the quantum dot at $r = 0$; (**b**) at $r = 0.5\ a.u.$; (**c**) at $r = 1.0\ a.u.$; (**d**) at $r = 1.5\ a.u.$ In the figure x' is the projection of $\mathbf{r}'$ on $\mathbf{r}$, i.e. $x' = r'\mathbf{i}_r \cdot \mathbf{i}_{r'}$, and y' is the projection of $\mathbf{r}'$ on the direction perpendicular to $\mathbf{r}$, i.e. $y' = r'[1 - (\mathbf{i}_r \cdot \mathbf{i}_{r'})^2]^{\frac{1}{2}}$

5.2.4 *Single-Particle Density Matrix* $\gamma(\mathbf{rr}')$

The single-particle density matrix $\gamma(\mathbf{rr}')$, a *nonlocal* quantal source, is defined as the expectation value

$$\gamma(\mathbf{rr}') = \langle\psi|\hat{\gamma}(\mathbf{rr}')|\psi\rangle, \tag{5.18}$$

where the complex single-particle density matrix operator [34, 35] is

$$\hat{\gamma}(\mathbf{rr}') = \hat{A} + i\hat{B}, \tag{5.19}$$

$$\hat{A} = \frac{1}{2}\sum_k \big[\delta(\mathbf{r}_k - \mathbf{r})T_k(\mathbf{a}) + \delta(\mathbf{r}_k - \mathbf{r}')T_k(-\mathbf{a})\big], \tag{5.20}$$

$$\hat{B} = -\frac{i}{2}\sum_k \big[\delta(\mathbf{r}_k - \mathbf{r})T_k(\mathbf{a}) - \delta(\mathbf{r}_k - \mathbf{r}')T_k(-\mathbf{a})\big], \tag{5.21}$$

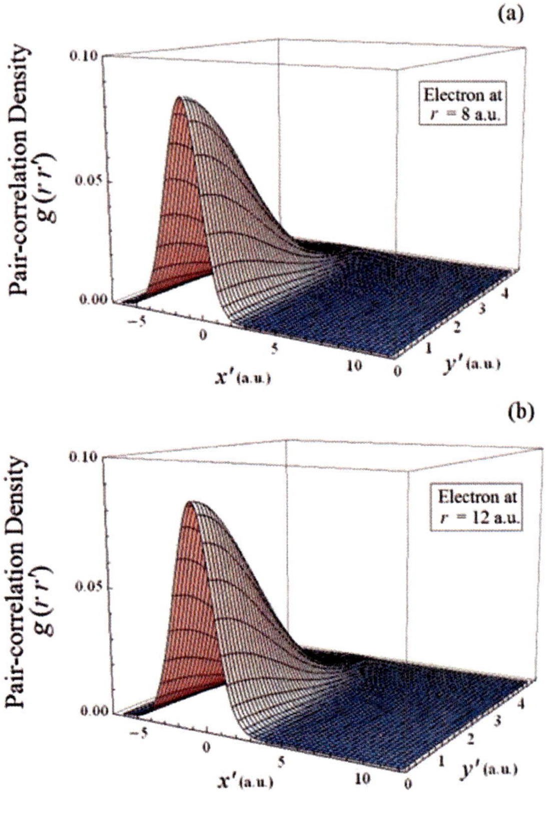

Fig. 5.11 The same as in Fig. 10 but for asymptotic electron positions: (**a**) at $r = 8.0\, a.u.$; (**b**) at $r = 12.0\, a.u.$

with $T_k(\mathbf{a})$ a translation operator such that $T_k(\mathbf{a})\psi(\ldots\mathbf{r}_k,\ldots) = \psi(\ldots\mathbf{r}_k + \mathbf{a},\ldots)$ and $\mathbf{a} = \mathbf{r}' - \mathbf{r}$. The operators $\hat{A}$ and $\hat{B}$ are Hermitian. The single-particle density matrix is the quantal source for all kinetic related properties [36] such as the kinetic energy tensor, the kinetic energy density, the kinetic field, the kinetic energy, and the paramagnetic current density (see Fig. 5.13).

In the panels of Fig. 5.13, the $\gamma(\mathbf{r}\mathbf{r}')$ for the triplet state is plotted as the positions $\mathbf{r}$ and $\mathbf{r}'$ change for (a) $\theta = \theta' = 0°$; (b) $\theta = 0°, \theta' = 45°$; (c) $\theta = 0°, \theta' = 60°$; (d) $\theta = 0°, \theta' = 90°$. The nonlocal nature of $\gamma(\mathbf{r}\mathbf{r}')$ is clearly evident as is shell structure. Observe the change in the shoulder of $\gamma(\mathbf{r}\mathbf{r}')$ as θ' changes from 0° to 90°. Also note that the $\gamma(\mathbf{r}\mathbf{r}')$ exhibits nodes as a consequence of the node in the wave function for this excited state (point #3e of Sect. 5.1). Although the wave function exhibits a node at the origin (point # 3a of Sect. 5.1), the $\gamma(\mathbf{r}\mathbf{r}')$ is finite there. For the cross sections for which $\mathbf{r} = \mathbf{r}'$, one obtains the density of Fig. 5.5 since $\gamma(\mathbf{r}\mathbf{r}) = \rho(\mathbf{r})$.

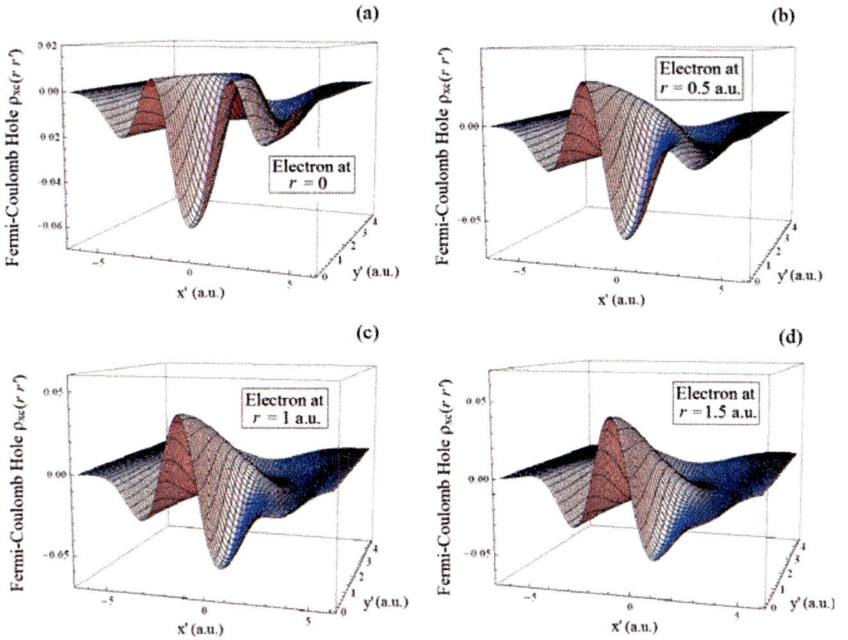

Fig. 5.12 Surface plot of the Fermi-Coulomb hole charge $\rho_{xc}(\mathbf{rr}')$ of the triplet 2^3S state of the quantum dot in a magnetic field for different electron positions located on the x-axis: (**a**) the center of the quantum dot at $r = 0$; (**b**) at $r = 0.5$ $a.u.$; (**c**) at $r = 1.0$ $a.u.$; (**d**) at $r = 1.5$ $a.u$. In the figure x' is the projection of $\mathbf{r}'$ on $\mathbf{r}$, i.e. $x' = r'\mathbf{i}_r \cdot \mathbf{i}_{r'}$, and y' is the projection of $\mathbf{r}'$ on the direction perpendicular to $\mathbf{r}$, i.e. $y' = r'[1 - (\mathbf{i}_r \cdot \mathbf{i}_{r'})^2]^{\frac{1}{2}}$

5.3 'Forces', Fields, and Energies

The 'forces' and fields derived from the quantal sources are described next. The contributions of the individual fields to the total energy E are given in Table 5.1. Due to the symmetry of the system, the fields are each conservative. Various analytical and semi-analytical expressions for the fields and components of the energy are given in Appendix B.

5.3.1 Electron-Interaction, Hartree, Pauli-Coulomb

The electron-interaction field $\mathcal{E}_{\text{ee}}(\mathbf{r})$ is obtained from its quantal source, the pair-correlation density $g(\mathbf{rr}')$, via Coulomb's Law, and may be written (see (5.17)) in terms of its Hartree $\mathcal{E}_{\text{H}}(\mathbf{r})$ and Pauli-Coulomb $\mathcal{E}_{\text{xc}}(\mathbf{r})$ components:

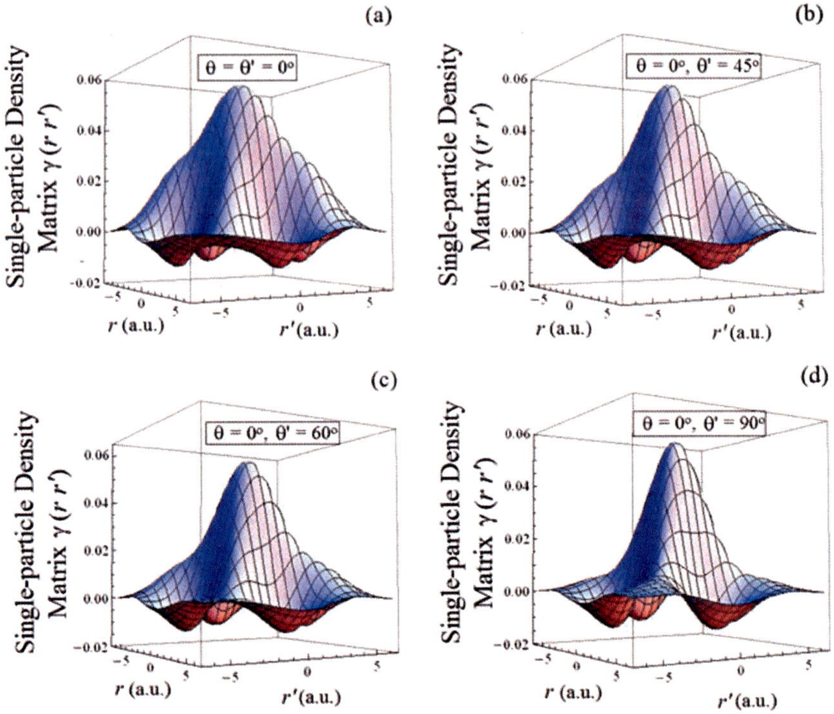

Fig. 5.13 The single particle density matrix $\gamma(\mathbf{rr}')$ for the triplet 2^3S state of the quantum dot in a magnetic field. The panels correspond to **(a)** $\theta = \theta' = 0°$; **(b)** $\theta = 0°, \theta' = 45°$; **(c)** $\theta = 0°, \theta' = 60°$; **(d)** $\theta = 0°, \theta' = 90°$

Table 5.1 Properties of the Triplet 2^3S state of the quantum dot in a magnetic field. The values are in effective atomic units $(a.u.)^\star$

Property	Value
T	0.615577
E_H	0.755497
E_{xc}	−0.501339
E_{ee}	0.254158
$E_{es} + E_{mag}$	0.742657
E	1.612391
$IP = E^{N-1} - E^N$	−1.343659
$\langle r^2 \rangle$	20.567403
$\langle r \rangle$	5.823553
$\langle 1/r \rangle$	1.041717
$\langle \delta(\mathbf{r}) \rangle$	0.0555377

$$\mathcal{E}_{ee}(\mathbf{r}) = \int \frac{g(\mathbf{r}\mathbf{r}')(\mathbf{r}-\mathbf{r}')}{|\mathbf{r}-\mathbf{r}'|^3} d\mathbf{r}' \tag{5.22}$$

$$= \mathcal{E}_{H}(\mathbf{r}) + \mathcal{E}_{xc}(\mathbf{r}), \tag{5.23}$$

where

$$\mathcal{E}_{H}(\mathbf{r}) = \int \frac{\rho(\mathbf{r}')(\mathbf{r}-\mathbf{r}')}{|\mathbf{r}-\mathbf{r}'|^3} d\mathbf{r}'; \ \mathcal{E}_{xc}(\mathbf{r}) = \int \frac{\rho_{xc}(\mathbf{r}\mathbf{r}')(\mathbf{r}-\mathbf{r}')}{|\mathbf{r}-\mathbf{r}'|^3} d\mathbf{r}'. \tag{5.24}$$

The fields may also be expressed in terms of their corresponding 'forces' $\mathbf{e}_{ee}(\mathbf{r})$, $\mathbf{e}_{H}(\mathbf{r})$, and $\mathbf{e}_{xc}(\mathbf{r})$:

$$\mathcal{E}_{ee}(\mathbf{r}) = \mathbf{e}_{ee}(\mathbf{r})/\rho(\mathbf{r}); \ \mathcal{E}_{H}(\mathbf{r}) = \mathbf{e}_{H}(\mathbf{r})/\rho(\mathbf{r}) \ ; \ \mathcal{E}_{xc}(\mathbf{r}) = \mathbf{e}_{xc}(\mathbf{r})/\rho(\mathbf{r}). \tag{5.25}$$

In Fig. 5.14(a, b) we plot the various 'forces' and fields, respectively. Shell structure is evident in the plots of both the 'forces' and fields. (For the 'force' $\mathbf{e}_{ee}(\mathbf{r})$ and field $\mathcal{E}_{ee}(\mathbf{r})$, the second shell becomes evident on an expanded scale.) As the quantal sources $g(\mathbf{r}\mathbf{r}')$, $\rho(\mathbf{r})$, $\rho_{xc}(\mathbf{r}\mathbf{r}')$ are all cylindrically symmetric for an electron position at the origin (see Figs. 5.5(a), 5.10(a), 5.12(a)), all the corresponding fields vanish there. The total charge of the pair-correlation density $g(\mathbf{r}\mathbf{r}')$ is 1, that of the Fermi-Coulomb hole $\rho_{xc}(\mathbf{r}\mathbf{r}')$ is -1, and that of the density $\rho(\mathbf{r})$ is 2. Since for asymptotic positions of the electron in the classically forbidden region, the nonlocal sources $g(\mathbf{r}\mathbf{r}')$ and $\rho_{xc}(\mathbf{r}\mathbf{r}')$ become essentially static charge distributions (see Fig. 5.11), and the density $\rho(\mathbf{r})$ is a static charge, the asymptotic structure of the fields as $r \to \infty$ is known exactly: $\mathcal{E}_{ee}(\mathbf{r}) \sim 1/r^2$, $\mathcal{E}_{H}(\mathbf{r}) \sim 2/r^2$, $\mathcal{E}_{xc}(\mathbf{r}) \sim -1/r^2$. That the decay of these fields is such is clearly evident in Fig. 5.14(b). Asymptotically, the 'forces' (see Fig. 5.14(a)) all vanish as their decay is faster than that of the density.

The electron-interaction E_{ee}, Hartree E_{H}, and Pauli-Coulomb E_{xc} energies are then obtained in integral virial form from the respective fields as

$$E_{ee} = \int \rho(\mathbf{r})\mathbf{r} \cdot \mathcal{E}_{ee}(\mathbf{r}) d\mathbf{r}, \tag{5.26}$$

$$E_{H} = \int \rho(\mathbf{r})\mathbf{r} \cdot \mathcal{E}_{H}(\mathbf{r}) d\mathbf{r}, \tag{5.27}$$

$$E_{xc} = \int \rho(\mathbf{r})\mathbf{r} \cdot \mathcal{E}_{xc}(\mathbf{r}) d\mathbf{r}. \tag{5.28}$$

These energy components are given in Table 5.1.

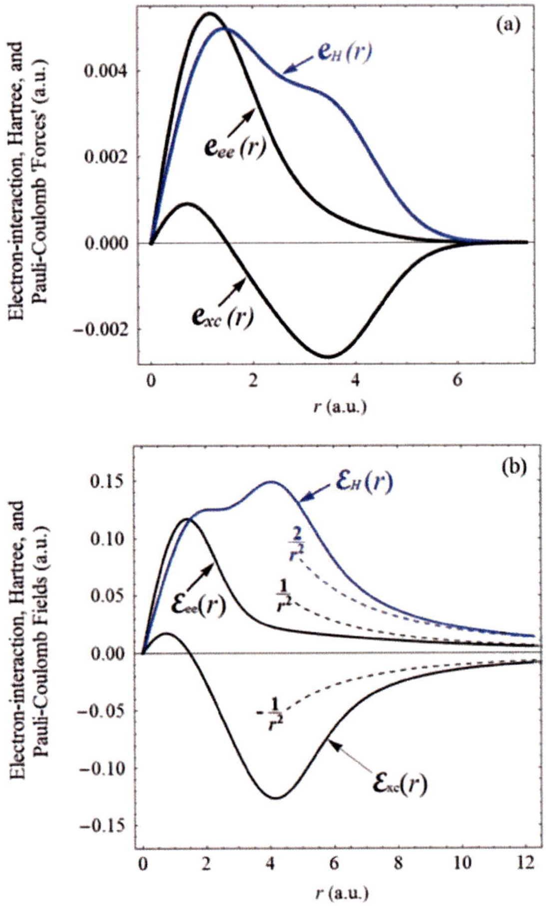

Fig. 5.14 (**a**) The electron-interaction $\mathbf{e}_{ee}(r)$, Hartree $\mathbf{e}_H(r)$, and Pauli-Coulomb $\mathbf{e}_{xc}(r)$ 'forces'. (**b**) The electron-interaction $\mathcal{E}_{ee}(r)$, Hartree $\mathcal{E}_H(r)$, and Pauli-Coulomb $\mathcal{E}_{xc}(r)$ fields. The functions $1/r^2, 2/r^2$, and $-1/r^2$ are also plotted as dashed lines

5.3.2 *Kinetic*

The quantal source for the kinetic 'force' $z(\mathbf{r})$, field $\mathcal{Z}(\mathbf{r})$, and energy T is the single-particle density matrix $\gamma(\mathbf{rr}')$. The field is defined in terms of the 'force' as

$$\mathcal{Z}(\mathbf{r}) = z(\mathbf{r})/\rho(\mathbf{r}), \tag{5.29}$$

where in Cartesian coordinates

$$z_\alpha(\mathbf{r}) = 2\sum_\beta \nabla_\beta t_{\alpha\beta}(\mathbf{r};\gamma), \tag{5.30}$$

and where the second-rank kinetic energy tensor

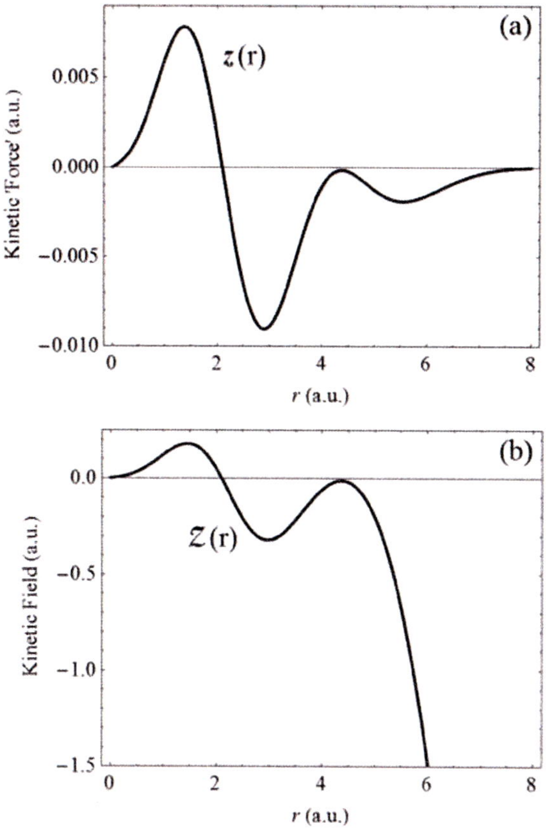

Fig. 5.15 The kinetic (**a**) 'force' $\mathbf{z}(r)$, and (**b**) field $\mathcal{Z}(r)$

$$t_{\alpha\beta}(\mathbf{r};\gamma) = \frac{1}{4}\left[\frac{\partial^2}{\partial r'_\alpha \partial r''_\beta} + \frac{\partial^2}{\partial r'_\beta \partial r''_\alpha}\right]\gamma(\mathbf{r}'\mathbf{r}'')\bigg|_{\mathbf{r}'=\mathbf{r}''=\mathbf{r}}. \tag{5.31}$$

The kinetic energy in terms of the field $\mathcal{Z}(\mathbf{r})$ is

$$T = -\frac{1}{2}\int \rho(\mathbf{r})\mathbf{r}\cdot\mathcal{Z}(\mathbf{r})d\mathbf{r}. \tag{5.32}$$

The kinetic 'force' $z(\mathbf{r})$ and field $\mathcal{Z}(\mathbf{r})$ are plotted in Fig. 5.15(a) and (b), respectively. Once again, shell structure is evident. Whilst the 'force' $z(\mathbf{r})$ decays and vanishes asymptotically, the field $\mathcal{Z}(\mathbf{r})$ is singular in this region. Both vanish at the origin. See Table 5.1 for the value of T. (For the derivation of the tensor $t_{\alpha\beta}(\mathbf{r};\gamma)$ and the kinetic 'force' $z(\mathbf{r})$, see Appendix C.)

(The singularity of the field $\mathcal{Z}(\mathbf{r})$ in the asymptotic region is canceled by the singularity in the differential density field $\mathcal{D}(\mathbf{r})$ discussed in the following subsection.)

5.3.3 *Differential Density*

The quantal source for the differential density 'force' $\mathbf{d}(\mathbf{r})$ and field $\boldsymbol{\mathcal{D}}(\mathbf{r})$ is the density $\rho(\mathbf{r})$. The field $\boldsymbol{\mathcal{D}}(\mathbf{r})$ is defined as

$$\boldsymbol{\mathcal{D}}(\mathbf{r}) = \mathbf{d}(\mathbf{r})/\rho(\mathbf{r}), \tag{5.33}$$

where

$$\mathbf{d}(\mathbf{r}) = -\frac{1}{4}\boldsymbol{\nabla}\nabla^2\rho(\mathbf{r}). \tag{5.34}$$

The 'force' $\mathbf{d}(\mathbf{r})$ and field $\boldsymbol{\mathcal{D}}(\mathbf{r})$ are plotted in Fig. 5.16(a) and (b), respectively. Their structure is similar to the kinetic case. The 'force' $\mathbf{d}(\mathbf{r})$ and field $\boldsymbol{\mathcal{D}}(\mathbf{r})$ exhibit shell structure, they both vanish at the origin, the 'force' decays asymptotically, whereas the field is singular in that region. There is no direct contribution of this field to the energy, however, its quantal source $\rho(\mathbf{r})$ is the source for the Hartree field $\boldsymbol{\mathcal{E}}_{\mathrm{H}}(\mathbf{r})$, and contributes to the energy through every contribution of the other energy components such as the electron-interaction E_{ee} (5.26), the kinetic T (5.32), the electrostatic E_{es} (5.46) and magnetostatic E_m (5.45) energies.

(As noted in the previous subsection, the singularity in the field $\boldsymbol{\mathcal{D}}(\mathbf{r})$ is canceled by the singularity in the kinetic field $\boldsymbol{\mathcal{Z}}(\mathbf{r})$.)

5.3.4 *Lorentz, Internal Magnetic, and External Electrostatic*

The quantal source for the Lorentz and internal magnetic 'forces' $(\boldsymbol{\ell}(\mathbf{r}), \boldsymbol{i}_m(\mathbf{r}))$ and fields $(\boldsymbol{\mathcal{L}}(\mathbf{r}), \boldsymbol{\mathcal{I}}_m(\mathbf{r}))$ is the physical current density $\mathbf{j}(\mathbf{r})$. The Lorentz field $\boldsymbol{\mathcal{L}}(\mathbf{r})$ is defined as

$$\boldsymbol{\mathcal{L}}(\mathbf{r}) = \boldsymbol{\ell}(\mathbf{r})/\rho(\mathbf{r}), \tag{5.35}$$

where

$$\boldsymbol{\ell}(\mathbf{r}) = \mathbf{j}(\mathbf{r}) \times \boldsymbol{\mathcal{B}}(\mathbf{r}), \tag{5.36}$$

or in Cartesian coordinates

$$\ell_\alpha(\mathbf{r}) = \sum_\beta \Big[j_\beta(\mathbf{r})\nabla_\alpha A_\beta(\mathbf{r}) - j_\beta(\mathbf{r})\nabla_\beta A_\alpha(\mathbf{r}) \Big]. \tag{5.37}$$

The internal magnetic field $\boldsymbol{\mathcal{I}}_m(\mathbf{r})$ is defined as

$$\boldsymbol{\mathcal{I}}_m(\mathbf{r}) = \boldsymbol{i}_m(\mathbf{r})/\rho(\mathbf{r}), \tag{5.38}$$

where in Cartesian coordinates

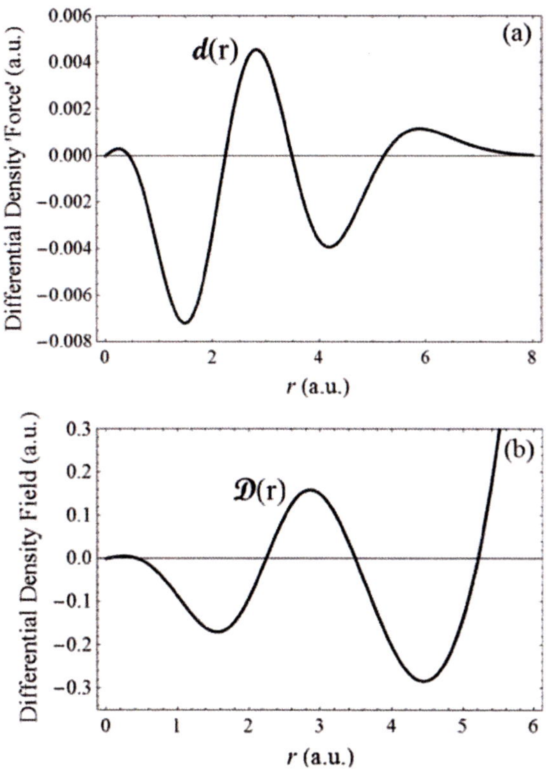

Fig. 5.16 The differential density (**a**) 'force' $\mathbf{d}(r)$, and (**b**) field $\mathcal{D}(r)$

$$i_{m,\alpha}(\mathbf{r}) = \sum_{\beta} \nabla_\beta I_{\alpha\beta}(\mathbf{r}), \tag{5.39}$$

and where the second-rank tensor

$$I_{\alpha\beta}(\mathbf{r}) = \left[j_\alpha(\mathbf{r}) A_\beta(\mathbf{r}) + j_\beta(\mathbf{r}) A_\alpha(\mathbf{r}) \right] - \rho(\mathbf{r}) A_\alpha(\mathbf{r}) A_\beta(\mathbf{r}). \tag{5.40}$$

As the physical current density $\mathbf{j}(\mathbf{r})$ is composed of its paramagnetic $\mathbf{j}_p(\mathbf{r})$, diamagnetic $\mathbf{j}_d(\mathbf{r})$, and magnetization $\mathbf{j}_m(\mathbf{r})$ components, it is then possible to determine the contributions of each component to the Lorentz $\boldsymbol{\ell}(\mathbf{r})$ and internal magnetic $\boldsymbol{i}_m(\mathbf{r})$ 'forces' and to the Lorentz $\boldsymbol{\mathcal{L}}(\mathbf{r})$ and internal magnetic $\boldsymbol{\mathcal{I}}_m(\mathbf{r})$ fields. Thus in Figs. 5.17, 5.18 and 5.19 panel (a) the corresponding Lorentz $\boldsymbol{\ell}_p(\mathbf{r})$, $\boldsymbol{\ell}_d(\mathbf{r})$, $\boldsymbol{\ell}_m(\mathbf{r})$, and internal magnetic $\boldsymbol{i}_{m,p}(\mathbf{r})$, $\boldsymbol{i}_{m,d}(\mathbf{r})$, and $\boldsymbol{i}_{m,m}(\mathbf{r})$ 'forces' are plotted for a Larmor frequency of $\omega_L = 0.1$. These 'forces' are all distinct and of the same order of magnitude. All the 'forces' vanish at the origin, exhibit shell structure, and vanish asymptotically. Observe, in particular, (see Fig. 5.19(a)), that the magnetization contribution of each electron, with one in the ground and the other in an excited state, to the Lorentz 'force' is about the same. Similarly, the contribution of each electron to the internal magnetic 'force' is also about the same. Hence, one may conclude that the contribution to these 'forces' due to the spin moment of each electron in the different shells is essentially the same. In panel (b) of these figures, the corresponding Lorentz $\mathcal{L}_p(\mathbf{r})$, $\mathcal{L}_d(\mathbf{r})$, $\mathcal{L}_m(\mathbf{r})$, and internal magnetic $\mathcal{I}_{m,p}(\mathbf{r})$, $\mathcal{I}_{m,d}(\mathbf{r})$, $\mathcal{I}_{m,m}(\mathbf{r})$ fields are plotted. The structure of each of these fields which vanish at the origin, are all different. Each field is also conservative. Observe, that in spite of each field being singular asymptotically, each separate sum $-[\mathcal{L}_p(\mathbf{r})+\mathcal{I}_{m,p}(\mathbf{r})]$, $-[\mathcal{L}_d(\mathbf{r})+\mathcal{I}_{m,d}(\mathbf{r})]$, and $-[\mathcal{L}_m(\mathbf{r})+\mathcal{I}_{m,m}(\mathbf{r})]$ results in a *linear* function. These linear functions too are plotted. Coincidentally, for the diamagnetic contribution (see Fig. 5.18(b), the sum $-[\mathcal{L}_d(\mathbf{r})+\mathcal{I}_{m,d}(\mathbf{r})]$ is equal to $\mathcal{I}_{m,d}(\mathbf{r})$.

The total Lorentz and internal magnetic 'forces $[\boldsymbol{\ell}(\mathbf{r}), \boldsymbol{i}_m(\mathbf{r})]$, and the corresponding Lorentz and internal magnetic fields $[\mathcal{L}(\mathbf{r}), \mathcal{I}_m(\mathbf{r})]$ are plotted in Fig. 5.20 for the value of the Larmor frequency of $\omega_L = 0.1$. Again, observe that these properties exhibit shell structure. The 'forces' Fig. 5.20(a) vanish at the origin and asymptotically in the classically forbidden region. The fields Fig. 5.20(b) vanish at the origin, but are singular asymptotically. On defining the field $\mathcal{M}(\mathbf{r})$ as being the sum

$$\mathcal{M}(\mathbf{r}) = -[\mathcal{L}(\mathbf{r})+\mathcal{I}_m(\mathbf{r})], \tag{5.41}$$

and based on the analysis of the individual components (Figs. 5.17(b), 5.18(b), 5.19(b)), it is observed that the field $\mathcal{M}(\mathbf{r})$ is a linear function as must be the case:

$$\mathcal{M}(\mathbf{r}) = -\omega_L^2 r\hat{\mathbf{i}}_r. \tag{5.42}$$

This linear function too is plotted in Fig. 5.20(b).

Also note that the field $\mathcal{M}(\mathbf{r})$ is conservative, i.e. $\nabla \times \mathcal{M}(\mathbf{r}) = 0$. Thus, one can define a *path-independent scalar* magnetic potential $v_m(\mathbf{r})$ such that (in general)

$$\mathcal{M}(\mathbf{r}) = -\nabla v_m(\mathbf{r})/e. \tag{5.43}$$

It follows from (5.42) and (5.43) that in (a.u.)*

$$v_m(\mathbf{r}) = \frac{1}{2}\omega_L^2 r^2. \tag{5.44}$$

Hence, the effect of the external magnetostatic field $\mathcal{B}(\mathbf{r})$ on the electronic system is that the magnetic potential energy of each electron $v_m(\mathbf{r})$ is harmonic.

As a consequence of (5.43), the contribution to the energy E_{mag} of the sum of the Lorentz and internal magnetic fields $\mathcal{M}(\mathbf{r})$ is

$$E_{\text{mag}} = \int \rho(\mathbf{r}) v_m d\mathbf{r}. \tag{5.45}$$

In a similar manner, as the external electrostatic field $\mathcal{E}(\mathbf{r}) = -\nabla v(\mathbf{r})/e$ is curl free, the contribution to the energy E_{es} due to this field is

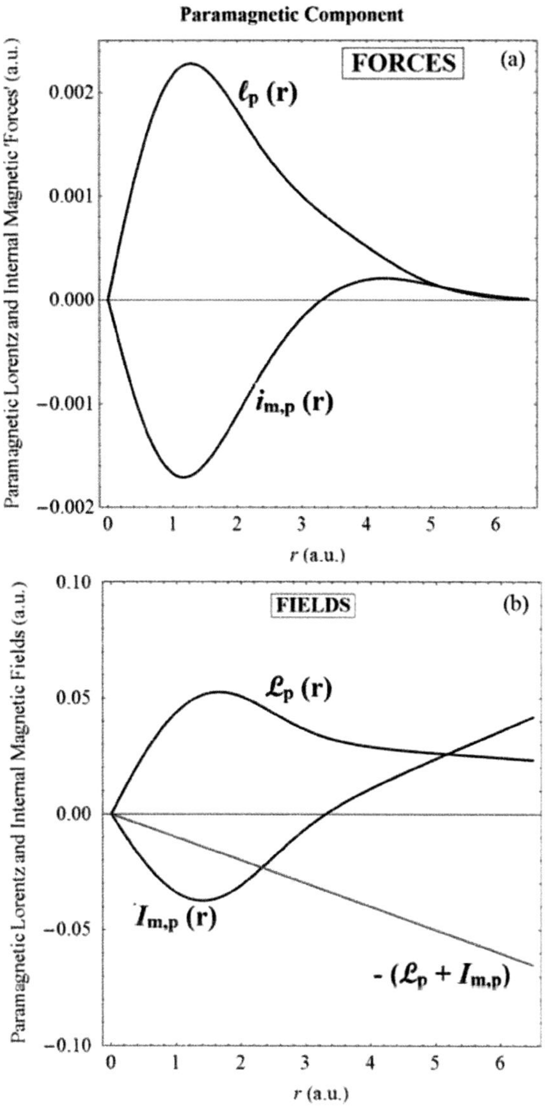

Fig. 5.17 The contribution of the paramagnetic current density $\mathbf{j}_p(\mathbf{r})$ to the Lorentz and internal magnetic (**a**) 'forces' $(\boldsymbol{\ell}_p(r), \boldsymbol{i}_m(r))$, and (**b**) fields $(\mathcal{L}_p(r), \mathcal{I}_{m,p}(r))$. The linear function $-(\mathcal{L}_p + \mathcal{I}_{m,p})$ is also plotted

$$E_{es} = \int \rho(\mathbf{r})v(\mathbf{r})d\mathbf{r}. \tag{5.46}$$

Thus, the sum of the electrostatic E_{es} and magnetostatic E_{mag} energies is

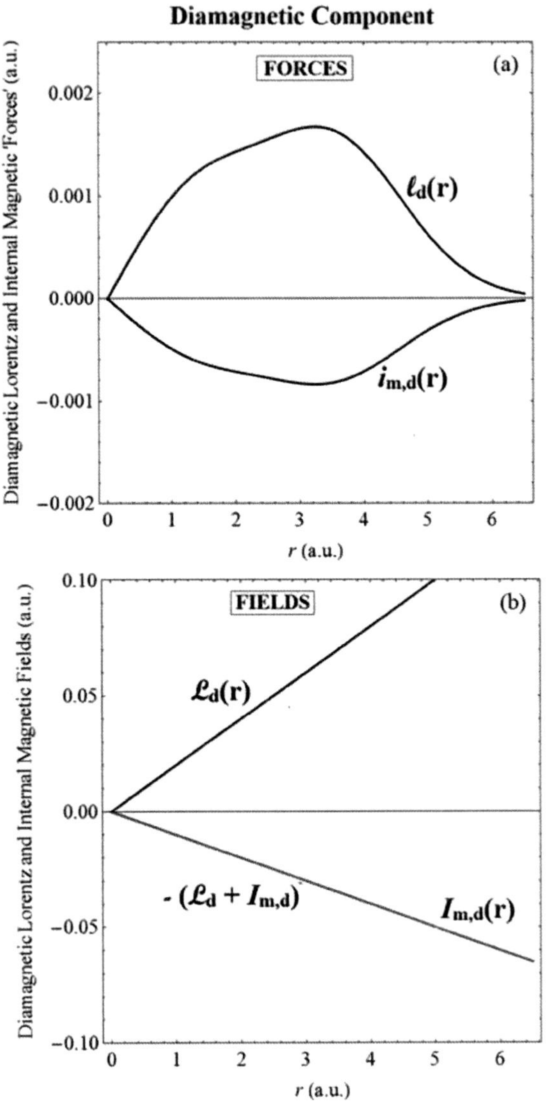

Fig. 5.18 The contribution of the diamagnetic current density $\mathbf{j}_d(\mathbf{r})$ to the Lorentz and internal magnetic (**a**) 'forces' $\boldsymbol{\ell}_d(r), \boldsymbol{i}_{m,d}(r)$ and (**b**) fields $(\mathcal{L}_d(r), \mathcal{I}_{m,d}(r))$. The linear function $-(\mathcal{L}_d + \mathcal{I}_{m,d})$ is equivalent to the function $\mathcal{I}_{m,d}(r)$

$$E_{es} + E_{\text{mag}} = \int \rho(\mathbf{r})\left\{\frac{1}{2}\left[\omega_0^2 + \omega_L^2\right]\right\} d\mathbf{r} \tag{5.47}$$

$$= \int \rho(\mathbf{r})\left[\frac{1}{2} k_{\text{eff}} r^2\right] d\mathbf{r}. \tag{5.48}$$

where $k_{\text{eff}} = \omega_0^2 + \omega_L^2 = 0.072217$ (see Sect. 5.1). The value of this sum of energies is given in Table 5.1.

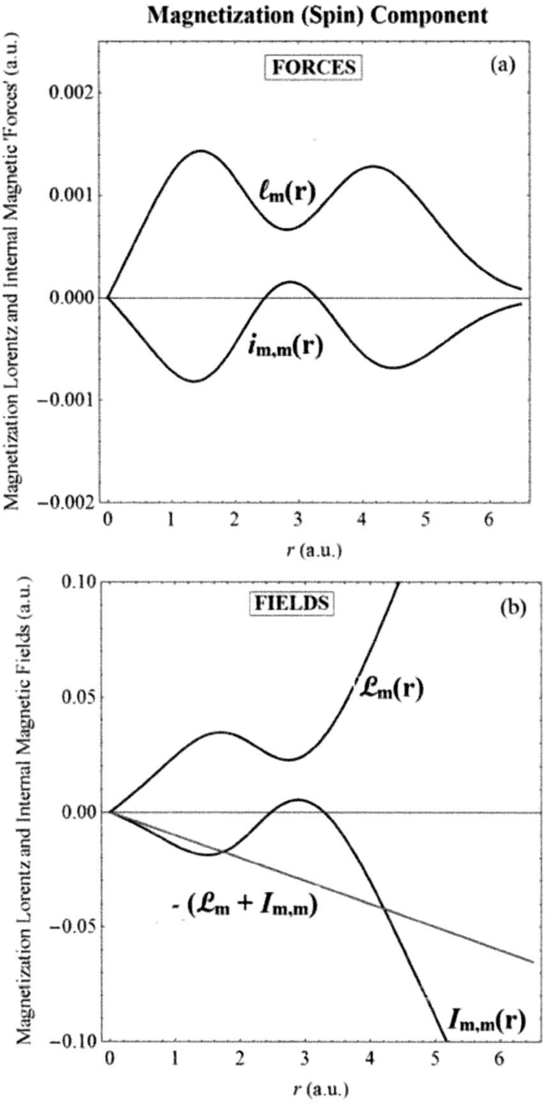

Fig. 5.19 The contribution of the magnetization current density $\mathbf{j}_m(\mathbf{r})$ to the Lorentz and internal magnetic (**a**) 'forces' $(\ell_m(r), i_{m,m}(r))$ and (**b**) fields $(\mathcal{L}_m(r), \mathcal{I}_{m,m}(r))$. The linear function $-(\mathcal{L}_m + \mathcal{I}_{m,m})$ is also plotted

5.4 Total Energy E and Ionization Potential IP

The total energy E of the quantum dot in its triplet 2^3S state is then the sum:

$$E = T + E_H + E_{xc} + E_{es} + E_{\text{mag}} = 1.612391(\text{a.u.})^*. \tag{5.49}$$

This value is consistent with the eigenvalue of the triplet state obtained by solution of the Schrödinger-Pauli equation (1) (see Chap. 9).

The corresponding ionization potential IP is defined as the negative of the energy difference between that of the quantum dot in its triplet state ($E^{N=2}$ of (5.49)) and the ion in its ground state ($E^{N=1}$):

$$IP = E^{N=1} - E^{N=2}, \tag{5.50}$$

where the energy of the ion is $E^{N=1} = \Omega(n+1); n = 0$. Note that the same effective frequency $\Omega = \sqrt{k_{\text{eff}}} = 0.2687322$ is employed in both terms. The value of this ionization potential is quoted in Table 5.1.

5.5 Expectation Values of Single-Particle Operators

In Table 5.1, the expectation values of the single-particle operators $\{\sum_i r_i^n; n = -1, 1, 2 \text{ and } \delta(\mathbf{r})\}$ for the quantum dot in its triplet state are also quoted. In terms of the density $\rho(\mathbf{r})$, the expectation

$$\langle\psi|\sum_i r_i^n|\psi\rangle = \int \rho(\mathbf{r}) r^n d\mathbf{r}. \tag{5.51}$$

These expectation values emphasize different regions of the quantum dot. They are also related to various properties of the 'artificial atom'. The expectation $\langle\delta(\mathbf{r})\rangle$ samples the deep interior of the dot, and is the value of the electron density $\rho(0)$ at the origin; the expectation $\langle r^2\rangle$ samples the far exterior of the dot, and is required for the determination of the diamagnetic susceptibility; the expectation $\langle r\rangle$ is the average size of the 'atom'. (For natural atoms, the expectation $\langle 1/r\rangle$ is required for the determination of the nuclear magnetic shielding constant; the electric field gradient, the magnetic dipole interaction, and spin-orbit coupling depends on the expectation $\langle 1/r^3\rangle$.)

5.6 Satisfaction of the 'Quantal Newtonian' First Law

The fields experienced by each electron must satisfy the 'Quantal Newtonian' First Law. According to the law, each electron experiences an external $\mathcal{F}^{\text{ext}}(\mathbf{r})$ and an internal $\mathcal{F}^{\text{int}}(\mathbf{r})$ field, the sum of which vanish:

$$\mathcal{F}^{\text{ext}}(\mathbf{r}) + \mathcal{F}^{\text{int}}(\mathbf{r}) = 0. \tag{5.52}$$

The external field is the sum of the binding electrostatic $\mathcal{E}(\mathbf{r})$ and Lorentz $\mathcal{L}(\mathbf{r})$ fields:

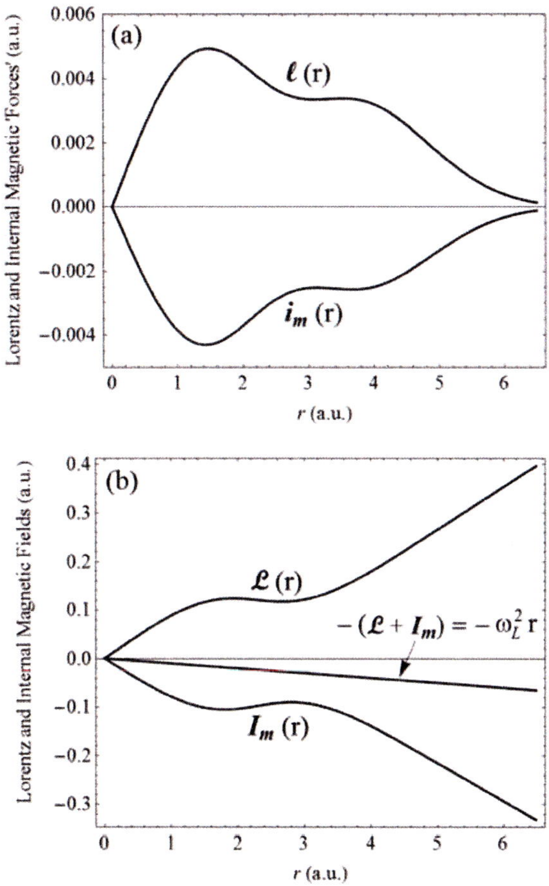

Fig. 5.20 The Lorentz and internal magnetic (**a**) 'forces' $(\boldsymbol{\ell}(r), \mathbf{i}_m(r))$, and (**b**) fields $(\mathcal{L}(r), \mathcal{I}_m(r))$. The linear function $\mathcal{M}(r) = -[\mathcal{L}(r) + \mathcal{I}_m(r)] = -\omega_L^2 r$ is also plotted

$$\mathcal{F}^{\text{ext}}(\mathbf{r}) = \mathcal{E}(\mathbf{r}) - \mathcal{L}(\mathbf{r}). \tag{5.53}$$

The internal field is a sum of the electron-interaction $\mathcal{E}_{ee}(\mathbf{r})$, kinetic $\mathcal{Z}(\mathbf{r})$, differential density $\mathcal{D}(\mathbf{r})$, and internal magnetic $\mathcal{I}_m(\mathbf{r})$ fields:

$$\mathcal{F}^{\text{int}}(\mathbf{r}) = \mathcal{E}_{ee}(\mathbf{r}) - \mathcal{Z}(\mathbf{r}) - \mathcal{D}(\mathbf{r}) - \mathcal{I}_m(\mathbf{r}). \tag{5.54}$$

The satisfaction of the First Law may be exhibited in two ways. The first is in terms of the effective force constant $k_{\text{eff}} = \omega_0^2 + \omega_L^2$ for which the law may be written as

$$-k_{\text{eff}} r = -[\omega_0^2 + \omega_L^2]r = -\mathcal{E}_{ee}(\mathbf{r}) + \mathcal{D}(\mathbf{r}) + \mathcal{Z}(\mathbf{r}). \tag{5.55}$$

In Fig. 5.21 the fields $\mathcal{E}_{ee}(\mathbf{r})$, $\mathcal{D}(\mathbf{r})$, and $\mathcal{Z}(\mathbf{r})$ are plotted. On summing these disparate fields according to (5.55) one obtains the linear function $-k_{\text{eff}} r$ where

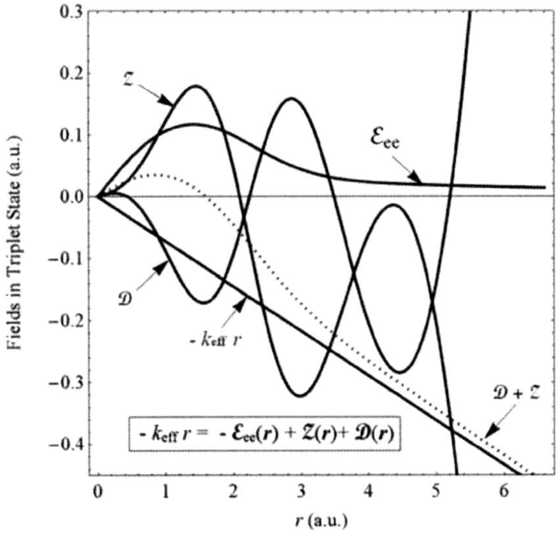

Fig. 5.21 The electron-interaction $\mathcal{E}_{ee}(r)$; kinetic $\mathcal{Z}(r)$; and differential density $\mathcal{D}(r)$ fields. Also plotted are the sum $\mathcal{D}(r) + \mathcal{Z}(r)$, and $-k_{\text{eff}} r$

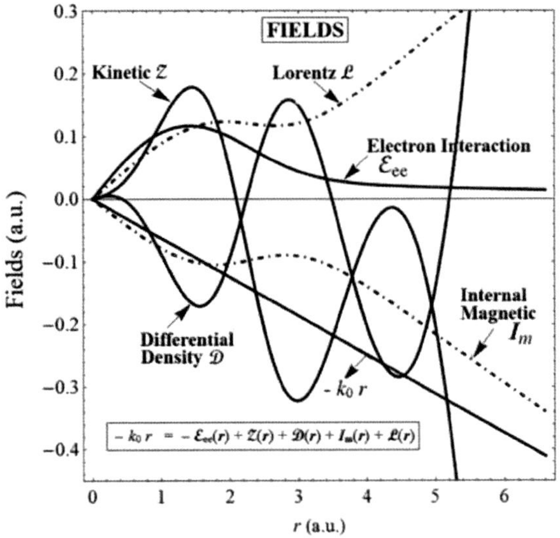

Fig. 5.22 The fields experienced by each electron: electron-interaction $\mathcal{E}_{ee}(r)$; kinetic $\mathcal{Z}(r)$; differential density $\mathcal{D}(r)$; Lorentz $\mathcal{L}(r)$; and internal magnetic $\mathcal{I}_m(r)$. The fields $\mathcal{L}(r)$ and $\mathcal{I}_m(r)$ are plotted for a value of the Larmor frequency of $\omega_L = 0.1$. Also plotted is the function $-k_0 r$

$k_{\text{eff}} = 0.072217$. This proves the law to be satisfied. Observe that the singularities in the differential density $\mathcal{D}(\mathbf{r})$ and kinetic $\mathcal{Z}(\mathbf{r})$ fields cancel so that their sum (see plot of $\mathcal{D}(\mathbf{r}) + \mathcal{Z}(\mathbf{r})$ in Fig. 5.21) approaches $-k_{\text{eff}} r$ asymptotically as the electron-interaction field $\mathcal{E}_{ee}(\mathbf{r})$ vanishes. (Note that the value of k_{eff} is fixed, and independent of the individual value of ω_0 and ω_L (see (5.55)).) (There are an infinite number of set of values of ω_0 and ω_L that lead to the same value of k_{eff}.)

The second way to demonstrate the 'Quantal Newtonian' First Law is in terms of the binding force constant $k_0 = \omega_0^2$. The corresponding equation of motion is

$$-k_0 r = -\mathcal{E}_{ee}(\mathbf{r}) + \mathcal{D}(\mathbf{r}) + \mathcal{Z}(\mathbf{r}) + \mathcal{I}_m(\mathbf{r}) + \mathcal{L}(\mathbf{r}). \quad (5.56)$$

In Fig. 5.22, in addition to the fields $\mathcal{E}_{ee}(\mathbf{r})$, $\mathcal{D}(\mathbf{r})$, and $\mathcal{Z}(\mathbf{r})$ we also plot the Lorentz $\mathcal{L}(\mathbf{r})$ and the internal magnetic field $\mathcal{I}_m(\mathbf{r})$ for a value of the Larmor frequency $\omega_L = 0.1$. On summing the disparate fields on the right hand side of (5.56), one obtains the linear function $-k_0 r$ with $k_0 = 0.062217$, as must be the case. Thus, all the fields experienced by each electron are such that they satisfy the 'Quantal Newtonian' First Law.

5.7 Self-Consistent Nature of the Schrödinger-Pauli Equation

The example of the triplet state of the quantum dot in a magnetic field can be employed to explain the intrinsic self-consistent nature of the Schrödinger-Pauli equation. In particular, to see how the external binding potential $v(\mathbf{r})$ is obtained self-consistently. Although a self-consistent calculation is not performed, the process of how one would proceed, and what the final iteration would comprise of, is explained.

Consider the Schrödinger-Pauli equation for the quantum dot written in its generalized form (in effective atomic units)

$$\hat{H}[\psi]\psi = E[\psi]\psi, \quad (5.57)$$

where the Hamiltonian $\hat{H}[\psi]$ with *unknown* binding potential $v[\psi](\mathbf{r})$ is

$$\hat{H}[\psi] = \frac{1}{2}\sum_{k=1}^{2}\big(\mathbf{p}_k + \mathbf{A}(\mathbf{r}_k)\big)^2 + \sum_{k=1}^{2}\mathcal{B}(\mathbf{r}_k)\cdot\mathbf{s}_k + \frac{1}{2}\sum_{k,\ell=1}^{2}{}' \frac{1}{|\mathbf{r}_k - \mathbf{r}_\ell|} + \sum_{k=1}^{2} v[\psi](\mathbf{r}_k), \quad (5.58)$$

with

$$v[\psi](\mathbf{r}) = \int_\infty^{\mathbf{r}} \big[\mathcal{E}_{ee}(\mathbf{r}') - \mathcal{D}(\mathbf{r}') - \mathcal{Z}(\mathbf{r}') - \mathcal{L}(\mathbf{r}') - \mathcal{I}_m(\mathbf{r}')\big]\cdot d\boldsymbol{\ell}'. \quad (5.59)$$

(The above equations are written in terms of the spatial part $\psi(\mathbf{r}_1\mathbf{r}_2)$ of the wave function.) In the symmetric gauge $\mathbf{A}(\mathbf{r}) = \frac{1}{2}\mathcal{B}(\mathbf{r}) \times \mathbf{r}$; $\mathcal{B}(\mathbf{r}) = \mathcal{B}\mathbf{i}_z$, with the assumption of cylindrical symmetry, let us assume the form of a trial input wave function to be

$$\psi(\mathbf{r}_1\mathbf{r}_2) = N e^{im\theta} e^{-\Omega(R^2+\frac{1}{4}u^2)}\big[u + c_2u^2 + c_3u^3 + c_4u^4\big] \quad (5.60)$$

where $\mathbf{r} = (\mathbf{r}_1 + \mathbf{r}_2)/2$; $\mathbf{u} = |\mathbf{r}_2 - \mathbf{r}_1|$; the angular momentum quantum number $m = 1$; the Larmor frequency $\omega_L = 0.1$; and $N, \Omega \equiv \sqrt{k_{\mathrm{eff}}}, c_2, c_3, c_4$ are *unknown* constants. The trial wave function is antisymmetric in an interchange of the vectors $\mathbf{r}_1$ and $\mathbf{r}_2$ as a result of the phase factor $e^{im\theta}$. As a consequence of cylindrical symmetry, let us assume all the individual fields are conservative.

For an assumed choice of the values of the constants, employ the approximate input $\psi(\mathbf{r}_1\mathbf{r}_2)$ to determine the various fields, and from them the potential $v[\psi](\mathbf{r})$. Substitute this $v[\psi](\mathbf{r})$ into (5.57) to solve for $\psi(\mathbf{r}_1\mathbf{r}_2)$ with new coefficients, and repeat the iterative procedure. At each iteration one also obtains the corresponding $E[\psi]$. (*We reiterate that what is meant by the functional* $v[\psi](\mathbf{r})$ *is that for each different* ψ, *one obtains a different function* $v[\psi](\mathbf{r})$.)

Suppose at the end of a particular iteration, the values of the coefficients turn out to be $N = 0.022466$, $c_2 = 0.33333$, $c_3 = -0.059108$, $c_4 = -0.015884$; $\Omega^2 \equiv k_{\mathrm{eff}} = 0.072217$. On substituting the $v(\mathbf{r})$ of (5.59) into the Schrödinger-Pauli equation (5.57) and solving, one obtains a wave function $\psi(\mathbf{r}_1\mathbf{r}_2)$ with the same coefficients. Thus, this constitutes the final iteration of the self-consistent procedure, thereby leading to the exact wave function $\psi(\mathbf{r}_1\mathbf{r}_2)$ and energy E. Hence, via the self-consistency procedure one obtains the Hamiltonian, the wave function, and eigen energy.

An examination of the final result of the sum of $-\boldsymbol{\mathcal{E}}_{\mathrm{ee}}(\mathbf{r})$, $\boldsymbol{\mathcal{D}}(\mathbf{r})$, and $\boldsymbol{\mathcal{Z}}(\mathbf{r})$ fields will show it to be a linear function of slope $-k_{\mathrm{eff}} = -0.072217$ (see Fig. 5.21), as follows

$$-\boldsymbol{\mathcal{E}}_{\mathrm{ee}}(\mathbf{r}) + \boldsymbol{\mathcal{D}}(\mathbf{r}) + \boldsymbol{\mathcal{Z}}(\mathbf{r}) = -k_{\mathrm{eff}} r. \tag{5.61}$$

As the individual fields are conservative, the sum of the magnetic fields $\boldsymbol{\mathcal{M}}(\mathbf{r}) = -[\boldsymbol{\mathcal{L}}(\mathbf{r}) + \boldsymbol{\mathcal{I}}_m(\mathbf{r})]$ is such that $\nabla \times \boldsymbol{\mathcal{M}}(\mathbf{r}) = 0$. Then these fields can be associated with a magnetic scalar potential $v_m(\mathbf{r})$ through

$$v_m[\psi](\mathbf{r}) = -\int_{\infty}^{\mathbf{r}} \boldsymbol{\mathcal{M}}(\mathbf{r}') \cdot d\boldsymbol{\ell}'. \tag{5.62}$$

As such, (5.59) can be rearranged to read

$$\begin{aligned} v[\psi](\mathbf{r}) + v_m[\psi](\mathbf{r}) &\equiv v_{\mathrm{eff}}[\psi](\mathbf{r}) \\ &= \int_{\infty}^{\mathbf{r}} \left[\boldsymbol{\mathcal{E}}_{\mathrm{ee}}(\mathbf{r}') - \boldsymbol{\mathcal{D}}(\mathbf{r}') - \boldsymbol{\mathcal{Z}}(\mathbf{r}')\right] \cdot d\boldsymbol{\ell}'. \end{aligned} \tag{5.63}$$

On substituting (5.61) in (5.63) the effective potential $v_{\mathrm{eff}}(\mathbf{r})$ turns out to be harmonic as

$$v_{\mathrm{eff}}(\mathbf{r}) = \frac{1}{2} k_{\mathrm{eff}} r^2, \tag{5.64}$$

with k_{eff} as given above.

A further examination of the final result of the field $\boldsymbol{\mathcal{M}}(\mathbf{r})$ (see Fig. 5.20) shows that the corresponding magnetic scalar potential $v_m(\mathbf{r})$ is also harmonic with Larmor frequency $\omega_L = 0.1$ as

$$v_m(\mathbf{r}) = \frac{1}{2}\omega_L^2 r^2. \quad (5.65)$$

Consequently, since both $v_{\text{eff}}(\mathbf{r})$ and $v_m(\mathbf{r})$ are harmonic, their difference which is the binding external electrostatic potential $v(\mathbf{r})$, must also be harmonic, with some angular frequency ω_0 as

$$v(\mathbf{r}) = \frac{1}{2}\omega_0^2 r^2, \quad (5.66)$$

where $\omega_0^2 = k_{\text{eff}} - \omega_L^2 = 0.062217$. Thus, via the self-consistent procedure, the unknown external potential $v(\mathbf{r})$ is determined to be harmonic. (Note that for a different value of the magnetic field $\boldsymbol{\mathcal{B}}(\mathbf{r})$ or equivalently Larmor frequency ω_L, one would also obtain a $v(\mathbf{r})$ that is harmonic, but with a different binding angular frequency ω_0. However, the value of the effective force constant k_{eff} will remain unchanged.)

Yet another way to understand how the self-consistency procedure leads to the binding potential is as follows. Following the final iteration, add up all the individual fields as on the right hand side of (5.59). Then one obtains the linear function $-k_0 r$ (see Fig. 5.22) with $k_0 = 0.24943$. This then leads to the binding potential $v(\mathbf{r})$ of (5.66).

References

1. E. Schrödinger, Ann. Physik **79**, 362 (1926); ibid. **79**, 489 (1926)
2. W. Pauli, Z. Physik **43**, 601 (1927)
3. V. Sahni, Int. J. Quantum Chem. **121**, e26556 (2021)
4. V. Sahni, X.-Y. Pan, Computation **5**, 15 (2017). https://doi.org/10.3390/computation5010015
5. V. Sahni, J. Comp. Chem. **39**, 1083–1089; https://doi.org/10.1002/jcc24888
6. M. Slamet, V. Sahni, Chem. Phys. **546**, 111073 (2021)
7. R.C. Ashoori, H.L. Stormer, J.S. Weiner, L.N. Pfeiffer, S.J. Pearton, K.W. Baldwin, K.W. West, Phys. Rev. Lett. **68**, 3088–3091 (1992)
8. R.C. Ashoori, Nature **379**, 413–419 (1996)
9. S.M. Reimann, M. Manninen, Rev. Mod. Phys. **74**, 1283–1342 (2002)
10. H. Saarikoski, S.M. Reimann, A. Harju, M. Manninen, Rev. Mod. Phys. **82**, 2785 (2010)
11. A. Kumar, S.E. Laux, F. Stern, Phys. Rev. B **42**, 5166–5175 (1990)
12. T.M. Henderson, K. Runge, R.J. Bartlett, Chem. Phys. Lett. **337**, 138–142 (2001)
13. T.M. Henderson, K. Runge, R.J. Bartlett, Phys. Rev. B **67**, 045320 (2003)
14. F. Pederiva, C.J. Umrigar, E. Lipparini, Phys. Rev. B **62**, 8120 (2000); Erratum, Phys. Rev. B **68**, 089901 (2003). https://doi.org/10.1103/PhysRevB.68.089901
15. M. Dineykhan, R.G. Nazmitdinov, Phys. Rev. B **55**, 13707–13714 (1997)
16. J.-L. Zhu, Z.-Q. Li, J.-Z. Yu, K. Ohno, Y. Kawazoe, Phys. Rev. B **55**, 15819–15823 (1997)
17. C. Yannouleas, U. Landman, Phys. Rev. Lett. **85**, 1726–1729 (2000)
18. R.G. Nazmitdinov, N.S. Simonovic, J.M. Rost, Phys. Rev. B **65**, 155307 (2002)
19. X. Lopez, J.M. Ugalde, L. Echevarría, E.V. Ludeña, Phys. Rev. A **74**, 042504 (2006)
20. N.S. Simonovic, R.G. Nazmitdinov, Phys. Rev. A **92**, 052322 (2015)
21. T. Yang, X.-Y. Pan, V. Sahni, Phys. Rev. A **83**, 042518 (2011)
22. D. Achan, L. Massa, V. Sahni, Phys. Rev. A **90**, 022502 (2014)

23. D. Achan, L. Massa, V. Sahni, Comp. Theor. Chem. **1035**, 14–18 (2014)
24. M. Slamet, V. Sahni, Comp. Theor. Chem. **1114**, 125–139 (2017)
25. M. Slamet, V. Sahni, Comp. Theor. Chem. **1138**, 140–157 (2018)
26. X.-Y. Pan, V. Sahni, J. Chem. Phys. **119**, 7083–7086 (2003)
27. W.A. Bingel, Z. Naturforsch. **18a**, 1249–1253 (1963)
28. R.T. Pack, W.B. Brown, J. Chem. Phys. **45**, 556–559 (1966)
29. W.A. Bingel, Theoret. Chim. Acta. (Berl) **8**, 54–61 (1967)
30. M. Taut, J. Phys. A **27**, 1045–1055 (1994); Corrigenda J. Phys. A **27**, 4723–4724 (1994)
31. M. Taut, J. Phys. Condens. Matter **12**, 3689–3710 (2000)
32. M. Taut, H. Eschrig, Z. Phys. Chem. **224**, 631–649 (2010)
33. M. Slamet, V. Sahni, Chem. Phys. **556**, 111453 (2022)
34. V. Sahni, J.B. Krieger, Phys. Rev. A. **11**, 409–417 (1975)
35. V. Sahni, J.B. Krieger, J. Gruenebaum, Phys. Rev. A **12**, 768–775 (1975)
36. M. Slamet, V. Sahni, Int. J. Quantum Chem. **119**, e25818 (2019). https://doi.org/10.1002/qua.25818

Chapter 6
Quantal Density Functional Theory: A Local Effective Potential Theory Complement to Schrödinger Theory

Abstract Quantal density functional theory (Q-DFT) is a *local effective potential* description of the Schrödinger theory of electrons. In such a description, the Hamiltonian is written such that the electron correlations due to the Pauli principle and Coulomb repulsion are incorporated into a local (multiplicative) potential energy term. This potential also includes the contributions of these correlations to the kinetic energy—the Correlation-Kinetic component. The energy is obtained from the properties of the model system. Stationary-state Q-DFT is a description of the mapping from a nondegenerate or degenerate ground or excited state of a system of electrons in a static electromagnetic field to one of noninteracting fermions experiencing the same external field and possessing the same electronic density $\rho(\mathbf{r})$ and physical current density $\mathbf{j}(\mathbf{r})$. The state of the model system is arbitrary. The mapping is in terms of 'classical' fields: an Electron-Interaction field representative of correlations due to the Pauli principle and Coulomb repulsion (which may be decomposed into its Hartree, Pauli and Coulomb components), and one descriptive of Correlation-Kinetic effects. The sources of these fields are quantum-mechanical expectation values of Hermitian operators. The local potential in which all these many-body effects are ensconced has a rigorous physical interpretation: It is the work done in the force of a conservative effective field which is the sum of the Electron-Interaction and Correlation-Kinetic fields. Within Q-DFT, the *separate* contributions to the energy due to the Pauli principle and Coulomb repulsion, as well as the Correlation-Kinetic energy are expressed in integral virial form in terms of the field representative of each property. The highest occupied eigenvalue of the differential equation of the model system of fermions, irrespective of its state, is the negative of the Ionization Potential. As such, Q-DFT constitutes a complement to Schrödinger theory. Properties of the model fermionic system arrived at via Q-DFT are discussed: ***(a)*** The non-uniqueness of the local electron-interaction potential in which the many-body correlations are embedded; ***(b)*** The proof that this non-uniqueness is solely due to Correlation-Kinetic effects; ***(c)*** The non-uniqueness of the model system wave function; ***(d)*** That it is solely the Correlation-Kinetic effects which contribute to the discontinuity in the local electron-interaction potential as the electron number passes an integer value. To elucidate Q-DFT, the mapping from two different 2-dimensional 2-electron semiconductor quantum dots one in a ground and the other in its first excited singlet 2^1S state, to one of noninteracting fermions in a ground state possessing the same

V. Sahni, *Schrödinger Theory of Electrons: Complementary Perspectives*,
Springer Tracts in Modern Physics 285,
https://doi.org/10.1007/978-3-030-97409-1_6

corresponding $\{\rho(\mathbf{r}), \mathbf{j}(\mathbf{r})\}$ is described. A brief description of the Q-DFT of the density amplitude is provided. Time-dependent (TD) Q-DFT of a system of electrons in a TD electromagnetic field, mapped to one of noninteracting fermions experiencing the same external field and possessing the same TD density $\rho(\mathbf{y})$ and physical current density $\mathbf{j}(\mathbf{y})$; $\mathbf{y} = \mathbf{r}t$, is described. In a manner similar to the stationary-state case, the local electron-interaction potential is the work done, at each instant of time, in the force of a conservative effective field which is the sum of the TD Electron-Interaction and Correlation-Kinetic fields.

Introduction

This chapter provides a brief description of the basic concepts underlying Quantal Density Functional theory (Q-DFT). Since the original ideas [1, 2] were formulated, there has been significant evolution in the understanding of the theory [3–31]. Two books [32, 33] have been written on the subject: *Quantal Density Functional Theory (2nd Edition)*, and *Quantal Density Functional Theory II: Approximation Methods and Applications*. These books are referred to in the chapter as ***QDFT*** and ***QDFT2***, respectively. Here the key ideas of the theory and the broader understanding it brings to local effective potential theory are presented. Examples to elucidate the theory are provided.

Stationary-State Quantal Density Functional Theory

We begin with a discussion of stationary-state Q-DFT. Q-DFT is a mathematically rigorous physics based description of *local effective potential* theory. The theory constitutes the mapping from an interacting system of electrons in static external electric and magnetic fields as described by Schrödinger or Schrödinger-Pauli theories to one of *noninteracting* fermions possessing the same *basic variables* of the electronic density $\rho(\mathbf{r})$ and physical current density $\mathbf{j}(\mathbf{r})$. (For the case when only an external electrostatic field is present, the mapping is such that the noninteracting fermions possess the same basic variable of the density $\rho(\mathbf{r})$.) *A basic variable of quantum mechanics is a gauge invariant property, knowledge of which determines the wave functions of the system.* The existence of the model fermionic system is an *assumption*. The description of this system is once again in terms of 'classical' fields and their quantal sources. The fields, in turn, obey the corresponding equation of motion or the *'Quantal Newtonian' First Law* for each model fermion. The theory encompasses *ground* and bound *excited* states, both *nondegenerate* and *degenerate*. In this chapter we describe the Q-DFT mapping from a *nondegenerate* ground or excited state of the interacting system.

Consider a system of N *spinless* interacting electrons in a static electromagnetic field : $\boldsymbol{\mathcal{E}}(\mathbf{r}) = -\boldsymbol{\nabla} v(\mathbf{r})/e$; $\boldsymbol{\mathcal{B}}(\mathbf{r}) = \boldsymbol{\nabla} \times \mathbf{A}(\mathbf{r})$, where $\{v(\mathbf{r}), \mathbf{A}(\mathbf{r})\}$ are the scalar and vector potentials . In units where $|e| = \hbar = m = c = 1$ (with the charge of the electron being $-e$), the Hamiltonian $\hat{H}$ is

$$\hat{H} = \hat{T}_A + \hat{V} + \hat{U}, \tag{6.1}$$

where the physical kinetic energy operator $\hat{T}_A$ is

$$\hat{T}_A = \frac{1}{2}\sum_k \left(\hat{\mathbf{p}}_k + \mathbf{A}(\mathbf{r}_k)\right)^2, \tag{6.2}$$

with $\hat{\mathbf{p}} = -i\nabla$, the canonical momentum operator. The electrostatic binding potential operator $\hat{V}$ is

$$\hat{V} = \sum_k v(\mathbf{r}_k), \tag{6.3}$$

and the electron-interaction potential operator $\hat{U}$ is

$$\hat{U} = \frac{1}{2}\sum_{k,\ell}{}' \frac{1}{|\mathbf{r}_k - \mathbf{r}_\ell|}. \tag{6.4}$$

The stationary-state Schrödinger equation is then

$$\hat{H}\Psi(\mathbf{X}) = E\Psi(\mathbf{X}), \tag{6.5}$$

where $\{\Psi(\mathbf{X}), E\}$ are the eigenfunctions and eigenvalues of the system; $\mathbf{X} = \mathbf{x}_1, \mathbf{x}_2, \ldots, \mathbf{x}_N$; $\mathbf{x} = \mathbf{r}\sigma$; $\{\mathbf{r}, \sigma\}$ being the spatial and spin coordinates of the electron. The description of this system of interacting electrons in terms of 'classical' fields and quantal sources is provided in Chap. 3.

The basic idea underlying Q-DFT is the mapping of the interacting system of electrons as defined by the Schrödinger equation (6.5) to one of *noninteracting* fermions possessing the same electronic density $\rho(\mathbf{r})$ and the physical current density $\mathbf{j}(\mathbf{r})$. (The rationale for this choice of gauge invariant properties will be explained below.) As the model fermions are noninteracting, the effective potential energy $v_s(\mathbf{r})$ of *each* such fermion is the *same*. The quantum-mechanical operator representative of this potential energy is thus a *local* or *multiplicative* operator just as is the scalar binding potential energy operator $v(\mathbf{r})$. The corresponding wave function of the model system is then a Slater determinant $\Phi\{\phi_k\}$ of the model fermion orbitals $\phi_k(\mathbf{x})$. The noninteracting fermion model system is referred to as the *S system*, *S* being the mnemonic for 'Slater' determinant.

What the mapping from the interacting to the noninteracting model *S* system then entails is the incorporation of the electron correlations of the former, *viz*. the correlations due to the Pauli principle and Coulomb repulsion, into the local effective potential $v_s(\mathbf{r})$. Further, as a consequence of the Heisenberg uncertainty principle, it is evident that the kinetic energy of the interacting electrons is different from those of the model fermions, both possessing the same density $\rho(\mathbf{r})$ and physical current density $\mathbf{j}(\mathbf{r})$. (To understand this, consider two same sized boxes, one containing N interacting electrons and the other N noninteracting fermions both having the same density $\rho(\mathbf{r})$. Now the effective volume available to the interacting and noninteracting particles is different. As a consequence, their momenta, and hence their kinetic energies are different.) Thus, the S system must additionally account for the difference in the kinetic energy between the interacting and noninteracting systems. The difference in kinetic energies is referred to as the *Correlation-Kinetic* effect. This effect too must then be accounted for by the local effective potential $v_s(\mathbf{r})$. In essence then, one must understand how the *nonlocality* of the interacting system is mapped into the *local* noninteracting model system. The mapping within Q-DFT explicitly accounts for each of these electron correlations. As such, Q-DFT provides a mathematically rigorous physical description of the local effective potential $v_s(\mathbf{r})$ which then generates via the corresponding Schrödinger equation the same $\{\rho(\mathbf{r}), \mathbf{j}(\mathbf{r})\}$ as that of the interacting electronic system.

The model S system Hamiltonian $\hat{H}_s$ is

$$\hat{H}_s = \hat{T}_A + \hat{V}_s = \sum_k \hat{h}_s(\mathbf{r}_k), \tag{6.6}$$

where $\hat{T}_A$ is defined by (6.2), and $\hat{V}_s$ is

$$\hat{V}_s = \sum_k v_s(\mathbf{r}_k). \tag{6.7}$$

Next one makes the *further assumption* that the external fields $\boldsymbol{\mathcal{E}}(\mathbf{r}) = -\boldsymbol{\nabla} v(\mathbf{r})/e$; $\boldsymbol{\mathcal{B}}(\mathbf{r}) = \boldsymbol{\nabla} \times \mathbf{A}(\mathbf{r})$ experienced by each model fermion is the *same* as for the interacting electrons. Consequently, the operator $v_s(\mathbf{r})$ may be written as

$$v_s(\mathbf{r}) = v(\mathbf{r}) + v_{ee}(\mathbf{r}), \tag{6.8}$$

with $v_{ee}(\mathbf{r})$ a *local effective electron-interaction potential* in which all the electron correlations are incorporated. Thus, the single model fermion Hamiltonian $\hat{h}_s(\mathbf{r})$ is

$$\hat{h}_s(\mathbf{r}) = \frac{1}{2}(\mathbf{p} + \mathbf{A}(\mathbf{r}))^2 + v(\mathbf{r}) + v_{ee}(\mathbf{r}), \tag{6.9}$$

and the corresponding S system Schrödinger equation is

$$\left[\frac{1}{2}(\mathbf{p} + \mathbf{A}(\mathbf{r}))^2 + v(\mathbf{r}) + v_{ee}(\mathbf{r})\right]\phi_k(\mathbf{x}) = \epsilon_k \phi_k(\mathbf{x}); \;\; k = 1, \ldots, N. \tag{6.10}$$

The wave function for the model fermions is the Slater determinant $\Phi\{\phi_k\}$ of the orbitals $\phi_k(\mathbf{x})$. The configurational state of the S system is *arbitrary* as discussed further below.

With the above assumption, the Q-DFT mapping corresponds to understanding how the many-body correlations are incorporated into the potential $v_{ee}(\mathbf{r})$. The assumption of existence of the scalar electron-interaction potential $v_{ee}(\mathbf{r})$ implies that there exists a corresponding *conservative effective field* $\boldsymbol{\mathcal{F}}^{\text{eff}}(\mathbf{r})$ such that $\boldsymbol{\mathcal{F}}^{\text{eff}}(\mathbf{r}) = -\boldsymbol{\nabla} v_{ee}(\mathbf{r})$. The effective field $\boldsymbol{\mathcal{F}}^{\text{eff}}(\mathbf{r})$ then fully defines the model S system.

Prior to describing the mapping within Q-DFT, we remark on various aspects of the theory. (These remarks are all valid in the absence of the magnetostatic field, i.e., for $\mathcal{B}(\mathbf{r}) = 0$, which constitutes a special case.)

Attributes of Quantal Density Functional Theory

We begin with attributes of stationary-state Q-DFT.

(a) Via the Q-DFT mapping to the model system possessing the same $\{\rho(\mathbf{r}), \mathbf{j}(\mathbf{r})\}$ as that of the interacting system, it is possible to determine properties of the latter not obtainable by direct solution of the Schrödinger equation (6.5). As described above, such a property is the contribution of electron correlations due to the Pauli principle and Coulomb repulsion to the kinetic energy – the *Correlation-Kinetic* energy T_c. Further, it is possible to obtain the *separate* contributions to the total energy E of the correlations due to the Pauli principle and Coulomb repulsion - the Pauli E_x and Coulomb E_c energies. The solution $\Psi(\mathbf{X})$ of the Schrödinger equation (6.5) accounts for both these correlations, but within Schrödinger theory, these correlations are not separable. (In quantum chemistry, the separation is accomplished in an approximate manner by performing a Hartree-Fock theory [34–36] calculation which then leads to the exchange energy - the contribution E_x^{HF} due to correlations arising from the Pauli principle (See Sect. 3.2.2). The difference between the total interacting system energy E and that of Hartree-Fock theory E^{HF} is referred to as the correlation energy. But the Hartree-Fock theory model, which only accounts for the correlations due to the Pauli principle, differs from the original fully-interacting system as its density $\rho(\mathbf{r})$ and physical current density $\mathbf{j}(\mathbf{r})$ are different. In the Q-DFT definitions, the model S system $\{\rho(\mathbf{r}), \mathbf{j}(\mathbf{r})\}$ are the same as those of the interacting electrons.) Finally, it is also possible to determine the ionization potential (or electron affinity) via Q-DFT. The highest occupied eigenvalue of the S system differential equation (6.10) is the negative of the ionization potential. (It requires two separate energy calculations to determine the ionization potential within Schrödinger theory: one for the charge-neutral and the other for the ionized system. (See the example of Chap. 5).) Thus, since via the mapping, it is possible to determine additional properties of the interacting system, Q-DFT *constitutes a complement to Schrödinger theory*.

(b) Q-DFT can also be viewed as an alternative *complementary* description of the interacting system as defined by the Schrödinger equation (6.5) because the model system possesses by definition the same density and physical current density $\{\rho(\mathbf{r}), \mathbf{j}(\mathbf{r})\}$, and can lead to the same energy E. (The energy E can be expressed in terms of the eigenvalues ϵ_k and the potential $v_{ee}(\mathbf{r})$ of the S system Schrödinger equation (6.10), and the *separate* Hartree E_H, Pauli E_x, Coulomb E_c, and Correlation-Kinetic T_c energies. (See Sect 6.1.5).) Furthermore, Q-DFT is a *local effective potential theory*, and thus the corresponding differential equation (6.10) is more amenable to numerical solution.

(c) The mapping via Q-DFT is mathematically rigorous, and involves the *'Quantal Newtonian' First Law* for the model system. As such the mapping is in terms of fields representative of the different electron correlations that must be accounted for by the S system. This then allows for *ad hoc* approximations based on these correlations. Thus, one could as a first step include only correlations due to the Pauli principle. Then include higher-order correlations such as Coulomb correlations

and Correlation-Kinetic effects. There is also a rigorous Many-body and Pseudo Møller-Plesset perturbation theory within Q-DFT. The lowest-order in the many-body perturbation theory corresponds to the correlations solely due to the Pauli principle. For these and other approximation methods and applications, the reader is referred to ***QDFT2***.

(d) The other *in-principle exact* local effective potential theories such as Kohn-Sham density functional theory (KS-DFT) [37] and the Optimized Potential Method (OPM) [38–40] are strictly *mathematical* in construct. They are *ground state* theories based on the variational principle for the energy [41]. KS-DFT, to be described in Chap. 7, is in terms of an energy functional $E[\rho]$ of the ground state density $\rho(\mathbf{r})$, with all the many-body effects incorporated into an *unknown* KS electron-interaction energy functional $E_{ee}^{KS}[\rho]$. The corresponding *local* electron-interaction potential $v_{ee}^{KS}(\mathbf{r})$ is then defined as its functional derivative $\delta E_{ee}^{KS}[\rho]/\delta\rho(\mathbf{r})$ taken with respect to the density. In the OPM, the local effective potential is obtained via a self-consistent solution of a differential and integral equation. As will be described in the sections to follow, the Q-DFT mapping is *physically* based. It is also mathematically rigorous, and in terms of the *separate* electron correlations that must be accounted for within local effective potential theory. Thus, Q-DFT provides a rigorous physical interpretation of KS-DFT and the OPM (see Chap. 5 of ***QDFT***). Q-DFT also provides a more fundamental understanding of various *ad hoc* approximations to the 'exchange-correlation' energy functional $E_{xc}^{KS}[\rho]$ of KS-DFT such as the Local Density Approximation (LDA). (Subtraction of the known Hartree energy functional $E_H[\rho]$ from $E_{ee}^{KS}[\rho]$ is the functional $E_{xc}^{KS}[\rho]$.) The LDA is the leading term of most approximate energy functionals in the literature. Q-DFT further explains why Slater theory [42, 43], another approximation scheme, is weakly founded. The reason is that the Slater 'exchange-correlation' potential does not represent the potential energy of an electron. (For the physical explanation of these approximations, see Chap. 10 of ***QDFT***).

(e) Interacting systems as represented within Hartree-Fock and Hartree theories can also be mapped via Q-DFT to model noninteracting fermionic systems possessing the same $\{\rho(\mathbf{r}), \mathbf{j}(\mathbf{r})\}$, and from which the corresponding total energies may be obtained. Hence, it is then possible to determine the Correlation-Kinetic effect contributions to the energy in these theories. This contribution is the difference between the Hartree-Fock theory kinetic energy and the kinetic energy of the model noninteracting system. In a similar manner, there is a Correlation-Kinetic contribution to the energy within Hartree theory. (For these mappings see Chap. 3 and ***QDFT***.)

Nonuniqueness of Local Electron-Interaction Potential $v_{ee}(\mathbf{r})$

Consider the case when only the external binding electrostatic field $\boldsymbol{\mathcal{E}}(\mathbf{r}) = -\boldsymbol{\nabla} v(\mathbf{r})/e$ is present, and that this field, and thus the binding potential $v(\mathbf{r})$, is known. Hence, instead of the local effective potential $v_s(\mathbf{r})$, one may focus on the electron-interaction potential $v_{ee}(\mathbf{r})$ instead (see (6.8)). Let us first consider the case of the mapping from a *nondegenerate ground* state of the interacting system. It is then possible to perform a Q-DFT mapping to a model noninteracting fermionic S system which possesses the *same* electronic density $\rho(\mathbf{r})$, and which is also in a *ground* state. There thus exists a corresponding potential function $v_{ee}(\mathbf{r})$. The highest occupied eigenvalue of the S system differential equation (6.10) is the negative of the first ionization potential. The constraint is that the electron number N be the same. However, it is also possible [22] to map to a model S system which is in a bound *excited* state configuration and which also possesses the *same ground state density* $\rho(\mathbf{r})$. There then exists a different potential function $v_{ee}(\mathbf{r})$. The highest occupied eigenvalue of this system of noninteracting fermions in the excited state is also the negative of the first ionization potential. *What the Q-DFT mapping reveals is that there exists an infinite number of local effective potentials* $v_s(\mathbf{r})$, *or equivalently electron-interaction potential functions* $v_{ee}(\mathbf{r})$, *that can generate the ground state density and ionization potential of the interacting system of electrons.*

Now let us consider an interacting system of electrons in a *nondegenerate excited* state with density $\rho^e(\mathbf{r})$. It is then possible [16, 30] to map this system to one of noninteracting fermions possessing the same density $\rho^e(\mathbf{r})$ but which is in a *ground* state. There thus exists a potential function $v_{ee}(\mathbf{r})$. Once again, the highest occupied eigenvalue of either system is the first ionization potential. (Interestingly, then the excited state density $\rho^e(\mathbf{r})$ that is obtained from a wave function which has nodes, may be reproduced by single-particle orbitals that are nodeless.) It is also possible [15] to map the interacting *excited* state system to a model S system in an *excited* state with the *same* or *different* configuration with corresponding different potentials $v_{ee}(\mathbf{r})$. Once again, irrespective of the state of the model system, its highest occupied eigenvalue is the first ionization potential. Therefore, *there exists an infinite number of local effective potentials* $v_s(\mathbf{r})$, *or equivalently electron-interaction potential functions* $v_{ee}(\mathbf{r})$, *that can generate the same density and ionization potential as a system of interacting electrons in an excited state.*

The *nonuniqueness* of the model S system is equally valid for the Q-DFT mapping from *degenerate ground* and *excited* states of the interacting system [19].

The mappings described are also valid for when a magnetic field $\boldsymbol{\mathcal{B}}(\mathbf{r})$ is present. However, the constraints on the model system are now that it possess the same electron number N, and the same orbital $\mathbf{L}$ and spin $\mathbf{S}$ angular momentum.

Finally, as will be shown, the contribution of the correlations due to the Pauli principle and Coulomb repulsion to the *infinite* number of local effective potentials is always the *same*. The difference between these potentials is *solely* due to the Correlation-Kinetic effects (see Sect. 6.1.7). Further, it is the Correlation-Kinetic effect via its definition that allows for the infinite number of potentials to be constructed [22].

Nonuniqueness of the Noninteracting System Wave Function

For the case when only an external field $\boldsymbol{\mathcal{E}}(\mathbf{r}) = -\boldsymbol{\nabla} v(\mathbf{r})/e$ is present it has been shown that in the mapping from a *ground* or *excited* state of the interacting system to an S system in an *excited* state, there is a *nonuniqueness* of the S system wave function [23, 24]. These wave functions all lead to the *same* density, thereby satisfying the sole requirement of reproducing the interacting system density. Thus, in Q-DFT, for the case when only an electric field $\boldsymbol{\mathcal{E}}(\mathbf{r})$ is present, there is no requirement that the configuration of the S system be the same as that of the interacting system. Thus, these S system wave functions are not constrained to be eigenfunctions of various spin-symmetry operators.

Rationale for Employing Basic Variables $\{\rho(\mathbf{r}), \mathbf{j}(\mathbf{r})\}$ for Equivalence

The rationale for the choice of the densities $\{\rho(\mathbf{r}), \mathbf{j}(\mathbf{r})\}$ for the equivalence of the interacting and noninteracting systems stems from the ground state theorem of Hohenberg and Kohn (HK) [44] and of its extension by Pan and Sahni (PS) [45] to the added presence of a *uniform* magnetic field. For a system of electrons in an external electrostatic field $\mathcal{E}(\mathbf{r}) = -\nabla v(\mathbf{r})/e$, and in a *nondegenerate ground state*, HK proved that knowledge of the ground state density $\rho(\mathbf{r})$ uniquely determines the external scalar potential $v(\mathbf{r})$ to within a constant. The constraint in the proof is that of fixed electron number N. As the kinetic $\hat{T}$ and electron-interaction potential $\hat{U}$ operators are assumed known, so thus is the Hamiltonian. Solution of the Schrödinger equation then leads to the eigenfunctions and eigenvalues of the system. Hence, the nondegenerate ground state density constitutes a *basic variable*. What PS proved was that in the added presence of a *uniform* magnetostatic field $\mathcal{B}(\mathbf{r}) = \nabla \times \mathbf{A}(\mathbf{r})$, knowledge of the nondegenerate ground state densities $\{\rho(\mathbf{r}), \mathbf{j}(\mathbf{r})\}$ uniquely determines the potentials $\{v(\mathbf{r}), \mathbf{A}(\mathbf{r})\}$ to within a constant and gradient of a scalar function, respectively. The constraints in the proof are that of fixed electron number N, orbital $\mathbf{L}$ and spin $\mathbf{S}$ angular momentum. The PS proof, which differs from that of HK, was for both spinless electrons and electrons with spin. Again, with the Hamiltonian now known, the solution of the corresponding Schrödinger and Schrödinger-Pauli equations then leads to the system eigenfunctions and eigenvalues. Hence, in the presence of a static electromagnetic field, the nondegenerate ground state densities $\{\rho(\mathbf{r}), \mathbf{j}(\mathbf{r})\}$ constitute the *basic variables*. The theorems of HK and PS are ground state theorems. Thus, within HK, the mapping is from an interacting system in a *ground* state to one of noninteracting fermions also in a *ground* state possessing the same density $\rho(\mathbf{r})$. This is the mapping performed, for example, in KS-DFT. However, within Q-DFT, the mapping to the model system with the same $\rho(\mathbf{r})$ or $\{\rho(\mathbf{r}), \mathbf{j}(\mathbf{r})\}$ is possible for ground, excited and degenerate states of the interacting system.

Discontinuity in the Local Electron-Interaction Potential $v_{ee}(\mathbf{r})$

The description thus far of the Q-DFT mapping from the interacting system of electrons in an external field $\mathcal{E}(\mathbf{r}) = -\nabla v(\mathbf{r})/e$ to one of noninteracting fermions with the same density $\rho(\mathbf{r})$ has been restricted to the case of integer number of electrons N. However, in order to understand [46, 47] the dissociation of molecules so that each fragment possesses integer charge, or to obtain [47–50] properties such as the band structure of semiconductors, the framework of the mapping must be extended to include the case of fractional charge $(N + \omega\,;\ 0 < \omega < 1)$. As a consequence, the electron-interaction potential $v_{ee}(\mathbf{r})$ exhibits a discontinuity Δ as the electron number passes through an integer value. Equivalently, as the fractional charge ω vanishes from above, the discontinuity is given as

$$\Delta = \lim_{\omega \downarrow 0} \left[v_{ee}^{N+\omega}(\mathbf{r}) - v_{ee}^{N}(\mathbf{r}) \right]. \tag{6.11}$$

It can be shown, and we refer the reader to ***QDFT*** and [11] for the details, that the electron correlations due to the Pauli principle and Coulomb repulsion do *not* contribute to the discontinuity in the limit as the fractional charge vanishes. *The discontinuity* Δ *arises solely as a consequence of Correlation-Kinetic effects.*

We emphasize the significance of Correlation-Kinetic effects in local effective potential theory. As noted above, the multiplicity of local effective potentials (or equivalently the potentials $v_{ee}(\mathbf{r})$) that can generate the same density $\rho(\mathbf{r})$ as that of the interacting system, is *solely* due to Correlation-Kinetic effects. The discontinuity Δ too is due to these effects. In the former, the contributions due to correlations arising from the Pauli principle and Coulomb repulsion to each potential remains the same. In the latter, correlations due to the Pauli principle and Coulomb repulsion play no role. Recently it has been shown that Correlation-Kinetic effects also contribute significantly [51, 52] to the total energy of low electron density Wigner systems. This low density regime is usually characterized by the fact that the Pauli and Coulomb correlation contributions to the energy are greater than the kinetic energy. *The Wigner regime can now also be characterized by a high Correlation-Kinetic energy.* In the limit of very low electron density, the zero-point energy of the electrons is the Correlation-Kinetic energy.

Degenerate State Quantal Density Functional Theory

Stationary-state Q-DFT has also been extended to the mapping from a *degenerate ground* or *excited* state of the N-electron interacting system in an electric field $\mathcal{E}(\mathbf{r}) = -\nabla v(\mathbf{r})/e$ to one of noninteracting fermions such that the equivalent density $\rho(\mathbf{r})$ and energy E are obtained [19]. (See also ***QDFT2***). The cases of both *pure-state* and *ensemble* v-representable densities have been developed. (A v-representable density is one obtained from a wave function that is a solution of the Schrödinger equation for interacting electrons having a potential energy $v(\mathbf{r})$.)

Arbitrariness of the Statistics of the Model System Particles

The Q-DFT mapping as discussed thus far is from the interacting system of electrons in a static electromagnetic field to one of *noninteracting fermions* possessing the same $\{\rho(\mathbf{r}), \mathbf{j}(\mathbf{r})\}$. However, the choice, and therefore the statistics, of the model system particles are entirely arbitrary. It is possible to map the interacting system of electrons via Q-DFT to one of *noninteracting bosons* possessing the same $\{\rho(\mathbf{r}), \mathbf{j}(\mathbf{r})\}$. This description is once again in terms of 'classical' fields and their quantal sources. The fields satisfy the corresponding model boson 'Quantal Newtonian' First Law. (See [18, 25], ***QDFT*** and ***QDFT2***.)

Time-dependent Quantal Density Functional Theory

Time-dependent (TD) Q-DFT [10, 12, 14, 29, 31] constitutes the mapping from an interacting system of electrons in the presence of a TD electromagnetic field as described by either Schrödinger or Schrödinger-Pauli theory to a model system of noninteracting fermions possessing the same TD basic variables of the density $\rho(\mathbf{y})$ and physical current density $\mathbf{j}(\mathbf{y})$, with $\mathbf{y} = \mathbf{r}t$. As in stationary-state Q-DFT, the existence of such a model system is an *assumption*. The mapping is in terms of 'classical' fields representative of properties of the system, and of their quantal sources. These fields obey the equation of motion or *'Quantal Newtonian' Second Law* for the noninteracting fermion.

(One could perform a Q-DFT map to a model system possessing *solely* the same electronic density $\rho(\mathbf{y})$ as that of the interacting system. This is the mapping done within traditional TD Runge-Gross DFT [53], the TD version of Kohn-Sham theory. However, in such a mapping, it then becomes evident from Q-DFT [10, 12, 14] that one must additionally account for the difference in the physical current densities of the interacting and noninteracting systems.)

For the interacting system, let us consider the simpler case of N spinless electrons in the presence of solely a TD external field $\boldsymbol{\mathcal{E}}(\mathbf{y}) = -\boldsymbol{\nabla} v(\mathbf{y})/e$ as described in Chap. 2. (The generalization to the added presence of a TD electromagnetic field $\boldsymbol{\mathcal{B}}(\mathbf{y}) = \boldsymbol{\nabla} \times \mathbf{A}(\mathbf{y}); \mathbf{E}(\mathbf{y}) = -\nabla\chi(\mathbf{y}) - \partial\mathbf{A}(y)/\partial t$ with $\chi(\mathbf{y})$ a scalar and $\mathbf{A}(\mathbf{y})$ a vector potential follows [29].) In atomic units with $e = \hbar = m = 1$, the Schrödinger equation is

$$\hat{H}(t)\Psi(\mathbf{Y}) = i\partial\Psi(\mathbf{Y})/\partial t, \tag{6.12}$$

where the Hamiltonian $\hat{H}(t)$ is

$$\hat{H}(t) = \frac{1}{2}\sum_k \hat{p}_k^2 + \frac{1}{2}\sum_{k,\ell}{}' \frac{1}{|\mathbf{r}_k - \mathbf{r}_\ell|} + \sum_k v(\mathbf{y}_k), \tag{6.13}$$

with the first two terms being the canonical kinetic and electron-interaction potential energy operators, and $\Psi(\mathbf{Y})$ is the wave function with $\mathbf{Y} = \mathbf{X}t$; $\mathbf{X} = \mathbf{x}_1, \ldots, \mathbf{x}_N$; $\mathbf{x} = \mathbf{r}\sigma$; $\mathbf{r}$ and σ the spatial and spin coordinates; $\mathbf{y}_k = \mathbf{r}_k t$; and $\hat{\mathbf{p}} = -i\boldsymbol{\nabla}$ is the canonical momentum operator. The equation is solved given an initial condition $\Psi(\mathbf{Y}_0)$; $\mathbf{Y}_0 = \mathbf{X}t_0$. This system is mapped via Q-DFT to one of noninteracting fermions possessing the same gauge invariant properties $\{\rho(\mathbf{y}), \mathbf{j}(\mathbf{y})\}$. The Hamiltonian for the corresponding model system is

$$\hat{H}_s = \sum_k \hat{h}_s(\mathbf{y}_k), \tag{6.14}$$

where

$$\hat{h}_s(\mathbf{y}) = -\frac{1}{2}\nabla^2 + v_s(\mathbf{y}). \tag{6.15}$$

With the *assumption* that the model fermions experience the *same* electric field $\boldsymbol{\mathcal{E}}(\mathbf{y})$, the local effective potential $v_s(\mathbf{y})$ may be written as

$$v_s(\mathbf{y}) = v(\mathbf{y}) + v_{ee}(\mathbf{y}), \tag{6.16}$$

where now all the many-body effects are incorporated into a *local electron-interaction potential* $v_{ee}(\mathbf{y})$. The corresponding model system Schrödinger equation is then

$$\left[\frac{1}{2}\hat{p}^2 + v_s(\mathbf{y})\right]\phi_k(\mathbf{y}) = i\partial\phi_k(\mathbf{y})/\partial t\ ; \quad k = 1, \ldots N. \tag{6.17}$$

With an appropriate initial condition, the S system wave function is obtained as the Slater determinant $\Phi\{\phi_k\}$ of the orbitals $\phi_k(\mathbf{y})$.

As a consequence of the above assumption, the Q-DFT mapping corresponds to understanding how all the many-electron correlations of the interacting system are incorporated into the local potential $v_{ee}(\mathbf{y})$. As another consequence, the correlations that must be accounted for are those of the Pauli principle, Coulomb repulsion, and Correlation-Kinetic effects. Once again, there must exist a *conservative* effective field $\mathcal{F}^{\text{eff}}(\mathbf{y}) = -\nabla v_{ee}(\mathbf{y})$ descriptive of the model system that allows for the same $\{\rho(\mathbf{y}), \mathbf{j}(\mathbf{y})\}$ to be generated.

As in stationary-state Q-DFT, the TD version allows the separation of the electron correlations due to the Pauli principle and Coulomb repulsion, and their separate contributions to the potential $v_{ee}(\mathbf{y})$ and the non-conserved energy $E(t)$. Additionally, it allows for the determination of the contributions of the Correlation-Kinetic effects to these properties. In this context, *TD Q-DFT then constitutes a complement to TD Schrödinger theory*. Of course, it may also be envisioned as a local effective potential theory description of Schrödinger theory. As such, and because the description of the local potential $v_{ee}(\mathbf{y})$ is in terms of fields, Q-DFT provides a rigorous physical interpretation of the action functionals $\mathcal{A}[\rho(\mathbf{y})]$ and their functional derivatives $\delta\mathcal{A}[\rho(\mathbf{y})]/\delta\rho(\mathbf{y})$ of TD Runge-Gross DFT [53]. (Runge-Gross DFT is the TD version of KS-DFT.) Finally, as stationary-state Schrödinger and Schrödinger-Pauli theories are special cases of the corresponding TD versions, stationary-state Q-DFT constitutes a special case of TD Q-DFT.

Rationale for Employing Basic Variables $\{\rho(\mathbf{y}), \mathbf{j}(\mathbf{y})\}$ for Equivalence

The rationale for employing $\{\rho(\mathbf{y}), \mathbf{j}(\mathbf{y})\}$ as the properties that the model system of noninteracting fermions must possess stems from the theorem of Runge and Gross [RG] [53]. The RG theorem is proved for an external electric field $\mathcal{E}(\mathbf{y}) = -\nabla v(\mathbf{y})/e$ for the case when the scalar potential $v(\mathbf{y})$ is Taylor expandable about some initial time. It is first proved that there is a one-to-one relationship between the potential $v(\mathbf{y})$ (to within an additive function of time) and the current density $\mathbf{j}(\mathbf{y})$. Then *employing this fact*, it is proved that there is a one-to-one relationship between $v(\mathbf{y})$ (to within an additive function of time) and the density $\rho(\mathbf{y})$. Thus, knowledge of both $\rho(\mathbf{y})$ and $\mathbf{j}(\mathbf{y})$ uniquely determine $v(\mathbf{y})$ to within an additive function of time. Then, since the kinetic $\hat{T}$ and electron-interaction potential $\hat{U}$ operators are assumed known, the Hamiltonian $\hat{H}(t)$ is known. Solution of the TD Schrödinger equation, with an appropriate initial condition, then leads to the wave function $\Psi(\mathbf{Y})$ of the system. As such, the density $\rho(\mathbf{y})$ and current density $\mathbf{j}(\mathbf{y})$ are *basic variables*.

Nomenclature

The rationale for the nomenclature of Quantal Density Functional Theory, now descriptive of the theory in the broad context as summarized above, remains the same as for its original coinage [54] as the quantum-mechanical interpretation of stationary-state Kohn-Sham density functional theory. The adjective *'quantal'* is employed because the sources of the fields that describe the noninteracting system are expectation values of Hermitian operators taken with respect to the interacting and noninteracting system wave functions. It is a *'density functional theory'* in that these wave functions are functionals of the ground state density. The original terminology is retained.

Stationary-state Q-DFT is described in the following section. These ideas are then elucidated by application of Q-DFT to the ground 1^1S and first excited singlet 2^1S states of a 2-electron 2-dimensional 'artificial atom' or semiconductor quantum dot in a magnetic field. A brief description of the Q-DFT of the density amplitude $\sqrt{\rho(\mathbf{r})}$ is provided. In the final section, the mapping within time-dependent Q-DFT is explained.

6.1 Stationary-State Quantal Density Functional Theory

In order to describe the mapping from the interacting system of N electrons in a static electromagnetic field as described by the Schrödinger theory equations (6.1)–(6.5) to one of noninteracting fermions – the S system – as described by the corresponding equations (6.6)–(6.10), it is necessary to first define the various quantal sources and fields descriptive of the model system. The S system *'Quantal Newtonian' First Law* then leads to the conservative effective field $\boldsymbol{\mathcal{F}}^{\text{eff}}(\mathbf{r})$ and corresponding local potential $v_{ee}(\mathbf{r})$ in which all the many-body effects are incorporated. This potential ensures the same density $\rho(\mathbf{r})$ and physical current density $\mathbf{j}(\mathbf{r})$ as that of the interacting system of electrons. The total energy E can then be defined in terms of the properties of the S system. Additionally, in this Section, other significant properties of the S system are described: the non-uniqueness of the potential $v_{ee}(\mathbf{r})$ is proved; a physical interpretation of the highest occupied eigenvalue provided; sum rules satisfied by the effective field $\boldsymbol{\mathcal{F}}^{\text{eff}}(\mathbf{r})$ stated.

6.1.1 Quantal Sources

The quantal sources of the S system are the *local* sources of the density $\rho(\mathbf{r})$ and physical current density $\mathbf{j}(\mathbf{r})$; the *nonlocal* sources of the Dirac single-particle density matrix $\gamma_s(\mathbf{rr}')$, the pair-correlation density $g_s(\mathbf{rr}')$ and from it the Fermi hole charge distribution $\rho_x(\mathbf{rr}')$. These properties are expectation values of Hermitian operators taken with respect to the Slater determinant $\Phi\{\phi_k\}$. The definition of the Fermi hole $\rho_x(\mathbf{rr}')$ in terms of the single-particle orbitals $\phi_k(\mathbf{r})$ thus appears naturally. As such it is then possible to define the *nonlocal* source of the Coulomb hole charge distribution $\rho_c(\mathbf{rr}')$. (Note that the subscript s is not employed in the definitions of $\rho(\mathbf{r})$ and $\mathbf{j}(\mathbf{r})$ as these densities are the same as for the interacting system.)

A. Electron Density $\rho(\mathbf{r})$

The electron density $\rho(\mathbf{r})$ is the expectation of the density operator $\hat{\rho}(\mathbf{r})$ of (2.12), and can be expressed in terms of the orbitals $\phi_k(\mathbf{r}\sigma)$:

$$\rho(\mathbf{r}) = \langle\Phi\{\phi_k\}|\hat{\rho}(\mathbf{r})|\Phi\{\phi_k\}\rangle = \sum_\sigma\sum_k \phi_k^*(\mathbf{r}\sigma)\phi_k(\mathbf{r}\sigma). \tag{6.18}$$

It satisfies the normalization condition

$$\int \rho(\mathbf{r})d\mathbf{r} = N, \tag{6.19}$$

and thus the constraint of equivalent electron number N.

B. Physical Current Density $\mathbf{j}(\mathbf{r})$

The physical current density $\mathbf{j}(\mathbf{r})$ is the expectation value of the current density operator of $\hat{\mathbf{j}}(\mathbf{r})$ of (3.17):

$$\mathbf{j}(\mathbf{r}) = \langle\Phi\{\phi_k\}|\hat{\mathbf{j}}(\mathbf{r})|\Phi\{\phi_k\}\rangle. \tag{6.20}$$

The current density $\mathbf{j}(\mathbf{r})$ too may be expressed in terms of the orbitals $\phi_k(\mathbf{r}\sigma)$. As the operator $\hat{\mathbf{j}}(\mathbf{r})$ may be decomposed into its paramagnetic $\hat{\mathbf{j}}_p(\mathbf{r})$ and diamagnetic $\hat{\mathbf{j}}_d(\mathbf{r})$ components (see (3.17)), the current density $\mathbf{j}(\mathbf{r})$ may be so decomposed. Thus,

$$\mathbf{j}(\mathbf{r}) = \mathbf{j}_{p,s}(\mathbf{r}) + \mathbf{j}_{d,s}(\mathbf{r}), \tag{6.21}$$

where

$$\mathbf{j}_{p,s}(\mathbf{r}) = \left\langle\Phi\{\phi_k\}|\frac{1}{2i}\sum_k\left[\nabla_{\mathbf{r}_k}\delta(\mathbf{r}_k-\mathbf{r}) + \delta(\mathbf{r}_k-\mathbf{r})\nabla_{\mathbf{r}_k}\right]|\Phi\{\phi_k\}\right\rangle \tag{6.22}$$

$$= \frac{1}{2i}\sum_\sigma\sum_k\left[\phi_k^*(\mathbf{r}\sigma)\nabla\phi_k(\mathbf{r}\sigma) - \phi_k(\mathbf{r}\sigma)\nabla\phi_k^*(\mathbf{r}\sigma)\right], \tag{6.23}$$

$$\mathbf{j}_{d,s} = \langle\Phi\{\phi_k\}|\hat{\rho}(\mathbf{r})\mathbf{A}(\mathbf{r})|\Phi\{\phi_k\}\rangle = \rho(\mathbf{r})\mathbf{A}(\mathbf{r}). \tag{6.24}$$

The physical current density satisfies the continuity constraint

$$\nabla\cdot\mathbf{j}(\mathbf{r}) = 0. \tag{6.25}$$

C. Dirac Spinless Single-Particle Density Matrix $\gamma_s(\mathbf{r}\mathbf{r}')$

The Dirac spinless single-particle density matrix $\gamma_s(\mathbf{r}\mathbf{r}')$ is the expectation value of the density matrix operator $\hat{\gamma}(\mathbf{r}\mathbf{r}')$ of (2.16). It too can be expressed in terms of the orbitals $\phi_k(\mathbf{r}\sigma)$:

$$\gamma_s(\mathbf{rr}') = \langle \Phi\{\phi_k\}|\hat{\gamma}(\mathbf{rr}')|\Phi\{\phi_k\}\rangle = \sum_\sigma \sum_k \phi_k^*(\mathbf{r}\sigma)\phi_k(\mathbf{r}'\sigma). \tag{6.26}$$

The properties of the Dirac density matrix are

$$\gamma_s(\mathbf{rr}) = \rho(\mathbf{r}), \tag{6.27}$$

$$\gamma_s(\mathbf{r}'\mathbf{r}) = \gamma_s^*(\mathbf{rr}'), \tag{6.28}$$

$$\int \gamma_s(\mathbf{rr}'')\gamma_s(\mathbf{r}''\mathbf{r}')d\mathbf{r}'' = \gamma_s(\mathbf{rr}'). \tag{6.29}$$

As a consequence of the equality in (6.29), the Dirac density matrix is said to be *idempotent*. Note that the Dirac density matrix $\gamma_s(\mathbf{rr}')$ differs from the interacting system density matrix $\gamma(\mathbf{rr}')$. It is only their diagonal matrix elements that are equivalent, i.e. $\gamma_s(\mathbf{rr}) = \gamma(\mathbf{rr}) = \rho(\mathbf{r})$.

D. Pair-Correlation Density $g_s(\mathbf{rr}')$, Fermi $\rho_x(\mathbf{rr}')$ and Coulomb $\rho_c(\mathbf{rr}')$ Hole Charge Distributions

The pair-correlation density $g_s(\mathbf{rr}')$ is the ratio of the expectation values of the pair-correlation $\hat{P}(\mathbf{rr}')$ and density $\hat{\rho}(\mathbf{r})$ operators of (2.28) and (2.12), respectively:

$$g_s(\mathbf{rr}') = \frac{\langle \Phi\{\phi_k\}|\hat{P}(\mathbf{rr}')|\Phi\{\phi_k\}\rangle}{\rho(\mathbf{r})}. \tag{6.30}$$

It satisfies the sum rule

$$\int g_s(\mathbf{rr}')d\mathbf{r}' = N - 1, \tag{6.31}$$

for arbitrary electron position at $\mathbf{r}$.

The pair-correlation density $g_s(\mathbf{rr}')$ may be decomposed into its *local* and *nonlocal* components as

$$g_s(\mathbf{rr}') = \rho(\mathbf{r}') + \rho_x(\mathbf{rr}'), \tag{6.32}$$

where $\rho_x(\mathbf{rr}')$ is the *Fermi* hole charge distribution. It represents the reduction in probability of two electrons of parallel spin approaching each other. Written as in (6.32), it is the reduction in density at $\mathbf{r}'$ due to an electron of parallel spin at $\mathbf{r}$. Its definition follows from the expectations of (6.30) as

$$\rho_x(\mathbf{rr}') = -\frac{\sum\limits_{\ell,k(\text{spin k=spin }\ell)} \phi_\ell^*(\mathbf{r})\phi_k^*(\mathbf{r}')\phi_\ell(\mathbf{r}')\phi_k(\mathbf{r})}{\sum\limits_\sigma \sum\limits_k \phi_k^*(\mathbf{r}\sigma)\phi_k(\mathbf{r}\sigma)} \tag{6.33}$$

$$= -\frac{|\gamma_s(\mathbf{rr}')|^2}{2\rho(\mathbf{r})}. \tag{6.34}$$

The Fermi hole $\rho_x(\mathbf{rr}')$ satisfies the sum rules

$$\int \rho_x(\mathbf{rr}')d\mathbf{r}' = -1, \tag{6.35}$$

$$\rho_x(\mathbf{rr}) = -\rho(\mathbf{r})/2, \tag{6.36}$$

$$\rho_x(\mathbf{rr}') \leq 0, \tag{6.37}$$

for each electron position at $\mathbf{r}$.

The Fermi hole $\rho_x(\mathbf{rr}')$ is representative of electron correlations due to the Pauli principle. Within Q-DFT one defines the *Coulomb* hole $\rho_c(\mathbf{rr}')$ representative of correlations due to Coulomb repulsion. It is a *nonlocal* quantal source. The Coulomb hole charge distribution at $\mathbf{r}'$ for an electron at $\mathbf{r}$ is defined as the difference between the pair-correlation densities of the interacting $g(\mathbf{rr}')$ and noninteracting $g_s(\mathbf{rr}')$ systems. This then reduces to the difference between the Fermi-Coulomb $\rho_{xc}(\mathbf{rr}')$ and Fermi $\rho_x(\mathbf{rr}')$ hole charge distributions. Thus,

$$\rho_c(\mathbf{rr}') = g(\mathbf{rr}') - g_s(\mathbf{rr}') \tag{6.38}$$

$$= \rho_{xc}(\mathbf{rr}') - \rho_x(\mathbf{rr}'). \tag{6.39}$$

Note that the equivalence of (6.39) follows only because the density $\rho(\mathbf{r})$ of the interacting and model S system are the same. As the total charge of the Fermi-Coulomb hole $\rho_{xc}(\mathbf{rr}')$ and that of the Fermi hole $\rho_x(\mathbf{rr}')$ are the same (see (2.36) and (6.35)), the sum rule satisfied by the Coulomb hole $\rho_c(\mathbf{rr}')$ is

$$\int \rho_c(\mathbf{rr}')d\mathbf{r}' = 0, \tag{6.40}$$

for each electron position at $\mathbf{r}$.

6.1.2 'Classical' Fields Experienced by Each Model Fermion

As was the case for the interacting system (see Sect. 3.1.1), it is meaningful to separate the fields experienced by each model fermion into its external $\mathcal{F}^{\text{ext}}(\mathbf{r})$ and internal $\mathcal{F}^{\text{int}}(\mathbf{r})$ components. This will then allow for the determination of the conservative effective field $\mathcal{F}^{\text{eff}}(\mathbf{r})$ defining the electron-interaction potential $v_{ee}(\mathbf{r})$ in which all the many-body effects are incorporated.

External Field $\mathcal{F}^{\text{ext}}(\mathbf{r})$

The external fields experienced by the noninteracting fermions are by assumption the same as those experienced by the interacting electrons. These fields are the following:

A. Electrostatic Binding Field $\mathcal{E}(\mathbf{r})$

The binding electrostatic field $\mathcal{E}(\mathbf{r})$ is defined as

$$\mathcal{E}(\mathbf{r}) = -\nabla v(\mathbf{r}), \tag{6.41}$$

where $v(\mathbf{r})$ is the external scalar potential.

B. Lorentz Field $\mathcal{L}(\mathbf{r})$

The Lorentz field $\mathcal{L}(\mathbf{r})$ is written in terms of the Lorentz 'force' $\boldsymbol{\ell}(\mathbf{r})$ and the density $\rho(\mathbf{r})$ as

$$\mathcal{L}(\mathbf{r}) = \boldsymbol{\ell}(\mathbf{r})/\rho(\mathbf{r}), \tag{6.42}$$

where

$$\boldsymbol{\ell}(\mathbf{r}) = \mathbf{j}(\mathbf{r}) \times \mathcal{B}(\mathbf{r}), \tag{6.43}$$

with the physical current density $\mathbf{j}(\mathbf{r})$ being the quantal source. Since $\mathcal{B}(\mathbf{r}) = \nabla \times \mathbf{A}(\mathbf{r})$, the components of the Lorentz 'force' may be written in Cartesian coordinates as

$$\ell_\alpha(\mathbf{r}) = \sum_{\beta=1}^{3} \left[j_\beta(\mathbf{r}) \nabla_\alpha A_\beta(\mathbf{r}) - j_\beta(\mathbf{r}) \nabla_\beta A_\alpha(\mathbf{r}) \right]. \tag{6.44}$$

(Note that as the S system is to be designed such that it possess the same density $\rho(\mathbf{r})$ and physical current density $\mathbf{j}(\mathbf{r})$ as that of the interacting system, and the magnetic field $\mathcal{B}(\mathbf{r})$ and the choice of gauge for the vector potential $\mathbf{A}(\mathbf{r})$ too are the same, the Lorentz field $\mathcal{L}(\mathbf{r})$ experienced by each model fermion is the same as that for the interacting system electron.)

The external field $\mathcal{F}^{\text{ext}}(\mathbf{r})$ is thus

$$\mathcal{F}^{\text{ext}}(\mathbf{r}) = \mathcal{E}(\mathbf{r}) - \mathcal{L}(\mathbf{r}) = -\nabla v(\mathbf{r}) - \mathcal{L}(\mathbf{r}). \tag{6.45}$$

This is rigorously the case as arrived at via the S system Schrödinger equation and the corresponding 'Quantal Newtonian' First Law. (See Appendix A).

Internal Field $\mathcal{F}_s^{\text{int}}(\mathbf{r})$

With the external field $\mathcal{F}^{\text{ext}}(\mathbf{r})$ defined as above, it is evident that the electron correlations due to the Pauli principle and Coulomb repulsion must be embedded in the S system *internal field* $\mathcal{F}_s^{\text{int}}(\mathbf{r})$. For the *interacting* system, these correlations are represented by the electron-interaction field $\mathcal{E}_{ee}(\mathbf{r})$ (see (2.46)). It turns out that this field plays a role in the mapping of these correlations to the model system. Further, by employing the corresponding electron-interaction field $\mathcal{E}_{ee,s}(\mathbf{r})$ of the S system, it is possible to decompose the field $\mathcal{E}_{ee}(\mathbf{r})$ into components that *separately* represent the correlations due to the Pauli principle and Coulomb repulsion. We begin by explaining this decomposition.

C. Electron-Interaction Field $\mathcal{E}_{ee}(\mathbf{r})$ and Its Hartree $\mathcal{E}_H(\mathbf{r})$, Pauli $\mathcal{E}_x(\mathbf{r})$, and Coulomb $\mathcal{E}_c(\mathbf{r})$ Components

The S system electron-interaction field $\mathcal{E}_{ee,s}(\mathbf{r})$ is obtained from its quantal source, the pair-correlation density $g_s(\mathbf{r}\mathbf{r}')$, via Coulomb's law as

$$\mathcal{E}_{ee,s}(\mathbf{r}) = \int \frac{g_s(\mathbf{rr}')(\mathbf{r}-\mathbf{r}')}{|\mathbf{r}-\mathbf{r}'|^3} d\mathbf{r}'. \tag{6.46}$$

The field may also be written in terms of the electron-interaction 'force' $\mathbf{e}_{ee,s}(\mathbf{r})$ and the (charge) density $\rho(\mathbf{r})$ as

$$\mathcal{E}_{ee,s}(\mathbf{r}) = \frac{\mathbf{e}_{ee,s}(\mathbf{r})}{\rho(\mathbf{r})}, \tag{6.47}$$

where $\mathbf{e}_{ee,s}(\mathbf{r})$ is determined via Coulomb's law from the S system pair-function $P_s(\mathbf{rr}')=\langle\Phi\{\phi_k\}|\hat{P}(\mathbf{rr}')|\Phi\{\phi_k\}\rangle$ which is the expectation value of the pair-correlation operator $\hat{P}(\mathbf{rr}')$ of (2.28) taken with respect to the Slater determinant $\Phi\{\phi_k\}$. Thus, the 'force' is

$$\mathbf{e}_{ee,s}(\mathbf{r}) = \int \frac{P_s(\mathbf{rr}')(\mathbf{r}-\mathbf{r}')}{|\mathbf{r}-\mathbf{r}'|^3} d\mathbf{r}'. \tag{6.48}$$

(The quantal source of the fields $\mathcal{E}_{ee,s}(\mathbf{r})$ may thus also be thought of as being the pair-function $P_s(\mathbf{rr}')$.)

On employing the decomposition of $g_s(\mathbf{rr}')$ as in (6.32), the field $\mathcal{E}_{ee,s}(\mathbf{r})$ may be decomposed into its Hartree $\mathcal{E}_H(\mathbf{r})$ and Pauli $\mathcal{E}_x(\mathbf{r})$ field components:

$$\mathcal{E}_{ee,s}(\mathbf{r}) = \mathcal{E}_H(\mathbf{r}) + \mathcal{E}_x(\mathbf{r}), \tag{6.49}$$

where

$$\mathcal{E}_H(\mathbf{r}) = \int \frac{\rho(\mathbf{r}')(\mathbf{r}-\mathbf{r}')}{|\mathbf{r}-\mathbf{r}'|^3} d\mathbf{r}', \tag{6.50}$$

and

$$\mathcal{E}_x(\mathbf{r}) - \int \frac{\rho_x(\mathbf{rr}')(\mathbf{r}-\mathbf{r}')}{|\mathbf{r}-\mathbf{r}'|^3} d\mathbf{r}'. \tag{6.51}$$

The quantal source of the Pauli field $\mathcal{E}_x(\mathbf{r})$ is the Fermi hole charge $\rho_x(\mathbf{rr}')$.

The *interacting* system electron-interaction field $\mathcal{E}_{ee}(\mathbf{r})$ is obtained from its quantal source, the pair-correlation density $g(\mathbf{rr}')$ (see (2.26)), via Coulomb's law as

$$\mathcal{E}_{ee}(\mathbf{r}) = \int \frac{g(\mathbf{rr}')(\mathbf{r}-\mathbf{r}')}{|\mathbf{r}-\mathbf{r}'|^3} d\mathbf{r}'. \tag{6.52}$$

On employing the definition of the Coulomb hole $\rho_c(\mathbf{rr}')$ of (6.39), the field $\mathcal{E}_{ee}(\mathbf{r})$ may be rewritten in terms of $\mathcal{E}_{ee,s}(\mathbf{r})$ and a Coulomb field $\mathcal{E}_c(\mathbf{r})$ as

$$\mathcal{E}_{ee}(\mathbf{r}) = \mathcal{E}_{ee,s}(\mathbf{r}) + \mathcal{E}_c(\mathbf{r}), \tag{6.53}$$

so that with (6.49) we have

$$\mathcal{E}_{ee}(\mathbf{r}) = \mathcal{E}_H(\mathbf{r}) + \mathcal{E}_x(\mathbf{r}) + \mathcal{E}_c(\mathbf{r}), \tag{6.54}$$

where

$$\mathcal{E}_c(\mathbf{r}) = \int \frac{\rho_c(\mathbf{r}\mathbf{r}')(\mathbf{r} - \mathbf{r}')}{|\mathbf{r} - \mathbf{r}'|^3} d\mathbf{r}'. \tag{6.55}$$

The quantal source of the Coulomb field $\mathcal{E}_c(\mathbf{r})$ is the Coulomb hole charge $\rho_c(\mathbf{r}\mathbf{r}')$. Thus, the electron-interaction field $\mathcal{E}_{ee}(\mathbf{r})$ may be decomposed into its Hartree $\mathcal{E}_H(\mathbf{r})$, Pauli $\mathcal{E}_x(\mathbf{r})$ and Coulomb $\mathcal{E}_c(\mathbf{r})$ components. The Pauli field $\mathcal{E}_x(\mathbf{r})$ accounts for correlations due to the Pauli principle, and the Coulomb field $\mathcal{E}_c(\mathbf{r})$ the correlations due to Coulomb repulsion beyond those incorporated into the Hartree field $\mathcal{E}_H(\mathbf{r})$.

As both the Fermi $\rho_x(\mathbf{r}\mathbf{r}')$ and Coulomb $\rho_c(\mathbf{r}\mathbf{r}')$ charge distributions are nonlocal, the corresponding Pauli $\mathcal{E}_x(\mathbf{r})$ and Coulomb $\mathcal{E}_c(\mathbf{r})$ fields are in general not conservative. The Hartree field $\mathcal{E}_H(\mathbf{r})$ derives from the density $\rho(\mathbf{r})$, a local quantal source, and is thus conservative.

D. Kinetic $\mathcal{Z}_s(\mathbf{r})$ and Correlation-Kinetic $\mathcal{Z}_{t_c}(\mathbf{r})$ Fields

The S system kinetic field $\mathcal{Z}_s(\mathbf{r})$ is defined in a manner similar to the kinetic field $\mathcal{Z}(\mathbf{r})$ of the interacting system (2.51), but its quantal source is the Dirac density matrix $\gamma_s(\mathbf{r}\mathbf{r}')$ instead of the interacting system nonidempotent density matrix $\gamma(\mathbf{r}\mathbf{r}')$. Thus, in terms of the kinetic 'force' $z_s(\mathbf{r}; [\gamma_s])$ and the (charge) density $\rho(\mathbf{r})$, the field $\mathcal{Z}_s(\mathbf{r})$ is

$$\mathcal{Z}_s(\mathbf{r}) = \frac{z_s(\mathbf{r}; [\gamma_s])}{\rho(\mathbf{r})}, \tag{6.56}$$

where the 'force' is defined by its component $z_{s,\alpha}(\mathbf{r})$ as

$$z_{s,\alpha}(\mathbf{r}) = 2 \sum_{\beta} \frac{\partial}{\partial r_\beta} t_{s,\alpha\beta}(\mathbf{r}), \tag{6.57}$$

with $t_{s,\alpha\beta}(\mathbf{r})$ the S system second-rank kinetic energy density tensor defined in Cartesian coordinates as

$$t_{s,\alpha\beta}(\mathbf{r}) = \frac{1}{4}\left[\frac{\partial^2}{\partial r'_\alpha \partial r''_\beta} + \frac{\partial^2}{\partial r'_\beta \partial r''_\alpha}\right] \gamma_s(\mathbf{r}'\mathbf{r}'')\bigg|_{\mathbf{r}'=\mathbf{r}''=\mathbf{r}}. \tag{6.58}$$

The *Correlation-Kinetic* field $\mathcal{Z}_{t_c}(\mathbf{r})$ is defined as the difference between the noninteracting and interacting system kinetic fields:

$$\mathcal{Z}_{t_c}(\mathbf{r}) = \mathcal{Z}_s(\mathbf{r}) - \mathcal{Z}(\mathbf{r}). \tag{6.59}$$

The field $\mathcal{Z}_{t_c}(\mathbf{r})$ is thus representative of the contribution to the interacting system kinetic field due to the correlations arising from the Pauli principle and Coulomb repulsion.

E. Differential Density Field $\mathcal{D}(\mathbf{r})$

The differential density field $\mathcal{D}(\mathbf{r})$ is defined in terms of the differential density 'force' $\mathbf{d}(\mathbf{r})$ and the (charge) density $\rho(\mathbf{r})$ as

$$\mathcal{D}(\mathbf{r}) = \frac{\mathbf{d}(\mathbf{r})}{\rho(\mathbf{r})}, \tag{6.60}$$

where

$$\mathbf{d}(\mathbf{r}) = -\frac{1}{4}\nabla\nabla^2\rho(\mathbf{r}). \tag{6.61}$$

As the quantal source of the field $\mathcal{D}(\mathbf{r})$ is the density $\rho(\mathbf{r})$, and the densities of the interacting and noninteracting systems are equivalent, this field is the same as that for the interacting system. It plays no role in the mapping to the S system.

F. Internal Magnetic Field $\mathcal{I}_m(\mathbf{r})$

In addition to the Lorentz field $\mathcal{L}(\mathbf{r})$, each noninteracting fermion also experiences an internal magnetic field $\mathcal{I}_m(\mathbf{r})$. The quantal source of this field is also the physical current density $\mathbf{j}(\mathbf{r})$. This field is defined in terms of an internal magnetic 'force' $\boldsymbol{i}_m(\mathbf{r})$ and the (charge) density $\rho(\mathbf{r})$ as

$$\mathcal{I}_m(\mathbf{r}) = \frac{\boldsymbol{i}_m(\mathbf{r})}{\rho(\mathbf{r})}. \tag{6.62}$$

The components of $\boldsymbol{i}_m(\mathbf{r})$ are defined as

$$\mathrm{i}_{m,\alpha}(\mathbf{r}) = \sum_{\beta=1}^{3} \nabla_\beta I_{\alpha\beta}(\mathbf{r};\mathbf{jA}), \tag{6.63}$$

where the second-rank tensor $I_{\alpha\beta}(\mathbf{r})$ in Cartesian coordinates is

$$I_{\alpha\beta}(\mathbf{r};\mathbf{jA}) = \left[j_\alpha(\mathbf{r})A_\beta(\mathbf{r}) + j_\beta(\mathbf{r})A_\alpha(\mathbf{r})\right] - \rho(\mathbf{r})A_\alpha(\mathbf{r})A_\beta(\mathbf{r}). \tag{6.64}$$

(Note that since the densities $\{\rho(\mathbf{r}), \mathbf{j}(\mathbf{r})\}$ of the S system are to be the same as that of the interacting system, and the magnetic field $\mathcal{B}(\mathbf{r})$ and choice of gauge for the vector potential $\mathbf{A}(\mathbf{r})$ are also the same, the internal field $\mathcal{I}_m(\mathbf{r})$ experienced by each model fermion too is the same.)

G. Effective Field $\mathcal{F}^{\mathrm{eff}}(\mathbf{r})$ Incorporating All Requisite Electron Correlations

The assumption of existence of the model S system, and thereby of a local effective electron-interaction potential $v_{ee}(\mathbf{r})$ in which all the many-body correlations are incorporated, implies that there must exist an effective field $\mathcal{F}^{\mathrm{eff}}(\mathbf{r})$ representative of these correlations. Hence, it follows that

$$\mathcal{F}^{\mathrm{eff}}(\mathbf{r}) = -\nabla v_{ee}(\mathbf{r}). \tag{6.65}$$

What then are the correlations that must be incorporated in $\mathcal{F}^{\text{eff}}(\mathbf{r})$? The S system electron-interaction field $\mathcal{E}_{ee,s}(\mathbf{r})$ includes only correlations due to the Pauli principle via the Pauli field $\mathcal{E}_x(\mathbf{r})$. These Pauli correlations are intrinsically incorporated. Correlations due to Coulomb repulsion beyond those included in the Hartree term $\mathcal{E}_H(\mathbf{r})$ are absent in $\mathcal{E}_{ee,s}(\mathbf{r})$, and must be accounted for. Further, as the kinetic energy of the S system differs from that of the interacting system, the contributions to the kinetic energy due to the correlations arising from the Pauli principle and Coulomb repulsion – the *Correlation-Kinetic* effects – must also be accounted for.

Thus, with all the electron correlations being incorporated into $\mathcal{F}^{\text{eff}}(\mathbf{r})$, the *internal field* experienced by each model fermion is then

$$\mathcal{F}_s^{\text{int}}(\mathbf{r}) = \mathcal{F}^{\text{eff}}(\mathbf{r}) - \mathcal{Z}_s(\mathbf{r}) - \mathcal{D}(\mathbf{r}) - \mathcal{I}_m(\mathbf{r}) \tag{6.66}$$

$$= -\nabla v_{ee}(\mathbf{r}) - \mathcal{Z}_s(\mathbf{r}) - \mathcal{D}(\mathbf{r}) - \mathcal{I}_m(\mathbf{r}). \tag{6.67}$$

In writing this expression, it is assumed that the density $\rho(\mathbf{r})$ and physical current density $\mathbf{j}(\mathbf{r})$ of the model fermions are the same as that of the interacting system.

The expression for $\mathcal{F}_s^{\text{int}}(\mathbf{r})$ is rigorously arrived from the S system Schrödinger differential equation (6.10) and the resulting *'Quantal Newtonian' First Law*. (See Appendix A). The manner by which the Pauli and Coulomb correlations and Correlation-Kinetic effects are incorporated into the effective field $\mathcal{F}^{\text{eff}}(\mathbf{r})$ will be explained in Sect. 6.1.4.

6.1.3 The S System 'Quantal Newtonian' First Law

As is the case for the interacting system of electrons, there exists a *'Quantal Newtonian' First Law* satisfied by each model fermion of the S system (see Appendix A). According to the law, the sum of the external $\mathcal{F}^{\text{ext}}(\mathbf{r})$ and internal $\mathcal{F}_s^{\text{int}}(\mathbf{r})$ fields vanish:

$$\mathcal{F}^{\text{ext}}(\mathbf{r}) + \mathcal{F}_s^{\text{int}}(\mathbf{r}) = 0, \tag{6.68}$$

where the fields $\mathcal{F}^{\text{ext}}(\mathbf{r})$ and $\mathcal{F}_s^{\text{int}}(\mathbf{r})$ are defined by (6.45) and (6.67), respectively.

From the law, the external scalar potential $v(\mathbf{r})$ may be expressed in terms of the various component fields of the S system as

$$v(\mathbf{r}) = \int_\infty^{\mathbf{r}} \mathcal{F}_s(\mathbf{r}') \cdot d\boldsymbol{\ell}', \tag{6.69}$$

where

$$\mathcal{F}_s(\mathbf{r}) = \mathcal{F}^{\text{eff}}(\mathbf{r}) - \mathcal{Z}_s(\mathbf{r}) - \mathcal{D}(\mathbf{r}) - \mathcal{I}_m(\mathbf{r}) - \mathcal{L}(\mathbf{r}), \tag{6.70}$$

with $\nabla \times \mathcal{F}_s(\mathbf{r}) = 0$. Thus, the potential $v(\mathbf{r})$ may be interpreted as the work done to move a model fermion from some reference point at infinity to its position at $\mathbf{r}$

in the force of a conservative field $\mathcal{F}_s(\mathbf{r})$. This is similar to the description of the potential $v(\mathbf{r})$ for the interacting system of electrons in terms of the corresponding fields experienced by each electron (see (3.23)).

6.1.4 *Effective Field $\mathcal{F}^{\text{eff}}(\mathbf{r})$ and Electron-Interaction Potential $v_{ee}(\mathbf{r})$*

We next determine the effective field $\mathcal{F}^{\text{eff}}(\mathbf{r})$ and the corresponding electron-interaction potential $v_{ee}(\mathbf{r})$ such that the density $\rho(\mathbf{r})$ and physical current density $\mathbf{j}(\mathbf{r})$ of the model S system are the same as those of the interacting system of electrons.

For the interacting system of electrons as defined by the Schrödinger equation (6.5), the statement of the *'Quantal Newtonian' First Law* is (see Sect. 3.1.1)

$$\mathcal{F}^{\text{ext}}(\mathbf{r}) + \mathcal{F}^{\text{int}}(\mathbf{r}) = 0, \tag{6.71}$$

where

$$\mathcal{F}^{\text{ext}}(\mathbf{r}) = \mathcal{E}(\mathbf{r}) - \mathcal{L}(\mathbf{r}) = -\nabla v(\mathbf{r}) - \mathcal{L}(\mathbf{r}) \tag{6.72}$$

and

$$\mathcal{F}^{\text{int}}(\mathbf{r}) = \mathcal{E}_{ee}(\mathbf{r}) - \mathcal{Z}(\mathbf{r}) - \mathcal{D}(\mathbf{r}) - \mathcal{I}_m(\mathbf{r}), \tag{6.73}$$

with the various fields defined in that section. The *'Quantal Newtonian' First Law* for the model S system is (see (6.68))

$$\mathcal{F}^{\text{ext}}(\mathbf{r}) + \mathcal{F}_s^{\text{int}}(\mathbf{r}) = 0, \tag{6.74}$$

where the expression for $\mathcal{F}^{\text{ext}}(\mathbf{r})$ is that of (6.72), and (see (6.67))

$$\mathcal{F}_s^{\text{int}}(\mathbf{r}) = -\nabla v_{ee}(\mathbf{r}) - \mathcal{Z}_s(\mathbf{r}) - \mathcal{D}(\mathbf{r}) - \mathcal{I}_m(\mathbf{r}). \tag{6.75}$$

One next *assumes* that the external field $\mathcal{F}^{\text{ext}}(\mathbf{r})$ experienced by each electron and model fermion are the *same*, i.e.. their scalar $v(\mathbf{r})$ and vector $\mathbf{A}(\mathbf{r})$ potentials are equivalent. This means the equivalence of the internal fields $\mathcal{F}^{\text{int}}(\mathbf{r})$ and $\mathcal{F}_s^{\text{int}}(\mathbf{r})$. Then, ensuring that the densities $\{\rho(\mathbf{r}), \mathbf{j}(\mathbf{r})\}$ of the interacting and noninteracting systems too are equivalent, one obtains on equating the corresponding internal fields that

$$\begin{aligned} -\nabla v_{ee}(\mathbf{r}) &= \mathcal{E}_{ee}(\mathbf{r}) + [\mathcal{Z}_s(\mathbf{r}) - \mathcal{Z}(\mathbf{r})] &(6.76)\\ &= \mathcal{E}_{ee}(\mathbf{r}) + \mathcal{Z}_{t_c}(\mathbf{r}), &(6.77)\end{aligned}$$

or equivalently that

$$\mathcal{F}^{\text{eff}}(\mathbf{r}) = \mathcal{E}_{ee}(\mathbf{r}) + \mathcal{Z}_{t_c}(\mathbf{r}), \tag{6.78}$$

with $\boldsymbol{\mathcal{E}}_{ee}(\mathbf{r})$ the electron-interaction and $\boldsymbol{\mathcal{Z}}_{t_c}(\mathbf{r})$ the Correlation-Kinetic fields. Thus the electron-interaction potential $v_{ee}(\mathbf{r})$ *is the work done to move a model fermion from some reference point at infinity to its position at* $\mathbf{r}$ *in the force of the conservative effective field* $\boldsymbol{\mathcal{F}}^{\mathrm{eff}}(\mathbf{r})$:

$$v_{ee}(\mathbf{r}) = -\int_{\infty}^{\mathbf{r}} \boldsymbol{\mathcal{F}}^{\mathrm{eff}}(\mathbf{r}') \cdot d\boldsymbol{\ell}'. \tag{6.79}$$

This work done is *path-independent* since $\nabla \times \boldsymbol{\mathcal{F}}^{\mathrm{eff}}(\mathbf{r}) = 0$. The vanishing of the curl implies (6.79) provided $\boldsymbol{\mathcal{F}}^{\mathrm{eff}}(\mathbf{r})$ is smooth in a simply connected region. (A function is smooth if it is continuous, differentiable, and has continuous first derivatives. By definition, a region is simply connected if any closed curve lying entirely within this region can shrink down to a point without leaving the region.) In general, the separate fields $\boldsymbol{\mathcal{E}}_{ee}(\mathbf{r})$ and $\boldsymbol{\mathcal{Z}}_{t_c}(\mathbf{r})$ are not conservative, but their sum is.

This then is the rigorous *physical* interpretation of the *local* electron-interaction potential $v_{ee}(\mathbf{r})$ that generates the S system orbitals $\phi_k(\mathbf{x})$ which in turn lead to the same densities $\{\rho(\mathbf{r}), \mathbf{j}(\mathbf{r})\}$ as of the interacting system. The mapping to the model S system accounts for the electron correlations due to the Pauli principle and Coulomb repulsion via the field $\boldsymbol{\mathcal{E}}_{ee}(\mathbf{r})$, and the contribution of these correlations to kinetic properties via the field $\boldsymbol{\mathcal{Z}}_{t_c}(\mathbf{r})$. Thus, in this mapping to the S system, all the many-electron correlations of the interacting system are explicitly accounted for.

Employing the decomposition of the electron-interaction field $\boldsymbol{\mathcal{E}}_{ee}(\mathbf{r})$ into its Hartree $\boldsymbol{\mathcal{E}}_H(\mathbf{r})$, Pauli $\boldsymbol{\mathcal{E}}_x(\mathbf{r})$, and Coulomb $\boldsymbol{\mathcal{E}}_c(\mathbf{r})$ components (see (6.54)), the effective field $\boldsymbol{\mathcal{F}}^{\mathrm{eff}}(\mathbf{r})$ may be written as

$$\boldsymbol{\mathcal{F}}^{\mathrm{eff}}(\mathbf{r}) = \boldsymbol{\mathcal{E}}_H(\mathbf{r}) + \boldsymbol{\mathcal{E}}_x(\mathbf{r}) + \boldsymbol{\mathcal{E}}_c(\mathbf{r}) + \boldsymbol{\mathcal{Z}}_{t_c}(\mathbf{r}). \tag{6.80}$$

Thus, the effective field $\boldsymbol{\mathcal{F}}^{\mathrm{eff}}(\mathbf{r})$ is expressed in terms of the *separate* electron correlations that must be accounted for by the model S system.

As the source $\rho(\mathbf{r})$ for the Hartree field $\boldsymbol{\mathcal{E}}_H(\mathbf{r})$ is a *local* (static) charge distribution, the field may be written as

$$\boldsymbol{\mathcal{E}}_H(\mathbf{r}) = -\boldsymbol{\nabla} W_H(\mathbf{r}), \tag{6.81}$$

where $W_H(\mathbf{r})$ is a scalar function. Since $\nabla \times \boldsymbol{\mathcal{E}}_H(\mathbf{r}) = 0$, the field $\boldsymbol{\mathcal{E}}_H(\mathbf{r})$ is conservative. Thus, the function $W_H(\mathbf{r})$, which equivalently is the work done in the field $\boldsymbol{\mathcal{E}}_H(\mathbf{r})$, may be expressed as

$$W_H(\mathbf{r}) = -\int_{\infty}^{\mathbf{r}} \boldsymbol{\mathcal{E}}_H(\mathbf{r}') \cdot d\boldsymbol{\ell}' \tag{6.82}$$

$$= \int \frac{\rho(\mathbf{r}')}{|\mathbf{r} - \mathbf{r}'|} d\mathbf{r}'. \tag{6.83}$$

The work done $W_H(\mathbf{r})$ of (6.82) is *path-independent.*

Hence, the potential energy $v_{ee}(\mathbf{r})$ of (6.79) may be written as

$$v_{ee}(\mathbf{r}) = W_H(\mathbf{r}) + \left(- \int_{\infty}^{\mathbf{r}} \left[\mathcal{E}_x(\mathbf{r}') + \mathcal{E}_c(\mathbf{r}') + \mathcal{Z}_{t_c}(\mathbf{r}') \right] \cdot d\boldsymbol{\ell}' \right). \tag{6.84}$$

For systems with symmetry such that the individual fields $\mathcal{E}_x(\mathbf{r})$, $\mathcal{E}_c(\mathbf{r})$ $\mathcal{Z}_{t_c}(\mathbf{r})$ are *separately* conservative, i.e. $\nabla \times \mathcal{E}_x(\mathbf{r}) = 0$; $\nabla \times \mathcal{E}_c(\mathbf{r}) = 0$; $\nabla \times \mathcal{Z}_{t_c}(\mathbf{r}) = 0$, the potential energy $v_{ee}(\mathbf{r})$ is the sum of the work done in these individual fields:

$$v_{ee}(\mathbf{r}) = W_H(\mathbf{r}) + W_x(\mathbf{r}) + W_c(\mathbf{r}) + W_{t_c}(\mathbf{r}), \tag{6.85}$$

where

$$W_x(\mathbf{r}) = - \int_{\infty}^{\mathbf{r}} \mathcal{E}_x(\mathbf{r}') \cdot d\boldsymbol{\ell}', \tag{6.86}$$

$$W_c(\mathbf{r}) = - \int_{\infty}^{\mathbf{r}} \mathcal{E}_c(\mathbf{r}') \cdot d\boldsymbol{\ell}', \tag{6.87}$$

$$W_{t_c}(\mathbf{r}) = - \int_{\infty}^{\mathbf{r}} \mathcal{Z}_{t_c}(\mathbf{r}') \cdot d\boldsymbol{\ell}'. \tag{6.88}$$

Each work done is separately *path-independent*. For systems of such symmetry, the contributions to the potential $v_{ee}(\mathbf{r})$ of the Pauli and Coulomb correlations, and Correlation-Kinetic effects, are therefore *separately* delineated. This separation on the basis of electron correlations will be explicated by example in Sect. 6.2.

6.1.5 Total Energy E in Terms of S System Properties

The total energy E of the interacting electronic system as defined by the Hamiltonian of (6.1)–(6.4) and the corresponding Schrödinger equation (6.5) may be expressed in terms of S system properties. The interacting system energy E in terms of the densities $\{\rho(\mathbf{r}), \mathbf{j}(\mathbf{r})\}$ and potentials $\{v(\mathbf{r}), \mathbf{A}(\mathbf{r})\}$ is

$$E = T + E_{ee} + \int \rho(\mathbf{r}) v(\mathbf{r}) d\mathbf{r} + \int \mathbf{j}(\mathbf{r}) \cdot \mathbf{A}(\mathbf{r}) d\mathbf{r} - \frac{1}{2} \int \rho(\mathbf{r}) A^2(\mathbf{r}) d\mathbf{r}, \tag{6.89}$$

where T is the canonical kinetic energy and E_{ee} the electron-interaction energy. Decomposing T into its noninteracting T_s and Correlation-Kinetic T_c components, the energy E may be written as

$$E = T_s + E_{ee} + \int \rho(\mathbf{r}) v(\mathbf{r}) d\mathbf{r} + \int \mathbf{j}(\mathbf{r}) \cdot \mathbf{A}(\mathbf{r}) d\mathbf{r} - \frac{1}{2} \int \rho(\mathbf{r}) A^2(\mathbf{r}) d\mathbf{r} + T_c, \tag{6.90}$$

where T_s is the expectation value

$$T_s = \langle \Phi\{\phi_k\}|\hat{T}|\Phi\{\phi_k\}\rangle = \sum_k \sum_\sigma \langle \phi_k(\mathbf{x})| -\frac{1}{2}\nabla^2|\phi_k(\mathbf{x})\rangle \tag{6.91}$$

$$= -\frac{1}{2}\int \rho(\mathbf{r})\mathbf{r}\cdot \boldsymbol{\mathcal{Z}}_s(\mathbf{r})d\mathbf{r}, \tag{6.92}$$

with $\boldsymbol{\mathcal{Z}}_s(\mathbf{r})$ the S system kinetic field (6.56). The Correlation-Kinetic energy component T_c is

$$T_c = \frac{1}{2}\int \rho(\mathbf{r})\mathbf{r}\cdot \boldsymbol{\mathcal{Z}}_{t_c}(\mathbf{r})d\mathbf{r}, \tag{6.93}$$

with $\boldsymbol{\mathcal{Z}}_{t_c}(\mathbf{r})$ the Correlation-Kinetic field (6.59). The electron-interaction energy E_{ee} is

$$E_{ee} = \int \rho(\mathbf{r})\mathbf{r}\cdot \boldsymbol{\mathcal{E}}_{ee}(\mathbf{r})d\mathbf{r}, \tag{6.94}$$

which on employing the decomposition of the electron-interaction field $\boldsymbol{\mathcal{E}}_{ee}(\mathbf{r})$ of (6.54) may be written as the sum of its Hartree E_H, Pauli E_x, and Coulomb E_c components:

$$E_{ee} = E_H + E_x + E_c, \tag{6.95}$$

where

$$E_H = \int \rho(\mathbf{r})\mathbf{r}\cdot \boldsymbol{\mathcal{E}}_H(\mathbf{r})d\mathbf{r}, \tag{6.96}$$

$$E_x = \int \rho(\mathbf{r})\mathbf{r}\cdot \boldsymbol{\mathcal{E}}_x(\mathbf{r})d\mathbf{r}, \tag{6.97}$$

$$E_c = \int \rho(\mathbf{r})\mathbf{r}\cdot \boldsymbol{\mathcal{E}}_c(\mathbf{r})d\mathbf{r}. \tag{6.98}$$

On multiplying the S system Schrödinger equation (6.10) by $\phi_k^\star(\mathbf{x})$, summing over all the fermions, and integrating over the spatial and spin coordinates, the kinetic energy T_s is obtained as

$$T_s = \sum_k \epsilon_k - \int \rho(\mathbf{r})v(\mathbf{r})d\mathbf{r} - \int \rho(\mathbf{r})v_{ee}(\mathbf{r})d\mathbf{r}$$
$$- \int \mathbf{j}(\mathbf{r})\cdot \mathbf{A}(\mathbf{r})d\mathbf{r} + \frac{1}{2}\int \rho(\mathbf{r})A^2(\mathbf{r})d\mathbf{r}, \tag{6.99}$$

with ϵ_k the eigenvalues of the differential equation (6.10).

On substituting for T_s from (6.99) into the expression for E of (6.90) one then obtains

$$E = \sum_k \epsilon_k - \int \rho(\mathbf{r}) v_{ee}(\mathbf{r}) d\mathbf{r} + E_{ee} + T_c \tag{6.100}$$

$$= \sum_k \epsilon_k - \int \rho(\mathbf{r}) v_{ee}(\mathbf{r}) d\mathbf{r} + E_H + E_x + E_c + T_c. \tag{6.101}$$

This expression for the total energy E in terms of the S system properties remains the same for the special case when $\mathcal{B}(\mathbf{r}) = 0$. In that case, the mapping from the interacting to the noninteracting model system is such that the density $\rho(\mathbf{r})$ and potential $v(\mathbf{r})$ of each system is the same.

Observe the similarity of this Q-DFT expression for E to those for Hartree-Fock (3.88) and Hartree (3.150) theories. Once again $E \neq \sum_k \epsilon_k$, but in this mapping *all* the electron correlations are accounted for, *viz*. those due to the Pauli principle, Coulomb repulsion and Correlation-Kinetic effects.

6.1.6 Sum Rules Satisfied by the Effective Field $\mathcal{F}^{\text{eff}}(\mathbf{r})$

The effective field $\mathcal{F}^{\text{eff}}(\mathbf{r})$ which incorporates all the many-body effects that the S system must account for, satisfies various sum rules. The effective field $\mathcal{F}^{\text{eff}}(\mathbf{r})$ is (see (6.76) - (6.78))

$$\mathcal{F}^{\text{eff}}(\mathbf{r}) = -\nabla v_{ee}(\mathbf{r}) = \mathcal{E}_{ee}(\mathbf{r}) + \mathcal{Z}_{t_c}(\mathbf{r}). \tag{6.102}$$

By operating on $\mathcal{F}^{\text{eff}}(\mathbf{r})$ by $\int d\mathbf{r} \rho(\mathbf{r})\mathbf{r}\cdot$, $\int d\mathbf{r} \rho(\mathbf{r})$, and $\int d\mathbf{r} \rho(\mathbf{r})\times$, the following sum rules are obtained. All that is required is that on application of these operators on the fields $\mathcal{E}_{ee}(\mathbf{r})$ and $\mathcal{Z}_{t_c}(\mathbf{r})$, the corresponding integrals vanish. (For the proofs, see Sect. 2.5).

Integral Virial Theorem

$$\int \rho(\mathbf{r}) \mathbf{r} \cdot \mathcal{F}^{\text{eff}}(\mathbf{r}) d\mathbf{r} = E_{ee} + 2T_c. \tag{6.103}$$

Zero Force Sum Rule

$$\int \rho(\mathbf{r}) \mathcal{F}^{\text{eff}}(\mathbf{r}) d\mathbf{r} = 0. \tag{6.104}$$

Zero Torque Sum Rule

$$\int \rho(\mathbf{r}) \mathbf{r} \times \mathcal{F}^{\text{eff}}(\mathbf{r}) d\mathbf{r} = 0. \tag{6.105}$$

6.1.7 Proof that Nonuniqueness of Effective Potential Energy is Solely Due to Correlation-Kinetic Effects

In the stationary-state Q-DFT mapping from the interacting system of electrons in an electromagnetic field to one of noninteracting fermions possessing the same densities $\{\rho(\mathbf{r}), \mathbf{j}(\mathbf{r})\}$ and potentials $\{v(\mathbf{r}), \mathbf{A}(\mathbf{r})\}$, the state of the model system is *arbitrary*. Hence, it is possible to map an interacting system in its *ground state* to a model S system also in its *ground state* or in *an excited state* with a *different* configuration. The corresponding effective electron-interaction potentials $v_{ee}(\mathbf{r})$ will then be different. In either case, the highest occupied eigenvalue ϵ_m is the negative of the ionization potential. Here we prove that the potentials $v_{ee}(\mathbf{r})$ of the different S systems differ *solely* in their Correlation-Kinetic components. The component due to the Pauli principle and Coulomb repulsion in the mapping remains the same.

Similarly, an interacting system in an *excited state* may be mapped to a model S system that is in an *excited state* of the *same* configuration, or an *excited state* with a *different* configuration, or even one in its *ground state*. In each case, the potential $v_{ee}(\mathbf{r})$ will be different. Again, the highest occupied eigenvalue ϵ_m will be the negative of the ionization potential. Once again, the potentials $v_{ee}(\mathbf{r})$ of the different S systems differ *solely* in their Correlation-Kinetic components, the component due to the Pauli principle and Coulomb repulsion remaining the same.

Consider the mapping from a *ground* or *excited* state of the interacting system of electrons in the presence of the electric and magnetic fields $\boldsymbol{\mathcal{E}}(\mathbf{r}) = -\boldsymbol{\nabla} v(\mathbf{r})/e$; $\boldsymbol{\mathcal{B}}(\mathbf{r}) = \boldsymbol{\nabla} \times \mathbf{A}(\mathbf{r})$, and with densities $\{\rho(\mathbf{r}), \mathbf{j}(\mathbf{r})\}$. Next consider two noninteracting systems S and S' experiencing the same electromagnetic field and possessing the same densities $\{\rho(\mathbf{r}), \mathbf{j}(\mathbf{r})\}$. For the S system, the Schrödinger equation is given by (6.10), and the electron-interaction potential $v_{ee}(\mathbf{r})$ in which all the many-electron effects are incorporated by (6.78) and (6.79).

For the S' system, the Schrödinger equation is

$$\left[\frac{1}{2}\big(\mathbf{p} + \mathbf{A}(\mathbf{r})\big)^2 + v(\mathbf{r}) + v'_{ee}(\mathbf{r})\right]\phi'_k(\mathbf{x}) = \epsilon'_k \phi'_k(\mathbf{x}); \quad k = 1, \ldots, N \tag{6.106}$$

with $v'_{ee}(\mathbf{r})$ the corresponding local electron-interaction potential. The 'Quantal Newtonian' first law satisfied by the model fermions of the S' system is

$$\boldsymbol{\mathcal{F}}^{\text{ext}}(\mathbf{r}) + \boldsymbol{\mathcal{F}}_s^{\text{int}'}(\mathbf{r}) = 0, \tag{6.107}$$

where the external field $\boldsymbol{\mathcal{F}}^{\text{ext}}(\mathbf{r})$ is defined by (6.45) and is the same as that experienced by the S system fermions. The internal field $\boldsymbol{\mathcal{F}}_s^{\text{int}'}(\mathbf{r})$ is

$$\boldsymbol{\mathcal{F}}_s^{\text{int}'}(\mathbf{r}) = -\boldsymbol{\nabla} v'_{ee}(\mathbf{r}) - \boldsymbol{\mathcal{Z}}'_s(\mathbf{r}) - \boldsymbol{\mathcal{D}}(\mathbf{r}) - \boldsymbol{\mathcal{I}}_m(\mathbf{r}), \tag{6.108}$$

with the fields $\boldsymbol{\mathcal{Z}}'_s(\mathbf{r})$, $\boldsymbol{\mathcal{D}}(\mathbf{r})$ and $\boldsymbol{\mathcal{I}}_m(\mathbf{r})$ as defined in Sect. 6.1.2.

On comparison of the S' system *'Quantal Newtonian' First Law* of (6.107) with that for the interacting system (6.71), one obtains that

$$v'_{ee}(\mathbf{r}) = -\int_{\infty}^{\mathbf{r}} \left[\mathcal{E}_{ee}(\mathbf{r}') + \mathcal{Z}'_{t_c}(\mathbf{r}')\right] \cdot d\boldsymbol{\ell}', \tag{6.109}$$

where the Correlation-Kinetic field $\mathcal{Z}'_{t_c}(\mathbf{r})$ is

$$\mathcal{Z}'_{t_c}(\mathbf{r}) = \mathcal{Z}'_s(\mathbf{r}) - \mathcal{Z}(\mathbf{r}). \tag{6.110}$$

Here $\mathcal{E}_{ee}(\mathbf{r})$ and $\mathcal{Z}(\mathbf{r})$ are the electron-interaction and kinetic fields of the interacting system, respectively, as defined in Sect. 2.2.

The difference between the $v_{ee}(\mathbf{r})$ and $v'_{ee}(\mathbf{r})$ of the S and S' systems is then

$$v_{ee}(\mathbf{r}) - v'_{ee}(\mathbf{r}) = -\int_{\infty}^{\mathbf{r}} \left[\mathcal{Z}_{t_c}(\mathbf{r}') - \mathcal{Z}'_{t_c}(\mathbf{r}')\right] \cdot d\boldsymbol{\ell}', \tag{6.111}$$

or equivalently

$$v_{ee}(\mathbf{r}) - v'_{ee}(\mathbf{r}) = -\int_{\infty}^{\mathbf{r}} \left[\mathcal{Z}_s(\mathbf{r}') - \mathcal{Z}'_s(\mathbf{r}')\right] \cdot d\boldsymbol{\ell}'. \tag{6.112}$$

The expression (6.111) for the difference is independent of the electron-interaction field $\mathcal{E}_{ee}(\mathbf{r})$ that accounts for Pauli and Coulomb correlations. Thus, its contribution to both $v_{ee}(\mathbf{r})$ and $v'_{ee}(\mathbf{r})$ is the same. The difference in the potentials $v_{ee}(\mathbf{r})$ and $v'_{ee}(\mathbf{r})$ depends on the difference between the Correlation-Kinetic fields $\mathcal{Z}_{t_c}(\mathbf{r})$ and $\mathcal{Z}'_{t_c}(\mathbf{r})$ of the S and S' systems. Or equivalently, the difference in the potentials is due to the difference in the kinetic fields $\mathcal{Z}_s(\mathbf{r})$ and $\mathcal{Z}'_s(\mathbf{r})$ of the S and S' systems. (See Chap. 5 of ***QDFT2*** for an illustrative example.)

6.1.8 *Physical Interpretation of Highest Occupied Eigenvalue* ϵ_m

Another property of the interacting system of electrons that may be determined via the mapping to the model S system is the first ionization potential I. It turns out that the highest occupied eigenvalue ϵ_m of the S system is the negative of the ionization potential I. To arrive at this interpretation, consider the N-electron system defined by the Hamiltonian $\hat{H}$ of (6.1) for a Coulombic external potential $v(\mathbf{r})$ and for the simpler case of $\mathcal{B}(\mathbf{r}) = 0$. As will be shown in Chap. 8, it is possible to determine the analytical solution to the corresponding Schrödinger equation in the classically forbidden region. With the asymptotic structure of the wave function known, the resulting asymptotic decay of the density is obtained as

$$\lim_{r\to\infty} \rho(\mathbf{r}) \sim \exp\big(-2\sqrt{2I}r\big), \tag{6.113}$$

where the ionization potential I is

$$I = E^{(N-1)} - E^{(N)}, \tag{6.114}$$

with $E^{(N)}$, $E^{(N-1)}$ the energies of the N-electron and the $(N-1)$-electron ion, respectively. The expression for the asymptotic structure of the density is valid for both ground and excited states.

Now *each* of the noninteracting fermions of the S system has the *same* potential $v_s(\mathbf{r})$ (see (6.8)). Hence, the asymptotic decay of each orbital $\phi_k(\mathbf{x})$ in the classically forbidden region depends on its corresponding eigenvalue ϵ_k. The orbital that decays the farthest is the highest occupied orbital $\phi_m(\mathbf{x})$, and depends on the highest occupied eigenvalue ϵ_m. Thus, the asymptotic structure of the density $\rho(\mathbf{r})$ of the S system is then

$$\lim_{r\to\infty} \rho(\mathbf{r}) = |\phi_m(x)|^2 \sim \exp\big(-2\sqrt{-2\epsilon_m}r\big). \tag{6.115}$$

This expression for the asymptotic density is valid whether the S system is in a ground or excited state.

A comparison of (6.113) and (6.115) then shows that the highest occupied eigenvalue of the S system is the negative of the first ionization potential:

$$\epsilon_m = -I. \tag{6.116}$$

Note that there is no such rigorous relationship between the remaining eigenvalues $\epsilon_k (k \neq m)$ and the other ionization potentials.

6.2 Application of Q-DFT to the Ground and First Excited Singlet State of a Quantum Dot in a Magnetic Field

We next elucidate the Q-DFT mapping from an interacting bound system of electrons in a magnetic field to one of *noninteracting fermions* possessing the same density $\rho(\mathbf{r})$ and physical current density $\mathbf{j}(\mathbf{r})$. The application is to the 2-electron 2-dimensional quantum dot in a magnetic field described previously in Chap. 5. Two mappings are performed: one from a 1^1S *ground state*, and the other from a first 2^1S *excited singlet state* of the quantum dot to two distinct 2-model-fermion S systems each in a 1^1S *ground state*. The constraints that in the mapping, the model S systems possess the same electron number N, orbital angular momentum $\mathbf{L}$, and spin angular momentum $\mathbf{S}$, is thus satisfied. This mapping [30], and the consequent S system properties: quantal sources, fields, potentials, and energies are described in detail. The results [28] of the mapping from the 1^1S ground state of the quantum dot to the S system in a 1^1S ground state are simply quoted.

The above mappings also demonstrate the following characteristics of Q-DFT:
(i) The mapping from the 2^1S *excited state* to one of model fermions in a 1^1S ground state demonstrates the *arbitrariness* of the *configuration* of the model system.
(ii) It turns out that each above mapping to a model system of 2 *noninteracting fermions* in their ground state, is equivalent to the corresponding Q-DFT mapping to a model B system [32, 33] of 2 *noninteracting bosons* also in their ground state possessing the same electronic density and physical current density as that of the interacting electronic system. (The differential equations for the model fermionic and bosonic systems turn out to be the same. The solutions in each case correspond to the density amplitude $\sqrt{\rho(\mathbf{r})}$.) This demonstrates the *arbitrariness* of the *statistics* of the particles of the model system. A brief description of the Q-DFT of the density amplitude is provided in Sect. 6.2.5 to illustrate the above equivalence.

6.2.1 Interacting Electronic System: The Quantum Dot

The stationary-state Schrödinger-Pauli theory equation for an N-electron semiconductor quantum dot in a magnetic field $\boldsymbol{\mathcal{B}}(\mathbf{r}) = \nabla \times \mathbf{A}(\mathbf{r})$ is

$$\hat{H}\Psi(\mathbf{X}) = E\Psi(\mathbf{X}), \tag{6.117}$$

where the Hamiltonian $\hat{H}$ is

$$\begin{aligned}\hat{H} &= \frac{1}{2m^\star}\sum_k \left(\hat{\mathbf{p}}_k + \frac{e}{c}\mathbf{A}(\mathbf{r}_k)\right)^2 + (g^\star\mu_B/\hbar)\sum_k \boldsymbol{\mathcal{B}}(\mathbf{r}_k)\cdot\mathbf{s}_k \\ &+ \frac{1}{2\epsilon}\sum_{k,\ell}{}' \frac{e^2}{|\mathbf{r}_k - \mathbf{r}_\ell|} + \frac{1}{2}m^\star\omega_0^2\sum_k r_k^2.\end{aligned} \tag{6.118}$$

Here the electrons are bound by a harmonic potential $v(\mathbf{r}) = \frac{1}{2}m^\star\omega_0^2 r^2$ with $m^\star$ the band effective mass and w_0 the binding frequency. The electron interaction is modified by the dielectric constant ϵ. The $g^\star$ is the corresponding gyromagnetic ratio; $\mu_B = e\hbar/2mc$ the Bohr magneton; and $\mathbf{s}$ the electron spin angular momentum vector operator.

Effective atomic units are employed: $e^2/\epsilon = \hbar = m^\star = c = 1$. The effective Bohr radius is $a_0^\star = a_0(m\epsilon/m^\star)$, where m is the free electron mass. The effective energy unit is $(a.u.)^\star = (a.u.)(m^\star/m\epsilon^2)$.

The Hamiltonian $\hat{H}$ for a 2-electron quantum dot in the *ground* and *first excited state* 2^1S state is then

$$\hat{H}_n = \sum_{k=1}^{2}\left\{\left(\hat{\mathbf{p}}_k + \mathbf{A}(\mathbf{r}_k)\right)^2 + \frac{1}{2}\omega_{0,n}^2 r_k^2\right\} + \frac{1}{|\mathbf{r}_1 - \mathbf{r}_2|}, \tag{6.119}$$

with $n = 0, 1$ indicating the ground and excited singlet states, respectively.

As was the case for the triplet 2^3S state (see Chap. 5), closed-form analytical solutions of the Schrödinger-Pauli equation.

$$\hat{H}_n \Psi_n(\mathbf{x}_1\mathbf{x}_2) = E_n \Psi_n(\mathbf{x}_1\mathbf{x}_2); \quad (n = 0, 1), \tag{6.120}$$

can be obtained for the ground and excited singlet states [28, 30, 55, 56] (see Chap. 9). In the symmetric gauge $\mathbf{A}(\mathbf{r}) = \frac{1}{2}\boldsymbol{\mathcal{B}}(\mathbf{r}) \times \mathbf{r}$ with the magnetic field in the z-direction: $\boldsymbol{\mathcal{B}}(\mathbf{r}) = B\mathbf{i}_z$, the wave functions are of the form

$$\Psi_n(\mathbf{x}_1\mathbf{x}_2) = \psi_n(\mathbf{r}_1\mathbf{r}_2)\chi_n(\boldsymbol{\sigma}_1\boldsymbol{\sigma}_2), \tag{6.121}$$

where the spatial component

$$\psi_n(\mathbf{r}_1\mathbf{r}_2) = C_n e^{-\Omega_n(R^2+\frac{1}{4}r^2)} f_n(r), \tag{6.122}$$

with C_n the normalization constant; $\mathbf{R} = (\mathbf{r}_1 + \mathbf{r}_2)/2$; $\mathbf{r} = \mathbf{r}_2 - \mathbf{r}_1$; $\Omega_n = \sqrt{k_{\text{eff}}^{(n)}}$; $k_{\text{eff}}^{(n)} = \Omega_n^2 = \omega_{0,n}^2 + \omega_L^2$; and $\omega_L = \mathcal{B}/2$ is the Larmor frequency. The functions $f_n(r)$ and the values of the various parameters are given below:

Ground State (n = 0)

$$\begin{aligned}
\Omega_0 &= \sqrt{k_{\text{eff}}^{(0)}} \;;\; k_{\text{eff}}^{(0)} = 1 \;;\; \Omega_0 = 1 \\
C_0 &= \Omega_0^{3/2} \Big/ \pi \left[2 + \Omega_0 + \sqrt{2\pi\Omega_0} \right]^{\frac{1}{2}} = 0.135646 \\
f_n(r) &= 1 + a_0 r \\
a_0 &= 1.000000.
\end{aligned}$$

Excited State (n = 1)

$$\begin{aligned}
\Omega_1 &= \sqrt{k_{\text{eff}}^{(1)}} \;;\; k_{\text{eff}}^{(1)} = 0.471716 \;;\; \Omega_1 = 0.686816 \\
C_1 &= 0.108563 \\
f_1(r) &= 1 + a_1 r + b_1 r^2 + c_1 r^3 \\
a_1 &= 1.000000 \;;\; b_1 = -0.26511 \;;\; c_1 = -0.182082.
\end{aligned}$$

The wave functions $\Psi_n(\mathbf{x}_1\mathbf{x}_2)$ are antisymmetric in an interchange of the two electrons. As both states considered are singlet states, the spin component $\chi_n(\sigma_1\sigma_2)$ is antisymmetric in an interchange of the spin coordinates σ_1 and σ_2. Hence, the spatial component $\psi_n(\mathbf{r}_1\mathbf{r}_2)$ is symmetric in an interchange of the spatial coordinates $\mathbf{r}_1$ and $\mathbf{r}_2$, i.e. $\psi_n(\mathbf{r}_1\mathbf{r}_2) = \psi_n(\mathbf{r}_2\mathbf{r}_1)$. The parity of $\psi_n(\mathbf{r}_1\mathbf{r}_2)$ is also *even* (see Sect. 8.3).

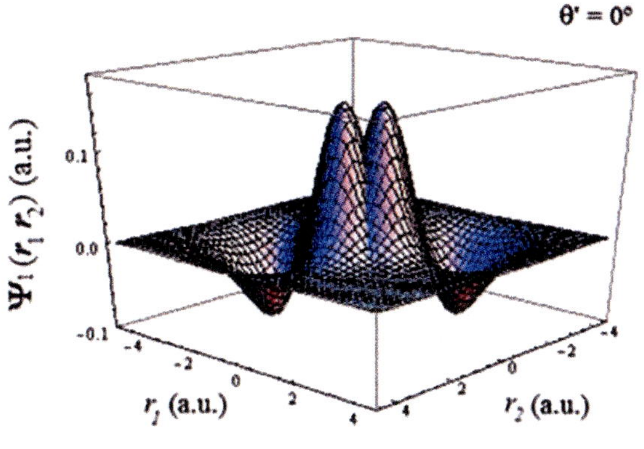

Fig. 6.1 Structure of the spatial component $\psi_1(\mathbf{r}_1\mathbf{r}_2)$ of the 2^1S state wave function of the quantum dot in a magnetic field. The angle between the vector $\mathbf{r}_1$ and $\mathbf{r}_2$ is $\theta' = 0°$. The vectors $\mathbf{r}_1$ and $\mathbf{r}_2$ are oriented along the positive and negative x-axis

The wave functions $\psi_n(\mathbf{r}_1\mathbf{r}_2)$ satisfy the electron-electron coalescence condition for dimensions $D = 2$. The $D \geq 2$ structure of the wave function for the coalescence of two particles of masses m_1, m_2 and charge Z_1, Z_2 is [33, 57]

$$\psi(\mathbf{r}_1\mathbf{r}_2, \ldots, \mathbf{r}_N) = \psi(\mathbf{r}_2\mathbf{r}_2, \mathbf{r}_3, \ldots, \mathbf{r}_N)\left(1 + \frac{2Z_2Z_2\mu}{D-1}r\right) + \mathbf{r}\cdot\mathbf{C}(\mathbf{r}_2, \mathbf{r}_3, \ldots, \mathbf{r}_N) + O(r^2), \tag{6.123}$$

where $\mu = m_1m_2/m_1 + m_2$, $r = |\mathbf{r}_2 - \mathbf{r}_1|$, and $\mathbf{C}$ an unknown vector. For these singlet states, the coalescence condition is a *cusp* coalescence condition. Note that the wave functions $\psi_n(\mathbf{r}_1\mathbf{r}_2)$ are both real. As such, the paramagnetic current density component $\mathbf{j}_p(\mathbf{r})$ of the physical current density $\mathbf{j}(\mathbf{r})$ vanishes. As both the ground and excited states are singlets, there is also no magnetization current density $\mathbf{j}_m(\mathbf{r})$. Thus, $\mathbf{j}(\mathbf{r})$ is the diamagnetic component $\mathbf{j}_d(\mathbf{r})$:

$$\mathbf{j}_n(\mathbf{r}) = \mathbf{j}_{d,n}(\mathbf{r}) = \rho_n(\mathbf{r})\mathbf{A}(\mathbf{r}), \tag{6.124}$$

where the density $\rho_n(\mathbf{r}) = \langle\Psi_n|\hat{\rho}(\mathbf{r})|\Psi_n\rangle$ with $\hat{\rho}(\mathbf{r})$ the density operator of (2.12).

The wave function $\psi_1(\mathbf{r}_1\mathbf{r}_2)$ for the singlet 2^1S state is plotted in Figs. 6.1, 6.2 and 6.3 for different angles $\theta' = 0°, 45°, 90°$ between the vectors $\mathbf{r}_1$ and $\mathbf{r}_2$, and for different orientations of these vectors. Observe the even parity of the wave function. Also observe the electron-electron coalescence cusp as exhibited by the wave function for $\mathbf{r}_1 = \mathbf{r}_2$ in Fig. 6.1. The cusp is particularly evident for $\mathbf{r}_1 = \mathbf{r}_2 = 0$ in this figure. The node of this excited state wave function is also evident.

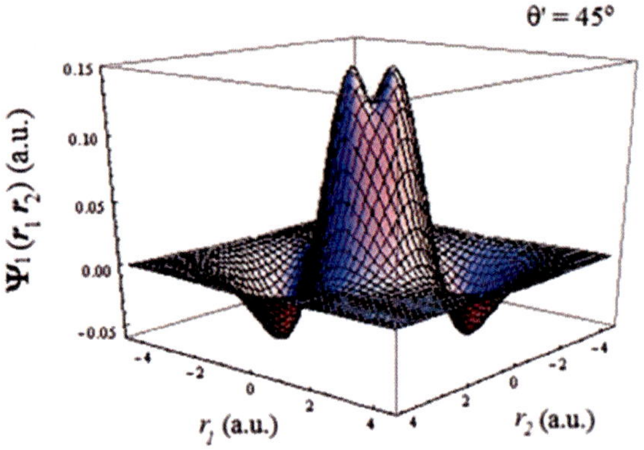

Fig. 6.2 The same as in Fig. 6.1 but for $\theta' = 45°$. Vector $\mathbf{r}_1$ is along the $\pm x$-axis, and vector $\mathbf{r}_2$ is at 45° from the $\pm x$-axis

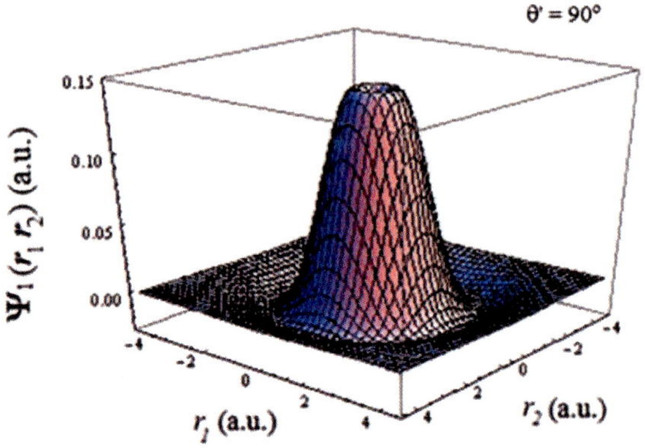

Fig. 6.3 The same as in Fig. 6.1 but for $\theta' = 90°$. Vector $\mathbf{r}_1$ is along the $\pm x$-axis, and vector $\mathbf{r}_2$ is along the $\pm y$-axis

6.2.2 Noninteracting Model Fermion System

We next map the interacting electronic system of the 2-electron quantum dot to a 2-model-fermion system in its ground state possessing the same density $\rho(\mathbf{r})$ and physical current density $\mathbf{j}(\mathbf{r})$. As the model system is in a ground state, the orbital $\mathbf{L}$ and spin $\mathbf{S}$ angular momentum is the same as that of the interacting system in either its ground or first excited singlet state.

The differential equation for the model system with the same potentials $\{v, \mathbf{A}\}$ as that of the quantum dot is

$$\left[\frac{1}{2}p^2 + \frac{1}{2}k_{\text{eff}}r^2 + v_{ee}(\mathbf{r})\right]\phi_k(\mathbf{x}) = \epsilon_k\phi_k(\mathbf{x}); \quad k = 1, 2. \tag{6.125}$$

(The superscript (n) indicating the ground $(n = 0)$ and first excited singlet $(n = 1)$ state of the interacting system is suppressed. Thus, $k_{\text{eff}} \equiv k_{\text{eff}}^{(n)} = \omega_{0,n}^2 + \omega_L^2$; $v_{ee}(\mathbf{r}) = v_{ee,n}(\mathbf{r})$; $\phi_k(\mathbf{x}) \equiv \phi_{k,n}(\mathbf{x})$; $n = 0, 1$.)

The corresponding wave function is the Slater determinant $\Phi\{\phi_k\}$ of the spin-orbitals $\phi_k(\mathbf{x})$ $(k = 1, 2)$ with

$$\phi_1(\mathbf{x}) = \psi(\mathbf{r})\alpha(\sigma) \ ; \ \ \phi_2(\mathbf{x}) = \psi(\mathbf{r})\beta(\sigma), \tag{6.126}$$

where the normalized $\psi(\mathbf{r})$ is the spatial part of the spin-orbital, and $\alpha(\sigma)$, $\beta(\sigma)$ the spin functions. Since the model fermions have opposite spin, the density $\rho(\mathbf{r})$ is

$$\begin{aligned}\rho(\mathbf{r}) &= \langle \Phi\{\phi_k\}|\hat{\rho}(\mathbf{r})|\Phi\{\phi_k\}\rangle \\ &= 2\psi^\star(\mathbf{r})\psi(\mathbf{r}).\end{aligned} \tag{6.127}$$

Thus, the model S system orbitals $\psi(\mathbf{r})$ are known in terms of the density $\rho(\mathbf{r})$ as

$$\psi(\mathbf{r}) = \sqrt{\frac{\rho(\mathbf{r})}{2}}. \tag{6.128}$$

As the wave functions $\Psi_n(\mathbf{x}_1\mathbf{x}_2)$ for the ground $(n = 0)$ and excited singlet $(n = 1)$ state of the interacting system are known, so are the corresponding densities $\rho_n(\mathbf{r})$. Thus, the ground state model fermion orbitals $\psi_n(\mathbf{r})$ of the two S systems are exactly known via (6.128).

Note also that although the wave function for the quantum dot in the excited state possesses a node (see Figs. 6.1, 6.2 and 6.3), the corresponding model system spatial orbitals, and hence the resulting Slater determinant, is nodeless.

Recall that the physical current density $\mathbf{j}(\mathbf{r})$ for both the ground and excited singlet state of the quantum dot is its diamagnetic component. Thus, $\mathbf{j}(\mathbf{r})$ is proportional to $\rho(\mathbf{r})$ (see (6.124)). Hence, for the model systems to reproduce the density $\rho(\mathbf{r})$ and physical current density $\mathbf{j}(\mathbf{r})$, all that is required is for the spatial orbitals $\psi(\mathbf{r})$ to reproduce the density $\rho(\mathbf{r})$.

Many properties of the mapping from the excited singlet state of the quantum dot to the model system in a ground state can be obtained in closed analytical or semianalytical form. The asymptotic structure of these properties in the classically forbidden region and near the center of the model system can also be obtained analytically. These expressions are given in Appendix D.

We next discuss the various properties.

6.2.3 Quantal Sources

Electron Density $\rho_1(\mathbf{r})$ and Physical Current Density $\mathbf{j}_1(\mathbf{r})$

The analytical expressions for the density $\rho_1(\mathbf{r})$ as obtained from the wave function $\psi_1(\mathbf{r}_1\mathbf{r}_2)$ of (6.122), and of its asymptotic structure in the classically forbidden and near the center of the quantum dot are given in Appendix D. The electron density $\rho_1(\mathbf{r})$ (which has cylindrical symmetry: $\rho_1(\mathbf{r}) = \rho_1(r)$), and the radial probability density

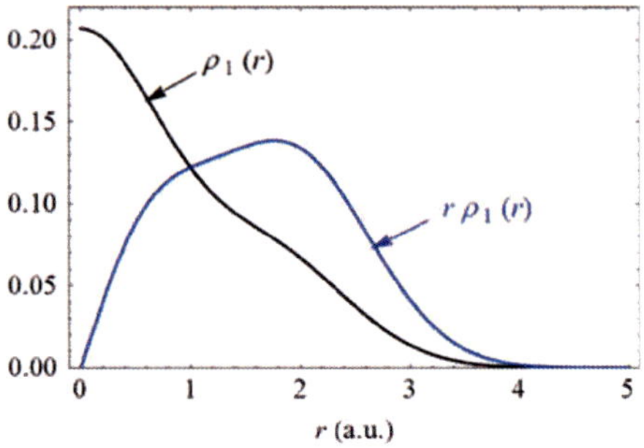

Fig. 6.4 The electron density $\rho_1(r)$ and the radial probability density $r\rho_1(r)$

$r\rho_1(r)$, are plotted in Fig. 6.4. Observe that the electron density does not exhibit a cusp at the center of the quantum dot but approaches it quadratically. The shell structure for this excited singlet state is evident in the plots of both the density $\rho_1(r)$ and the radial probability density $r\rho_1(r)$. As the wavefunction $\psi_1(\mathbf{r}_1\mathbf{r}_2)$ is real, and because there is no magnetization current density component, the physical current density $\mathbf{j}_1(\mathbf{r})$ is its diamagnetic component so that $\mathbf{j}_1(\mathbf{r}) = \rho_1(\mathbf{r})\mathbf{A}(\mathbf{r})$. As such the current density $\mathbf{j}_1(\mathbf{r})$ is not plotted. It satisfies the continuity condition $\boldsymbol{\nabla} \cdot \mathbf{j}_1(\mathbf{r}) = 0$.

***Density Matrices: Single-Particle** $\gamma(\mathbf{rr}')$; **Dirac** $\gamma_s(\mathbf{rr}')$; **Correlation-Kinetic** $\gamma_{t_c}(\mathbf{rr}')$*

The single-particle density matrix $\gamma(\mathbf{rr}')$, whose semi-analytical expression is given in Appendix D, is plotted in Fig. 6.5. The different panels correspond to (a) $\theta = 0°, \theta' = 0°$; (b) $\theta = 0°, \theta' = 60°$; (c) $\theta = 0°, \theta' = 90°$. The *nonlocal* structure is quite evident from these panels. In particular, note the change in the structure of the shoulder. Observe the existence of the nodes of $\gamma(\mathbf{rr}')$, a consequence of the node in the wave function $\psi_1(\mathbf{r}_1\mathbf{r}_2)$. In each panel of Fig. 6.5, shell structure is also clearly evident. The density matrix $\gamma(\mathbf{rr}')$ is finite at the center of the quantum dot irrespective of the direction in which the center is approached. The diagonal matrix element $\gamma(\mathbf{rr})$ is the density $\rho_1(\mathbf{r})$ plotted in Fig. 6.4. This too is evident in Fig. 6.5.

The analytical expression for the Dirac density matrix $\gamma_s(\mathbf{rr}')$ for the model fermion system in a ground state as given in Appendix D is plotted in Fig. 6.6. Observe that $\gamma_s(\mathbf{rr}')$ is cylindrically symmetric. As the matrix is for a ground state, and since the orbitals $\psi(\mathbf{r})$ are proportional to the density $\rho_1(\mathbf{r})$ (see (6.128)), there are no nodes, and the matrix is positive definite. It exhibits shell structure.

Finally, the quantal source for the Correlation-kinetic effects $\gamma_{t_c}(\mathbf{rr}') = \gamma_s(\mathbf{rr}') - \gamma(\mathbf{rr}')$ is plotted in Fig. 6.7. The different panels once again correspond to (a) $\theta = 0°, \theta' = 0°$; (b) $\theta = 0°, \theta' = 60°$; (c) $\theta = 0°, \theta' = 90°$. Its *nonlocal* structure is evident. As the Dirac density matrix $\gamma_s(\mathbf{rr}')$ is positive, so is the Correlation-Kinetic source $\gamma_{t_c}(\mathbf{rr}')$. As a consequence, the Correlation-Kinetic energy T_c is positive (See Table 6.1). Also observe that the magnitude of $\gamma_{t_c}(\mathbf{rr}')$ is a significant fraction of $\gamma(\mathbf{rr}')$. As such the Correlation-Kinetic energy T_c is a substantial fraction of the kinetic energy T (see Table 6.1).

Fig. 6.5 The single-particle density matrix $\gamma(\mathbf{r}\mathbf{r}')$ for the first excited singlet state 2^1S of the quantum dot in a magnetic field. The panels correspond to (**a**) $\theta = 0°, \theta' = 0°$; (**b**) $\theta = 0°, \theta' = 60°$; (**c**) $\theta = 0°, \theta' = 90°$. This is the *nonlocal* quantal source for the interacting system kinetic field $\mathcal{Z}(\mathbf{r})$

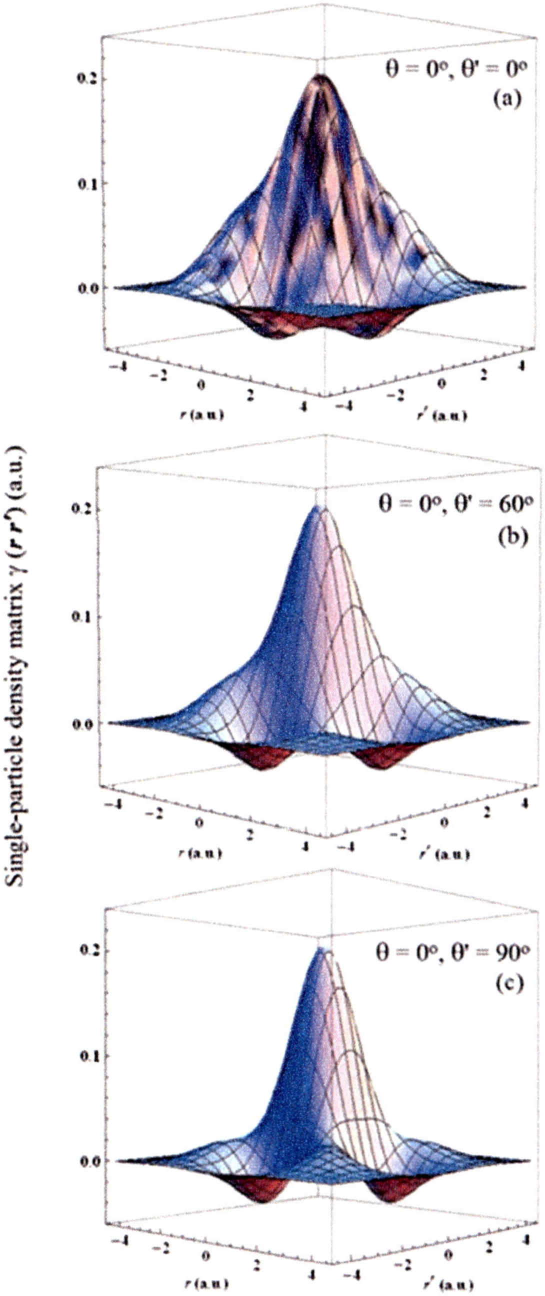

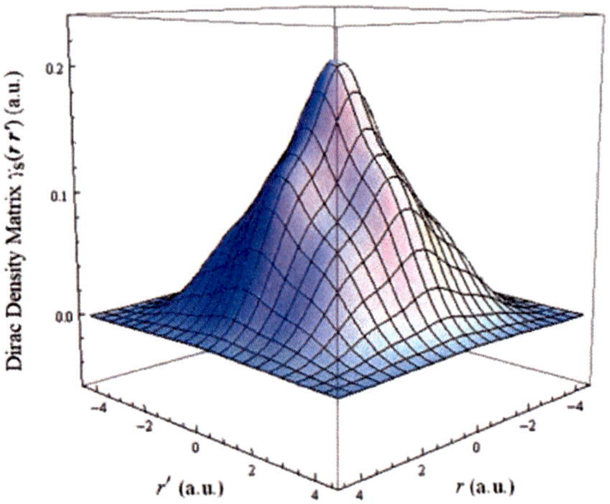

Fig. 6.6 The Dirac density matrix $\gamma_s(\mathbf{rr}')$ for the S system in its ground state possessing the same density $\rho_1(\mathbf{r})$ as that of the first excited singlet 2^1S state of the quantum dot in a magnetic field. This is the quantal source for the noninteracting system kinetic field $\boldsymbol{\mathcal{Z}}_s(\mathbf{r})$

Pair-Correlation Density $g_s(\mathbf{rr}')$; ***Fermi*** $\rho_x(\mathbf{rr}')$; ***and Coulomb Hole*** $\rho_c(\mathbf{rr}')$ ***Charge Distributions***

The model S system pair-correlation density $g_s(\mathbf{rr}')$ of (6.32) is

$$g_s(\mathbf{rr}') = \rho_1(\mathbf{r}')/2. \tag{6.129}$$

In local effective potential theory, it is customary to define a Fermi hole $\rho_x(\mathbf{rr}')$ for a two model-fermion system for which the fermions have opposite spin. Thus, employing the decomposition of the pair-correlation density $g_s(\mathbf{rr}')$ of (6.32), the expression for the Fermi hole is

$$\rho_x(\mathbf{rr}') = -\rho_1(\mathbf{r}')/2. \tag{6.130}$$

This is a *local* charge distribution independent of the electron position. The model system Coulomb hole $\rho_c(\mathbf{rr}')$ is the difference between the Fermi-Coulomb $\rho_{xc}(\mathbf{rr}')$ and Fermi $\rho_x(\mathbf{rr}')$ hole distributions. Thus,

$$\rho_c(\mathbf{rr}') = \rho_{xc}(\mathbf{rr}') + \rho_1(\mathbf{r}')/2, \tag{6.131}$$

which is a *nonlocal* charge distribution whose structure depends upon the electron position. The Fermi-Coulomb hole $\rho_{xc}(\mathbf{rr}')$ in turn is obtained from the interacting system pair-correlation density $g(\mathbf{rr}')$ whose expression is given in Appendix D. Thus, as the Fermi hole $\rho_x(\mathbf{rr}')$ is a local charge distribution, the intrinsic nonlocality of the interacting system as exhibited via its Fermi-Coulomb hole $\rho_{xc}(\mathbf{rr}')$, is now reflected via the nonlocality of the Coulomb hole $\rho_c(\mathbf{rr}')$.

In Fig. 6.8(a) the cross-sections through the Fermi-Coulomb $\rho_{xc}(\mathbf{rr}')$, Fermi $\rho_x(\mathbf{rr}')$, and Coulomb $\rho_c(\mathbf{rr}')$ holes for an electron at the center of the quantum dot at $r = 0$ are plotted. The electron position is indicated by the arrow. Observe that

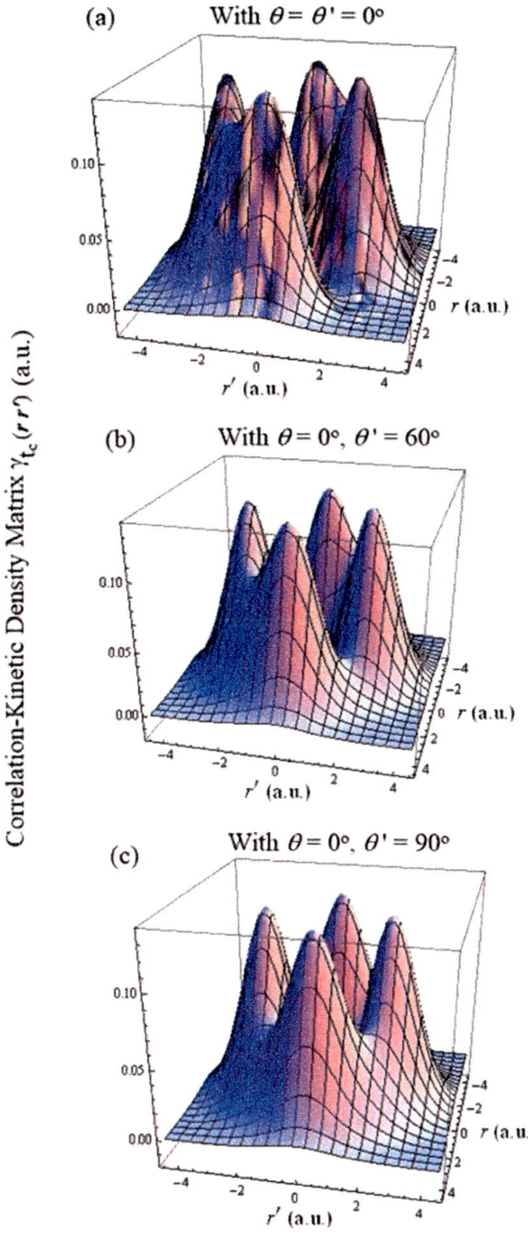

Fig. 6.7 The Correlation-Kinetic density matrix $\gamma_{t_c}(\mathbf{rr}')$. This is the *nonlocal* quantal source for the Correlation-Kinetic field $\mathcal{Z}_{t_c}(\mathbf{r})$

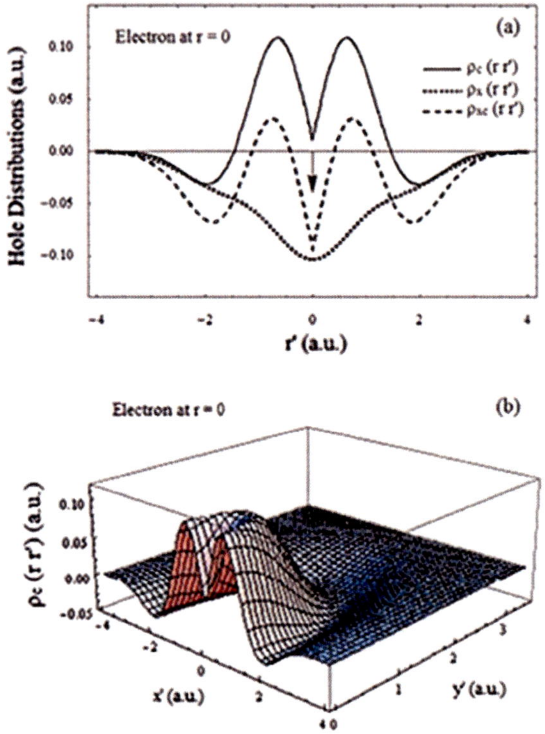

Fig. 6.8 (**a**) Cross-sections through the Fermi-Coulomb $\rho_{xc}(\mathbf{rr}')$, Fermi $\rho_x(\mathbf{rr}')$, and Coulomb $\rho_c(\mathbf{rr}')$ holes for an electron at the center of the quantum dot at $r = 0$. The electron position indicated by the arrow. (**b**) Surface plot of the Coulomb hole $\rho_c(\mathbf{rr}')$ for an electron at the same position. Here x' is the projection of $\mathbf{r}'$ on $\mathbf{r}$, i.e., $x' = r'\hat{\mathbf{r}} \cdot \hat{\mathbf{r}}'$, and y' is the projection of $\mathbf{r}'$ on the direction perpendicular to $\mathbf{r}$, i.e., $y' = r'[1 - (\hat{\mathbf{r}} \cdot \hat{\mathbf{r}}')^2]^{\frac{1}{2}}$

for this electron position, these charge distributions are all cylindrically symmetric about the z-axis. (As a consequence, the corresponding fields vanish at this point.) Also observe how the electron-electron coalescence condition on the wave function is exhibited in the corresponding Fermi-Coulomb and Coulomb hole distributions. A better pictorial sense of the full structure of the Coulomb hole $\rho_c(\mathbf{rr}')$ is obtained via Fig. 6.8(b) in which a surface plot of the hole for the same electron position is presented. Here too, observe the electron-electron coalescence cusp at the electron position. (For this plot, x' is the projection of $\mathbf{r}'$ on $\mathbf{r}$, i.e., $x' = r'\hat{\mathbf{r}} \cdot \hat{\mathbf{r}}'$, and y' is the projection of $\mathbf{r}'$ on the direction perpendicular to $\mathbf{r}$, i.e., $y' = r'[1 - (\hat{\mathbf{r}} \cdot \hat{\mathbf{r}}')^2]^{\frac{1}{2}}$.)

In the panels (a) of Figs. 6.9, 6.10, 6.11 and 6.12, cross sections of the Coulomb hole $\rho_c(\mathbf{rr}')$ in different directions $\theta' = 0°, 45°, 90°$ with respect to the center of quantum dot - electron direction are plotted. The electron positions considered are at $r = 0.5, 1, 2, 20$ $a.u.$. In the panels (b), the surface plots of the Coulomb hole $\rho_c(\mathbf{rr}')$ for the same electron positions are shown. Observe the nonlocal (dynamic) structure of the Coulomb hole, and the fact that it is not symmetrical about the electron position. In Figs. 6.9 and 6.10, the electron-electron coalescence cusp is also clearly exhibited. The cusp is too weak to be exhibited in Figs. 6.11 and 6.12 on the scale of these figures. Note that for asymptotic positions of the electron (Fig. 6.12), the Coulomb hole is becoming essentially cylindrically symmetric and static. For far

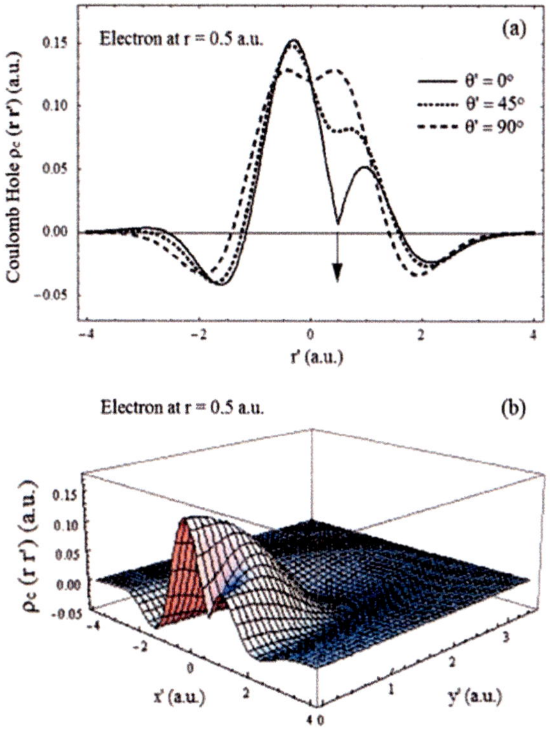

Fig. 6.9 (**a**) Cross sections through the Coulomb hole $\rho_c(\mathbf{rr}')$ in different directions corresponding to $\theta' = 0°, 45°, 90°$ with respect to the center of the quantum dot - electron direction. The electron is at $r = 0.5\ a.u.$ as indicated by the arrow. (**b**) Surface plot of the Coulomb hole for an electron at the same position

asymptotic positions of the electron in the classically forbidden region, the hole is symmetric and static.

6.2.4 *Fields, Potentials, Energies, and Eigenvalues*

Fields

The analytical expressions for the electron-interaction field $\mathcal{E}_{ee}(\mathbf{r})$, and of its asymptotic structure in the classically forbidden region and near the center of the quantum dot, are given in Appendix D. Its Hartree $\mathcal{E}_H(\mathbf{r})$, Pauli $\mathcal{E}_x(\mathbf{r})$, and Coulomb $\mathcal{E}_c(\mathbf{r})$ components are plotted in Fig. 6.13. The structure of these fields is predictive and understandable based on the structure of the corresponding quantal sources. Observe that the Hartree field $\mathcal{E}_H(\mathbf{r})$ is positive, and the Pauli field $\mathcal{E}_x(\mathbf{r})$ negative throughout space. This is a result of the density $\rho_1(\mathbf{r})$ being positive and the Fermi hole charge $\rho_x(\mathbf{rr}')$ negative. As a result of the sum rule $\int \rho_c(\mathbf{rr}')d\mathbf{r}' = 0$, the Coulomb field is both positive and negative. Recall that the electron density $\rho_1(\mathbf{r})$, Fermi $\rho_x(\mathbf{rr}')$, and Coulomb $\rho_c(\mathbf{rr}')$ holes are all cylindrically symmetric for an electron position at the center. Hence, the fields $\mathcal{E}_H(\mathbf{r})$, $\mathcal{E}_x(\mathbf{r})$, $\mathcal{E}_c(\mathbf{r})$ must all vanish there. In proceeding

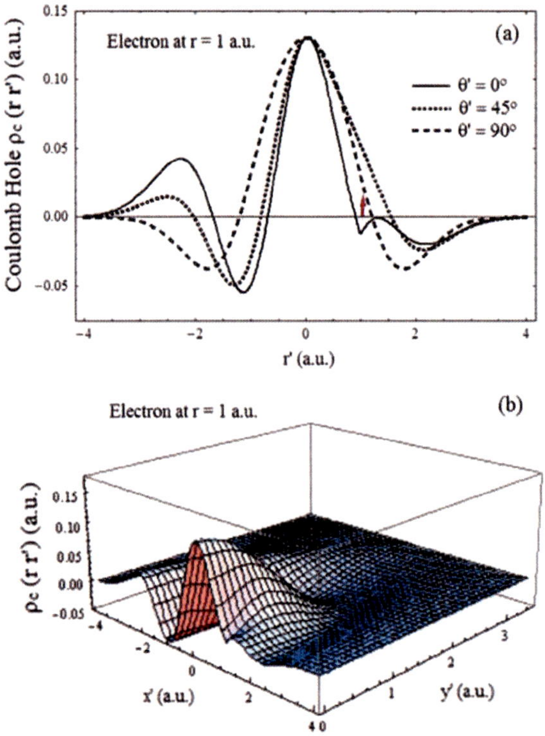

Fig. 6.10 The same as in Fig. 6.9 except that the electron is at $r = 1\ a.u.$

away from the center, each field is seen to exhibit shell structure, as must be the case. Finally, as the density $\rho_1(\mathbf{r})$ and Fermi hole $\rho_x(\mathbf{rr}')$ are static charge distributions satisfying the constraints $\int \rho_1(\mathbf{r})d\mathbf{r} = 2$ and $\int \rho_x(\mathbf{rr}')d\mathbf{r}' = -\frac{1}{2}\int \rho_1(\mathbf{r}')d\mathbf{r}' = -1$, the asymptotic structure of the Hartree $\mathcal{E}_H(\mathbf{r})$ and Pauli $\mathcal{E}_x(\mathbf{r})$ fields in the classically forbidden region to leading order is

$$\mathcal{E}_H(\mathbf{r}) \underset{r\to\infty}{\sim} 2/r^2, \tag{6.132}$$

$$\mathcal{E}_x(\mathbf{r}) \underset{r\to\infty}{\sim} -1/r^2. \tag{6.133}$$

The asymptotic structure of the Coulomb field $\mathcal{E}_c(\mathbf{r})$ to leading order turns out to be

$$\mathcal{E}_c(\mathbf{r}) \underset{r\to\infty}{\sim} -10.1/r^4. \tag{6.134}$$

The merging of the above structures with the functions $2/r^2$, $-1/r^2$, and $-10.1/r^4$, respectively, is clearly evident in Fig. 6.13.

For the interacting $\mathcal{Z}(\mathbf{r})$ and noninteracting $\mathcal{Z}_s(\mathbf{r})$ system kinetic fields, the quantal sources are the single particle density matrix $\gamma(\mathbf{rr}')$ and Dirac density matrix

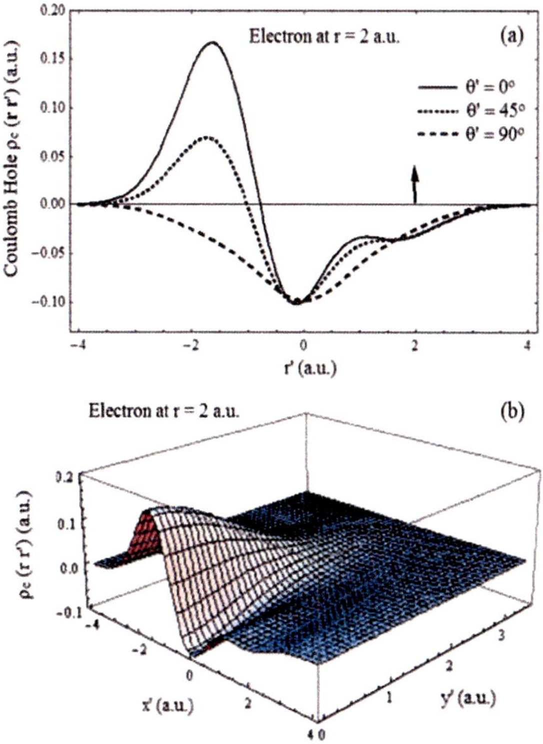

Fig. 6.11 The same as in Fig. 6.9 except that the electron is at $r = 2\ a.u.$

$\gamma_s(\mathbf{rr}')$, respectively. The expressions for the kinetic energy tensor for the interacting $t_{\alpha\beta}(\mathbf{r};\gamma)$ and noninteracting $t_{s,\alpha\beta}(\mathbf{r};\gamma_s)$ systems, and those for the corresponding 'forces' $z_\alpha(\mathbf{r};\gamma)$ and $z_{\alpha,s}(\mathbf{r};\gamma_s)$, as well as the asymptotic structures of the 'forces' are given in Appendix D. The derivation of the interacting system kinetic energy tensor $t_{\alpha\beta}(\mathbf{r};\gamma)$ and the corresponding kinetic 'force' $z_\alpha(\mathbf{r};\gamma)$ is similar to that for the triplet 2^3S state given in Appendix C and is not repeated. The kinetic 'forces' $z(r)$ and $z_s(r)$ are plotted in Fig. 6.14(a). The Correlation-Kinetic field $\mathcal{Z}_{t_c}(\mathbf{r})$ is given in Fig. 6.14(b). As both $z(r)$ and $z_s(r)$ vanish at the origin, so does the Correlation-Kinetic field $\mathcal{Z}_{t_c}(\mathbf{r})$. This field exhibits shell structure, and decays asymptotically in the classically forbidden region as a positive function:

$$\mathcal{Z}_{t_c}(\mathbf{r}) \underset{r\to\infty}{\sim} 9/r^3. \tag{6.135}$$

Observe that the field $\mathcal{Z}_{t_c}(\mathbf{r})$ decays more slowly than the Coulomb field $\mathcal{E}_c(\mathbf{r})$.

A comparison of all the fields is made in Fig. 6.15. Observe that the magnitude of the Correlation-Kinetic field $\mathcal{Z}_{t_c}(\mathbf{r})$ is much larger than those of the components of the electron-interaction field $\mathcal{E}_{ee}(\mathbf{r})$. This will then reflect on the respective magnitudes of the corresponding energies as discussed below in Table 6.1.

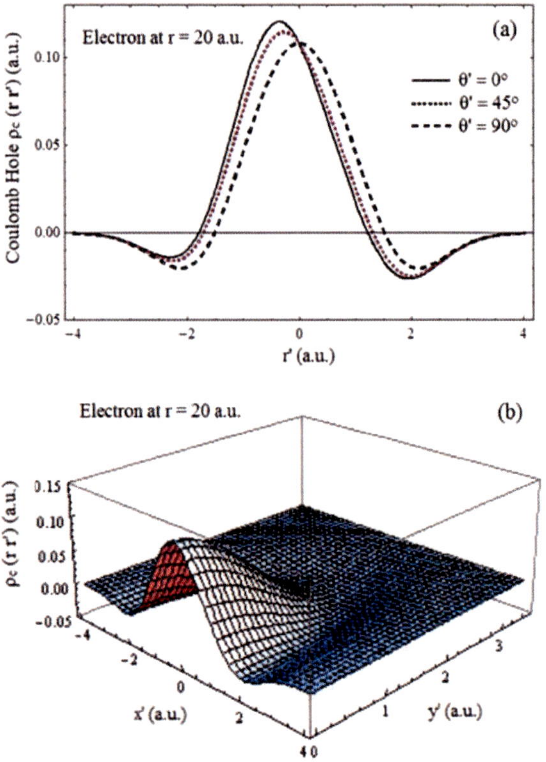

Fig. 6.12 The same as in Fig. 6.9 except that the electron is at $r = 20\ a.u.$

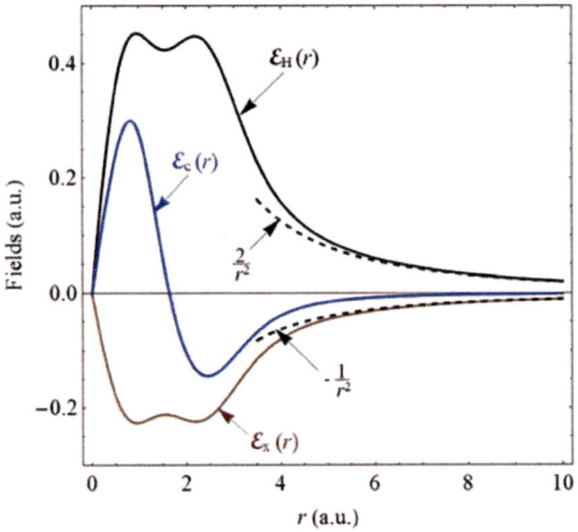

Fig. 6.13 The Hartree $\mathcal{E}_H(r)$, Pauli $\mathcal{E}_x(r)$, and Coulomb $\mathcal{E}_c(r)$ fields. The functions $2/r^2$ and $-1/r^2$ and $-10.1/r^4$ are also plotted in the asymptotic region. The Coulomb field decays as $-10.1/r^4$ to leading order

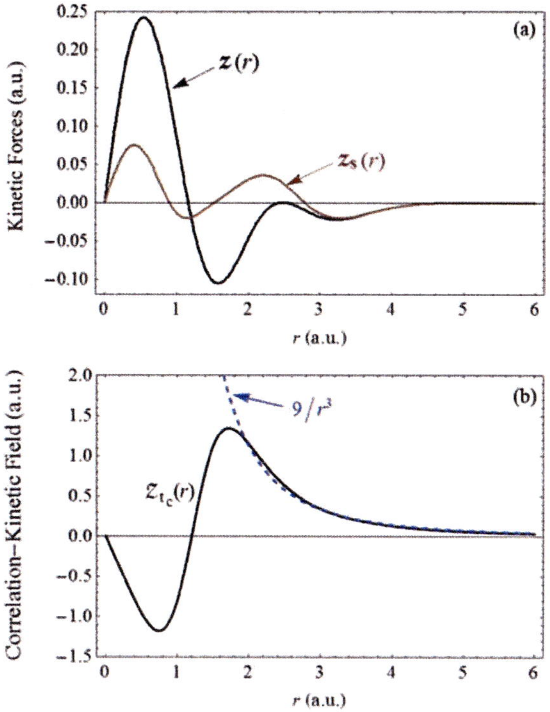

Fig. 6.14 (**a**) Kinetic 'forces' $z(r)$, $z_s(r)$ for the interacting and noninteracting systems, respectively. (**b**) The correlation-kinetic field $\mathcal{Z}_{t_c}(r)$. Its asymptotic decay is of $9/r^3$ to leading order

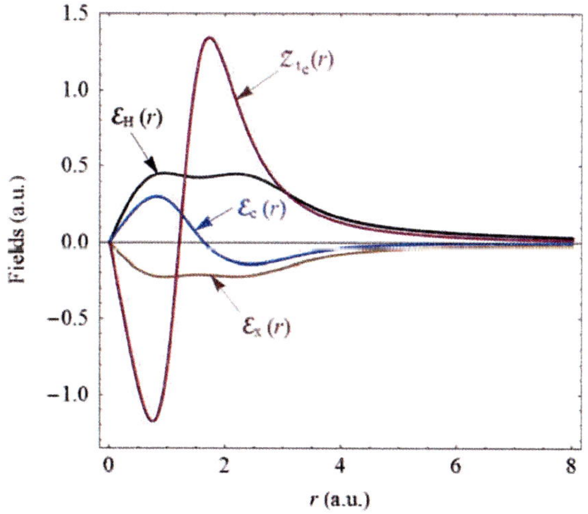

Fig. 6.15 A comparison of the Hartree $\mathcal{E}_H(r)$, Pauli $\mathcal{E}_x(r)$, and Coulomb $\mathcal{E}_c(r)$, and correlation-kinetic $\mathcal{Z}_{t_c}(\mathbf{r})$ fields

Table 6.1 Quantal density functional theory (Q-DFT) properties [30] (Column 1) (in effective atomic units) of the model system in its *ground state* that reproduces the density, physical current density, orbital and spin angular momentum, and total energy E of the quantum dot in a magnetic field in its 2^1S state with effective force constant $k_{\text{eff}} = 0.471716$ (Column 2). Expectation values of single-particle operators are also quoted. For comparison, the corresponding properties [28, 32] for the QDFT mapping from a ground state of the quantum dot in a magnetic field with effective force constant $k_{\text{eff}} = 1$ to a model system in its ground state are shown (Column 3)

Property	Value in (a.u.)*			
	Q-DFT mapping from			
	$2^1S \to 1^1S$		$1^1S \to 1^1S$	
	$k_{\text{eff}} = 0.471716$		$k_{\text{eff}} = 1$	
E	3.434066		3.000000	
E_{ext}	1.566907		1.295400	
E_{ee}	0.600476		0.818401	
E_H	1.343218		1.789832	
E_x	−0.671609		−0.894916	
	$\frac{\lvert E_x\rvert}{E} = 20\%$		$\frac{\lvert E_x\rvert}{E} = 30\%$	
E_c	−0.071133		−0.076515	
	$\frac{\lvert E_c\rvert}{E} = 2.1\%$		$\frac{\lvert E_c\rvert}{E} = 2.6\%$	
T	1.266683		0.886199	
T_s	0.338856		0.780987	
T_c	0.927827		0.105212	
	$\frac{T_c}{T} = 73\%$	$\frac{T_c}{E} = 27\%$	$\frac{T_c}{T} = 12\%$	$\frac{T_c}{E} = 3.5\%$
ϵ	2.746374		2.000000	
	$\frac{\epsilon}{E} = 80\%$		$\frac{\epsilon}{E} = 67\%$	
$\langle\delta(\mathbf{r})\rangle = \rho(0)$	0.207299		0.436132	
$\langle r\rangle$	3.276767		2.037894	
$\langle r^2\rangle$	6.643441		2.590800	
$\langle 1/r\rangle$	1.930010		2.996873	

Potentials

As a consequence of cylindrical symmetry, the electron-interaction field $\boldsymbol{\mathcal{E}}_{ee}(\mathbf{r})$, and its Hartree $\boldsymbol{\mathcal{E}}_H(\mathbf{r})$, Pauli $\boldsymbol{\mathcal{E}}_x(\mathbf{r})$, and Coulomb $\boldsymbol{\mathcal{E}}_c(\mathbf{r})$ components, and the Correlation-Kinetic field $\boldsymbol{\mathcal{Z}}_{t_c}(\mathbf{r})$, are all conservative. Hence, the local effective electron-interaction potential $v_{ee}(\mathbf{r})$ of (6.79) can be expressed as a sum of its Hartree $W_H(\mathbf{r})$, Pauli $W_x(\mathbf{r})$, Coulomb $W_c(\mathbf{r})$, and Correlation-Kinetic $W_{t_c}(\mathbf{r})$ components (see (6.85)). Each of these components of the potential is the work done in the corresponding field. A comparison of these potentials is made in Fig. 6.16. Their structure follows from that of the fields. For example, since the Pauli field $\boldsymbol{\mathcal{E}}_x(\mathbf{r})$ is negative, so is the potential $W_x(\mathbf{r})$. As $\boldsymbol{\mathcal{E}}_x(\mathbf{r})$ decays asymptotically as $-1/r^2$, the potential $W_x(\mathbf{r})$ decays as $-1/r$. The Coulomb field $\boldsymbol{\mathcal{E}}_c(\mathbf{r})$ is both positive and negative, and thus so is the Coulomb potential $W_c(\mathbf{r})$ although it is principally negative, and so on. The asymptotic structure of the various potentials near the center and

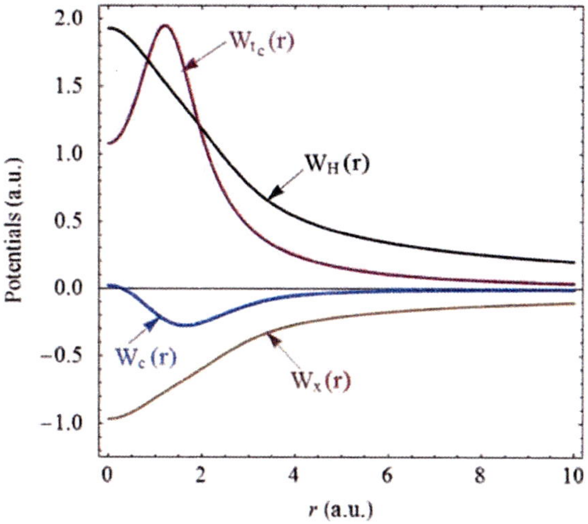

Fig. 6.16 A comparison of the Hartree $W_H(r)$, Pauli $W_x(r)$, Coulomb $W_c(r)$, and Correlation-Kinetic $W_{t_c}(r)$ potentials

classically forbidden region is given in Appendix D. Since the electron-interaction field approaches the center linearly, the corresponding potential $W_{ee}(\mathbf{r})$ approaches the center quadratically. It is interesting that although the Correlation-Kinetic field $\mathcal{Z}_{t_c}(\mathbf{r})$ is both positive and negative, the Correlation-Kinetic potential $W_{t_c}(\mathbf{r})$ is positive throughout space. (This is also the case of the mapping from a ground state of the quantum dot to a model system in a ground state [28, 32]. Thus, in the model system, since $W_{t_c}(\mathbf{r})$ is positive, $W_x(\mathbf{r})$ negative, and $W_c(\mathbf{r})$ essentially negative throughout space, Correlation-Kinetic effects cancel to a great degree the Pauli and Coulomb correlations.

The local electron-interaction potential $v_{ee}(\mathbf{r})$ in which all the many-body effects of the interacting system are incorporated, and its electron-interaction $W_{ee}(\mathbf{r})$ (the sum of its Hartree, Pauli and Coulomb components), and Correlation-Kinetic $W_{t_c}(\mathbf{r})$ components are plotted in Fig. 6.17. It is the potential $v_{ee}(\mathbf{r})$ which then generates the orbitals which in turn lead to the same density $\rho(\mathbf{r})$ and physical current density $\mathbf{j}(\mathbf{r})$ as that of the interacting quantum dot system.

Energies

The Q-DFT determined properties of the model system in its ground state that reproduce the densities $\{\rho_1(\mathbf{r}), \mathbf{j}_1(\mathbf{r})\}$ and energy E of the singlet 2^1S state of the quantum dot are given in Table 6.1. The Hartree E_H, Pauli E_x, Coulomb E_c, and Correlation-Kinetic T_c energies are obtained from the corresponding fields. The expectation values of the single-particle operators $\hat{O} = \delta(\mathbf{r}), r, r^2, 1/r$, are also quoted. Analytical expressions for the various energy components and the expectations are provided in Appendix D. (For purposes of comparison, included in Table 6.1 are the Q-DFT properties of the mapping from the ground 1^1S state of the quantum dot to the model system also in a 1^1S ground state [28, 32]).

As expected, the Pauli E_x and Coulomb E_c energies as a fraction of the total energy E for the 2^1S state of the quantum dot are less than those of the 1^1S state but

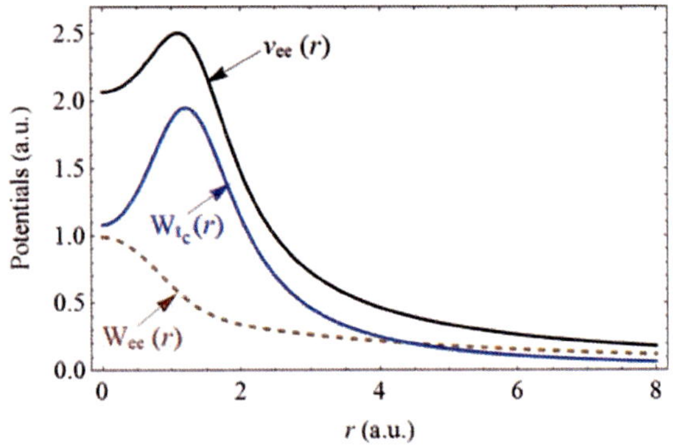

Fig. 6.17 The potential $v_{ee}(r)$, and its electron-interaction $W_{ee}(r)$ and correlation-kinetic $W_{t_c}(r)$ components

of the same order of magnitude: 20% and 2.1%, respectively, as opposed to 30% and 2.6% (see Table 6.1). Also observe that since the quantum dots are 2-dimensional low density 'artificial atoms', Correlation-Kinetic effects play a significant role (see Table 6.1). Thus, for the mapping $2^1S \to 1^1S$, the ratio $T_c/T = 73\%$ and $T_c/E = 27\%$. (Hence, the Correlation-Kinetic contribution to the total energy E is greater than that of even the Pauli energy E_x). For the mapping $1^1S \to 1^1S$, these ratios are $T_c/T = 12\%$ and $T_c/E = 3.5\%$. The increase in the significance of Correlation-Kinetic effects is foreseen in the structure of the Correlation-Kinetic field $\mathcal{Z}_{t_c}(\mathbf{r})$ (see Fig. 6.15). That the excited 2^1S state is a lower density system than the ground 1^1S state, and that therefore Correlation-Kinetic effects are more significant, is also clearly seen by a comparison of the expectations $\langle r \rangle$ and $\langle \delta(\mathbf{r}) \rangle$.

The values for the total energy E as obtained by the Q-DFT mapping (see (6.100)) of both the ground and singlet states of the quantum dot is the same as obtained via the expectation value of the Hamiltonian $\hat{H}$ of (6.119) taken with respect to the wave functions $\psi_n(\mathbf{r}_1\mathbf{r}_2)$; $n = 0, 1$ of (6.122). (The determination of the eigenvalue ϵ which is required in the Q-DFT expression for the energy E of (6.100) is discussed in the following subsection.)

Eigenvalues

For the mapping from either the ground ($n = 0$) or first excited singlet ($n = 1$) state, the single ground state eigenvalue $\epsilon^{(n)}$ of the S system may be obtained from the corresponding differential equation (6.125). Thus, employing (6.128) we have

$$\epsilon^{(n)} = \frac{1}{2}\frac{\nabla^2\sqrt{\rho(\mathbf{r})}}{\sqrt{\rho(\mathbf{r})}} + \frac{1}{2}k_{\text{eff}}^{(n)}r^2 + v_{ee}^{(n)}(\mathbf{r}); \quad n = 0, 1. \tag{6.136}$$

This expression is valid for arbitrary electron position. As each term on the right hand side is known, the choice of any electron position will suffice to determine $\epsilon^{(n)}$. Alternatively, the $\epsilon^{(n)}$ could be obtained by taking the limit of the right hand side as $r \to \infty$. In this limit, $v_{ee}^{(n)}(r) = 0$, and the analytical expressions for $\sqrt{\rho(\mathbf{r})}$ are known (See Appendix D). Then, since $\nabla^2 = \partial^2/\partial r^2 + (1/r)\partial/\partial r$, one may obtain the eigenvalues. The values of the ϵ for the two mappings are given in Table 6.1.

Now in Sect. 6.1.8, it was shown that the highest occupied eigenvalue of the model S system is equivalent to the negative of the ionization potential IP for Hamiltonians (6.1) in which the binding potential $v(\mathbf{r})$ was Coulombic. In the mappings discussed, the binding potential $v(\mathbf{r})$ is harmonic. It is interesting to note that nonetheless, the eigenvalues in each case of the present mappings $\epsilon^{(n)} = -IP = -[E_n^{N=1} - E_n^{N=2}]$, where $E_n^{N=2}$ is the energy of the quantum dot, and $E_n^{N=1} = (m+1)\Omega_n$; $m = 0$, is the energy of the one-electron quantum dot in its ground state but with the same effective frequency Ω_n.

6.2.5 *Quantal Density Functional Theory of the Density Amplitude*

As noted in the Introduction, the bound interacting electronic system of energy E may be mapped [25, 32, 58, 59] into one of *noninteracting bosons* - the model B system - possessing the same density $\rho(\mathbf{r})$ and physical current density $\mathbf{j}(\mathbf{r})$, and from which the energy E could be obtained. The solution to the corresponding B system differential equation is the density amplitude $\sqrt{\rho(\mathbf{r})}$, and its eigenvalue μ is the negative of the ionization potential. As the bosons are noninteracting, their potential $v_B(\mathbf{r})$ is also a *local* or multiplicative operator. For the different methods of deriving the B system differential equation, and the Q-DFT mapping to the B system in terms of fields and quantal sources, we refer the reader to Chap. 6 of ***QDFT***. The advantage of mapping to the model bosonic system is that the differential equation needs be solved self-consistently only *once* to obtain the density $\rho(\mathbf{r})$. In contrast, on mapping to the model fermionic system, the corresponding differential equation must be solved for *each* of the N orbitals from which the density $\rho(\mathbf{r})$ is then obtained.

For a system of N electrons in an external electromagnetic field: $\mathcal{E}(\mathbf{r}) = -\nabla v(\mathbf{r})/e$; $\mathcal{B}(\mathbf{r}) = \nabla \times \mathbf{A}(\mathbf{r})$, where $\{v(\mathbf{r}), \mathbf{A}(\mathbf{r})\}$ are the scalar and vector potentials, the B system differential equation in units where $|e| = \hbar = m = c = 1$ (with the charge of the electron being $-e$), is

$$\left[\frac{1}{2}(\hat{\mathbf{p}} + \mathbf{A}(\mathbf{r}))^2 + v_B(\mathbf{r})\right]\sqrt{\rho(\mathbf{r})} = \mu\sqrt{\rho(\mathbf{r})}, \tag{6.137}$$

where

$$v_B(\mathbf{r}) = v(\mathbf{r}) + v_{ee}^B(\mathbf{r}), \tag{6.138}$$

and $v_{ee}^B(\mathbf{r})$ the *local* potential in which all the many-electron correlations of the interacting system are incorporated. It is assumed that the model bosons are subject to the same electromagnetic field. Within Q-DFT, the potential $v_{ee}^B(\mathbf{r})$ is the work done to move the model boson from infinity to its position at $\mathbf{r}$ in the force of an effective field $\boldsymbol{\mathcal{F}}_B^{\text{eff}}(\mathbf{r})$. For the definition of this field, and for the expressions for the energy E of the interacting electronic system in terms of the B system components such as the bosonic kinetic T_B and Correlation-Kinetic T_c^B energy see ***QDFT***.

For the mapping from either the ground or first excited 2^1S state of the 2-electron quantum dot, the corresponding differential equation for the B system is

$$\left[\frac{1}{2}\mathbf{p}^2 + \frac{1}{2}k_{\text{eff}}r^2 + v_{ee}^B(\mathbf{r})\right]\sqrt{\rho(\mathbf{r})} = \mu\sqrt{\rho(\mathbf{r})}, \tag{6.139}$$

where k_{eff} is the appropriate effective force constant corresponding to the ground or excited state (see Sect. 6.2.1). The potential $v_{ee}^B(\mathbf{r})$ turns out to be equal to the potential $v_{ee}(\mathbf{r})$ of (6.79) for the noninteracting fermionic system. This is a consequence of the fact that the orbitals $\phi_k(\mathbf{r})$ of the S system differential equation (6.125) are proportional to the density amplitude $\sqrt{\rho(\mathbf{r})}$. Hence, the B and S system differential equations (6.139) and (6.125) are equivalent. Therefore the B and S system kinetic energies are equivalent: $T_B = T_s$, as are the corresponding Correlation-Kinetic energies: $T_c^B = T_c$. The energy E is thus also the same.

Hence, the mapping from an interacting 2-electron natural or 'artificial' atom, either in a ground or any excited singlet state, to one of noninteracting fermions in a ground state possessing the same $\{\rho(\mathbf{r}), \mathbf{j}(\mathbf{r})\}$, is entirely equivalent to the mapping to one of noninteracting bosons with also the same $\{\rho(\mathbf{r}), \mathbf{j}(\mathbf{r})\}$.

6.3 Time-Dependent Quantal Density Functional Theory

The case of TD Q-DFT we consider, as noted in the Introduction, is that of a system (see (6.12)) of N electrons in a TD external field $\boldsymbol{\mathcal{F}}^{\text{ext}}(\mathbf{y}) = \boldsymbol{\mathcal{E}}(\mathbf{y}) = -\boldsymbol{\nabla} v(\mathbf{y})/e$; $(\mathbf{y} = \mathbf{r}t)$. This system is then mapped to one of noninteracting fermions – the S system – whose density $\rho_s(\mathbf{y})$ and physical current density $\mathbf{j}_s(\mathbf{y})$ are the same as those of the interacting system, i.e. $\{\rho_s(\mathbf{y}), \mathbf{j}_s(\mathbf{y})\} = \{\rho(\mathbf{y}), \mathbf{j}(\mathbf{y})\}$, where $\{\rho(\mathbf{y}), \mathbf{j}(\mathbf{y})\}$ are defined as the expectations of the density $\hat{\rho}(\mathbf{r})$ and current density $\hat{\mathbf{j}}(\mathbf{r})$ operators taken with respect to $\Psi(\mathbf{y})$ (see Sect. 2.1). The S system densities $\{\rho_s(\mathbf{y}), \mathbf{j}_s(\mathbf{y})\}$ are the expectations of these operators taken with respect to the Slater determinant $\Phi\{\phi_k(\mathbf{y})\}$. With the model fermions subject to the *same* external field $\boldsymbol{\mathcal{F}}^{\text{ext}}(\mathbf{y})$, the only correlations that the model S system must then account for are those of due to the Pauli principle, Coulomb repulsion, and Correlation-Kinetic effects. Hence, paralleling the stationary-state case, it is the TD *conservative* effective field $\boldsymbol{\mathcal{F}}^{\text{eff}}(\mathbf{y})$ in which these correlations are incorporated that must first be determined. The resulting *local* electron-interaction potential $v_{ee}(\mathbf{y})$ of (6.16) in which these many-body effects are embedded, is then the work done in this effective field.

6.3.1 The S System 'Quantal Newtonian' Second Law

In order to obtain the expression for the local potential $v_{ee}(\mathbf{y})$, the S system *'Quantal Newtonian' Second Law* must first be derived (see Appendix A). The statement of the second law is the following: each model fermion experiences both the external $\mathcal{F}^{\text{ext}}(\mathbf{y})$ and an internal $\mathcal{F}_s^{\text{int}}(\mathbf{y})$ field. The response of the model fermion to these fields is then represented by a current-density field $\mathcal{J}_s(\mathbf{y})$. Thus, (in units of $e = \hbar = m = 1$)

$$\mathcal{F}^{\text{ext}}(\mathbf{y}) + \mathcal{F}_s^{\text{int}}(\mathbf{y}) = \mathcal{J}_s(\mathbf{y}), \tag{6.140}$$

with

$$\mathcal{F}^{\text{ext}}(\mathbf{y}) = -\nabla v(\mathbf{y}), \tag{6.141}$$

$$\mathcal{F}_s^{\text{int}}(\mathbf{y}) = -\nabla v_{ee}(\mathbf{y}) - \mathcal{Z}_s(\mathbf{y}) - \mathcal{D}(\mathbf{y}), \tag{6.142}$$

$$\mathcal{J}_s(\mathbf{y}) = \frac{1}{\rho_s(\mathbf{y})}\frac{\partial \mathbf{j}_s(\mathbf{y})}{\partial t} = \frac{1}{\rho(\mathbf{y})}\frac{\partial \mathbf{j}(\mathbf{y})}{\partial t} = \mathcal{J}(\mathbf{y}), \tag{6.143}$$

where we have employed the equality $\{\rho_s(\mathbf{y}), \mathbf{j}_s(\mathbf{y})\} = \{\rho(\mathbf{y}), \mathbf{j}(\mathbf{y})\}$. Hence, the differential density $\mathcal{D}(\mathbf{y})$ and current-density $\mathcal{J}(\mathbf{y})$ fields are those of the interacting system. The S system kinetic field $\mathcal{Z}_s(\mathbf{y})$ is obtained from its quantal source, the TD Dirac density matrix $\gamma_s(\mathbf{y}\mathbf{y}') = \sum_k \phi_k^*(\mathbf{y})\phi_k(\mathbf{y}')$; $(\mathbf{y}' = \mathbf{r}'t)$, in the usual manner via the TD kinetic energy tensor $t_{s,\alpha\beta}(\mathbf{y})$ (see (6.57), (6.58)). The law is derived so as to satisfy the continuity equation: $\nabla \cdot \mathbf{j}(\mathbf{y}) + \partial\rho(\mathbf{y})/\partial t = 0$.

6.3.2 Effective Field $\mathcal{F}^{\text{eff}}(\mathbf{y})$ and Electron-Interaction Potential $v_{ee}(\mathbf{y})$

The *conservative* effective field $\mathcal{F}^{\text{eff}}(\mathbf{y}) = -\nabla v_{ee}(\mathbf{y})$ of the model fermion S system, in which all the many-body effects are incorporated, ensures the equivalence of the model and interacting system densities $\{\rho(\mathbf{y}), \mathbf{j}(\mathbf{y})\}$. The field $\mathcal{F}^{\text{eff}}(\mathbf{y})$ is obtained by comparison of the *'Quantal Newtonian' Second Laws* of the interacting electronic and model fermionic systems.

The statement of the *'Quantal Newtonian' Second Law* for the interacting system, (see Sect. 2.4), is that at each instant of time t, the sum of the external $\mathcal{F}^{\text{ext}}(\mathbf{y})$ and internal $\mathcal{F}^{\text{int}}(\mathbf{y})$ fields experienced by each electron must equal the electron response current-density field $\mathcal{J}(\mathbf{y})$:

$$\mathcal{F}^{\text{ext}}(\mathbf{y}) + \mathcal{F}^{\text{int}}(\mathbf{y}) = \mathcal{J}(\mathbf{y}), \tag{6.144}$$

where

$$\mathcal{F}^{\text{ext}}(\mathbf{y}) = -\nabla v(\mathbf{y}), \tag{6.145}$$
$$\mathcal{F}^{\text{int}}(\mathbf{y}) = \mathcal{E}_{ee}(\mathbf{y}) - \mathcal{Z}(\mathbf{y}) - \mathcal{D}(\mathbf{y}), \tag{6.146}$$

with $\mathcal{E}_{ee}(\mathbf{y})$, $\mathcal{Z}(\mathbf{y})$, $\mathcal{D}(\mathbf{y})$, the electron-interaction, kinetic, and differential density fields, respectively.

As the external $\mathcal{F}^{\text{ext}}(\mathbf{y})$ and the response current-density $\mathcal{J}(\mathbf{y})$ fields experienced by the interacting and model system fermions are the same, the effective field $\mathcal{F}^{\text{eff}}(\mathbf{y})$ is obtained by equating the corresponding internal fields $\mathcal{F}^{\text{int}}(\mathbf{y})$ and $\mathcal{F}_s^{\text{int}}(\mathbf{y})$. Hence, one obtains

$$\mathcal{F}^{\text{eff}}(\mathbf{y}) = -\nabla v_{ee}(\mathbf{y}) = \mathcal{E}_{ee}(\mathbf{y}) + \mathcal{Z}_{t_c}(\mathbf{y}), \tag{6.147}$$

where the Correlation-Kinetic field $\mathcal{Z}_{t_c}(\mathbf{y})$ is defined as

$$\mathcal{Z}_{t_c}(\mathbf{y}) = \mathcal{Z}_s(\mathbf{y}) - \mathcal{Z}(\mathbf{y}). \tag{6.148}$$

Thus, the local potential $v_{ee}(\mathbf{y})$ has the following physical interpretation: *At each instant of time t, it is the work done to move a model fermion from some reference point at infinity to its position at $\mathbf{r}$ in the force of the conservative effective field $\mathcal{F}^{\text{eff}}(\mathbf{y})$:*

$$v_{ee}(\mathbf{y}) = -\int_{\infty}^{\mathbf{r}} \mathcal{F}^{\text{eff}}(\mathbf{y}') \cdot d\boldsymbol{\ell}'. \tag{6.149}$$

Since, $\nabla \times \mathcal{F}^{\text{eff}}(\mathbf{y}) = 0$, the work done $v_{ee}(\mathbf{y})$, at each instant of time t, is *path-independent*.

It is also evident from the expression for $\mathcal{F}^{\text{eff}}(\mathbf{y})$ of (6.147) that the potential $v_{ee}(\mathbf{y})$ explicitly accounts for electron correlations due to the Pauli principle and Coulomb repulsion via the electron-interaction field $\mathcal{E}_{ee}(\mathbf{y})$, and Correlation-Kinetic effects through the Correlation-Kinetic field $\mathcal{Z}_{t_c}(\mathbf{y})$. The field $\mathcal{E}_{ee}(\mathbf{y})$ can be decomposed into its Hartree $\mathcal{E}_H(\mathbf{y})$, Pauli $\mathcal{E}_x(\mathbf{y})$, and Coulomb $\mathcal{E}_c(\mathbf{y})$ components as described for the stationary-state case so that

$$\mathcal{F}^{\text{eff}}(\mathbf{y}) = \mathcal{E}_H(\mathbf{y}) + \mathcal{E}_x(\mathbf{y}) + \mathcal{E}_c(\mathbf{y}) + \mathcal{Z}_{t_c}(\mathbf{y}). \tag{6.150}$$

Thus, the effective field $\mathcal{F}^{\text{eff}}(\mathbf{y})$, and thereby the potential $v_{ee}(\mathbf{y})$, is expressed in terms of the *separate* Hartree, Pauli, Coulomb, and Correlation-Kinetic components. The Hartree field $\mathcal{E}_H(\mathbf{y})$, for each instant of time t, is conservative. This is because, for each instant of time t, the density $\rho(\mathbf{y})$ is a *local* charge distribution. The potential $v_{ee}(\mathbf{y})$ may thus be written as

$$v_{ee}(\mathbf{y}) = W_H(\mathbf{y}) + \left(- \int_{\infty}^{\mathbf{r}} \left[\mathcal{E}_x(\mathbf{y}') + \mathcal{E}_c(\mathbf{y}') + \mathcal{Z}_{t_c}(\mathbf{y}') \right] \cdot d\boldsymbol{\ell}' \right), \tag{6.151}$$

where

$$W_H(\mathbf{y}) = \int \frac{\rho(\mathbf{y}')}{|\mathbf{r} - \mathbf{r}'|} d\mathbf{r}'. \tag{6.152}$$

The expressions for the non-conserved energy $E(t)$ and Ehrenfest's Theorem, in terms of the S system properties, and various sum rules involving the corresponding effective field $\boldsymbol{\mathcal{F}}^{\mathrm{eff}}(\mathbf{y})$ are the following:

Energy

$$E(t) = T_s(t) + \int \rho(\mathbf{y})v(\mathbf{y})d\mathbf{r} + E_{ee}(t) + T_c(t), \tag{6.153}$$

where

$$T_s(t) = \sum_k \langle \phi_k(\mathbf{y})| -\frac{1}{2}\nabla^2 |\phi_k(\mathbf{y})\rangle = -\frac{1}{2}\int \rho(\mathbf{y})\mathbf{r}\cdot \boldsymbol{\mathcal{Z}}_s(\mathbf{y})d\mathbf{r}, \tag{6.154}$$

$$T_c(t) = \frac{1}{2}\int \rho(\mathbf{y})\mathbf{r}\cdot \boldsymbol{\mathcal{Z}}_{t_c}(\mathbf{y})d\mathbf{r}, \tag{6.155}$$

$$E_{ee}(t) = \int \rho(\mathbf{y})\mathbf{r}\cdot \boldsymbol{\mathcal{E}}_{ee}(\mathbf{y})d\mathbf{r}. \tag{6.156}$$

Ehrenfest's Theorem

$$\int \rho(\mathbf{y})\left[\boldsymbol{\mathcal{F}}^{\mathrm{ext}}(\mathbf{y}) - \boldsymbol{\mathcal{J}}_s(\mathbf{y})\right]d\mathbf{r} = 0. \tag{6.157}$$

Integral Virial Theorem

$$\int \rho(\mathbf{y})\mathbf{r}\cdot \boldsymbol{\mathcal{F}}^{\mathrm{eff}}(\mathbf{y})d\mathbf{r} = E_{ee}(t) + 2T_c(t). \tag{6.158}$$

Zero Force Sum Rule

$$\int \rho(\mathbf{y})\boldsymbol{\mathcal{F}}^{\mathrm{eff}}(\mathbf{y})d\mathbf{r} = 0. \tag{6.159}$$

Zero Torque Sum Rule

$$\int \rho(\mathbf{y})\mathbf{r}\times \boldsymbol{\mathcal{F}}^{\mathrm{eff}}(\mathbf{y})d\mathbf{r} = 0. \tag{6.160}$$

TD Q-DFT thus constitutes an alternative description of the interacting TD system defined by the Schrödinger equation (6.12). This description is in terms of, and elucidates, the separate electron correlations – Pauli, Coulomb, Correlation-Kinetic – that exist in the system. Such a separation into the individual correlation components is not possible via the TD Schrödinger equation itself. In that context, TD Q-DFT constitutes a complement to TD Schrödinger theory. It is also a local effective potential theory description of TD Schrödinger theory.

References

1. M.K. Harbola, V. Sahni, Phys. Rev. Lett. **62**, 489 (1989)
2. V. Sahni, M.K. Harbola, Int. J. Quantum Chem. **24**, 569 (1990)
3. V. Sahni, M. Slamet, Phys. Rev. B **48**, 1910 (1993)
4. M.K. Harbola, M. Slamet, V. Sahni, Phys. Lett. A **157**, 60 (1991)
5. M. Slamet, V. Sahni, M.K. Harbola, Phys. Rev. A **49**, 809 (1994)
6. V. Sahni, Phys. Rev. A **55**, 1846 (1997)
7. V. Sahni, Top. Curr. Chem. **182**, 1 (1996)
8. Z. Qian, V. Sahni, Phys. Rev. A **57**, 2527 (1998)
9. Z. Qian, V. Sahni, Phys. Lett. A **247**, 303 (1998)
10. Z. Qian, V. Sahni, Phys. Lett. A **248**, 393 (1998)
11. Z. Qian, V. Sahni, Phys. Rev. B **62**, 16364 (2000)
12. Z. Qian, V. Sahni, Int. J. Quantum Chem. **78**, 341 (2000)
13. Z. Qian, V. Sahni, Int. J. Quantum Chem. **80**, 555 (2000)
14. Z. Qian, V. Sahni, Phys. Rev. A **63**, 042508 (2001)
15. V. Sahni, L. Massa, R. Singh, M. Slamet, Phys. Rev. Lett. **87**, 113002 (2001)
16. M. Slamet, V. Sahni, Int. J. Quantum Chem. **85**, 436 (2001)
17. V. Sahni, in *Electron Correlations and Materials Properties* 2. ed. by A. Gonis, N. Kioussis, M. Ciftan (Kluwer Academic/Plenum Publishers, New York, 2002)
18. X.-Y. Pan, V. Sahni, Phys. Rev. A **67**, 012501 (2003)
19. V. Sahni, X.-Y. Pan, Phys. Rev. Lett. **90**, 123001 (2003)
20. M. Slamet, R. Singh, L. Massa, V. Sahni, Phys. Rev. A **68**, 042504 (2003)
21. X.-Y. Pan, V. Sahni, J. Chem. Phys. **120**, 5642 (2004)
22. V. Sahni, M. Slamet, Int. J. Quantum Chem. **100**, 858 (2004)
23. V. Sahni, M. Slamet, Int. J. Quantum Chem. **106**, 3087 (2006)
24. V. Sahni, M. Slamet, X.-Y. Pan, J. Chem. Phys. **126**, 204106 (2007)
25. X.-Y. Pan, V. Sahni, Phys. Rev. A **80**, 022506 (2009)
26. V. Sahni, in *Proceedings of the 26th International Colloquium on Group Theoretical Methods in Physics*. ed. by J.L. Birman, S. Catto, B. Nicolescu (Canopus Publishers, 2009)
27. X.-Y. Pan, V. Sahni, in *Theoretical and Computational Developments in Modern Density Functional Theory*, ed. by A.K. Roy (Nova Science Publishers, New York, 2012)
28. T. Yang, X.-Y. Pan, V. Sahni, Phys. Rev. A **83**, 042518 (2011)
29. V. Sahni, X.-Y. Pan, T. Yang, Computation **4**, 30 (2016). https://doi.org/10.3390/computation4030030
30. M. Slamet, V. Sahni, Comput. Theor. Chem. **1114**, 125 (2017)
31. X.-Y. Pan, V. Sahni, Computation **6**, 25 (2018). https://doi.org/10.3390/computation6010025
32. V. Sahni, *Quantal Density Functional Theory*, 2nd edn. (Springer, Berlin, Heidelberg, 2016) (Referred to as ***QDFT***)
33. V. Sahni, *Quantal Density Functional Theory II: Approximation Methods and Applications* (Springer, Berlin, Heidelberg, 2010) (Referred to as ***QDFT2***)
34. J.C. Slater, Phys. Rev. **34**, 1293 (1929)
35. V. Fock, Z. Phys. **61**, 126 (1930)
36. J.C. Slater, Phys. Rev. **35**, 210 (1930)
37. W. Kohn, L.J. Sham, Phys. Rev. **140**, A1133 (1965)
38. R.T. Sharp, G.K. Horton, Phys. Rev. **30**, 317 (1953)
39. J.D. Talman, W.F. Shadwick, Phys. Rev. A **14**, 36 (1976)
40. E. Engel, S.H. Vosko, Phys. Rev. A **47**, 2800 (1993)
41. B.L. Moiseiwitsch, *Variational Principles* (Wiley, London, 1966)
42. J.C. Slater, Phys. Rev. **81**, 385 (1951)
43. J.C. Slater, T.M. Wilson, J.H. Wood, Phys. Rev. **179**, 28 (1969)
44. P. Hohenberg, W. Kohn, Phys. Rev. **136**, B864 (1964)
45. X.-Y. Pan, V. Sahni, J. Chem. Phys. **143**, 174105 (2015)

46. J.P. Perdew, R.G. Parr, M. Levy, J.L. Balduz, Phys. Rev. Lett. **49**, 1691 (1982)
47. J.P. Perdew, in *Density Functional Methods in Physics*, ed. by R.M. Dreizler, J. da Providencia. NATO ASI series, Series B: Physics, Vol. 123 (Plenum, New York, 1985)
48. J.P. Perdew, M. Levy, Phys. Rev. Lett. **51**, 1884 (1983)
49. L.J. Sham, M. Schlüter, Phys. Rev. Lett. **51**, 1888 (1983)
50. L.J. Sham, M. Schlüter, Phys. Rev. B **32**, 3883 (1985)
51. D. Achan, L. Massa, V. Sahni, Comp. Theor. Chem. **1035**, 14 (2014)
52. D. Achan, L. Massa, V. Sahni, Phys. Rev. A **90**, 022502 (2014)
53. E. Runge, E.K.U. Gross, Phys. Rev. Lett. **52**, 997 (1984)
54. V. Sahni, Bull. Am. Phys. Soc. **44**, 1902 (1999)
55. M. Taut, J. Phys. A **27**, 1045 (1994); Corrigenda J. Phys. A **27**, 4723 (1994)
56. M. Taut, H. Eschrig, Z. Phys. Chem. **224**, 631 (2010)
57. X.-Y. Pan, V. Sahni, J. Chem. Phys. **119**, 7083 (2003)
58. N.H. March, *Electron Density Theory of Atoms and Molecules* (Academic Press, London, 1992)
59. M. Levy, J.P. Perdew, V. Sahni, Phys. Rev. A **30**, 2745 (1984)

Chapter 7
Modern Density Functional Theory

Abstract This chapter is on the fundamental aspects of stationary-state Hohenberg and Kohn density functional theory; its extension to the added presence of a uniform magnetic field by Pan and Sahni to a density and physical current density functional theory for both spinless electrons and electrons with a spin moment; and Runge and Gross time-dependent density functional theory. The different proofs of the basic theorems, new insights into these proofs, the generalizations of the theorems via density preserving unitary transformations, and Corollaries to the theorems are described. The perspectives these theories bring to the Schrödinger theory of electrons is discussed. The Kohn and Sham density functional theory constitutes the mapping, in both the stationary-state and temporal cases, to model systems of noninteracting fermions possessing the same density as the interacting electronic system. In the Kohn-Sham version of local effective potential theory, the definitions of properties are in terms of undefined energy or action functionals of the density, with the corresponding potentials defined in terms of their functional derivatives. The potentials may equally well be defined in terms of the requisite self-consistent equations. These functionals and functional derivatives are provided a rigorous physical interpretation via Quantal Density Functional Theory. These descriptions are in terms of fields representative of electron correlations that the model system must account for. In the stationary-state case, these correlations are those due to the Pauli principle, Coulomb repulsion, and Correlation-Kinetic effects. In the temporal case, there exists additionally a field representative of Correlation-Current-Density effects.

Introduction

This chapter is on modern stationary-state Hohenberg-Kohn [1] and Kohn-Sham [2] density functional theory (DFT); the extension by Pan-Sahni [3] to the added presence of a uniform magnetic field; time-dependent Runge-Gross [4, 5] DFT; and the rigorous physical interpretations of these theories as afforded by quantal density functional theory (Q-DFT) [6, 7]. The emphasis of the chapter is on the fundamental

V. Sahni, *Schrödinger Theory of Electrons: Complementary Perspectives*,
Springer Tracts in Modern Physics 285,
https://doi.org/10.1007/978-3-030-97409-1_7

aspects of DFT, the new insights achieved, and the perspectives it brings to the Schrödinger theory of electrons.

Kohn-Sham DFT, both stationary-state and temporal, is a local effective potential theory just as the Q-DFT described in Chap. 6. It is basically a mathematical construct with no formal prescription on how the physics of electron correlations is embodied within it. In its approximate form, with the physics incorporated extrinsically via the satisfaction of various sum rule constraints and scaling laws, the theory is nonetheless extensively employed in electronic structure calculations. We do not delve here into the numerous approximation schemes that exist in the literature. A focus of this chapter is to explain Kohn-Sham DFT from the more fundamental perspective of the various electron correlations ensconced within it.

For the standard description of modern DFT, the material precursory to the Hohenberg-Kohn Theorems, and for the purpose of comparison with the present chapter, the reader is referred to the original texts on the subject [8–10] and to more recent texts [11, 12].

Another local effective potential theory employed extensively in its approximate form, and the immediate precursor to stationary-state Kohn-Sham DFT, is Slater [13, 14] theory. For an understanding of the physics underlying Slater theory, see Chap. 10 of ***QDFT*** [6].

(The books on Quantal Density Functional Theory, [6] and [7] are referred to in the chapter as ***QDFT*** and ***QDFT2***, respectively.)

7.1 Paths to the Hamiltonian

In 1964, Hohenberg and Kohn (HK) [1] proposed a new path to the determination of the Hamiltonian $\hat{H}$ of an electronic system, and thereby to its eigenfunctions $\Psi_n(\mathbf{X})$ and eigenvalues E_n.

Prior to explaining the *HK path* [15], consider a system of N electrons in an electrostatic field $\boldsymbol{\mathcal{E}}(\mathbf{r}) = -\boldsymbol{\nabla} v(\mathbf{r})/e$. (When one speaks of a system of electrons, the kinetic energy $\hat{T}$ and electron-interaction potential energy $\hat{U}$ operators are *assumed known* and *fixed*. The field $\boldsymbol{\mathcal{E}}(\mathbf{r})$, or the potential $v(\mathbf{r})$, is 'external' to the system of the N electrons. The functional form of the field $\boldsymbol{\mathcal{E}}(\mathbf{r})$, or potential $v(\mathbf{r})$, is *arbitrary*. It is this potential $v(\mathbf{r})$ that is *changeable*.) The stationary-state Schrödinger equation for this system in atomic units ($e = \hbar = m = 1$) is then

$$\hat{H}\Psi_n(\mathbf{X}) = E_n\Psi_n(\mathbf{X}). \tag{7.1}$$

The Hamiltonian $\hat{H}$ of the system is the sum of the kinetic $\hat{T}$, electron-interaction potential $\hat{U}$, and binding potential $\hat{V}$ operators:

$$\hat{T} = \frac{1}{2}\sum_k \hat{\mathbf{p}}_k^2; \quad \hat{U} = \frac{1}{2}\sum_{k,\ell}{}' \frac{1}{|\mathbf{r}_k - \mathbf{r}_\ell|}; \quad \hat{V} = \sum_k v(\mathbf{r}_k), \tag{7.2}$$

with $\hat{\mathbf{p}} = -i\nabla$, the canonical momentum operator. Traditionally, for a particular physical system, the potential $v(\mathbf{r})$ is *assumed known*. For natural atoms, the binding potential is Coulombic: $v(\mathbf{r}) = -Z/r$, with Z the atomic number; for 'artificial atoms' or quantum dots it is harmonic: $v(\mathbf{r}) = \frac{1}{2}kr^2$, with k the force constant. The $\{\Psi_n(\mathbf{X}), E_n\}$ are the antisymmetric N electron eigenfunctions and eigenenergies, respectively; $\mathbf{X} = \mathbf{x}_1 \dots \mathbf{x}_N$; $\mathbf{x} = \mathbf{r}\sigma$ with $\mathbf{r}$ and σ the spatial and spin coordinates. With each term of the Hamiltonian $\hat{H}$ defined, the $\{\Psi_n(\mathbf{X}), E_n\}$ are obtained by direct analytical or numerical solution of the Schrödinger differential equation (7.1).

The Hohenberg-Kohn Path

The *HK path* to the determination of the Hamiltonian $\hat{H}$ and the consequent eigenfunctions and eigenenergies $\{\Psi_n(\mathbf{X}), E_n\}$ of the system, is based on the *First Hohenberg-Kohn Theorem*. The theorem is proved for a *nondegenerate ground state*. The proof is for *arbitrary* field $\mathcal{E}(\mathbf{r})$ or potential $v(\mathbf{r})$, *arbitrary* electron number N, but for *fixed* N. The fundamental property on which the theorem is based is the electronic density $\rho(\mathbf{r})$, the expectation value of the density operator $\hat{\rho}(\mathbf{r})$ of (2.12):

$$\rho(\mathbf{r}) = \langle \Psi(\mathbf{X}) | \hat{\rho}(\mathbf{r}) | \Psi(\mathbf{X}) \rangle, \tag{7.3}$$

with the constraint that it integrates to the electron number N:

$$\int \rho(\mathbf{r}) d\mathbf{r} = N. \tag{7.4}$$

(The lack of subscripts on $\Psi(\mathbf{X})$ or $\rho(\mathbf{r})$ indicate a nondegenerate ground state.)

It is proved in the first HK theorem that knowledge of the nondegenerate ground state density $\rho(\mathbf{r})$ uniquely determines the external potential $v(\mathbf{r})$ to within an additive constant C. With the kinetic $\hat{T}$ and electron-interaction $\hat{U}$ operators known, the Hamiltonian $\hat{H}$ of the system is then known to within the constant C. The solution of the corresponding Schrödinger equation then leads to the eigenfunctions and eigenenergies $\{\Psi_n(\mathbf{X}), E_n\}$ of the system. The HK path in equation form, with the constant C chosen to be zero, is then

$$\rho(\mathbf{r}) \quad \rightarrow \quad v(\mathbf{r}) \quad \rightarrow \quad \hat{H}. \tag{7.5}$$

The proof of the First HK theorem from which the HK path is derived is provided in the following section Sect. 7.2. (There are other fundamental understandings of Schrödinger theory arrived at via the First HK Theorem that will be discussed in subsequent sections.) The proof of the Theorem is for *v-representable* densities $\rho(\mathbf{r})$. These are densities obtained from wave functions of the interacting system Schrödinger equation (7.1) with an external potential $v(\mathbf{r})$. The initial and key component of the proof is that there is a bijective or *one-to-one* relationship between the potential $v(\mathbf{r})$ and the *nondegenerate ground state* wave function $\Psi(\mathbf{X})$. What this means is that for each potential $v(\mathbf{r})$ there exists one-and-only- one ground state wave function $\Psi(\mathbf{X})$, *and* that for each ground state wave function $\Psi(\mathbf{X})$ which is a solution of a Schrödinger equation of the form of (7.1), there can only be one-and-only-one potential $v(\mathbf{r})$. *Employing this fact*, it is then proved that there is a bijective or *one-to-one* relationship between the ground state wave function $\Psi(\mathbf{X})$ and the ground state density $\rho(\mathbf{r})$. The meaning of this bijectivity follows in a similar vein as before. Hence, knowledge of $\rho(\mathbf{r})$ determines $v(\mathbf{r})$, and the HK path of (7.5) follows.

The HK path from the ground state density $\rho(\mathbf{r})$ to the Hamiltonian $\hat{H}$, is the basis for the concept of a *basic variable* in quantum mechanics. *A basic variable is a gauge invariant property of an electronic system whose knowledge determines the wave functions and eigenenergies of that system.* Hence the nondegenerate ground state density $\rho(\mathbf{r})$ constitutes a basic variable.

It is to be noted that the bijective relationship between the potential $v(\mathbf{r})$ and the nondegenerate ground state wave function $\Psi(\mathbf{X})$, which is critical to the HK path, can be proved *only* for the case of an external electrostatic field $\mathcal{E}(\mathbf{r})$. In the added presence of a magnetostatic field $\mathcal{B}(\mathbf{r}) = \nabla \times \mathbf{A}(\mathbf{r})$, with $\mathbf{A}(\mathbf{r})$ the vector potential, the relationship between the potentials $\{v(\mathbf{r}), \mathbf{A}(\mathbf{r})\}$ and the nondegenerate ground state wave function $\Psi(\mathbf{X})$ is fundamentally different. As such the proof of the corresponding HK path in the presence of a uniform magnetic field differs. The path as proved by Pan and Sahni [3] is discussed in a subsection below, with the proof provided in Sect. 7.6.

The E. Bright Wilson Path

A different way of determining the Hamiltonian $\hat{H}$ from the ground state density $\rho(\mathbf{r})$ for potentials $v(\mathbf{r})$ that are singular at the nucleus such as the Coulomb potential is due to E. Bright Wilson [16]. Accordingly, *knowledge of the density* $\rho(\mathbf{r})$ leads on integration to the electron number N (see (7.4)). The cusps in the density at the nuclei satisfy the *electron-nucleus cusp coalescence constraint* (see [17–22] and Chap. 8) whereby the positions of the nuclei and their charge Z are determined. The cusp coalescence constraint in 3 dimensions for an arbitrary state Ψ of a system of N charged particles as particles 1 and 2 at $\mathbf{r}_1, \mathbf{r}_2$ coalesce (with the spin index suppressed) is

$$\begin{aligned}\Psi(\mathbf{r}_1, \mathbf{r}_2, \ldots \mathbf{r}_N) &= \Psi(\mathbf{r}_2, \mathbf{r}_2, \mathbf{r}_3, \ldots \mathbf{r}_N)(1 + \zeta r_{12}) \\ &\quad + \mathbf{r}_{12} \cdot \mathbf{C}(\mathbf{r}_2, \mathbf{r}_3, \ldots \mathbf{r}_N).\end{aligned} \tag{7.6}$$

Here $r_{12} = |\mathbf{r}_1 - \mathbf{r}_2|$, $\mathbf{r}_{12} = \mathbf{r}_1 - \mathbf{r}_2$, $\mathbf{C}(\mathbf{r}_2, \mathbf{r}_3, \ldots \mathbf{r}_N)$ an undetermined vector, and $\zeta = Z$ the nuclear charge. The corresponding differential form of the cusp coalescence constraint may be determined from this expression. In terms of the spherical averaged density $\rho(r)$, as the electron at $\mathbf{r}_1$ approaches the nucleus of charge Z at the origin ($\mathbf{r}_2 = 0$), the cusp condition is

$$\lim_{r\to 0} \frac{d\rho(r)}{dr} = -2Z\,\rho(r=0). \tag{7.7}$$

From the cusps in the density, the charge Z and the positions of the nuclei are known. With the electron number N; the charge Z; the positions of the nuclei; the operators $\hat{T}$, $\hat{U}$, and $\hat{V}$; now known, the Hamiltonian $\hat{H}$ is then explicitly defined. Note that this argument is not applicable to external potentials $v(\mathbf{r})$ that are not singular at the nucleus because then there is no *electron-nucleus cusp* constraint on the wave function and hence the density. However, it is valid for *arbitrary* state because the cusp coalescence constraint is. The Wilson reasoning ought not to be equated to the HK path, as is commonly assumed to be the case. The reason for this is twofold: one, because in the HK path, the density $\rho(\mathbf{r})$ first uniquely determines the potential *function* $v(\mathbf{r})$. There is no such equivalent step in the Wilson path; two, the HK proof is *general* in that the potential $v(\mathbf{r})$ is *arbitrary* and not restricted to being singular at the nucleus. It is worth reiterating that the Wilson path *assumes knowledge* of the ground state density $\rho(\mathbf{r})$ as being the basic variable. The fact that $\rho(\mathbf{r})$ is the basic variable is due to the First HK theorem.

The Percus-Levy-Lieb Path

The Percus-Levy-Lieb (PLL) [23–26] *constrained-search* path also begins with the *assumption of knowledge* that the basic variable is the ground state density $\rho(\mathbf{r})$. A search, constrained to all antisymmetric wave functions Ψ_ρ that generate $\rho(\mathbf{r})$, is performed to determine the true ground state Ψ. The manner of the search is by application of the variational principle for the energy [27]. As will be shown in Sect. 7.3.3, of all the antisymmetric functions Ψ_ρ that yield the ground state density $\rho(\mathbf{r})$, the true wave function Ψ is the one that yields the density $\rho(\mathbf{r})$ and obtains the infimum of the expectation value of the operators $\hat{T} + \hat{U}$:

$$\inf_{\Psi_\rho \to \rho} \langle \Psi_\rho | \hat{T} + \hat{U} | \Psi_\rho \rangle. \tag{7.8}$$

The notation $\inf_{\Psi_\rho \to \rho}$ means that one searches for the smallest (infimum) value of the expectation $\langle \hat{T} + \hat{U} \rangle$ taken with respect to all antisymmetric functions Ψ_ρ that yield $\rho(\mathbf{r})$. (The set $\{\Psi_\rho\}$ is a subset of all functions $\{\Psi\}$ that could be employed in the expectation value.) Now, according to Levy [26], Ψ cannot be an eigenfunction of more than one $\hat{H}$ with a multiplicative potential $v(\mathbf{r})$. This statement, though correct, lacks specificity about how $v(\mathbf{r})$ is determined from Ψ, and hence how $\hat{H}$ is obtained. However, once the ground state wave function Ψ is known, the potential $v(\mathbf{r})$ is known because the functional $v[\Psi](\mathbf{r})$ is *exactly known* via the 'Quantal Newtonian' First Law (QNFL) (see Chap. 3). (Of course, the First HK theorem also shows that $v(\mathbf{r}) = v[\Psi](\mathbf{r})$, but the functional dependence of $v(\mathbf{r})$ on Ψ is not provided.) Thus, with the operators $\hat{T}$ and $\hat{U}$ assumed known, $\rho(\mathbf{r})$ determines $\hat{H}$ uniquely to within an additive constant. The original description of the constrained-search path according to Levy is then

$$\{\Psi_\rho\} \;\rightarrow\; \rho(\mathbf{r}) \;\rightarrow\; \Psi \;\xrightarrow{\text{(not defined)}}\; v(\mathbf{r}) \rightarrow \hat{H}. \tag{7.9}$$

But as noted above, the *explicit* dependence of $v(\mathbf{r})$ on Ψ is known according to the QNFL. Thus, every step of the constrained-search path is now rigorously defined. The path, more precisely, is then

$$\begin{array}{ccccc} & & & \text{(QNFL)} & \\ \{\Psi_\rho\} \;\rightarrow\; \rho(\mathbf{r}) \;\rightarrow\; \Psi & & \rightarrow & & v[\Psi](\mathbf{r}) \rightarrow \;\hat{H}. \end{array} \tag{7.10}$$

There are two other points of note with regard to the PLL path: ***(i)*** The original proof is for *N-representable densities* $\rho(\mathbf{r})$. That is densities obtained from N-particle antisymmetric functions. Note, however, that the density $\rho(\mathbf{r})$ contains no information about the Pauli principle, so that the density could correspond to that of a fermionic or bosonic system. Hence, the functions $\{\Psi_\rho\}$ need not be restricted to being antisymmetric. They could equally well be symmetric or lack a symmetry. Thus, the constrained search path is valid for a broader class of functions. Unlike the HK path, the constrained search path is not restricted to v-representable densities; ***(ii)*** When degeneracies exist, i.e. when more than one function Ψ gives the same ground state energy E, the constrained search path of (7.9) or (7.10) is still valid. Hence, the PLL path encompasses the case of degenerate ground states.

The Hohenberg-Kohn Path in a Uniform Magnetic Field

In the added presence of a magnetostatic field $\boldsymbol{\mathcal{B}}(\mathbf{r}) = \boldsymbol{\nabla} \times \mathbf{A}(\mathbf{r})$, and depending on whether the electrons are spinless or have a spin moment, the corresponding Hamiltonians, respectively, are

$$\hat{H}_{spinless} = \hat{T}_A + \hat{U} + \hat{V}, \tag{7.11}$$

$$\hat{H}_{spin} = \hat{T}_F + \hat{U} + \hat{V}, \tag{7.12}$$

where

$$\hat{T}_A = \frac{1}{2}\sum_k \left(\hat{\mathbf{p}}_{k,phys}\right)^2 = \frac{1}{2}\sum_k \left(\hat{\mathbf{p}}_k + (1/c)\mathbf{A}(\mathbf{r}_k)\right)^2, \tag{7.13}$$

$$\hat{T}_F = \frac{1}{2}\sum_k \left(\boldsymbol{\sigma}_k \cdot \hat{\mathbf{p}}_{k,phys}\right)\left(\boldsymbol{\sigma}_k \cdot \hat{\mathbf{p}}_{k,phys}\right) \tag{7.14}$$

$$= \frac{1}{2}\sum_k \left(\hat{\mathbf{p}}_k + (1/c)\mathbf{A}(\mathbf{r}_k)\right)^2 + (1/c)\sum_k \boldsymbol{\mathcal{B}}(\mathbf{r}_k)\cdot\mathbf{s}_k. \tag{7.15}$$

In (7.13), (7.14), the physical momentum $\hat{\mathbf{p}}_{phys} = (\hat{\mathbf{p}} + (1/c)\mathbf{A}(\mathbf{r}))$, the canonical momentum $\hat{\mathbf{p}} = -i\boldsymbol{\nabla}$, and $\mathbf{s} = \frac{1}{2}\boldsymbol{\sigma}$ with $\mathbf{s}$ the electron spin angular momentum operator and $\boldsymbol{\sigma}$ the spin matrix. The corresponding Schrödinger and Schrödinger-Pauli equations are then, respectively,

$$\hat{H}_{spinless}\Psi(\mathbf{X}) = E\Psi(\mathbf{X}), \tag{7.16}$$

$$\hat{H}_{spin}\Psi(\mathbf{X}) = E\Psi(\mathbf{X}). \tag{7.17}$$

According to the First HK Theorem, in the case when only an electrostatic field $\boldsymbol{\mathcal{E}}(\mathbf{r}) = -\boldsymbol{\nabla} v(\mathbf{r})/e$ is present, the relationship between the scalar potential $v(\mathbf{r})$

and the nondegenerate ground state wave function $\Psi(\mathbf{X})$ is *one-to-one*. However, in the added presence of a magnetic field, the relationship between the potentials $\{v(\mathbf{r}), \mathbf{A}(\mathbf{r})\}$ and the nondegenerate ground state $\Psi(\mathbf{X})$ can be *many-to-one* [28–31] and even *infinite-to-one* [32, 33]:

$$\begin{array}{lcc} \{v(\mathbf{r}), \mathbf{A}(\mathbf{r})\} & \searrow & \\ \{v'(\mathbf{r}), \mathbf{A}'(\mathbf{r})\} & \longrightarrow & \\ \{v''(\mathbf{r}), \mathbf{A}''(\mathbf{r})\} & \longrightarrow & \Psi(\mathbf{X}) \\ \cdot & & \\ \cdot & \nearrow & \\ \cdot & & \end{array} \tag{7.18}$$

Hence, a proof of the HK path in the presence of a magnetic field is not possible along the lines of the First HK Theorem. What Pan and Sahni [3] proved, taking into account the *many-to-one* relationship between $\{v(\mathbf{r}), \mathbf{A}(\mathbf{r})\}$ and $\Psi(\mathbf{X})$, is that there is a *one-to-one* or bijective relationship between the potentials $\{v(\mathbf{r}), \mathbf{A}(\mathbf{r})\}$ and the nondegenerate ground state density $\rho(\mathbf{r})$ and physical current density $\mathbf{j}(\mathbf{r})$. That is, the knowledge of the nondegenerate ground state $\{\rho(\mathbf{r}), \mathbf{j}(\mathbf{r})\}$ determines the potentials $\{v(\mathbf{r}), \mathbf{A}(\mathbf{r})\}$ to within an additive constant C and the gradient of a scalar function $\nabla\chi$. The proof is for both spinless electrons and electrons with spin. (For spinless electrons, the current density $\mathbf{j}(\mathbf{r})$ is the sum of its paramagnetic and diamagnetic components. For electrons with spin, there is an added magnetization (spin) current density component to $\mathbf{j}(\mathbf{r})$.) As the magnetic field constitutes an added degree of freedom, there is an additional constraint beyond that of fixed electron number N. For spinless electrons, the added constraint is that of fixed canonical angular momentum $\mathbf{L}$. For electrons with spin, there is the added constraint of fixed spin angular momentum $\mathbf{S}$. With the canonical momentum $\hat{\mathbf{p}}$ and electron-interaction $\hat{U}$ operators assumed known, and with the potentials $\{v(\mathbf{r}), \mathbf{A}(\mathbf{r})\}$ now also known, the Schrödinger and Schrödinger-Pauli Hamiltonians are defined. Hence, the HK path in the presence of both an electric and magnetic field is

$$\{\rho(\mathbf{r}), \mathbf{j}(\mathbf{r})\} \rightarrow \{v(\mathbf{r}), \mathbf{A}(\mathbf{r})\} \rightarrow \hat{H}_{spinless} \text{ or } \hat{H}_{spin}. \tag{7.19}$$

Solution of the corresponding differential equations then leads to the eigenfunctions $\Psi_n(\mathbf{X})$ and eigenvalues E_n of the system. Thus, in the presence of a magnetic field, the *basic variables* are the gauge invariant properties $\{\rho(\mathbf{r}), \mathbf{j}(\mathbf{r})\}$.

The Self-consistent Path

Although all the above paths to determine the Hamiltonian $\hat{H}$ are distinct, there is one point of commonality between them. They all proceed from knowledge of the properties that constitute the basic variables. These are either the nondegenerate ground state density $\rho(\mathbf{r})$, or the $\{\rho(\mathbf{r}), \mathbf{j}(\mathbf{r})\}$ with $\mathbf{j}(\mathbf{r})$ the *physical* current density. There is, however, another way [34–36] of deriving the Hamiltonian $\hat{H}$ that does not depend on the concept of a basic variable. As explained in Chap. 3, recognition of the fact that the Hamiltonian $\hat{H}$ is an *exactly known universal functional of the eigenfunctions* $\Psi_n(\mathbf{X})$, i.e. $\hat{H} = \hat{H}[\Psi_n(\mathbf{X})]$, allows it, and the corresponding eigenfunctions $\Psi_n(\mathbf{X})$ and eigenvalues E_n, to be determined self-consistently. The Schrödinger equation (7.1) is first written in its generalized form as

$$\hat{H}[\Psi_n]\Psi_n(\mathbf{X}) = E_n[\Psi_n]\Psi_n(\mathbf{X}). \tag{7.20}$$

In this form, the external potential $v(\mathbf{r})$ is expressed as the *exact functional* of the eigenfunctions $\Psi_n(\mathbf{X})$, i.e. $v(\mathbf{r}) = v[\Psi_n](\mathbf{r})$. With the kinetic $\hat{T}$ and electron-interaction $\hat{U}$ operators *known*, it is the potential $v(\mathbf{r})$, and therefore the Hamiltonian $\hat{H}$, together with the $\{\Psi_n(\mathbf{X}), E_n\}$ that are determined by the self-consistent solution of the generalized Schrödinger equation. This is the case irrespective of whether the potential $v(\mathbf{r})$ is *known* or *unknown*.

7.2 The First Hohenberg-Kohn Theorem

The First Hohenberg-Kohn Theorem is derived for the physical system defined by the Schrödinger equation of (7.1). The statement of the Theorem is the following:

Theorem 1 *The nondegenerate ground state density $\rho(\mathbf{r})$ determines the external field $\mathcal{E}(\mathbf{r})$, or equivalently the external potential $v(\mathbf{r})$ to within a trivial additive constant.*

Proof The theorem is proved for *nondegenerate* v-representable ground state densities $\rho(\mathbf{r})$, *arbitrary* external field $\mathcal{E}(\mathbf{r})$ or potential $v(\mathbf{r})$, and for *fixed* electron number N. For a *specific* external field $\mathcal{E}(\mathbf{r})$ or potential $v(\mathbf{r})$, the Schrödinger equation of (7.1) is solved to determine the ground state wave function $\Psi(\mathbf{X})$. This defines the map C between the different potentials $v(\mathbf{r})$ and the corresponding nondegenerate ground state wave functions $\Psi(\mathbf{X})$ (see Fig. 7.1 with the arrow pointing from left to right). The different ground state wave functions $\Psi(\mathbf{X})$ then lead to the different ground state densities $\rho(\mathbf{r})$ according to (7.3). This establishes the map D between the wave functions $\Psi(\mathbf{X})$ and densities $\rho(\mathbf{r})$ (see Fig. 7.1 with the arrow going from left to right). The combination (CD) of the maps C and D then maps each potential energy $v(\mathbf{r})$ to a ground state density $\rho(\mathbf{r})$.

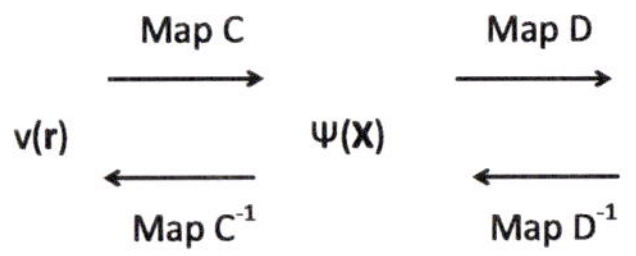

Fig. 7.1 The maps C and C^{-1} between the potentials $v(\mathbf{r})$ and the ground state wave functions $\Psi(\mathbf{X})$, and the maps D and D^{-1} between the wave functions $\Psi(\mathbf{X})$ and the ground state densities $\rho(\mathbf{r})$

The statement of the First Theorem is that the map (CD) is invertible. In other words, the inverse map $(\text{CD})^{-1}$ exists and thereby ensures that knowledge of the ground state density $\rho(\mathbf{r})$ uniquely determines the external potential $v(\mathbf{r})$ to within an *additive* constant, thus determining the Hamiltonian $\hat{H}$. To prove the invertibility of map (CD), the separate inverse maps C^{-1} and D^{-1} must exist (see Fig. 7.1 with the arrows pointing from right to left). That is, for each ground state wave function $\Psi(\mathbf{X})$ there exists one-and-only-one external potential function $v(\mathbf{r})$. And for each ground state density $\rho(\mathbf{r})$, there exists one-and-only-one ground state wave function $\Psi(\mathbf{X})$.

Part 1 of Proof To prove the invertibility C^{-1} of map C, what needs to be proved is that two different external potential operators $\hat{V}$ and $\hat{V}'$ that differ by more than a constant, i.e. $V - V' \neq$ constant, must lead to different ground state wave functions $\Psi(\mathbf{X})$ and $\Psi'(\mathbf{X})$.

The Schrödinger equations corresponding to the operators $\hat{V}$ and $\hat{V}'$ are

$$\hat{H}\Psi(\mathbf{X}) = (\hat{T} + \hat{U} + \hat{V})\Psi(\mathbf{X}) = E\Psi(\mathbf{X}), \tag{7.21}$$

and

$$\hat{H}'\Psi'(\mathbf{X}) = (\hat{T} + \hat{U} + \hat{V}')\Psi'(\mathbf{X}) = E'\Psi'(\mathbf{X}), \tag{7.22}$$

with E and E' the respective ground state energies.

Now if one makes the assumption that $\Psi(\mathbf{X}) = \Psi'(\mathbf{X})$, then on subtraction of the above equations, one obtains

$$(\hat{V} - \hat{V}')\Psi(\mathbf{X}) = (E - E')\Psi(\mathbf{X}). \tag{7.23}$$

As the operators $\hat{V}$ and $\hat{V}'$ are multiplicative operators, the above equation reduces to

$$\hat{V} - \hat{V}' = E - E'. \tag{7.24}$$

As $(E - E')$ is a constant, (7.24) contradicts the assumption that $\hat{V}$ and $\hat{V}'$ must differ by more than a constant. *Thus, it is proved that for every ground state $\Psi(\mathbf{X})$ there corresponds only one potential function $v(\mathbf{r})$*, and the inverse map C^{-1} is established. Consequently, a bijective or one-to-one relationship between $v(\mathbf{r})$ and the nondegenerate ground state $\Psi(\mathbf{X})$ is established.

Note 1 Although the Theorem establishes the bijective relationship between $v(\mathbf{r})$ and $\Psi(\mathbf{X})$, it does not provide the explicit dependence of $v(\mathbf{r})$ on $\Psi(\mathbf{X})$. That *explicit relationship for arbitrary* state is provided by the 'Quantal Newtonian' First Law as discussed further in Sect. 7.2.2.

Note 2 *The bijectivity between the potentials $v(\mathbf{r})$ and the wave functions $\Psi(\mathbf{X})$ of maps C and C^{-1} is critical to the proof of the invertibility D^{-1} of map D.* Let us consider that there exist two nondegenerate ground state wave functions $\Psi(\mathbf{X})$ and

$\Psi'(\mathbf{X})$ with $\Psi(\mathbf{X}) \neq \Psi'(\mathbf{X})$ generated from different $v(\mathbf{r})$ and $v'(\mathbf{r})$, respectively. *This is only possible because of the bijectivity between* $v(\mathbf{r})$ *and* $\Psi(\mathbf{X})$. That is, there is only one $\Psi(\mathbf{X})$ for each $v(\mathbf{r})$. (If there were many $v(\mathbf{r})$ that generated the same $\Psi(\mathbf{X})$, the above statement of the existence of $\Psi(\mathbf{X})$ and $\Psi'(\mathbf{X})$ could not be made.) The statement that there exists a $\Psi(\mathbf{X})$ and a $\Psi'(\mathbf{X})$ is then employed to prove that $\rho(\mathbf{r}) \neq \rho'(\mathbf{r})$.

Part 2 of Proof From the variational principle for the energy [27] we have

$$E = \langle \Psi(\mathbf{X}) | \hat{H} | \Psi(\mathbf{X}) \rangle < \langle \Psi'(\mathbf{X}) | \hat{H} | \Psi'(\mathbf{X}) \rangle. \tag{7.25}$$

Note 3 The inequality of (7.25) is justified because we are considering nondegenerate ground states. To see this, recall that according to the variational principle, for $\Psi(\mathbf{X}) \neq \Psi'(\mathbf{X})$, the energy $E \leq \langle \Psi'(\mathbf{X}) | \hat{H} | \Psi'(\mathbf{X}) \rangle$. Thus, for the equality $E = \langle \Psi'(\mathbf{X}) | \hat{H} | \Psi'(\mathbf{X}) \rangle$, then $\hat{H}\Psi'(\mathbf{X}) = E\Psi'(\mathbf{X})$. This contradicts the assumption that the ground state is nondegenerate.

Rewriting the last term of (7.25) as

$$\begin{aligned} \langle \Psi'(\mathbf{X}) | \hat{H} | \Psi'(\mathbf{X}) \rangle &= \langle \Psi'(\mathbf{X}) | \hat{T} + \hat{U} + \hat{V}' + \hat{V} - \hat{V}' | \Psi'(\mathbf{X}) \rangle \\ &= \langle \Psi'(\mathbf{X}) | \hat{H}' | \Psi'(\mathbf{X}) \rangle + \langle \Psi'(\mathbf{X}) | \hat{V} - \hat{V}' | \Psi'(\mathbf{X}) \rangle \\ &= E' + \int \rho'(\mathbf{r}) [v(\mathbf{r}) - v'(\mathbf{r})] d\mathbf{r}, \end{aligned} \tag{7.26}$$

we see that (7.25) becomes

$$E < E' + \int \rho'(\mathbf{r}) [v(\mathbf{r}) - v'(\mathbf{r})]. \tag{7.27}$$

In a similar manner we have as a result of the variational principle

$$E' = \langle \Psi'(\mathbf{X}) | \hat{H}' | \Psi'(\mathbf{X}) \rangle \ < \ \langle \Psi(\mathbf{X}) | \hat{H}' | \Psi(\mathbf{X}) \rangle, \tag{7.28}$$

so that in this case we obtain

$$E' < E + \int \rho(\mathbf{r}) [v'(\mathbf{r}) - v(\mathbf{r})] d\mathbf{r}. \tag{7.29}$$

Note 4 Note that the densities $\rho(\mathbf{r})$ and $\rho'(\mathbf{r})$ are v-representable as they are obtained from the solutions of the Schrödinger equations (7.21) and (7.22), respectively.

On adding the two inequalities of (7.27) and (7.29) together with the assumption that $\rho(\mathbf{r}) = \rho'(\mathbf{r})$, then leads to the contradiction

$$E + E' < E + E'. \tag{7.30}$$

This proves that for each nondegenerate ground state density $\rho(\mathbf{r})$, there exists one-and-only-one ground state wave function $\Psi(\mathbf{X})$ which gives rise to this density, and hence the inverse map D^{-1} is established. A bijective relationship between the nondegenerate ground state $\Psi(\mathbf{X})$ and the density $\rho(\mathbf{r})$ is thus proved.

Having proved the existence of the inverse maps C^{-1} and D^{-1}, the inverse map $(CD)^{-1}$ ensures that there is a one-to-one correspondence between ground state densities $\rho(\mathbf{r})$ and external potentials $v(\mathbf{r})$. That is, for each nondegenerate ground state density $\rho(\mathbf{r})$ there exists only *one* external potential $v(\mathbf{r})$ that leads to this density. The First HK theorem is thus proved.

Note 5 Once again, although the First HK Theorem establishes a bijective relationship between the nondegenerate ground state $\Psi(\mathbf{X})$ and the density $\rho(\mathbf{r})$, the dependence of $\Psi(\mathbf{X})$ on $\rho(\mathbf{r})$ is not provided. Hence, for the inverse map D^{-1}, there exists *no explicit* functional relationship whereby $\Psi(\mathbf{X})$ can be determined from $\rho(\mathbf{r})$. There is, however, a related question that can be addressed: Of the infinite number of functions $\Psi_\rho(\mathbf{X})$ that integrate to the true density $\rho(\mathbf{r})$, how does one distinguish the true ground state wave function $\Psi(\mathbf{X})$? The answer to this question is in the description (see Sect. 7.1) of the Percus-Levy-Lieb path from the density $\rho(\mathbf{r})$ to the Hamiltonian $\hat{H}$: One performs a variational search over all function $\Psi_\rho(\mathbf{X})$ that yield the density $\rho(\mathbf{r})$ and obtains the infimum of the expectation value of the sum of the operators $\hat{T}$ and $\hat{U}$. The proof of this statement is provided in the Sect. 7.2.2 on Inverse Maps.

7.2.1 The Gunnarsson-Lundqvist Theorem for Excited States

The proof of the First HK Theorem relies on *(i)* the one-to-one relationship between the potential $v(\mathbf{r})$ and the wave function $\Psi(\mathbf{X})$, and *(ii)* on the variational principle for the energy. It follows that in addition to the ground state, the proof of the First HK Theorem is then also valid for those excited states for which these conditions are met. This is the Gunnarsson-Lundqvist Theorem [37, 38].

Consider, for example, the 2-electron Helium atom or Quantum Dot with their corresponding binding potentials $v(\mathbf{r})$. The symmetry of the excited states such as 2^3S, 2^1P, 2^3P, 3^1D, 3^3D, differs from that of the ground state. For these states, there exists only one wave function for each state. Hence, there exists a one-to-one relationship between the potential $v(\mathbf{r})$ and the corresponding wave function. Further, for states of such symmetry, the variational principle is applicable because for these states the subsidiary condition that the wave function be orthogonal to the eigenfunctions of all lower states is automatically satisfied. This is because two states that have a different value of the total orbital $\mathbf{L}$ or total spin $\mathbf{S}$ angular momentum or both are automatically orthogonal. Hence, the First HK Theorem is valid for such states. In other words, the excited state density $\rho^e(\mathbf{r})$ for such states uniquely determines the potential $v(\mathbf{r})$ to within an additive constant. The Theorem is not valid for the excited singlet 2^1S state.

7.2.2 The Inverse Maps C^{-1} and D^{-1}

The Maps C from the external potentials $v(\mathbf{r})$ to the ground state wave functions $\Psi(\mathbf{X})$, and of the Maps D from the wave functions $\Psi(\mathbf{X})$ to the ground state densities $\rho(\mathbf{r})$ (see Fig. 7.1), are well defined. One solves the Schrödinger equation (7.1) for the ground state to obtain $\Psi(\mathbf{X})$ for a particular $v(\mathbf{r})$ for Map C, and then the corresponding ground state density $\rho(\mathbf{r})$ from $\Psi(\mathbf{X})$ via its definition of (7.4). Here we address the inverse Maps C^{-1} and D^{-1}. In other words, the paths from the ground state wave function $\Psi(\mathbf{X})$ to the external potential $v(\mathbf{r})$, and that from the ground state density $\rho(\mathbf{r})$ to the wave function $\Psi(\mathbf{X})$.

Inverse Map C^{-1}

The path from knowledge of a ground state wave function $\Psi(\mathbf{X})$ to the potential $v(\mathbf{r})$ is given by the 'Quantal Newtonian' First Law of (2.130). It provides the functional dependence of $v(\mathbf{r})$ on $\Psi(\mathbf{X})$. Accordingly, the potential $v(\mathbf{r})$ is the work done to move an electron from some reference point at infinity to its position at $\mathbf{r}$ in the force of the *conservative* internal field $\mathcal{F}^{\mathrm{int}}[\Psi(\mathbf{X})](\mathbf{r})$ experienced by each electron:

$$v(\mathbf{r}) = v[\Psi](\mathbf{r}) = \int_{\infty}^{\mathbf{r}} \mathcal{F}^{\mathrm{int}}[\Psi](\mathbf{r}') \cdot d\boldsymbol{\ell}', \tag{7.31}$$

where

$$\mathcal{F}^{\mathrm{int}}[\Psi(\mathbf{X})](\mathbf{r}) = \mathcal{E}_{ee}[\Psi](\mathbf{r}) - \mathcal{Z}[\Psi](\mathbf{r}) - \mathcal{D}[\Psi](\mathbf{r}), \tag{7.32}$$

with the electron-interaction $\mathcal{E}_{ee}[\Psi](\mathbf{r})$, kinetic $\mathcal{Z}[\Psi](\mathbf{r})$ and differential density $\mathcal{D}[\Psi](\mathbf{r})$ fields as defined in Sect. 2.8.3. These fields are *known functionals* of the wave function $\Psi(\mathbf{X})$. They are derived from quantal sources which are expectation values of Hermitian operators taken with respect to the wave function $\Psi(\mathbf{X})$. Furthermore $\nabla \times \mathcal{F}^{\mathrm{int}}[\Psi(\mathbf{X})](\mathbf{r}) = 0$. Thus, the functional dependence of $v(\mathbf{r})$ on $\Psi(\mathbf{X})$ is known.

Note 1 The definition of the potential $v(\mathbf{r})$ of (7.31) is not restricted to the ground state $\Psi(\mathbf{X})$. It is valid for arbitrary state $\Psi_n(\mathbf{X})$.

For completeness the definitions of the fields $\mathcal{E}_{ee}[\Psi](\mathbf{r})$, $\mathcal{Z}[\Psi](\mathbf{r})$, $\mathcal{D}[\Psi](\mathbf{r})$ in terms of the corresponding 'forces' $\mathbf{e}_{ee}[\Psi](\mathbf{r})$, $z[\Psi](\mathbf{r})$, $\mathbf{d}[\Psi](\mathbf{r})$ and the density $\rho(\mathbf{r})$ are given below. The respective quantal sources are the pair function $P(\mathbf{r}\mathbf{r}')$, the single-particle density matrix $\gamma(\mathbf{r}\mathbf{r}')$, and the density $\rho(\mathbf{r})$ (see Sect. 2.1).

The electron-interaction field $\mathcal{E}_{ee}[\Psi](\mathbf{r})$:

$$\mathcal{E}_{ee}[\Psi](\mathbf{r}) = \frac{\mathbf{e}_{ee}[\Psi](\mathbf{r})}{\rho(\mathbf{r})}\ ; \quad \mathbf{e}_{ee}[\Psi](\mathbf{r}) = \int \frac{P(\mathbf{r}\mathbf{r}')(\mathbf{r}-\mathbf{r}')}{|\mathbf{r}-\mathbf{r}'|^3} d\mathbf{r}'. \tag{7.33}$$

The kinetic field $\mathcal{Z}[\Psi](\mathbf{r})$:

$$\mathcal{Z}[\Psi](\mathbf{r}) = \frac{\mathbf{z}[\Psi](\mathbf{r})}{\rho(\mathbf{r})}; \quad z_{\alpha}[\Psi](\mathbf{r}) = 2\sum_{\beta} \frac{\partial}{\partial r_{\beta}} t_{\alpha\beta}(\mathbf{r}; [\Psi]), \tag{7.34}$$

with the kinetic energy tensor

$$t_{\alpha\beta}(\mathbf{r}) = \frac{1}{4}\left[\frac{\partial^2}{\partial r'_\alpha \partial''_\beta} + \frac{\partial^2}{\partial r'_\beta \partial''_\alpha}\right]\gamma(\mathbf{r}'\mathbf{r}'')\bigg|_{\mathbf{r}'=\mathbf{r}''=\mathbf{r}}. \tag{7.35}$$

The differential density field $\mathcal{D}[\Psi](\mathbf{r})$:

$$\mathcal{D}[\Psi](\mathbf{r}) = \frac{\mathbf{d}[\Psi](\mathbf{r})}{\rho(\mathbf{r})}; \quad \mathbf{d}[\Psi](\mathbf{r}) = -\frac{1}{4}\nabla\nabla^2\rho(\mathbf{r}). \tag{7.36}$$

The path from the wavefunction $\Psi(\mathbf{X})$ to the potential $v(\mathbf{r})$ or equivalently the Map C^{-1} is thus explicitly defined.

Note 2 It is also possible to determine the operator $\hat{V}$ by inversion of the Schrödinger equation: $\hat{V} = [(\hat{T} + \hat{U})\Psi(\mathbf{X})]/\Psi(\mathbf{X})$.

Inverse Map D^{-1}

In principle, knowledge of the density $\rho(\mathbf{r})$ determines the wave function $\Psi(\mathbf{X})$. However, unlike the explicit functional dependence of the potential $v(\mathbf{r})$ on the wave function $\Psi(\mathbf{X})$ (see (7.31)), the functional dependence of $\Psi(\mathbf{X})$ on $\rho(\mathbf{r})$ is not known. The difficulty lies in that the mapping required is from a one-particle property of the density $\rho(\mathbf{r})$ to the many-particle wave function $\Psi(\mathbf{X})$. As there exist many functions $\Psi_\rho(\mathbf{X})$ that integrate to $\rho(\mathbf{r})$, it is possible to determine the true $\Psi(\mathbf{X})$ from amongst these functions. One performs a search constrained to the functions $\Psi_\rho(\mathbf{X})$.

From the variational principle for the energy, we have

$$\langle\Psi_\rho(\mathbf{X})|\hat{H}|\Psi_\rho(\mathbf{X})\rangle \geq \langle\Psi(\mathbf{X})|\hat{H}|\Psi(\mathbf{X})\rangle = E, \tag{7.37}$$

or equivalently

$$\langle\Psi_\rho(\mathbf{X})|\hat{T} + \hat{U}|\Psi_\rho(\mathbf{X})\rangle + \int \rho(\mathbf{r})v(\mathbf{r})d\mathbf{r} \geq \langle\Psi(\mathbf{X})|\hat{T} + \hat{U}|\Psi(\mathbf{X})\rangle + \int \rho(\mathbf{r})v(\mathbf{r})d\mathbf{r}, \tag{7.38}$$

which reduces to

$$\langle\Psi_\rho(\mathbf{X})|\hat{T} + \hat{U}|\Psi_\rho(\mathbf{X})\rangle \geq \langle\Psi(\mathbf{X})|\hat{T} + \hat{U}|\Psi(\mathbf{X})\rangle. \tag{7.39}$$

Thus, of all the functions $\Psi_\rho(\mathbf{X})$ that yield the ground state density $\rho(\mathbf{r})$, the true ground state wave function $\Psi(\mathbf{X})$ is that which minimizes the expectation value $\langle\hat{T} + \hat{U}\rangle$. This constrained variational search can then be construed as a mechanism for the inverse mapping D^{-1} from $\rho(\mathbf{r})$ to $\Psi(\mathbf{X})$.

Note 3 Although the constrained variational search involves only the operators $\hat{T}$ and $\hat{U}$, there is nonetheless an implicit dependence on the potential $v(\mathbf{r})$. This follows from the First HK theorem according to which knowledge of the density $\rho(\mathbf{r})$ uniquely determines the potential $v(\mathbf{r})$ to within an additive constant. The search is thus for a particular physical system.

Note 4 Methods for constructing N-particle anti-symmetric functions $\Psi_\rho(\mathbf{X})$ that yield $\rho(\mathbf{r})$ are described by Gilbert [39], Harriman [40], and Cioslowski [41].

Note 5 It is also possible to construct [42] antisymmetric functions $\Psi_\rho(\mathbf{X})$ that are functionals of functions χ, i.e. $\Psi_\rho(\mathbf{X}) = \Psi_\rho[\chi](\mathbf{X})$ which also reproduce a given density $\rho(\mathbf{r})$.

7.2.3 Generalization of the First Hohenberg-Kohn Theorem via Density Preserving Unitary Transformations

From the proof of the First HK Theorem, it follows that the ground state wave function $\Psi(\mathbf{X})$ is a functional of the density $\rho(\mathbf{r})$, i.e. $\Psi(\mathbf{X}) = \Psi[\rho]$. In their paper [1] HK state: "Thus, $v(\mathbf{r})$ is (to within a constant) a unique functional of $\rho(\mathbf{r})$; since, in turn, $v(\mathbf{r})$ fixes $\hat{H}$ we see that *the full many-particle ground state is a unique functional of* $\rho(\mathbf{r})$." (Our emphases). This statement is the HK path described by (7.5). (The fact that the wave function $\Psi(\mathbf{X})$ is a functional of $\rho(\mathbf{r})$ also follows from the bijective maps D and D^{-1} provided that the Maps C and C^{-1} are first established.) The statement of HK though correct is incomplete. As is known in quantum mechanics, the density $\rho(\mathbf{r})$ is a *gauge invariant* property, whereas the Hamiltonian $\hat{H}$ and the wave function $\Psi(\mathbf{X})$ are *gauge variant*. Hence, the wave function $\Psi(\mathbf{X})$ when written *solely* as a functional of the density $\rho(\mathbf{r})$ must then be *gauge invariant* in contradiction to the tenet of quantum mechanics. To address this issue, one performs [43] a *density preserving* unitary transformation involving a gauge function $\alpha(\mathbf{R})$ with $\mathbf{R} = \mathbf{r}_1, \ldots, \mathbf{r}_N$. As the physical system, and the energy E, remain unchanged in the transformation, the HK proof is equally valid for the transformed Hamiltonian $\hat{H}'(\mathbf{R})$. As a consequence of the transformation, however, the wave function $\Psi(\mathbf{X})$ is then seen to be a functional of both the density $\rho(\mathbf{r})$ as well as the gauge function $\alpha(\mathbf{R})$, i.e. $\Psi(\mathbf{X}) = \Psi[\rho(\mathbf{r}); \alpha(\mathbf{R})]$. It is the presence of the gauge function $\alpha(\mathbf{R})$ that allows the wave function written as a functional to be gauge variant. As will be shown, the original Hamiltonian $\hat{H}(\mathbf{R})$, and thus the original statement of HK, is recovered as a special case when $\alpha(\mathbf{R}) = \alpha$, a constant.

Consider the physical system defined by the Hamiltonian $\hat{H}(\mathbf{R})$ of (7.2) (with the electron-interaction operator $\hat{U}$ replaced by the symbol $\hat{W}$). The unitary operator employed is

$$\hat{U} = e^{j\alpha(\mathbf{R})}, \tag{7.40}$$

so that the transformed wave function $\Psi'(\mathbf{X})$ is

$$\Psi'(\mathbf{X}) = \hat{U}^\dagger \Psi(\mathbf{X}). \tag{7.41}$$

The transformed density $\rho'(\mathbf{r})$ is then

$$\rho'(\mathbf{r}) = \langle \Psi'(\mathbf{X}) | \hat{\rho}(\mathbf{r}) | \Psi'(\mathbf{X}) \rangle = \rho(\mathbf{r}). \tag{7.42}$$

The unitary transformation thus *preserves the density*.

The transformed Hamiltonian $\hat{H}'(\mathbf{R})$ is

$$\hat{H}'(\mathbf{R}) = \hat{U}^{\dagger}\hat{H}\hat{U}, \tag{7.43}$$

so that the transformed Schrödinger equation is

$$\hat{H}'(\mathbf{R})\Psi'(\mathbf{X}) = E'\Psi'(\mathbf{X}), \tag{7.44}$$

with $E' = E = \langle\Psi(\mathbf{X})|\hat{H}(\mathbf{R})|\Psi(\mathbf{X})\rangle$.

We next obtain the transformed Hamiltonian $\hat{H}'(\mathbf{R})$. From (7.43)

$$\hat{H}'(\mathbf{R}) = e^{-j\alpha(\mathbf{R})}\sum_i\Big(-\frac{1}{2}\nabla_i^2\Big)e^{j\alpha(\mathbf{R})} + \hat{W} + \hat{V}. \tag{7.45}$$

Since the commutator

$$[\nabla^2, e^{j\alpha}] = \nabla^2 e^{j\alpha} - e^{j\alpha}\nabla^2, \tag{7.46}$$

the Hamiltonian $\hat{H}'(\mathbf{R})$ is

$$\hat{H}'(\mathbf{R}) = -\frac{1}{2}\sum_i\{e^{-j\alpha(\mathbf{R})}[\nabla_i^2, e^{j\alpha(\mathbf{R})}] + \nabla_i^2\} + \hat{W} + \hat{V} \tag{7.47}$$

or

$$\hat{H}'(\mathbf{R}) = \hat{H}(\mathbf{R}) - \frac{1}{2}\sum_i\{e^{-j\alpha(\mathbf{R})}[\nabla_i^2, e^{j\alpha(\mathbf{R})}]\}. \tag{7.48}$$

Solving for the commutator in (7.48), the transformed Hamiltonian $\hat{H}'(\mathbf{R})$ may be written as

$$\hat{H}'(\mathbf{R}) = \frac{1}{2}\sum_i\big(\hat{\mathbf{p}}_i + \hat{\mathbf{A}}_i\big)^2 + \hat{W} + \hat{V}, \tag{7.49}$$

with

$$\hat{\mathbf{A}}_i = \nabla_i\alpha(\mathbf{R}). \tag{7.50}$$

Note that the choice of gauge function $\alpha(\mathbf{R})$ in the unitary transformation is *arbitrary*. (It is chosen so as to preserve the density.) That is, it could be some $\alpha_k(\mathbf{R})$, in which case the transformed wave function would be $\Psi_k'(\mathbf{X}) = exp\big(i\alpha_k(\mathbf{R})\big)\Psi(\mathbf{X})$, and the transformed Hamiltonian $\hat{H}_k'(\mathbf{R})$ to be

$$\hat{H}_k'(\mathbf{R}) = \frac{1}{2}\sum_i\big(\hat{\mathbf{p}}_i + \hat{\mathbf{A}}_{ik}\big)^2 + \hat{W} + \hat{V}, \tag{7.51}$$

with

$$\hat{\mathbf{A}}_{ik} = \nabla_i \alpha_k(\mathbf{R}). \tag{7.52}$$

As the transformations are unitary, each transformed Hamiltonian $\hat{H}'_k(\mathbf{R})$ corresponds to the *same physical system* as that represented by the Hamiltonian $\hat{H}(\mathbf{R})$. Thus, the proof of bijectivity between the density $\rho(\mathbf{r})$ and the Hamiltonian $\hat{H}(\mathbf{R})$ is equally valid for the more general Hamiltonians $\hat{H}'_k(\mathbf{R})$. This generalizes the First HK Theorem to such Hamiltonians. As a consequence, we learn that the wave function $\Psi(\mathbf{X})$ must be a functional of both the density $\rho(\mathbf{r})$ as well as the gauge function $\alpha(\mathbf{R})$, i.e. $\Psi(\mathbf{X}) = \Psi[\rho(\mathbf{r}); \alpha(\mathbf{R})]$. It is the presence of the gauge function $\alpha(\mathbf{R})$ which ensures that the wave function written as a functional is gauge variant.

Finally, as the choice of gauge function $\alpha(\mathbf{R})$ is arbitrary, the choice of $\alpha(\mathbf{R}) = \alpha$, a constant, or $\alpha(\mathbf{R}) = 0$ is equally valid. This choice is consistent with the HK statement quoted above. However, it is now understood that therein lies implicitly the assumption of vanishing of the gauge function, i.e. $\Psi(\mathbf{X}) = \Psi[\rho(\mathbf{r}); \alpha(\mathbf{R}) = 0]$.

Writing the gauge function $\Psi(\mathbf{X})$ as a functional of the density $\Psi[\rho]$, the expectation value of any operator $\hat{O}$ is then a unique functional of the density:

$$\langle \hat{O} \rangle = O[\rho(\mathbf{r})] = \langle \Psi[\rho] | \hat{O} | \Psi[\rho] \rangle. \tag{7.53}$$

Thus, the energy $\hat{H}$ is a functional of the density: $E = E[\rho(\mathbf{r})]$. However, as the functional dependence of $\Psi(\mathbf{X})$ on $\rho(\mathbf{r})$ is unknown, the unique expectation value functionals $O[\rho]$ are unknown.

7.2.4 Corollary to the First Hohenberg-Kohn Theorem

For a system of N electrons in an external field $\mathcal{E}(\mathbf{r}) = -\nabla v(\mathbf{r})/e$, the First HK Theorem proves that the relationship between the Hamiltonian $\hat{H}$ of the system (see (7.1)) and the nondegenerate ground state density $\rho(\mathbf{r})$ is one-to-one or bijective and therefore invertible. That is, the Hamiltonian $\hat{H}$ uniquely determines the density $\rho(\mathbf{r})$ via solution of the Schrödinger equation. In turn, knowledge of the density $\rho(\mathbf{r})$ determines the external potential $v(\mathbf{r})$, and thereby the Hamiltonian $\hat{H}$, to within an *arbitrary additive constant* C, i.e. $\hat{H} + C$. The Hamiltonians $\hat{H} + C$ correspond to the *same physical system* as that represented by $\hat{H}$: all properties are the same with the exception of the energy E which differs by the constant C. The relationship between the Hamiltonian $\hat{H}$ and the density may then be stated in equation form as

$$\hat{H} \text{ or } \hat{H} + C \leftrightarrow \rho(\mathbf{r}), \tag{7.54}$$

with the two-sided arrow indicative of the bijectivity.

It is, however, possible [44] to construct a set of Hamiltonians $\{\hat{H}\}$ that represent *different physical systems* whose Hamiltonians differ by an *intrinsic constant* C, but which all have the *same* density $\rho(\mathbf{r})$. By intrinsic constant is meant one that is inherent to the system Hamiltonian and not extrinsically additive. It is the constant C which helps distinguish between the different Hamiltonians of the set $\{\hat{H}\}$. As the density $\rho(\mathbf{r})$ is the same, it cannot determine each external potential energy $v(\mathbf{r})$ uniquely, and thereby cannot distinguish between the Hamiltonians of the set $\{\hat{H}\}$. The relationship between the Hamiltonians $\{\hat{H}\}$ and the density $\rho(\mathbf{r})$ in equation form is then

$$\{\hat{H}\} \to \rho(\mathbf{r}), \tag{7.55}$$

with the right-directed arrow indicating the lack of invertibility.

The statement of the Corollary to the First HK Theorem is as follows:

Corollary *Degenerate time-independent Hamiltonians $\{\hat{H}\}$ that represent different physical systems, but which each differ by a constant C, and yet possess the same density $\rho(\mathbf{r})$, cannot be distinguished on the basis of the First Hohenberg-Kohn Theorem.*

The set of Hamiltonians $\{\hat{H}\}$ with the above properties are those of the Hooke's species (see Fig. 7.2). The species comprise of two electrons (interacting Coulombically) coupled harmonically to a variable number $\mathcal{N}$ of nuclei. The electrons are coupled to each nuclei with different spring constants k_j, $j = 1, \ldots, \mathcal{N}$. The species comprise of the Hooke's atom: $\mathcal{N} = 1$, atomic number $Z = 2$, spring constant k; Hooke's molecule: $\mathcal{N} = 2$, atomic number of each nuclei $Z = 1$, spring constants k_1 and k_2; and the Hooke's positive molecular ions: $\mathcal{N} > 2$, atomic number of each nuclei $Z = 1$, spring constants $k_1, k_2, k_3, \ldots, k_{\mathcal{N}}$.

The Hamiltonians $\hat{H}_{\mathcal{N}}$ for the Hooke's species (in atomic units) is

$$\hat{H}_{\mathcal{N}} = \hat{T} + \hat{U} + \hat{V}_{\mathcal{N}}, \tag{7.56}$$

where $\hat{T}$ the kinetic energy operator is

$$\hat{T} = -\frac{1}{2}\sum_{i=1}^{2} \nabla_i^2, \tag{7.57}$$

$\hat{U}$ the electron-interaction potential operator is

$$\hat{U} = \frac{1}{|\mathbf{r}_1 - \mathbf{r}_2|}, \tag{7.58}$$

and $\hat{V}_{\mathcal{N}}$ the external potential operator is

$$\hat{V}_{\mathcal{N}} = \sum_{i=1}^{2} v_{\mathcal{N}}(\mathbf{r}_i), \tag{7.59}$$

with the external potential $v_{\mathcal{N}}(\mathbf{r})$ being harmonic:

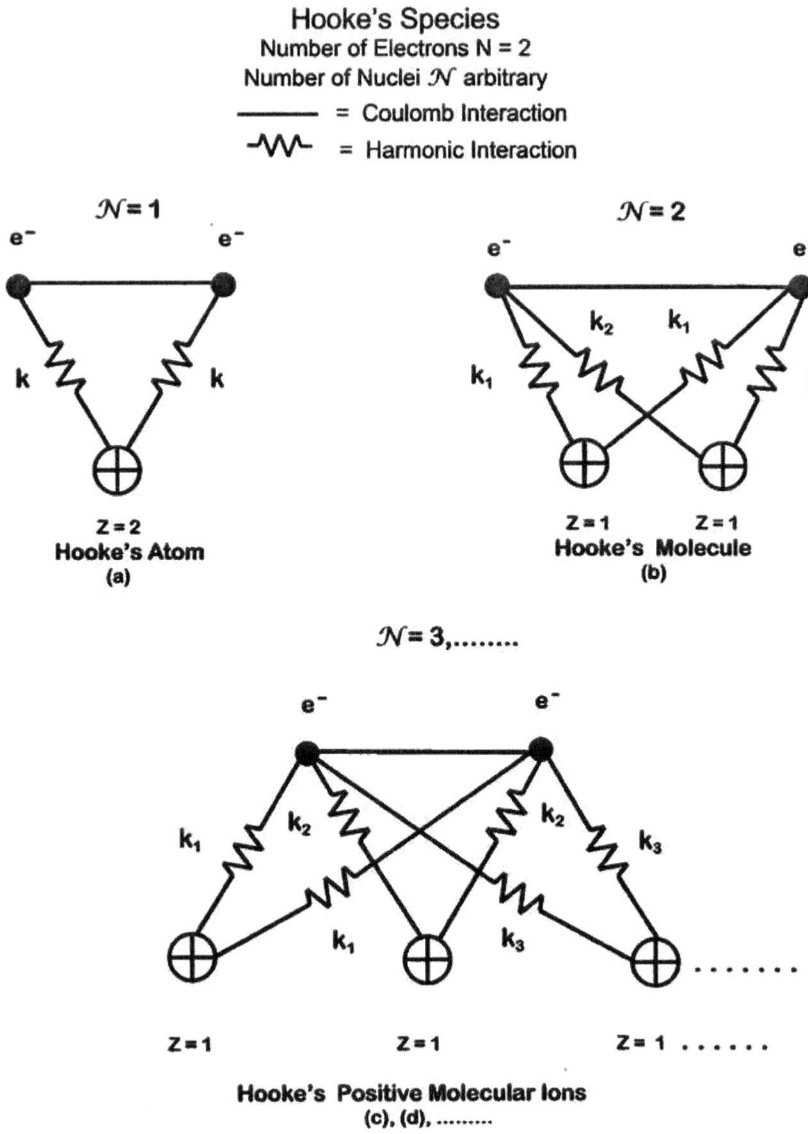

Fig. 7.2 The Hooke's species comprises of two electrons and an arbitrary number $\mathcal{N}$ of nuclei, the interaction between the electrons is Coulombic, and that between electrons and nuclei is harmonic with spring constant $k, k_1, \ldots, k_{\mathcal{N}}$: (**a**) Hooke's atom; (**b**) Hooke's molecule; (**c**), (**d**), ... Hooke's positive molecular ions. Here $\mathcal{N}$ is the number of nuclei, Z the nuclear charge, e^- the electronic charge. Note that each element of the species corresponds to a different physical system

$$v_{\mathcal{N}}(\mathbf{r}) = \frac{1}{2}\sum_{j=1}^{\mathcal{N}} k_j(\mathbf{r} - \mathbf{R}_j)^2. \tag{7.60}$$

Here $\mathbf{r}_1$ and $\mathbf{r}_2$ are the electron positions, and $\mathbf{R}_j (j = 1, \ldots, \mathcal{N})$ the positions of the nuclei.

Each element of the Hooke's species represents a *different* physical system. Thus, for example, the Hamiltonian for Hooke's atom is

$$\hat{H}_a = -\frac{1}{2}\nabla_1^2 - \frac{1}{2}\nabla_2^2 + \frac{1}{|\mathbf{r}_1 - \mathbf{r}_2|} + \frac{1}{2}k[(\mathbf{r}_1 - \mathbf{R}_1)^2 + (\mathbf{r}_2 - \mathbf{R}_1)^2], \tag{7.61}$$

and that of Hooke's molecule is

$$\hat{H}_m = -\frac{1}{2}\nabla_1^2 - \frac{1}{2}\nabla_2^2 + \frac{1}{|\mathbf{r}_1 - \mathbf{r}_2|} + \frac{1}{2}\left\{k_1\left[(\mathbf{r}_1 - \mathbf{R}_1)^2 + (\mathbf{r}_2 - \mathbf{R}_1)^2\right]\right.$$
$$\left. + \, k_2\left[(\mathbf{r}_1 - \mathbf{R}_2)^2 + (\mathbf{r}_2 - \mathbf{R}_2)^2\right]\right\}, \tag{7.62}$$

where $k \neq k_1 \neq k_2$, and so on for the various Hooke's positive molecular ions with $\mathcal{N} > 2$.

For the Hooke's species, however, the external potential energy operator $\hat{V}_{\mathcal{N}}$ which is

$$\hat{V}_{\mathcal{N}} = \frac{1}{2}\sum_{j=1}^{\mathcal{N}}[k_j(\mathbf{r}_1 - \mathbf{R}_j)^2 + k_j(\mathbf{r}_2 - \mathbf{R}_j)^2], \tag{7.63}$$

may be rewritten as

$$\hat{V}_{\mathcal{N}}(\mathbf{r}) = \left(\frac{1}{2}\sum_{j=1}^{\mathcal{N}} k_j\right)[(\mathbf{r}_1 - \mathbf{a})^2 + (\mathbf{r}_2 - \mathbf{a})^2] + C(\{k\}, \{\mathbf{R}\}, \mathcal{N}), \tag{7.64}$$

where the translation vector $\mathbf{a}$ is

$$\mathbf{a} = \sum_{j=1}^{\mathcal{N}} k_j \mathbf{R}_j \bigg/ \sum_{j=1}^{\mathcal{N}} k_j, \tag{7.65}$$

and the constant C is

$$C = b - d \tag{7.66}$$

with

$$b = \sum_{j=1}^{\mathcal{N}} k_j \mathbf{R}_j^2. \tag{7.67}$$

$$d = \left(\sum_{j=1}^{\mathcal{N}} k_j \mathbf{R}_j \right)^2 \Bigg/ \sum_{j=1}^{\mathcal{N}} k_j \, , \tag{7.68}$$

or

$$C = \frac{1}{2} \sum_{i \neq j}^{\mathcal{N}} k_i k_j \left(\mathbf{R}_i - \mathbf{R}_j \right)^2 \Bigg/ \sum_{j=1}^{\mathcal{N}} k_j \, . \tag{7.69}$$

From (7.64) it is evident that the Hamiltonians $\hat{H}_{\mathcal{N}}$ of the Hooke's species are those of a Hooke's atom ($\sum_{j=1}^{\mathcal{N}} k_j = k$), (to within a constant $C(\{k\}, \{\mathbf{R}\}, \mathcal{N})$), whose center of mass is at $\mathbf{a}$. The constant C which depends upon the spring constants $\{k\}$, the positions of the nuclei $\{\mathbf{R}\}$, and the number $\mathcal{N}$ of the nuclei, differs from a trivial additive constant in that it is an *intrinsic* part of *each* Hamiltonian $\hat{H}_{\mathcal{N}}$, and distinguishes between the different elements of the species. It does so because the constant $C(\{k\}, \{\mathbf{R}\}, \mathcal{N})$ contains physical information about the system such as the positions $\{\mathbf{R}\}$ of the nuclei.

Now according to the First HK theorem, the ground state density determines the external potential energy, and hence the Hamiltonian, to within a constant. Since the density of *each* element of the Hooke's species is that of the Hooke's atom, it can only determine the Hamiltonian of a Hooke's atom and not the constant $C(\{k\}, \{\mathbf{R}\}, \mathcal{N})$. Therefore, it cannot determine the Hamiltonian $\hat{H}_{\mathcal{N}}$ for $\mathcal{N} > 1$. This is reflected by the fact that the density of the elements of the Hooke's species does not satisfy the electron-nucleus coalescence cusp condition. (It is emphasized that although the 'degenerate Hamiltonians' of the Hooke's species have a ground state wavefunction and density that corresponds to that of a Hooke's atom, each element of the species represents a *different* physical system. Thus, for example, a neutron diffraction experiment on the Hooke's molecule and Hooke's positive molecular ions would *all* give different results).

It is also possible to construct a Hooke's species such that the density of each element is the *same*. This is most readily seen for the case when the center of mass is moved to the origin of the coordinate system, i.e.. for $\mathbf{a} = 0$. This requires, from (7.65), the product of the spring constants and the coordinates of the nuclei satisfy the condition

$$\sum_{j=1}^{\mathcal{N}} k_j \mathbf{R}_j = 0 \, , \tag{7.70}$$

so that the external potential energy operator is then

$$v_{\mathcal{N}}(\mathbf{r}) = \frac{1}{2} \sum_{j=1}^{\mathcal{N}} k_j \mathbf{r}^2 + \frac{1}{2} \sum_{j=1}^{\mathcal{N}} k_j \mathbf{R}_j^2 \, , \tag{7.71}$$

where $\mathbf{r}$ is the distance to the origin. If the sum $\sum_{j=1}^{\mathcal{N}} k_j$ is then adjusted to equal a particular value of the spring constant k of Hooke's atom:

$$\sum_{j=1}^{\mathcal{N}} k_j = k \ , \tag{7.72}$$

then the Hamiltonian $\hat{H}_{\mathcal{N}}$ of any element of the species may be rewritten as

$$\hat{H}_{\mathcal{N}}(\{k\}, \{\mathbf{R}\}, \mathcal{N}) = \hat{H}_a(k) + C(\{k\}, \{\mathbf{R}\}, \mathcal{N}) \ , \tag{7.73}$$

where $\hat{H}_a(k)$ is the Hooke's atom Hamiltonian and the constant $C(\{k\}, \{\mathbf{R}\}, \mathcal{N})$ is

$$C(\{k\}, \{\mathbf{R}\}, \mathcal{N}) = \sum_{j=1}^{\mathcal{N}} k_j \mathbf{R}_j^2 \ . \tag{7.74}$$

The solution of the Schrödinger equation and the corresponding density for *each* element of the species are therefore the *same*.

As an example, again consider the case of Hooke's molecule and atom. For Hooke's atom $\mathcal{N} = 1$, $\mathbf{R}_1 = 0$ and let us assume $k = \frac{1}{4}$. Thus, the external potential energy operator is

$$v_a(\mathbf{r}) = \frac{1}{2}kr^2 = \frac{1}{8}r^2 \ . \tag{7.75}$$

For this choice of k, the singlet ground state solution of the time-independent Schrödinger equation ($\hat{H}_{\mathcal{N}}\Psi = E_{\mathcal{N}}\Psi$) is analytical and given by:

$$\Psi(\mathbf{r}_1\mathbf{r}_2) = De^{-y^2/2}e^{-r^2/8}(1 + r/2) \ , \tag{7.76}$$

where $\mathbf{r} = \mathbf{r}_1 - \mathbf{r}_2$, $\mathbf{y} = (\mathbf{r}_1 + \mathbf{r}_2)/2$, and $D = 1/[2\pi^{5/4}(5\sqrt{\pi} + 8)^{1/2}]$. The corresponding ground state density $\rho(\mathbf{r})$ is

$$\rho(\mathbf{r}) = \frac{\pi\sqrt{2\pi}}{r} D^2 e^{-r^2/2}\{7r + r^3 + (8/\sqrt{2\pi})re^{-r^2/2} + 4(1 + r^2)\mathrm{erf}(r/\sqrt{2})\} \ , \tag{7.77}$$

where

$$\mathrm{erf}(x) = \frac{2}{\sqrt{\pi}} \int_0^x e^{-z^2} dz \tag{7.78}$$

is the error function.

For the Hooke's molecule, $\mathcal{N} = 2$, $\mathbf{R}_1 = -\mathbf{R}_2$, and we choose $k_1 = k_2 = \frac{1}{8}$, so that the external potential energy operator is

$$v_m(\mathbf{r}) = \frac{1}{8}r^2 + \frac{1}{16}(R_1^2 + R_2^2) = \frac{1}{8}r^2 + \frac{1}{8}R^2 \ , \tag{7.79}$$

where $|\mathbf{R}_1| = R$. Thus, the Hamiltonian for Hooke's molecule differs from that of Hooke's atom by only the constant $\frac{1}{8}R^2$, thereby leading to the *same* ground state wave function and density. However, the ground state energy of the two elements of the species differ by $\frac{1}{8}R^2$.

The above example demonstrating the equivalence of the density of the Hooke's atom and molecule is for a specific value of the spring constant k for which the wavefunction happens to be analytical. However, this conclusion is valid for arbitrary value of k for which solutions of the Schrödinger equation exist but are not necessarily analytical. For example, if we assume that for each element of the species ($\mathcal{N} \geq 2$), all the spring constants k_j, $j = 1, 2, \ldots, \mathcal{N}$ are the same and designated by k', then for the three values of k for the Hooke's atom corresponding to $k = \frac{1}{4}, \frac{1}{2}, 1$, the values of k' for which the Hooke's molecule and molecular ion ($\mathcal{N} = 3$) wavefunctions are the same are $k' = \frac{1}{8}, \frac{1}{12}$; $k' = \frac{1}{4}, \frac{1}{6}$; $k' = \frac{1}{2}, \frac{1}{3}$, respectively.

Thus, for the case where the elements of the Hooke's species are all made to have the *same* ground state density $\rho(\mathbf{r})$, the density cannot, on the basis of the First HK theorem, distinguish between the different physical elements of the species.

We conclude by noting that the Hooke's species *does not constitute a counter example* to the First HK Theorem. The reason for this is that the proof of the Theorem is *independent* of whether the constant C is additive or intrinsic. The Hamiltonians still differ by a constant C. A counter example would be one in which Hamiltonians that differ by more than a constant C have the same density $\rho(\mathbf{r})$.

7.3 The Second Hohenberg-Kohn Theorem

Having established via the First Hohenberg-Kohn Theorem that the ground state energy E is a functional of the nondegenerate ground state density $\rho(\mathbf{r})$, i.e. $E = E[\rho]$, the statement of the Second Theorem is as follows:

Theorem 2 *The nondegenerate ground state density $\rho(\mathbf{r})$ can be determined from the ground state energy functional $E[\rho]$ via the variational principle for the energy by variation of the density.*

Proof The ground state energy E as a functional of the density $\rho(\mathbf{r})$ is

$$E = E[\rho] = \langle \Psi[\rho]|\hat{H}|\Psi[\rho]\rangle. \tag{7.80}$$

Consider a v-representable ground state density $\tilde{\rho}(\mathbf{r}) \neq \rho(\mathbf{r})$. From the First HK Theorem, knowledge of the density $\tilde{\rho}(\mathbf{r})$ determines the corresponding wave function functional $\tilde{\Psi}[\tilde{\rho}(\mathbf{r})]$. From the variational principle for the energy, it follows that

$$\begin{aligned} \tilde{E} = E[\tilde{\rho}] = \langle \tilde{\Psi}[\tilde{\rho}(\mathbf{r})]|\hat{H}|\tilde{\Psi}[\tilde{\rho}(\mathbf{r})]\rangle \; &> \; E \text{ for } \tilde{\rho}(\mathbf{r}) \neq \rho(\mathbf{r}) \\ &= E \text{ for } \tilde{\rho}(\mathbf{r}) = \rho(\mathbf{r}). \end{aligned} \tag{7.81}$$

Thus, the ground state density $\rho(\mathbf{r})$ can be obtained by minimization of the functional $E[\rho]$ for arbitrary variations $\rho(\mathbf{r}) + \delta\rho(\mathbf{r})$ of v-representable densities with the constraint $\int \delta\rho(\mathbf{r})d\mathbf{r} = 0$ to ensure particle number conservation. Introducing a Lagrange multiplier μ to ensure this constraint ($\int \rho(\mathbf{r})d\mathbf{r} = N$), the stationary point is achieved via the variational principle at the vanishing of the first-order variation:

$$\delta\Big\{E[\rho] - \mu \int \rho(\mathbf{r})d\mathbf{r}\Big\} = 0. \tag{7.82}$$

Equivalently, the ground state density $\rho(\mathbf{r})$ and the Lagrange multiplier μ may be obtained from the self-consistent solution of the Euler-Lagrange equation

$$\frac{\delta E[\rho]}{\delta\rho(\mathbf{r})} = \mu, \tag{7.83}$$

together with the constraint of electron number N. This proves Theorem 2.

Separating out the external potential energy component, the ground state energy functional $E[\rho]$ may be written as

$$E[\rho] = \int \rho(\mathbf{r})v(\mathbf{r})d\mathbf{r} + F_{\mathrm{HK}}[\rho], \tag{7.84}$$

where the functional

$$F_{\mathrm{HK}}[\rho] = \langle\Psi[\rho]|\hat{T} + \hat{U}|\Psi[\rho]\rangle. \tag{7.85}$$

The functional $F_{\mathrm{HK}}[\rho]$, depends only on the kinetic $\hat{T}$ and electron-interaction potential $\hat{U}$ operators. It is universal in that it is the same functional for all electronic systems. It is a functional of v-representable densities. However, as the functional dependence of $\Psi(\mathbf{X})$ on $\rho(\mathbf{r})$ is unknown, the functional $F_{\mathrm{HK}}[\rho]$ is *unknown*.

7.3.1 Physical Interpretation of Lagrange Multiplier μ

The Lagrange multiplier μ in the Euler-Lagrange equation (7.83) has the physical interpretation of being the chemical potential. The chemical potential $\mu(N)$ is a number that depends on the electron number N. It represents the change in energy $E^{(N)}$ with respect to N:

$$\mu(N) = \frac{\partial E^{(N)}}{\partial N}\,. \tag{7.86}$$

If $\rho^{(N)}(\mathbf{r})$ is the solution of (7.83) for an N-electron system with ground state energy $E[\rho^{(N)}]$, then the energy difference

$$E^{(N+\epsilon)} - E^{(N)} = E[\rho^{(N+\epsilon)}] - E[\rho^{(N)}]$$
$$= \int \frac{\delta E[\rho]}{\delta \rho(\mathbf{r})}\bigg|_{\rho^{(N)}} (\rho^{(N+\epsilon)}(\mathbf{r}) - \rho^{(N)}(\mathbf{r}))d\mathbf{r} . \quad (7.87)$$

Employing (7.83), the right hand side reduces to

$$= \mu(N) \int (\rho^{(N+\epsilon)}(\mathbf{r}) - \rho^{(N)}(\mathbf{r}))d\mathbf{r}$$
$$= \mu(N)(N + \epsilon - N) = \mu(N)\epsilon , \quad (7.88)$$

so that $\lim_{\epsilon\to 0}(E^{(N+\epsilon)} - E^{(N)})/\epsilon = \mu(N)$, which is the desired result.

7.3.2 *The Primacy of Electron Number N*

In the description of an electronic system, the kinetic energy $\hat{T}$ and electron-interaction potential $\hat{U}$ operators are assumed known and kept *fixed*. It is the external binding potential operator $\hat{V} = \sum_i v(\mathbf{r}_i)$ that is *arbitrary*, and it is the potential $v(\mathbf{r})$ that is descriptive of different physical systems. For example, in natural atoms, $v(\mathbf{r})$ is Coulombic, whereas for 'artificial atoms' or quantum dots $v(\mathbf{r})$ is harmonic. Each physical system is also comprised of N electrons. Thus, the ground state energy E is a functional of the electron number N and the external potential $v(\mathbf{r})$:

$$E = \langle \Psi(\mathbf{X})|\hat{H}_N|\Psi(\mathbf{X})\rangle \quad (7.89)$$
$$= E[N, v], \quad (7.90)$$

with $\hat{H}_N$ the N-electron Hamiltonian.

Now, according to the First HK Theorem, there is a one-to-one correspondence between the v-representable nondegenerate ground state densities $\rho(\mathbf{r})$ and the external potentials $v(\mathbf{r})$ to within an additive constant C: $\rho(\mathbf{r}) \leftrightarrow v(\mathbf{r}) + C$. This Theorem is proved *for fixed* N. Thus, in the energy functional of (7.90), the $v(\mathbf{r})$ may be replaced by $\rho(\mathbf{r})$. Further, the v-representable variational densities employed in the Second HK Theorem all integrate to the *same electron number* N. Equivalently, in the self-consistent solution of the Euler-Lagrange equation (7.83), the constancy of electron number N must be imposed to determine the chemical potential μ. Hence, in the energy functional of (7.90), the N may be replaced by $\int \rho(\mathbf{r})d\mathbf{r}$. Therefore, we may write

$$E = E[N, v] = E\left[\int \rho(\mathbf{r})d\mathbf{r}, \rho(\mathbf{r})\right]. \quad (7.91)$$

In HK theory, however, it is the *sole* dependence of the energy E on the density $\rho(\mathbf{r})$ that is emphasized. But as explained above, the role of the electron number N in HK density functional theory is primary [45]. One must know N prior to proving the Theorems and solving the Euler-Lagrange equation for the density $\rho(\mathbf{r})$ and the chemical potential $\mu(N)$.

7.3.3 *The Percus-Levy-Lieb Constrained-Search Proof*

According to the Second Hohenberg-Kohn Theorem, the nondegenerate ground state energy $E[\rho]$ is obtained for the true ground state density $\rho(\mathbf{r})$ via the variational principle for arbitrary variations of v-representable densities. The Percus-Levy-Lieb (PLL) proof expands the domain of applicability to all functions $\{\Psi_\rho(\mathbf{X})\}$ that integrate to the true density $\rho(\mathbf{r})$, as well as to degenerate states. As noted in Sect. 7.1, the functions $\{\Psi_\rho(\mathbf{X})\}$ are not even required to be N-representable. *The PLL proof requires the a priori knowledge that the basic variable is the ground state density* $\rho(\mathbf{r})$. That information is provided by the First HK Theorem. In that context, the PLL proof is not as fundamental as the two Hohenberg-Kohn Theorems.

With the energy functional $E[\rho]$ written as in (7.84), a comparison with (7.39) shows that the universal functional $F_{\text{HK}}[\rho]$ may be defined as

$$F_{\text{HK}}[\rho] = \inf_{\Psi_\rho \to \rho} \langle \Psi_\rho | \hat{T} + \hat{U} | \Psi_\rho \rangle. \tag{7.92}$$

This infimum can be shown to be a minimum [25].

The PLL proof is for *arbitrary* functions $\{\Psi_\rho(\mathbf{X})\}$ such that the corresponding densities $\rho(\mathbf{r})$ satisfy the conditions of nonnegativity, normalization to N electron number, and continuity. The variational principle for the energy E for all N-particle functions $\{\Psi_\rho(\mathbf{X})\}$ is

$$E = \inf_{\Psi_\rho} \langle \Psi_\rho(\mathbf{X}) | \hat{T} + \hat{U} + \hat{V} | \Psi_\rho(\mathbf{X}) \rangle. \tag{7.93}$$

This expression for the energy E may be written as two nested infima as

$$E = \inf_{\rho(\mathbf{r})} \left[\inf_{\Psi_\rho \to \rho} \langle \Psi_\rho(\mathbf{X}) | \hat{T} + \hat{U} | \Psi_\rho(\mathbf{X}) \rangle + \int \rho(\mathbf{r}) v(\mathbf{r}) d\mathbf{r} \right], \tag{7.94}$$

where the inner infimum is restricted to all the functions $\{\Psi_\rho(\mathbf{X})\}$ that *yield a given density* $\rho(\mathbf{r})$, and the outer infimum is a search over *all* densities $\rho(\mathbf{r})$. Separating the external potential energy component, the energy on employing (7.92) is then

$$E = \inf_{\rho(\mathbf{r})} \left[F_{\text{HK}}[\rho] + \int \rho(\mathbf{r}) v(\mathbf{r}) d\mathbf{r} \right] \tag{7.95}$$

$$= \inf_{\rho(\mathbf{r})} E[\rho], \tag{7.96}$$

with $E[\rho]$ defined by (7.84). The search in (7.96) is over *all* densities $\rho(\mathbf{r})$ obtained from the N-particle functions $\{\Psi_\rho(\mathbf{X})\}$ some of which integrate to the true $\rho(\mathbf{r})$. It is evident that the above reasoning is equally valid if the energy E is degenerate.

7.3.4 Comment on the Constrained-Search Definition of the Functional $F_{\mathrm{HK}}[\rho]$

The constrained-search definition of the universal functional $F_{\mathrm{HK}}[\rho]$ of (7.92) is based on the a priori assumption that the basic variable is the ground state density $\rho(\mathbf{r})$. Further, that this density, corresponding to a physical system, is known. The statement of this definition is that it is a search over all functions $\{\Psi_\rho(\mathbf{X})\}$ (that yield the true ground state density $\rho(\mathbf{r})$) to determine the smallest (infimum) value of the expectation $< \Psi_\rho(\mathbf{X})|\hat{T} + \hat{U}|\Psi_\rho(\mathbf{X}) >$. This expectation value depends upon the kinetic $\hat{T}$ and electron-interaction potential $\hat{U}$ operators, and is independent of the external potential $v(\mathbf{r})$. The functions $\{\Psi_\rho(\mathbf{X})\}$, however, all yield the ground state density $\rho(\mathbf{r})$. But from the First HK Theorem, knowledge of the ground state density $\rho(\mathbf{r})$ uniquely determines the potential $v(\mathbf{r})$ to within a constant C. Thus, the potential $v(\mathbf{r})$, or equivalently the physical system it corresponds to, is known. There is hence an *implicit* dependence on $v(\mathbf{r})$ in this constrained-search definition of the functional $F_{\mathrm{HK}}[\rho]$.

7.4 Kohn-Sham Density Functional Theory

The First Hohenberg-Kohn theorem establishes that the nondegenerate ground state density $\rho(\mathbf{r})$ of a system of N electrons in an external field $\boldsymbol{\mathcal{E}}(\mathbf{r}) = -\boldsymbol{\nabla} v(\mathbf{r})/e$ constitutes a basic variable of quantum mechanics. The Second Hohenberg-Kohn Theorem then informs that the corresponding ground state energy $E[\rho]$ may be obtained by the variational principle for arbitrary variation of v-representable densities via solution of the resulting Euler Lagrange equation. The domain of v-representable densities is then expanded via the Percus-Levy-Lieb constrained-search to encompass the broader class of N-representable densities.

Kohn-Sham density functional theory (KS-DFT) is a local effective potential theory based on the Hohenberg-Kohn theorems. Hence, it is a ground state theory. KS-DFT is an alternate description of the mapping from a *nondegenerate ground state* of the interacting system to one of *noninteracting fermions* also in its *ground state* possessing the same density $\rho(\mathbf{r})$. This is the *sole* mapping permitted by KS-DFT. In further contrast to Q-DFT, its description of the mapping is strictly *mathematical* in that it is in terms of energy functionals of the density, with potentials being defined in terms of functional derivatives of the energy functionals taken with respect to the density. As Q-DFT is a physical description of local effective potential theory in terms of the various electron correlations, it provides a rigorous physical interpretation of the energy functionals and functional derivatives of KS-DFT. This physical description is provided in the following section. As in Q-DFT, the existence of the model system of noninteracting fermions—the S system—is an *assumption*. The meaning of the assumption is that the interacting system v-representable densities are also *noninteracting v-representable*. But, as was the case for the interacting system, the weaker constraint of N-representability suffices.

The assumption of existence of the model S system allows one to directly write down the Hamiltonian $\hat{H}_s$ for the noninteracting fermions, and the corresponding single-particle Schrödinger equation for each model fermion. The further assumption is that the model fermions are subject to the same field $\mathcal{E}(\mathbf{r})$. Thus, the Hamiltonian is

$$\hat{H}_s = \hat{T} + \hat{V}_s = \sum_i \hat{h}_s(\mathbf{r}_i), \tag{7.97}$$

$$\hat{T} = -\frac{1}{2}\sum_i \nabla_i^2, \tag{7.98}$$

$$\hat{V}_s = \sum_i v_s(\mathbf{r}_i), \tag{7.99}$$

$$\hat{h}_s(\mathbf{r}) = -\frac{1}{2}\nabla^2 + v_s(\mathbf{r}), \tag{7.100}$$

$$v_s(\mathbf{r}) = v(\mathbf{r}) + v_{ee}(\mathbf{r}). \tag{7.101}$$

The local effective potential is $v_s(\mathbf{r})$, with all the many-body effects incorporated into the electron-interaction potential $v_{ee}(\mathbf{r})$. The Schrödinger equation for each model fermion is

$$\hat{h}_s(\mathbf{r})\phi_i(\mathbf{x}) = \epsilon_i\phi_i(\mathbf{x}); \quad i = 1, \ldots, N. \tag{7.102}$$

The wave function is consequently a Slater determinant $\Phi\{\phi_i\}$ of the N *lowest* orbitals $\phi_i(\mathbf{x})$. The ground state density $\rho(\mathbf{r})$ which is the same as that of the interacting system, is the expectation value

$$\rho(\mathbf{r}) = \langle\Phi\{\phi_i\}|\hat{\rho}(\mathbf{r})|\Phi\{\phi_i\}\rangle = \sum_\sigma\sum_i |\phi_i(\mathbf{r}\sigma)|^2, \tag{7.103}$$

with $\hat{\rho}(\mathbf{r})$ the density operator of (2.12). From the First Hohenberg-Kohn Theorem, it follows that knowledge of the density $\rho(\mathbf{r})$ uniquely determines the effective potential $v_s(\mathbf{r})$ (map $(CD)^{-1}$ for the S system) to within a constant, and hence its electron-interaction component $v_{ee}(\mathbf{r})$. As the kinetic energy operator $\hat{T}$ of the model fermions is assumed known, the Hamiltonian $\hat{H}_s$ is then fully defined. Solution of the Schrödinger equation (7.102) then leads to the Slater determinant wave function $\Phi\{\phi_i\}$. Thus, the wave function $\Phi\{\phi_i\}$ and the orbitals $\phi_i(\mathbf{x})$ are functionals of the ground state density $\rho(\mathbf{r})$, i.e. $\phi_i(\mathbf{x}) = \phi_i[\rho]$. The kinetic energy T_s of the model fermions is therefore a functional of $\rho(\mathbf{r})$:

$$T_s = T_s[\rho] = \sum_\sigma \sum_i \langle \phi_i(\mathbf{r}\sigma; [\rho])| -\frac{1}{2}\nabla_i^2 |\phi_i(\mathbf{r}\sigma; [\rho])\rangle. \tag{7.104}$$

The energy functional $E[\rho]$ of the interacting system (see (7.84)), may then be written in terms of $T_s[\rho]$. By adding and subtracting $T_s[\rho]$ from the expression (7.84), the energy $E[\rho]$ may be expressed as

$$E[\rho] = T_s[\rho] + \int \rho(\mathbf{r})v(\mathbf{r})d\mathbf{r} + E_{ee}^{\mathrm{KS}}[\rho]. \tag{7.105}$$

where

$$E_{ee}^{\mathrm{KS}}[\rho] = F_{\mathrm{HK}}[\rho] - T_s[\rho], \tag{7.106}$$

which then defines the KS-DFT electron-interaction energy functional $E_{ee}^{\mathrm{KS}}[\rho]$. It follows from (7.105) that all the many-body effects are now incorporated in the energy functional $E_{ee}^{\mathrm{KS}}[\rho]$. By expressing $T_s[\rho]$ in terms of the eigenvalues ϵ_i of the S system Schrödinger equation, one may write the energy $E[\rho]$ in terms of these eigenvalues. By multiplying (7.102) by $\phi_i^\star(\mathbf{r}\sigma)$, summing over all the model fermions, and intergrating over the spatial and spin coordinates leads to

$$T_s[\rho] = \sum_i \epsilon_i - \int \rho(\mathbf{r})v(\mathbf{r})d\mathbf{r} - \int \rho(\mathbf{r})v_{ee}(\mathbf{r})d\mathbf{r}. \tag{7.107}$$

On substituting for $T_s[\rho]$ from (7.107) into (7.105) we have

$$E[\rho] = \sum_i \epsilon_i - \int \rho(\mathbf{r})v_{ee}(\mathbf{r})d\mathbf{r} + E_{ee}^{\mathrm{KS}}[\rho]. \tag{7.108}$$

This expression for the energy is entirely in terms of the S system properties.

All that remains to fully describe the model system then is to define the electron-interaction potential $v_{ee}(\mathbf{r})$ that appears in the S system Schrödinger equation (7.102) and the corresponding expression for the energy $E[\rho]$ of (7.108). This is arrived at via the Second Hohenberg-Kohn Theorem by application of the variational principle for the energy in terms of arbitrary variations $\rho(\mathbf{r}) + \delta\rho(\mathbf{r})$ of the density $\rho(\mathbf{r})$. Thus, at the vanishing of the first-order variation of the energy functional, we have

$$\delta E[\rho] = E[\rho + \delta\rho] - E[\rho] \tag{7.109}$$
$$= \int \frac{\delta E[\rho]}{\delta\rho(\mathbf{r})}\delta\rho(\mathbf{r})d\mathbf{r} \tag{7.110}$$
$$= \delta T_s[\rho] + \int [v(\mathbf{r}) + v_{ee}(\mathbf{r})]\delta\rho(\mathbf{r}) \tag{7.111}$$
$$= 0, \tag{7.112}$$

where $v_{ee}(\mathbf{r})$ is the *functional derivative* of $E_{ee}^{\mathrm{KS}}[\rho]$:

$$v_{ee}(\mathbf{r}) = \frac{\delta E_{ee}^{\mathrm{KS}}[\rho]}{\delta\rho(\mathbf{r})}. \tag{7.113}$$

This is the KS-DFT definition of the potential $v_{ee}(\mathbf{r})$ of local effective potential theory.

The first-order variation $\delta T_s[\rho]$ in (7.111) is obtained from $T_s[\rho]$ of (7.104) for arbitrary variations $\phi_i(\mathbf{x}) + \delta\phi_i(\mathbf{x})$ of the orbitals $\phi_i(\mathbf{x})$ which lead to variations of the density $\rho(\mathbf{r}) + \delta\rho(\mathbf{r})$. On employing the Schrödinger equation (7.102), the normalization of the orbitals $\phi_i(\mathbf{x})$, and the neglect of second-order terms, one obtains

$$\delta T_s[\rho] = -\int v_s(\mathbf{r})\delta\rho(\mathbf{r})d\mathbf{r}. \tag{7.114}$$

Substitution of (7.114) into (7.111) then leads to

$$\int \big[- v_s(\mathbf{r}) + v(\mathbf{r}) + v_{ee}(\mathbf{r})\big]\delta\rho(\mathbf{r})d\mathbf{r} = 0. \tag{7.115}$$

As the variations $\delta\rho(\mathbf{r})$ are arbitrary, one recovers (7.101) with the electron-interaction potential $v_{ee}(\mathbf{r})$ now defined as the functional derivative of (7.113). With this definition of $v_{ee}(\mathbf{r})$, the Schrödinger equation (7.102) is referred to as the *Kohn-Sham Equation*.

In KS-DFT, the KS equation (7.102) is solved self-consistently for the orbitals $\phi_i(\mathbf{x})$ and eigenvalues ϵ_i. From the orbitals, the density $\rho(\mathbf{r})$ is determined via (7.103) and the kinetic energy T_s from (7.104). The energy $E[\rho]$ is then determined either from the expressions of (7.105) or (7.108).

In the KS-DFT energy expression of (7.105), $T_s[\rho]$ is the kinetic energy of *noninteracting* fermions whose density is the true ground state density $\rho(\mathbf{r})$. The contribution of the electron correlations, (due to the Pauli principle and Coulomb repulsion), to the kinetic energy $T_c[\rho]$—the Correlation-Kinetic component—is subsumed into the functional $E_{ee}^{\mathrm{KS}}[\rho]$. *Thus, the KS electron-interaction energy functional* $E_{ee}^{\mathrm{KS}}[\rho]$ *and its functional derivative* $v_{ee}(\mathbf{r})$ *are representative of electron correlations due to the Pauli principle, Coulomb repulsion, and Correlation-Kinetic effects.*

As the Hartree or Coulomb self-energy functional $E_{\mathrm{H}}[\rho]$ of the density $\rho(\mathbf{r})$ is known:

$$E_{\mathrm{H}}[\rho] = \frac{1}{2}\iint \frac{\rho(\mathbf{r})\rho(\mathbf{r}')}{|\mathbf{r}-\mathbf{r}'|} d\mathbf{r}d\mathbf{r}', \tag{7.116}$$

the functional $E_{ee}^{\mathrm{KS}}[\rho]$ is customarily partitioned as

$$E_{ee}^{\mathrm{KS}}[\rho] = E_{\mathrm{H}}[\rho] + E_{xc}^{\mathrm{KS}}[\rho], \tag{7.117}$$

which defines the KS 'exchange-correlation' energy functional. It follows from (7.113) that the electron-interaction potential $v_{ee}(\mathbf{r})$ within KS-DFT is then

$$v_{ee}(\mathbf{r}) = v_{\mathrm{H}}(\mathbf{r}) + v_{xc}(\mathbf{r}), \tag{7.118}$$

where the Hartree potential energy $v_{\mathrm{H}}(\mathbf{r})$ is the functional derivative

$$v_{\mathrm{H}}(\mathbf{r}) = \frac{\delta E_{\mathrm{H}}[\rho]}{\delta\rho(\mathbf{r})} = \int \frac{\rho(\mathbf{r}')}{|\mathbf{r}-\mathbf{r}'|} d\mathbf{r}', \tag{7.119}$$

and the KS 'exchange-correlation' potential $v_{xc}(\mathbf{r})$ the functional derivative

$$v_{xc}(\mathbf{r}) = \frac{\delta E_{xc}^{\mathrm{KS}}[\rho]}{\delta\rho(\mathbf{r})}. \tag{7.120}$$

Thus, the KS energy functional $E_{xc}^{\mathrm{KS}}[\rho]$ and its functional derivative $v_{xc}(\mathbf{r})$ are representative of electron correlations due to the Pauli principle, and Coulomb repulsion beyond those of the Hartree term, and those of the Correlation-Kinetic effects.

The functional $E_{\mathrm{xc}}^{\mathrm{KS}}[\rho]$ is usually further partitioned into its KS 'exchange' $E_{\mathrm{x}}^{\mathrm{KS}}[\rho]$ and KS 'correlation' $E_{c}^{\mathrm{KS}}[\rho]$ energy functional components. Thus

$$E_{xc}^{\mathrm{KS}}[\rho] = E_{\mathrm{x}}^{\mathrm{KS}}[\rho] + E_{\mathrm{c}}^{\mathrm{KS}}[\rho] \;, \tag{7.121}$$

so that the KS 'exchange-correlation' potential energy $v_{\mathrm{xc}}(\mathbf{r})$ is

$$v_{\mathrm{xc}}(\mathbf{r}) = v_{\mathrm{x}}(\mathbf{r}) + v_{\mathrm{c}}(\mathbf{r}) \;, \tag{7.122}$$

where the KS 'exchange' potential energy $v_{\mathrm{x}}(\mathbf{r})$ is defined as the functional derivative

$$v_{\mathrm{x}}(\mathbf{r}) = \frac{\delta E_{\mathrm{x}}^{\mathrm{KS}}[\rho]}{\delta\rho(\mathbf{r})} \;, \tag{7.123}$$

and the KS 'correlation' potential energy $v_{\mathrm{c}}(\mathbf{r})$ as the functional derivative

$$v_{\mathrm{c}}(\mathbf{r}) = \frac{\delta E_{\mathrm{c}}^{\mathrm{KS}}[\rho]}{\delta\rho(\mathbf{r})} \;. \tag{7.124}$$

The partitioning of $E_{\mathrm{xc}}^{\mathrm{KS}}[\rho]$ into its 'exchange' $E_{\mathrm{x}}^{\mathrm{KS}}[\rho]$ and 'correlation' $E_{\mathrm{c}}^{\mathrm{KS}}[\rho]$ energy components is based on the *ad hoc* choice that $E_{\mathrm{x}}^{\mathrm{KS}}[\rho]$ is given by the Hartree-Fock theory *expression* (3.72) for the exchange energy, but with the orbitals $\phi_i(\mathbf{x})$ of the S system differential equation (7.102) employed instead. Thus, it is assumed that

$$E_x^{\mathrm{KS}}[\rho] = \frac{1}{2}\iint \frac{\rho(\mathbf{r})\rho_x(\mathbf{rr}')}{|\mathbf{r}-\mathbf{r}'|}d\mathbf{r}d\mathbf{r}', \tag{7.125}$$

where $\rho_x(\mathbf{rr}')$ is the S system Fermi hole. The KS 'exchange' energy thus defined is a functional of the ground state density $\rho(\mathbf{r})$ because the orbitals $\phi_i(\mathbf{X})$ are such functionals. This choice for $E_x^{\mathrm{KS}}[\rho]$ then defines the KS 'correlation' energy functional $E_c^{KS}[\rho]$.

The KS-DFT energy functionals $E_{\mathrm{ee}}^{\mathrm{KS}}[\rho]$, $E_{\mathrm{xc}}^{\mathrm{KS}}[\rho]$, $E_{\mathrm{H}}[\rho]$, $E_{\mathrm{x}}^{\mathrm{KS}}[\rho]$, $E_{\mathrm{c}}^{\mathrm{KS}}[\rho]$, and their functional derivatives $v_{\mathrm{ee}}(\mathbf{r})$, $v_{\mathrm{xc}}(\mathbf{r})$, $v_{\mathrm{H}}(\mathbf{r})$, $v_{\mathrm{x}}(\mathbf{r})$, $v_{\mathrm{c}}(\mathbf{r})$, respectively, satisfy [46] the following integral virial theorems:

$$E_{\mathrm{ee}}^{\mathrm{KS}}[\rho] + \int \rho(\mathbf{r})\mathbf{r}\cdot\nabla v_{\mathrm{ee}}(\mathbf{r})d\mathbf{r} = -T_{\mathrm{c}}[\rho] \leq 0\,, \tag{7.126}$$

$$E_{\mathrm{H}}[\rho] + \int \rho(\mathbf{r})\cdot\nabla v_{\mathrm{H}}(\mathbf{r}) = 0, \tag{7.127}$$

$$E_{\mathrm{xc}}^{\mathrm{KS}}[\rho] + \int \rho(\mathbf{r})\mathbf{r}\cdot\nabla v_{\mathrm{xc}}(\mathbf{r})d\mathbf{r} = -T_{\mathrm{c}}[\rho] \leq 0\,, \tag{7.128}$$

$$E_{\mathrm{x}}^{\mathrm{KS}}[\rho] + \int \rho(\mathbf{r})\mathbf{r}\cdot\nabla v_{\mathrm{x}}(\mathbf{r})d\mathbf{r} = 0\,, \tag{7.129}$$

$$E_{\mathrm{c}}^{\mathrm{KS}}[\rho] + \int \rho(\mathbf{r})\mathbf{r}\cdot\nabla v_{\mathrm{c}}(\mathbf{r})d\mathbf{r} = -T_{\mathrm{c}}[\rho] \leq 0\,. \tag{7.130}$$

Since in KS-DFT, the mapping is from a ground state of the interacting system to an S system also in its ground state, the value of $T_c \geq 0$. (Within Q-DFT, it is also possible to map to an S system in an excited state, in which case $T_c < 0$. For such an example, see ***QDFT2***.)

The functional $T_s]\rho]$ satisfies the sum rule

$$2T_s[\rho] = \int \rho(\mathbf{r})\mathbf{r}\cdot\nabla v_s(\mathbf{r})d\mathbf{r}, \tag{7.131}$$

where $v_s(\mathbf{r})$ is the effective potential of the model fermions as defined by (7.101).

Finally, the highest occupied eigenvalue ϵ_m of the KS differential equation has the physical interpretation of being the removal energy. For the proof see Sect. 6.1.8.

7.5 Physical Interpretation of Kohn-Sham Theory

The physical interpretation of the KS-DFT energy functionals and functional derivatives follows by comparison with the corresponding expressions for the energy E in terms of the S system properties, and the electron-interaction potential $v_{ee}(\mathbf{r})$ of Q-DFT as described in Chap. 6. (There the description of Q-DFT included the presence of a magnetic field $\mathcal{B}(\mathbf{r}) = \nabla \times \mathbf{A}(\mathbf{r})$. The Q-DFT expressions become equivalent to those of KS-DFT on putting the vector potential $\mathbf{A}(\mathbf{r})$ to zero in the equations of the former.)

A comparison of the Q-DFT expression for the energy E of (6.100) (valid for both ground and excited states) to that of the KS-DFT ground state energy functional $E[\rho]$ of (7.108) leads to

$$E_{\mathrm{ee}}^{\mathrm{KS}}[\rho] = E_{ee} + T_c. \tag{7.132}$$

Here E_{ee} is the quantum-mechanical electron-interaction energy expressed in integral virial form in terms of the electron-interaction field $\mathcal{E}_{ee}(\mathbf{r})$ of (6.52) as

$$E_{\mathrm{ee}} = \int \rho(\mathbf{r})\mathbf{r} \cdot \mathcal{E}_{ee}(\mathbf{r})d\mathbf{r}, \tag{7.133}$$

and T_c the Correlation-Kinetic energy defined in terms of the Correlation-Kinetic field $\mathcal{Z}_{t_c}(\mathbf{r})$ as

$$T_c = \frac{1}{2}\int \rho(\mathbf{r})\mathbf{r} \cdot \mathcal{Z}_{t_c}(\mathbf{r})d\mathbf{r}, \tag{7.134}$$

with

$$\mathcal{Z}_{t_c}(\mathbf{r}) = \mathcal{Z}_s(\mathbf{r}) - \mathcal{Z}(\mathbf{r}), \tag{7.135}$$

and where $\mathcal{Z}_s(\mathbf{r})$ and $\mathcal{Z}(\mathbf{r})$ are the S and interacting system kinetic energy fields, respectively (see Sect. 6.1.2). *Thus, it is explicitly shown that the energy functional $E_{ee}^{\mathrm{KS}}[\rho]$ is comprised of the sum of contributions due to the correlations arising from the Pauli principle and Coulomb repulsion, and those due to Correlation-Kinetic effects.* Further, the separate components E_{ee} and T_c of this partition of $E_{ee}^{\mathrm{KS}}[\rho]$ are precisely defined via Q-DFT. Note that the fields $\mathcal{E}_{ee}(\mathbf{r})$ and $\mathcal{Z}_{t_c}(\mathbf{r})$ are not necessarily separately conservative. Their sum always is. However, the respective expressions for the energy components in terms of these fields are valid whether or not the fields are conservative.

Equating the Q-DFT and KS-DFT expressions (6.79) and (7.113) for the electron-interaction potential energy we have

$$v_{ee}(\mathbf{r}) = \frac{\partial E_{ee}^{\mathrm{KS}}[\rho]}{\partial \rho(\mathbf{r})} = -\int_{\infty}^{\mathbf{r}} \mathcal{F}^{\mathrm{eff}}(\mathbf{r}') \cdot d\boldsymbol{\ell}', \tag{7.136}$$

with

$$\mathcal{F}^{\mathrm{eff}}(\mathbf{r}) = \mathcal{E}_{ee}(\mathbf{r}) + \mathcal{Z}_{t_c}(\mathbf{r}). \tag{7.137}$$

Hence, the physical interpretation of the functional derivative $\delta E_{ee}^{\mathrm{KS}}[\rho]/\delta\rho(\mathbf{r})$ is that of the work done to bring the model fermion from some reference point at infinity to its position at $\mathbf{r}$ *in the force of the conservative effective field $\mathcal{F}^{\mathrm{eff}}(\mathbf{r})$. Since $\nabla \times \mathcal{F}^{\mathrm{eff}}(\mathbf{r}) = 0$, the work done is path independent.* Once again, the contribution of electron correlations due to the Pauli principle and Coulomb repulsion, and those of Correlation kinetic effects, to the functional derivative $v_{ee}(\mathbf{r})$ are *explicitly* defined via Q-DFT.

If the electron-interaction $\mathcal{E}_{ee}(\mathbf{r})$ and Correlation-Kinetic $\mathcal{Z}_{t_c}(\mathbf{r})$ fields are separately conservative, i.e. if $\nabla \times \mathcal{E}_{ee}(\mathbf{r}) = 0$ and $\nabla \times \mathcal{Z}_{t_c}(\mathbf{r}) = 0$, then the functional derivative $v_{ee}(\mathbf{r})$ may be written as

$$v_{ee}(\mathbf{r}) = \frac{\partial E_{ee}^{\mathrm{KS}}[\rho]}{\partial\rho(\mathbf{r})} = W_{ee}(\mathbf{r}) + W_{t_c}(\mathbf{r}), \tag{7.138}$$

where

$$W_{ee}(\mathbf{r}) = -\int_{\infty}^{\mathbf{r}} \mathcal{E}_{ee}(\mathbf{r}') \cdot d\boldsymbol{\ell}' \,; \quad W_{t_c}(\mathbf{r}) = -\int_{\infty}^{\mathbf{r}} \mathcal{Z}_{t_c}(\mathbf{r}') \cdot d\boldsymbol{\ell}', \tag{7.139}$$

and where $W_{ee}(\mathbf{r})$ and $W_{t_c}(\mathbf{r})$ are, respectively, the work done in the fields $\mathcal{E}_{ee}(\mathbf{r})$ and $\mathcal{Z}_{t_c}(\mathbf{r})$. As each field is conservative, each work done is *path-independent* and represents a potential.

From the above two definitions of the potential $v_{ee}(\mathbf{r})$, one mathematical and the other physical, we have

$$\nabla v_{\mathrm{ee}}(\mathbf{r}) = \nabla\left(\frac{\delta E_{\mathrm{ee}}^{\mathrm{KS}}[\rho]}{\delta\rho(\mathbf{r})}\right) = -\mathcal{F}^{\mathrm{eff}}(\mathbf{r})\,, \tag{7.140}$$

or, equivalently employing (7.132) and (7.136) that

$$\nabla\left(\frac{\delta E_{\mathrm{ee}}}{\delta\rho(\mathbf{r})} + \frac{\delta T_{\mathrm{c}}}{\delta\rho(\mathbf{r})}\right) = -(\mathcal{E}_{\mathrm{ee}}(\mathbf{r}) + \mathcal{Z}_{\mathrm{t_c}}(\mathbf{r}))\,. \tag{7.141}$$

This equation relates the functional derivatives of E_{ee} and T_{c} to the component fields $\mathcal{E}_{\mathrm{ee}}(\mathbf{r})$ and $\mathcal{Z}_{\mathrm{t_c}}(\mathbf{r})$. (The energies E_{ee} and T_c are also functionals of the density $\rho(\mathbf{r})$ since the wave function $\Psi(\mathbf{X})$ and the orbitals $\Phi\{\phi_i\}$ are.) Note, however, that

$$\nabla\left(\frac{\delta E_{\mathrm{ee}}}{\delta\rho(\mathbf{r})}\right) \neq -\mathcal{E}_{\mathrm{ee}}(\mathbf{r})\,, \tag{7.142}$$

and

$$\nabla\left(\frac{\delta T_{\mathrm{c}}}{\delta\rho(\mathbf{r})}\right) \neq -\mathcal{Z}_{\mathrm{t_c}}(\mathbf{r})\,. \tag{7.143}$$

These inequalities hold whether or not the fields $\mathcal{E}_{\text{ee}}(\mathbf{r})$ and $\mathcal{Z}_{t_c}(\mathbf{r})$ are conservative. The *equality* of the functional derivatives to the fields is that given by (7.136) or (7.141).

Since the pair-correlation density may also be written as $g(\mathbf{rr}') = \rho(\mathbf{r}') + \rho_{\text{xc}}(\mathbf{rr}')$, where $\rho_{\text{xc}}(\mathbf{rr}')$ is the Fermi-Coulomb hole charge, the electron-interaction field $\mathcal{E}_{\text{ee}}(\mathbf{r})$ of (6.52) may be expressed as the sum of its Hartree $\mathcal{E}_{\text{H}}(\mathbf{r})$ and Pauli-Coulomb $\mathcal{E}_{\text{xc}}(\mathbf{r})$ components:

$$\mathcal{E}_{\text{ee}}(\mathbf{r}) = \mathcal{E}_{\text{H}}(\mathbf{r}) + \mathcal{E}_{\text{xc}}(\mathbf{r}) \ , \tag{7.144}$$

where

$$\mathcal{E}_{\text{H}}(\mathbf{r}) = \int \frac{\rho(\mathbf{r}')(\mathbf{r} - \mathbf{r}')}{|\mathbf{r} - \mathbf{r}'|^3} d\mathbf{r}' \ \text{ and } \ \mathcal{E}_{\text{xc}}(\mathbf{r}) = \int \frac{\rho_{\text{xc}}(\mathbf{rr}')(\mathbf{r} - \mathbf{r}')}{|\mathbf{r} - \mathbf{r}'|^3} d\mathbf{r}' \ . \tag{7.145}$$

As the field $\mathcal{E}_{\text{H}}(\mathbf{r})$ is due to a static or local charge distribution $\rho(\mathbf{r})$, it may be expressed as

$$\mathcal{E}_{\text{H}}(\mathbf{r}) = -\boldsymbol{\nabla} W_{\text{H}}(\mathbf{r}) \ , \tag{7.146}$$

with the scalar potential energy $W_{\text{H}}(\mathbf{r})$ being

$$W_{\text{H}}(\mathbf{r}) = \int \frac{\rho(\mathbf{r}')}{|\mathbf{r} - \mathbf{r}'|} d\mathbf{r}' \ . \tag{7.147}$$

Since $\nabla \times \mathcal{E}_{\text{H}}(\mathbf{r}) = 0$, the potential energy $W_{\text{H}}(\mathbf{r})$ is equivalently the work done in the conservative field $\mathcal{E}_{\text{H}}(\mathbf{r})$:

$$W_{\text{H}}(\mathbf{r}) = -\int_{\infty}^{\mathbf{r}} \mathcal{E}_{\text{H}}(\mathbf{r}') \cdot d\boldsymbol{\ell}' \ . \tag{7.148}$$

A comparison of the functional derivative $v_{\text{H}}(\mathbf{r}) = \delta E_{\text{H}}[\rho]/\delta\rho(\mathbf{r})$ of (7.119) and (7.147) shows that

$$v_{\text{H}}(\mathbf{r}) = W_{\text{H}}(\mathbf{r}) \ . \tag{7.149}$$

Thus, the physical interpretation of the functional derivative $\delta E_{\text{H}}[\rho]/\delta\rho(\mathbf{r})$ is that it is the work done to move a model fermion from its reference point at infinity to its position at $\mathbf{r}$ *in the force of the conservative field* $\mathcal{E}_{\text{H}}(\mathbf{r})$. Equivalently

$$\boldsymbol{\nabla}\left(\frac{\delta E_{\text{H}}[\rho]}{\delta\rho(\mathbf{r})}\right) = -\mathcal{E}_{\text{H}}(\mathbf{r}) \ . \tag{7.150}$$

The Hartree energy functional $E_{\text{H}}[\rho]$ of (7.116), which is the energy of self-interaction of the density, may also be expressed in terms of the Hartree field $\mathcal{E}_{\text{H}}(\mathbf{r})$ as

$$E_{\text{H}} = \int \rho(\mathbf{r})\mathbf{r} \cdot \mathcal{E}_{\text{H}}(\mathbf{r}) d\mathbf{r} \ . \tag{7.151}$$

Again, employing the partitioning of the pair-correlation density $g(\mathbf{rr}')$ into its local density $\rho(\mathbf{r})$ and nonlocal Fermi-Coulomb hole $\rho_{xc}(\mathbf{rr}')$ components, we can write the quantum-mechanical electron-interaction energy E_{ee} as

$$E_{\mathrm{ee}} = E_{\mathrm{H}} + E_{\mathrm{xc}} \; , \tag{7.152}$$

where E_{xc} is the Pauli-Coulomb energy. Thus, the KS electron-interaction energy functional (7.132) is

$$E_{\mathrm{ee}}^{\mathrm{KS}} = E_{\mathrm{H}} + E_{\mathrm{xc}} + T_{\mathrm{c}} \; . \tag{7.153}$$

Comparison with (7.151) then defines the KS 'exchange-correlation' energy functional in terms of the Pauli and Coulomb correlations and Correlation-Kinetic effects as

$$E_{\mathrm{xc}}^{\mathrm{KS}}[\rho] = E_{\mathrm{xc}} + T_{\mathrm{c}} \; , \tag{7.154}$$

where E_{xc} is expressed in terms of the Pauli-Coulomb field $\boldsymbol{\mathcal{E}}_{\mathrm{xc}}(\mathbf{r})$ as

$$E_{\mathrm{xc}} = \int \rho(\mathbf{r})\mathbf{r} \cdot \boldsymbol{\mathcal{E}}_{\mathrm{xc}}(\mathbf{r}) d\mathbf{r} \; , \tag{7.155}$$

and with T_{c} as previously defined by (7.134).

The KS 'exchange-correlation' potential $v_{\mathrm{xc}}(\mathbf{r})$ or functional derivative $\delta E_{xc}^{\mathrm{KS}}[\rho]/\delta\rho$ is the work done to bring the model fermion from a reference point at infinity to its position at $\mathbf{r}$ in the conservative field $\boldsymbol{\mathcal{F}}_{\mathrm{xct_c}}(\mathbf{r})$:

$$v_{\mathrm{xc}}(\mathbf{r}) = \frac{\delta E_{\mathrm{xc}}^{\mathrm{KS}}[\rho]}{\delta\rho(\mathbf{r})} = -\int_{\infty}^{\mathbf{r}} \boldsymbol{\mathcal{F}}_{\mathrm{xct_c}}(\mathbf{r}') \cdot d\boldsymbol{\ell}' \; , \tag{7.156}$$

where

$$\boldsymbol{\mathcal{F}}_{\mathrm{xct_c}}(\mathbf{r}) = \boldsymbol{\mathcal{E}}_{\mathrm{xc}}(\mathbf{r}) + \boldsymbol{\mathcal{Z}}_{\mathrm{t_c}}(\mathbf{r}) \; . \tag{7.157}$$

This follows from (7.136) using the fact that the Hartree field $\boldsymbol{\mathcal{E}}_{\mathrm{H}}(\mathbf{r})$ is conservative so that $\boldsymbol{\nabla} \times \boldsymbol{\mathcal{F}}_{\mathrm{xct_c}}(\mathbf{r}) = 0$. Equivalently,

$$\boldsymbol{\nabla} v_{\mathrm{xc}}(\mathbf{r}) = \boldsymbol{\nabla}\left(\frac{\delta E_{\mathrm{xc}}^{\mathrm{KS}}[\rho]}{\delta\rho(\mathbf{r})}\right) = -(\boldsymbol{\mathcal{E}}_{\mathrm{xc}}(\mathbf{r}) + \boldsymbol{\mathcal{Z}}_{\mathrm{t_c}}(\mathbf{r})) \; . \tag{7.158}$$

Thus, the KS 'exchange-correlation' energy functional $E_{\mathrm{xc}}^{\mathrm{KS}}[\rho]$ and its functional derivative $v_{\mathrm{xc}}(\mathbf{r})$ can be expressed in terms of the Pauli-Coulomb $\boldsymbol{\mathcal{E}}_{\mathrm{xc}}(\mathbf{r})$ and Correlation-Kinetic $\boldsymbol{\mathcal{Z}}_{\mathrm{t_c}}(\mathbf{r})$ fields. Hence, the dependence of the functional $E_{\mathrm{xc}}^{\mathrm{KS}}[\rho]$ and its derivative $v_{\mathrm{xc}}(\mathbf{r})$ on the *separate* electron correlations due to the Pauli principle, Coulomb repulsion, and Correlation-Kinetic effects is *explicitly* defined within the framework of Q-DFT.

Substituting (7.154) into (7.158) leads to

$$\nabla\left(\frac{\delta E_{\mathrm{xc}}}{\delta\rho(\mathbf{r})}+\frac{\delta T_{\mathrm{c}}}{\delta\rho(\mathbf{r})}\right)=-(\boldsymbol{\mathcal{E}}_{\mathrm{xc}}(\mathbf{r})+\boldsymbol{\mathcal{Z}}_{t_{\mathrm{c}}}(\mathbf{r}))\,, \tag{7.159}$$

which relates the functional derivative of the quantum-mechanical exchange-correlation E_{xc} and Correlation-Kinetic T_{c} energies to the fields $\boldsymbol{\mathcal{E}}_{\mathrm{xc}}(\mathbf{r})$ and $\boldsymbol{\mathcal{Z}}_{t_{\mathrm{c}}}(\mathbf{r})$ that give rise to them, respectively. Again, irrespective of whether the field $\boldsymbol{\mathcal{E}}_{\mathrm{xc}}(\mathbf{r})$ is conservative or not

$$\nabla\left(\frac{\delta E_{\mathrm{xc}}}{\delta\rho(\mathbf{r})}\right)\neq-\boldsymbol{\mathcal{E}}_{\mathrm{xc}}(\mathbf{r})\,. \tag{7.160}$$

Thus, we see that the mathematical entities of KS-DFT, *viz.* the electron-interaction energy functional $E_{\mathrm{ee}}^{\mathrm{KS}}[\rho]$, its Hartree $E_{\mathrm{H}}[\rho]$ and 'exchange-correlation' $E_{\mathrm{xc}}^{\mathrm{KS}}[\rho]$ components, and their respective functional derivatives $v_{\mathrm{ee}}(\mathbf{r})$, $v_{\mathrm{H}}(\mathbf{r})$, and $v_{xc}(\mathbf{r})$ can all be afforded a rigorous physical interpretation.

7.5.1 Electron Correlations in Kohn-Sham 'Exchange' and 'Correlation'

It is also possible to understand the correlations that contribute to the KS 'exchange' $E_{x}^{\mathrm{KS}}[\rho]$ and 'correlation' $E_{c}^{\mathrm{KS}}[\rho]$ energy functionals and their respective functional derivatives $v_x(\mathbf{r})$ and $v_c(\mathbf{r})$, and to provide a rigorous physical interpretation of these potentials [47, 48]. This is accomplished via adiabatic coupling constant (λ) perturbation theory [49]. (The respective energy functionals and their functional derivatives are related by the sum rules of (7.129) and (7.130)). We summarize the conclusions here, and refer the reader to Chap. 5 of ***QDFT*** for the detailed proofs.

Kohn-Sham 'Exchange'

(***a***) The KS 'exchange energy functional $E_{x}^{\mathrm{KS}}[\rho]$ and its functional derivative $v_x(\mathbf{r})$ are representative of Pauli correlations and *lowest-order* $O(\lambda)$ Correlation-Kinetic effects. (This is not evident in the sum rule (7.129).)

(***b***) The functional derivative $v_x(\mathbf{r})$ has the rigorous physical interpretation of the work done to move the model fermion in a *conservative* field representative of these correlations.

(***c***) The lowest-order $O(\lambda)$ Correlation-Kinetic effects do not contribute directly to the energy $E_{x}^{\mathrm{KS}}[\rho]$ (see sum rule (7.129)). Their contribution is indirect via the potential $v_x(\mathbf{r})$.

Kohn-Sham 'Correlation'

(***a***) The KS 'correlation' energy $E_{c}^{\mathrm{KS}}[\rho]$ and functional derivative $v_c(\mathbf{r})$ (see sum rule (7.130)) commence in *second-order* $O(\lambda^2)$.

(***b***) Both Coulomb correlations and Correlation-Kinetic effects contribute to the second and higher order of the potentials $v_c(\mathbf{r})$. Thus, lowest-order $O(\lambda)$ Coulomb correlations and second-order $O(\lambda^2)$ Correlation-Kinetic effects contribute to the leading (second-order) 'correlation' potential $v_{c,2}(\mathbf{r})$.

(*c*) In each order, the 'correlation' potential is the work done to move the model fermion in a *conservative* field representative of Coulomb correlations and Correlation-Kinetic effects.

(*d*) Only second-order $O(\lambda^2)$ Correlation-Kinetic effects contribute to the KS energy $E_c^{\mathrm{KS}}[\rho]$ to leading order $O(\lambda^2)$.

(*e*) For energy terms beyond the leading order, both Coulomb correlations and Correlation-Kinetic effects contribute.

Thus, KS 'exchange' is not *solely* representative of Pauli correlations. It includes lowest-order Correlation-Kinetic effects. And, KS 'correlation' which proceeds only at second-order, is representative of Coulomb correlations and second- and higher-order Correlation-Kinetic effects.

7.5.2 *Definitions of the Correlation Energy*

As neither the 'exchange' $E_x^{\mathrm{KS}}[\rho]$ nor the 'correlation' $E_c^{\mathrm{KS}}[\rho]$ energy functionals in the partitioning of $E_{xc}^{\mathrm{KS}}[\rho]$ of (7.121) are known, the definition of the 'correlation' energy depends upon the choice of the 'exchange' energy component employed. In practice, the 'exchange' energy functional $E_x^{\mathrm{KS}}[\rho]$ is replaced *ad hoc* by the Hartree-Fock theory *expression* for the exchange energy (3.72). However, in this expression, it is the orbitals $\phi_i(\mathbf{x})$ of the S system differential equation (7.102) that are employed instead. Thus, with

$$E_x^{\mathrm{KS}}[\rho] = \frac{1}{2}\iint \frac{\rho(\mathbf{r})\rho_x(\mathbf{r}\mathbf{r}')}{|\mathbf{r}-\mathbf{r}'|} d\mathbf{r}d\mathbf{r}', \tag{7.161}$$

where $\rho_x(\mathbf{r}\mathbf{r}')$ is the S system Fermi hole, then defines $E_c^{\mathrm{KS}}[\rho]$. This definition differs from that employed in quantum chemistry (QC). There the correlation energy $E_c^{\mathrm{QC}}[\rho]$ is the difference between the ground state energy $E[\rho]$ and the Hartree-Fock theory value of the energy $E_{\mathrm{HF}}[\rho]$:

$$E_c^{\mathrm{QC}}[\rho] = E[\rho] - E_{\mathrm{HF}}[\rho], \tag{7.162}$$

where

$$E_{\mathrm{HF}}[\rho] = \langle \Phi_{\mathrm{HF}} | \hat{T} + \hat{U} + \hat{V} | \Phi_{\mathrm{HF}} \rangle, \tag{7.163}$$

and where Φ_{HF}, the Hartree-Fock theory wave function, is that single Slater determinant that minimizes the expectation value of (7.163) with no further restrictions.

There is yet another definition [50, 51] of the correlation energy $E_c'[\rho]$ within the context of local effective potential theory. This is the difference between the ground state energy $E[\rho]$, and the energy $E_{xo}[\rho]$ obtained within an 'exchange-only' (xo) local effective potential energy calculation:

$$E'_c[\rho] = E[\rho] - E_{xo}[\rho]. \tag{7.164}$$

The energy $E_{xo}[\rho]$ is defined [50] as the expectation value

$$E_{xo}[\rho] = \langle \Phi' | \hat{T} + \hat{U} + \hat{V} | \Phi' \rangle, \tag{7.165}$$

where Φ' is that single Slater determinant which is constrained to be the ground state of some noninteracting Hamiltonian of the form $\hat{T} + \sum_i v'_s(\mathbf{r}_i)$, and which *simultaneously* minimizes the expectation value of (7.165). (This is the energy as determined within the 'exchange-only' Optimized Potential Method [52, 53] and ***QDFT***). Note that the Kohn-Sham $\Phi\{\phi_i\}$, Hartree-Fock theory $\Phi_{\rm HF}\{\phi_i\}$, and $\Phi'\{\phi_i\}$ Slater determinants are different, and thus lead to different densities and values for the correlation energy. The following inequalities with regard to the correlation energies can be proved [51]:

$$E_c^{\rm KS}[\rho] < E'_c[\rho] < E_c^{\rm QC}[\rho]. \tag{7.166}$$

7.5.3 Electron Correlations in Approximate Kohn-Sham Theory

The Kohn-Sham energy functionals $E_{ee}^{\rm KS}[\rho]$, $E_{xc}^{\rm KS}[\rho]$, $E_x^{\rm KS}[\rho]$, $E_c^{\rm KS}[\rho]$ of the density $\rho(\mathbf{r})$ are unknown, and must be approximated. Hence, it is in its approximate form that KS theory is applied. Since the advent of the KS paper in 1965, however, much effort has been expended in constructing accurate energy functionals, and this in general has proved successful. As such, the Kohn-Sham version of DFT is extensively employed in present-day electronic structure calculations. Herein are a few remarks on approximate KS-DFT.

In approximating the various 'exchange-correlation', 'exchange' and 'correlation' energy functionals, one is in essence approximating the manner in which the electron correlations are represented. Hence the Hamiltonian, and thereby the physical system is approximated. The rigor of the Hohenberg-Kohn theorems is thus lost, and the bounds on the ground state energy are no longer rigorous. The results for the energy could thus lie below the true nonrelativistic value [54].

There is yet another point to note about the electron correlations in an approximate energy functional. In the construction of say an approximate 'exchange-correlation' functional $E_{xc}^{\rm KS-approx}[\rho]$, it is *assumed* that the electron correlations of the system are those represented by this approximate energy functional. The corresponding approximate functional derivative $v_{xc}^{approx}(\mathbf{r}) = \delta E_{xc}^{\rm KS-approx}[\rho]/\delta\rho(\mathbf{r})$ is thus also *assumed* to be representative of the same electron correlations. However, it could turn out that assumptions made, within the framework of KS-DFT, as to the correlations in both the approximate functional $E_{xc}^{\rm KS-approx}[\rho]$ and its derivative $v_{xc}^{approx}(\mathbf{r})$ are both incorrect.

As an example of this, consider the Local Density Approximation (LDA) for the 'exchange-correlation' energy functional $E_{xc}^{LDA}[\rho]$ and our understanding of it within KS-DFT. This functional is the leading term in the majority of approximate 'exchange-correlation' energy functionals employed in the literature. The approximation treats each electron point of the *nonuniform* electron gas as if it were *uniform* defined by the *local* value of the density $\rho(\mathbf{r})$ at that point. As a consequence, the correlations at each electron position are those of a *uniform* electron gas. The corresponding LDA Fermi-Coulomb hole is then a spherically symmetric charge distribution $\rho_{xc}^{sph-sym}(\mathbf{rr}')$ about each electron position at $\mathbf{r}$. The energy $E_{xc}^{LDA}[\rho]$ is the energy of interaction between the density $\rho(\mathbf{r})$ and hole charge distribution $\rho_{xc}^{homog}(\mathbf{rr}')$. It is then *assumed* that the potential or functional derivative $v_{xc}^{LDA}(\mathbf{r}) = \delta E_{xc}^{LDA}[\rho]/\delta\rho(\mathbf{r})$, a monotonic function, is representative of these same correlations. That this cannot be the case becomes evident [55, 56] if one determines the electric field due to the LDA Fermi-Coulomb hole charge at each electron position, and from it the potential. Since the LDA Fermi-Coulomb hole is spherically symmetrical about each electron position, the field at each point in space vanishes. Therefore, the potential, which is the work done in this field, too vanishes or is constant. But, as noted, the functional derivative $\delta E_{xc}^{LDA}[\rho]/\delta\rho(\mathbf{r})$ exists and is a monotonic function. Obviously, there is an inconsistency.

The understanding of the true electron correlations within the LDA is arrived at via Q-DFT. A brief explanation is provided here, but the reader is referred to ***QDFT*** and [57–60] for the details. It turns out that the *true* Fermi-Coulomb hole in the LDA is comprised of the spherically symmetric component $\rho_{xc}^{sph-sym}(\mathbf{rr}')$, and an *asymmetric* part $\rho_{xc}^{asym}(\mathbf{rr}')$ which is proportional to the *gradient of the density* $\boldsymbol{\nabla}\rho(\mathbf{r})$ at each electron position. That is, the correct Fermi-Coulomb hole in the LDA is $\rho_{xc}^{LDA}(\mathbf{rr}') = \rho_{xc}^{sph-sym}(\mathbf{rr}') + \rho_{xc}^{asym}(\mathbf{rr}')$. Only the spherically symmetric component $\rho_{xc}^{sph-sym}(\mathbf{rr}')$ contributes to the energy $E_{xc}^{LDA}[\rho]$. The asymmetric part $\rho_{xc}^{asym}(\mathbf{rr}')$ leads to an electric field. The work done in this field $W_{xc}^{LDA}(\mathbf{r})$ is the same as the expression for the functional derivative $v_{xc}^{LDA}(\mathbf{r})$. Thus, the LDA samples not only the electron density $\rho(\mathbf{r})$ at each electron position but also its gradient $\boldsymbol{\nabla}\rho(\mathbf{r})$ at each point. Hence, the true electron correlations within the LDA are far more accurate than that understood to be the case within the context of KS-DFT.

7.6 The Hohenberg-Kohn Theorems in a Uniform Magnetic Field

We next provide the Pan and Sahni [3] proofs of the First and Second Hohenberg-Kohn (HK) theorems for a system of N *spinless electrons* in an electrostatic field $\boldsymbol{\mathcal{E}}(\mathbf{r}) = -\boldsymbol{\nabla}v(\mathbf{r})/e$ in the added presence of a uniform magnetic field $\boldsymbol{\mathcal{B}}(\mathbf{r}) = \boldsymbol{\nabla} \times \mathbf{A}(\mathbf{r})$. On the basis of the remarks made in Sect. 7.1, it is evident that the proof must differ from that of the original HK theorems. The underlying reason for this (see (7.18)) is that the relationship between the potentials $\{v(\mathbf{r}), \mathbf{A}(\mathbf{r})\}$ and the nondegenerate ground state wave function $\Psi(\mathbf{X})$ can be *many-to-one* and even *infinite-to-one*. Hence, there is no equivalent of the Maps C and C^{-1} in this case. Any proof of the theorems in the presence of a magnetic field must account for this many-to-one relationship. The proof of the First HK Theorem reveals that the basic variables, knowledge of which determines the potentials $\{v(\mathbf{r}), \mathbf{A}(\mathbf{r})\}$, and hence the Hamiltonian $\hat{H}$, are the nondegenerate ground state density and *physical* current density $\{\rho(\mathbf{r}), \mathbf{j}(\mathbf{r})\}$. The proof is valid for $(v, \mathbf{A})$-representable densities. Finally, because the presence of the magnetic field constitutes an added degree of freedom, there exists an additional constraint to that of fixed electron number N. For spinless electrons, the added constraint is that of fixed canonical angular momentum $\mathbf{L}$. The Second HK Theorem then determines the densities $\{\rho(\mathbf{r}), \mathbf{j}(\mathbf{r})\}$ by solution of the corresponding Euler-Lagrange equations with the constraints of fixed electron number and canonical angular momentum together with the satisfaction of the equation of continuity. With the knowledge that the basic variables are $\{\rho(\mathbf{r}), \mathbf{j}(\mathbf{r})\}$, a Percus-Levy-Lieb constrained-search-type proof then extends these theorems to N-representable densities, and to degenerate states.

The proof that $\{\rho(\mathbf{r}), \mathbf{j}(\mathbf{r})\}$ constitute the basic variables is then extended to the case of *electrons with spin*. There is in this case the additional constraint of fixed spin angular momentum $\mathbf{S}$.

Finally, we address the proofs of theories which claim properties other than the gauge invariant physical current density $\mathbf{j}(\mathbf{r})$ to be basic variables, properties such as the gauge variant paramagnetic $\mathbf{j}_p(\mathbf{r})$ and magnetization $\mathbf{m}(\mathbf{r})$ current densities.

7.6.1 The First Hohenberg-Kohn Theorem: Case 1: Spinless Electrons

The statement of the first Hohenberg-Kohn Theorem for spinless electrons in a uniform magnetic field is the following:

Theorem 3 *For electrons in an external electrostatic field and a uniform magnetostatic field, and for fixed electron number N and orbital angular momentum $\mathbf{L}$, the nondegenerate ground state density $\rho(\mathbf{r})$ and physical current density $\mathbf{j}(\mathbf{r})$, determine the external scalar $v(\mathbf{r})$ and vector $\mathbf{A}(\mathbf{r})$ potentials to within an additive constant and the gradient of a scalar function, respectively.*

Proof The Schrödinger equation for spinless electrons is given by (7.16). The corresponding Hamiltonian $\hat{H}_{spinless}$ of (7.11) for fixed electron number N and canonical angular momentum $\mathbf{L}$ may be expressed in terms of the density $\hat{\rho}(\mathbf{r})$ and physical current density $\hat{\mathbf{j}}(\mathbf{r})$ operators as

$$\hat{H}_{spinless} = \hat{T} + \hat{W} + \hat{V} + \frac{1}{c}\int \hat{\mathbf{j}}(\mathbf{r}) \cdot \mathbf{A}(\mathbf{r}) d\mathbf{r} - \frac{1}{2c^2}\int \hat{\rho}(\mathbf{r}) A^2(\mathbf{r}) d\mathbf{r}, \quad (7.167)$$

where $\hat{\mathbf{j}}(\mathbf{r})$ is the sum of the paramagnetic $\hat{\mathbf{j}}_p(\mathbf{r})$ and diamagnetic $\hat{\mathbf{j}}_d(\mathbf{r})$ components:

$$\hat{\mathbf{j}}_p(\mathbf{r}) = \frac{1}{2}\sum_k \left[\hat{\mathbf{p}}_k\delta(\mathbf{r}_k - \mathbf{r}) + \delta(\mathbf{r}_k - \mathbf{r})\hat{\mathbf{p}}_k\right], \tag{7.168}$$

$$\hat{\mathbf{j}}_d(\mathbf{r}) = \frac{1}{c}\hat{\rho}(\mathbf{r})\mathbf{A}(\mathbf{r}), \tag{7.169}$$

and where

$$\hat{\rho}(\mathbf{r}) = \sum_k \delta(\mathbf{r}_k - \mathbf{r}), \tag{7.170}$$

with $\hat{p}_k = -i\nabla_{\mathbf{r}_k}$ the canonical momentum operator. The proof is for $(v, \mathbf{A})$-representable densities $\{\rho(\mathbf{r}), \mathbf{j}(\mathbf{r})\}$.

Let us then consider two different physical systems $\{v, \mathbf{A}\}$ and $\{v', \mathbf{A}'\}$ that generate different nondegenerate ground state wave functions Ψ and Ψ'. We assume the gauges of the unprimed and primed systems to be the same. Let us further assume that these systems lead to the *same* nondegenerate ground state $\{\rho(\mathbf{r}), \mathbf{j}(\mathbf{r})\}$. We prove this cannot be the case.

From the variational principle for the energy for a nondegenerate ground state

$$E = \langle\Psi|\hat{H}_{spinless}|\Psi\rangle \quad < \quad \langle\Psi'|\hat{H}_{spinless}|\Psi'\rangle. \tag{7.171}$$

Now the term on the right hand side of the inequality may be written as

$$\begin{aligned}\langle\Psi'|\hat{H}_{spinless}|\Psi'\rangle &= \langle\Psi'|\hat{T} + \hat{U} + \hat{V}' + \frac{1}{c}\int \hat{\mathbf{j}}'(\mathbf{r})\cdot\mathbf{A}'(\mathbf{r})d\mathbf{r} - \frac{1}{2c^2}\int \hat{\rho}(\mathbf{r})A'^2(\mathbf{r})d\mathbf{r}|\Psi'\rangle \\ &+ \langle\Psi'|\hat{V} - \hat{V}'|\Psi'\rangle \\ &+ \langle\Psi'|\frac{1}{c}\int [\hat{\mathbf{j}}(\mathbf{r})\cdot\mathbf{A}(\mathbf{r}) - \hat{\mathbf{j}}'(\mathbf{r})\cdot\mathbf{A}'(\mathbf{r})]d\mathbf{r}|\Psi'\rangle \\ &- \frac{1}{2c^2}\langle\Psi'|\int \hat{\rho}(\mathbf{r})[A^2(\mathbf{r}) - A'^2(\mathbf{r})]d\mathbf{r}|\Psi'\rangle.\end{aligned} \tag{7.172}$$

For the primed system, the physical current density operator is

$$\hat{\mathbf{j}}'(\mathbf{r}) = \hat{\mathbf{j}}_p(\mathbf{r}) + \frac{1}{c}\hat{\rho}(\mathbf{r})\mathbf{A}'(\mathbf{r}), \tag{7.173}$$

so that

$$\mathbf{j}'(\mathbf{r}) = \langle\Psi'|\hat{\mathbf{j}}'(\mathbf{r})|\Psi'\rangle = \mathbf{j}'_p(\mathbf{r}) + \frac{1}{c}\rho'(\mathbf{r})\mathbf{A}'(\mathbf{r}). \tag{7.174}$$

Employing the original assumption that Ψ and Ψ' lead to the same $\rho(\mathbf{r})$, we have

$$\langle\Psi'|\hat{\mathbf{j}}(\mathbf{r})|\Psi'\rangle = \mathbf{j}'_p(\mathbf{r}) + \frac{1}{c}\rho(\mathbf{r})\mathbf{A}(\mathbf{r}). \tag{7.175}$$

and

$$\langle\Psi'|\hat{\mathbf{j}}'(\mathbf{r})|\Psi'\rangle = \mathbf{j}'_p(\mathbf{r}) + \frac{1}{c}\rho(\mathbf{r})\mathbf{A}'(\mathbf{r}). \quad (7.176)$$

Therefore

$$\langle\Psi'|\int\hat{\mathbf{j}}(\mathbf{r})\cdot\mathbf{A}(\mathbf{r})d\mathbf{r}|\Psi'\rangle = \int\mathbf{j}'_p(\mathbf{r})\cdot\mathbf{A}(\mathbf{r})d\mathbf{r} + \frac{1}{c}\int\rho(\mathbf{r})A^2(\mathbf{r})d\mathbf{r}, \quad (7.177)$$

and

$$\langle\Psi'|\int\hat{\mathbf{j}}'(\mathbf{r})\cdot\mathbf{A}'(\mathbf{r})d\mathbf{r}|\Psi'\rangle = \int\mathbf{j}'_p(\mathbf{r})\cdot\mathbf{A}'(\mathbf{r})d\mathbf{r} + \frac{1}{c}\int\rho(\mathbf{r})A'^2(\mathbf{r})d\mathbf{r}, \quad (7.178)$$

so that in (7.172) the term

$$\begin{aligned}&\frac{1}{c}\langle\Psi'|\int[\hat{\mathbf{j}}(\mathbf{r})\cdot\mathbf{A}(\mathbf{r}) - \hat{\mathbf{j}}'(\mathbf{r})\cdot\mathbf{A}'(\mathbf{r})]d\mathbf{r}|\Psi'\rangle\\ &= \frac{1}{c}\int\mathbf{j}'_p(\mathbf{r})\cdot[\mathbf{A}(\mathbf{r}) - \mathbf{A}'(\mathbf{r})]d\mathbf{r} + \frac{1}{c^2}\int\rho(\mathbf{r})[A^2(\mathbf{r}) - A'^2(\mathbf{r})]d\mathbf{r}. \quad (7.179)\end{aligned}$$

Finally, employing again that $\rho'(\mathbf{r}) = \rho(\mathbf{r})$, the last term of (7.172) is

$$\begin{aligned}&\frac{1}{2c^2}\langle\Psi'|\int\hat{\rho}(\mathbf{r})[A^2(\mathbf{r}) - A'^2(\mathbf{r})]d\mathbf{r}|\Psi'\rangle\\ &= \frac{1}{2c^2}\int\rho(\mathbf{r})[A^2(\mathbf{r}) - A'^2(\mathbf{r})]d\mathbf{r}. \quad (7.180)\end{aligned}$$

Therefore, the inequality of (7.171) is

$$\begin{aligned}E < E' + \int\rho(\mathbf{r})[v(\mathbf{r}) - v'(\mathbf{r})]d\mathbf{r} + \frac{1}{c}\int\mathbf{j}'_p(\mathbf{r})\cdot[\mathbf{A}(\mathbf{r}) - \mathbf{A}'(\mathbf{r})]d\mathbf{r}\\ + \frac{1}{2c^2}\int\rho(\mathbf{r})[A^2(\mathbf{r}) - A'^2(\mathbf{r})]d\mathbf{r}. \quad (7.181)\end{aligned}$$

On interchanging the primed and unprimed quantities,

$$\begin{aligned}E' < E + \int\rho(\mathbf{r})[v'(\mathbf{r}) - v(\mathbf{r})]d\mathbf{r} + \frac{1}{c}\int\mathbf{j}_p(\mathbf{r})\cdot[\mathbf{A}'(\mathbf{r}) - \mathbf{A}(\mathbf{r})]d\mathbf{r}\\ + \frac{1}{2c^2}\int\rho(\mathbf{r})[A'^2(\mathbf{r}) - A^2(\mathbf{r})]d\mathbf{r}. \quad (7.182)\end{aligned}$$

On adding the previous two equations one obtains the inequality

$$E + E' < E + E' + \frac{1}{c}\int[\mathbf{j}'_p(\mathbf{r}) - \mathbf{j}_p(\mathbf{r})]\cdot[\mathbf{A}(\mathbf{r}) - \mathbf{A}'(\mathbf{r})]d\mathbf{r}. \quad (7.183)$$

The inequality of (7.183) is a general result.

Consider next the third term on the right hand side of (7.183). With $\boldsymbol{\mathcal{B}}(\mathbf{r}) = \mathcal{B}\hat{\mathbf{i}}_z$, $\boldsymbol{\mathcal{B}}'(\mathbf{r}) = \mathcal{B}'\hat{\mathbf{i}}_z$, and the symmetric gauge $\mathbf{A}(\mathbf{r}) = \frac{1}{2}\boldsymbol{\mathcal{B}} \times \mathbf{r}$; $\mathbf{A}'(\mathbf{r}) = \frac{1}{2}\boldsymbol{\mathcal{B}}' \times \mathbf{r}$, this term is

$$I = \int \big[\mathbf{j}'_p(\mathbf{r}) - \mathbf{j}_p(\mathbf{r})\big] \cdot \big[\mathbf{A}(\mathbf{r}) - \mathbf{A}'(\mathbf{r})\big] d\mathbf{r} \tag{7.184}$$

$$= \int \big[\mathbf{j}'_p(\mathbf{r}) - \mathbf{j}_p(\mathbf{r})\big] \cdot \big[\frac{1}{2}\Delta\mathbf{B} \times \mathbf{r}\big] d\mathbf{r} \tag{7.185}$$

$$= \frac{1}{2}\Delta\mathbf{B} \cdot \int \mathbf{r} \times \big[\mathbf{j}'_p(\mathbf{r}) - \mathbf{j}_p(\mathbf{r})\big] d\mathbf{r}, \tag{7.186}$$

where $\Delta\mathbf{B} = (B - B')\hat{i}_z$. First consider the integral

$$I_1 = \int \mathbf{r} \times \mathbf{j}_p(\mathbf{r}) d\mathbf{r} \tag{7.187}$$

$$= -\frac{i}{2}\sum_k \int \mathbf{r} \times \Big\{\int \Psi^\star(\mathbf{X})\big[\nabla_{\mathbf{r}_k}\delta(\mathbf{r} - \mathbf{r}_k)$$

$$+\delta(\mathbf{r} - \mathbf{r}_k)\nabla_{\mathbf{r}_k}\big]\Psi(\mathbf{X})d\mathbf{X}\Big\}d\mathbf{r} \tag{7.188}$$

$$= -\frac{i}{2}\sum_k \int d\mathbf{X} \int d\mathbf{r}\Psi^\star(\mathbf{X})\big[\mathbf{r} \times \nabla_{\mathbf{r}_k}\delta(\mathbf{r} - \mathbf{r}_k)$$

$$+\delta(\mathbf{r} - \mathbf{r}_k)\mathbf{r} \times \nabla_{\mathbf{r}_k}\big]\Psi(\mathbf{X}). \tag{7.189}$$

Next consider the second integral of I_1 of (7.189):

$$I_{12} = -\frac{i}{2}\sum_k \int d\mathbf{X}\Psi^\star(\mathbf{X})\big[\int d\mathbf{r}\delta(\mathbf{r} - \mathbf{r}_k)\mathbf{r} \times \nabla_{\mathbf{r}_k}\big]\Psi(\mathbf{X}) \tag{7.190}$$

$$= -\frac{i}{2}\sum_k \int d\mathbf{X}\Psi^\star(\mathbf{X})\mathbf{r}_k \times \nabla_{\mathbf{r}_k}\Psi(\mathbf{X}) \tag{7.191}$$

$$= \frac{1}{2}\int d\mathbf{X}\Psi^\star(\mathbf{X})\big(\sum_k \mathbf{r}_k \times \mathbf{p}_k\big)\Psi(\mathbf{X}) \tag{7.192}$$

$$= \frac{1}{2}\int d\mathbf{X}\Psi^\star(\mathbf{X})\sum_k \hat{\mathbf{L}}_k\Psi(\mathbf{X}) \tag{7.193}$$

$$= \frac{1}{2}\mathbf{L}, \tag{7.194}$$

where $\hat{\mathbf{L}}_k = \mathbf{r}_k \times \hat{\mathbf{p}}_k$ is the canonical orbital angular momentum operator, and $\mathbf{L}$ the total canonical orbital angular momentum defined by (7.193).

The first integral of I_1 of (7.189) is

$$I_{11} = -\frac{i}{2}\sum_k \int d\mathbf{X} \int d\mathbf{r} \Psi^\star(\mathbf{X})\mathbf{r} \times \nabla_{\mathbf{r}_k}\delta(\mathbf{r} - \mathbf{r}_k)\Psi(\mathbf{X}) \tag{7.195}$$

$$= -\frac{i}{2}\sum_k \int d\mathbf{X} \int d\mathbf{r} \Psi^\star(\mathbf{X})\epsilon_{\alpha\beta\gamma}\frac{\partial}{\partial r_{k\gamma}}\big(r_\beta \delta(\mathbf{r} - \mathbf{r}_k)\Psi(\mathbf{X})\big). \tag{7.196}$$

On integrating the inner integral by parts and dropping the surface term, one obtains

$$I_{11} = -\frac{i}{2}\sum_k \int d\mathbf{X}\Big[- \epsilon_{\alpha\beta\gamma} \int d\mathbf{r} \frac{\partial \Psi^\star(\mathbf{X})}{\partial r_{k\gamma}} r_\beta \delta(\mathbf{r} - \mathbf{r}_k)\Psi(\mathbf{X})\Big] \tag{7.197}$$

$$= -\frac{i}{2}\sum_k \int d\mathbf{X}\Big[- \epsilon_{\alpha\beta\gamma} \frac{\partial \Psi^\star(\mathbf{X})}{\partial r_{k\gamma}} r_{k\beta}\Psi(\mathbf{X})\Big]. \tag{7.198}$$

On integrating by parts again, one obtains to

$$I_{11} = -\frac{i}{2}\sum_k \epsilon_{\alpha\beta\gamma} \int d\mathbf{X}\Psi^\star(\mathbf{X})\frac{\partial}{\partial r_{k\gamma}}\big(r_{k\beta}\Psi(\mathbf{X})\big) \tag{7.199}$$

$$= -\frac{i}{2}\sum_k \int d\mathbf{X}\Psi^\star(\mathbf{X})\ (\mathbf{r}_k \times \nabla_{\mathbf{r}_k})\Psi(\mathbf{X}) \tag{7.200}$$

$$= \frac{1}{2}\mathbf{L}. \tag{7.201}$$

Hence, the integral I of (7.186) is

$$I = \frac{1}{2}\Delta\mathbf{B} \cdot (\mathbf{L}' - \mathbf{L}). \tag{7.202}$$

If one imposes the constraint that the total canonical orbital angular momentum is *fixed* so that $\mathbf{L} = \mathbf{L}'$, then the integral I of (7.186) vanishes. Hence, (7.183) reduces to the contradiction

$$E + E' < E + E'. \tag{7.203}$$

What this means is that the original assumption Ψ and Ψ' differ is erroneous, and that there can exist a $\{v, \mathbf{A}\}$ and a $\{v', \mathbf{A}'\}$ with the same nondegenerate ground state wave function. The fact that $\Psi = \Psi'$ means that

$$\rho(\mathbf{r})\big|_\Psi = \rho'(\mathbf{r})\big|_{\Psi'} \tag{7.204}$$

However, the corresponding physical current densities are not the same:

$$\mathbf{j}(\mathbf{r})\big|_\Psi \neq \mathbf{j}'(\mathbf{r})\big|_{\Psi'}. \tag{7.205}$$

This is because the diamagnetic components are not the same

$$\mathbf{j}_d(\mathbf{r})\big|_{\Psi} \neq \mathbf{j}'_d(\mathbf{r})\big|_{\Psi'}, \tag{7.206}$$

if one hews to the original assumption that $\mathbf{A}(\mathbf{r})$ is different from $\mathbf{A}'(\mathbf{r})$. This proves that the assumption that there exists a different $\{v', \mathbf{A}'\}$ (with the same N and $\mathbf{L}$) that leads to the same $\{\rho, \mathbf{j}\}$ as that due to $\{v, \mathbf{A}\}$ is incorrect. *This step takes into account the fact that there could exist many* $\{v, \mathbf{A}\}$ *that lead to the same nondegenerate ground state* Ψ. Hence, there exists only one $\{v, \mathbf{A}\}$ for fixed N and $\mathbf{L}$ that generates a nondegenerate ground state $\{\rho, \mathbf{j}\}$. The one-to-one relationship between $\{\rho, \mathbf{j}\}$ and $\{v, \mathbf{A}\}$ is therefore proved.

With the kinetic $\hat{T}$ and electron-interaction $\hat{U}$ operators known, knowledge of $\{\rho(\mathbf{r}), \mathbf{j}(\mathbf{r})\}$ determines the potentials $\{v(\mathbf{r}), \mathbf{A}(\mathbf{r})\}$ and thereby the Hamiltonian $\hat{H}_{spinless}$. Solution of the Schrödinger equation (7.16) then determines the eigenfunctions $\Psi(\mathbf{X})$. The wave functions $\Psi(\mathbf{X})$ are thus functionals of $\{\rho(\mathbf{r}), \mathbf{j}(\mathbf{r})\}$. By performing a density and physical current density *preserving* unitary transformation of the Schrödinger equation (7.16), it can be shown (see *QDFT*) that the wave function must also be a functional of a gauge function $\alpha(\mathbf{R})$. Thus, the wave functions $\Psi(\mathbf{X})$ are functionals of $\{\rho, \mathbf{j}, \alpha\}$, i.e. $\Psi(\mathbf{X}) = \Psi[\rho, \mathbf{j}, \alpha]$. As the properties $\{\rho(\mathbf{r}), \mathbf{j}(\mathbf{r})\}$ are each gauge invariant, it is the presence of the gauge function $\alpha(\mathbf{R})$ that ensures that the wave function functional is gauge variant. However, as the physical system remains unchanged for different gauge functions, the choice of vanishing gauge function is equally valid.

That the properties $\{\rho(\mathbf{r}), \mathbf{j}(\mathbf{r})\}$ for fixed orbital angular momentum $\mathbf{L}$ are the basic variables is most readily seen for the case of $N = 1$. Writing the wave function in polar form

$$\psi(\mathbf{r}) = \Phi(\mathbf{r})e^{i\Theta(\mathbf{r})}, \tag{7.207}$$

with $\Phi(\mathbf{r})$, $\Theta(\mathbf{r})$ real valued, we see that $\rho(\mathbf{r}) = \Phi^2(\mathbf{r})$ and $\mathbf{j}(\mathbf{r}) = \mathbf{j}_p(\mathbf{r}) + \mathbf{j}_d(\mathbf{r})$, $\mathbf{j}_p(\mathbf{r}) = \rho(\mathbf{r})\nabla\Theta(\mathbf{r})$, $\mathbf{j}_d(\mathbf{r}) = \frac{1}{c}\rho(\mathbf{r})\mathbf{A}(\mathbf{r})$, so that $\frac{1}{c}\mathbf{A}(\mathbf{r}) = \frac{\mathbf{j}(\mathbf{r})}{\rho(\mathbf{r})} - \nabla\Theta(\mathbf{r})$. Thus knowledge of $\{\rho(\mathbf{r}), \mathbf{j}(\mathbf{r})\}$ determines the potential $\mathbf{A}(\mathbf{r})$ to within the gradient of a scalar function. Employing this and the fact that one can perform a gauge transformation to eliminate the phase, the one-electron Schrödinger equation can be written as

$$\left[\frac{1}{2}\left(\hat{\mathbf{p}} + \frac{\mathbf{j}(\mathbf{r})}{\rho(\mathbf{r})}\right)^2 + v(\mathbf{r}) - E\right]\rho(\mathbf{r}) = 0, \tag{7.208}$$

from which the potential $v(\mathbf{r})$ can be obtained to within the constant E since $\{\rho(\mathbf{r}), \mathbf{j}(\mathbf{r})\}$ are known. (This example is given in [61] but without the understanding of the added constraint on the angular momentum.) What is interesting about this example is that the $\{\rho(\mathbf{r}), \mathbf{j}(\mathbf{r})\}$ as basic variables are not restricted to being solely the nondegenerate ground state densities. The above arguments are equally valid for $\{\rho(\mathbf{r}), \mathbf{j}(\mathbf{r})\}$ corresponding to any state and angular momentum.

7.6.2 *The First Hohenberg-Kohn Theorem: Case 2: Electrons With Spin*

The statement of the first Hohenberg-Kohn Theorem for electrons with spin in a uniform magnetic field is the following:

Theorem 3 *For electrons in an external electrostatic field and a uniform magnetostatic field, and for fixed electron number N, orbital angular momentum $\mathbf{L}$, and spin angular momentum S, the nondegenerate ground state density $\rho(\mathbf{r})$ and physical current density $\mathbf{j}(\mathbf{r})$, determine the external scalar $v(\mathbf{r})$ and vector $\mathbf{A}(\mathbf{r})$ potentials to within an additive constant and the gradient of a scalar function, respectively.*

Proof For electrons with spin, the differential equation descriptive of the system is the Schrödinger-Pauli equation (7.17), with the corresponding Hamiltonian $\hat{H}_{spin}$ given by (7.12). The form of this Hamiltonian in terms of the density $\hat{\rho}(\mathbf{r})$ and physical current density $\hat{\mathbf{j}}(\mathbf{r})$ operators is the same as $\hat{H}_{spinless}$ for spinless electrons of (7.167), i.e.

$$\hat{H}_{spin} = \hat{T} + \hat{W} + \hat{V} + \frac{1}{c}\int \hat{\mathbf{j}}(\mathbf{r}) \cdot \mathbf{A}(\mathbf{r})d\mathbf{r} - \frac{1}{2c^2}\int \hat{\rho}(\mathbf{r})A^2(\mathbf{r})d\mathbf{r}. \qquad (7.209)$$

The sole difference in $\hat{H}_{spin}$ lies in the expression for the physical current density operator $\hat{\mathbf{j}}$. In addition to the paramagnetic $\hat{\mathbf{j}}_p(\mathbf{r})$ (7.168) and diamagnetic $\hat{\mathbf{j}}_d(\mathbf{r})$ (7.169) components, there is now a magnetization (spin) current density $\hat{\mathbf{j}}_m(\mathbf{r})$ component:

$$\hat{\mathbf{j}}_m(\mathbf{r}) = -c\boldsymbol{\nabla} \times \hat{\mathbf{m}}(\mathbf{r}), \qquad (7.210)$$

where the magnetization density operator $\hat{\mathbf{m}}(\mathbf{r})$ is defined in terms of the spin angular momentum vector $\mathbf{s}$ as

$$\hat{\mathbf{m}}(\mathbf{r}) = -\frac{1}{c}\sum_k \mathbf{s}_k\delta(\mathbf{r}_k - \mathbf{r}). \qquad (7.211)$$

As the Hamiltonian $\hat{H}_{spin}$ and $\hat{H}_{spinless}$ are of the same form, one would then expect that for electrons with spin the basic variables would also be the nondegenerate ground state density $\rho(\mathbf{r})$ and the physical current density $\mathbf{j}(\mathbf{r})$. And further, that in addition to the constraints of fixed electron number N and orbital angular momentum $\mathbf{L}$, there would now also be the requirement of fixed spin angular momentum $\mathbf{S}$. In order to prove the above statements, we rewrite the Hamiltonian $\hat{H}_{spin}$ of (7.12) in yet another form as the sum of the Hamiltonian $\hat{H}_{spinless}$ and the term involving the spin moment of the electron. Thus,

$$\hat{H}_{spin} = \hat{H}_{spinless} + \frac{1}{c}\sum_k \mathcal{B}(\mathbf{r}_k)\cdot\mathbf{s}_k \tag{7.212}$$

$$= \hat{T} + \hat{U} + \hat{V} + \frac{1}{c}\int \hat{\mathbf{j}}(\mathbf{r})\cdot\mathbf{A}(\mathbf{r})d\mathbf{r}$$

$$-\frac{1}{2c^2}\int \hat{\rho}(\mathbf{r})A^2(\mathbf{r})d\mathbf{r} - \int \hat{\mathbf{m}}(\mathbf{r})\cdot\mathcal{B}(\mathbf{r})d\mathbf{r}, \tag{7.213}$$

with $\hat{\mathbf{m}}(\mathbf{r})$ defined by (7.211). Note that the operator $\hat{\mathbf{j}}(\mathbf{r})$ in (7.213) is the sum of its paramagnetic $\hat{\mathbf{j}}_p(\mathbf{r})$ and diamagnetic $\hat{\mathbf{j}}_d(\mathbf{r})$ components as in (7.168) and (7.169). Once again let us consider two different physical systems $\{v, \mathbf{A}\}$ and $\{v', \mathbf{A}'\}$ that generate different nondegenerate wave functions Ψ and Ψ' but lead to the *same* nondegenerate ground state $\{\rho(\mathbf{r}), \mathbf{j}(\mathbf{r})\}$. Again, we prove this cannot be the case. With these assumptions, the inequality of (7.183) is replaced by

$$E + E' < E + E' + \int [\mathbf{j}'_p(\mathbf{r}) - \mathbf{j}_p(\mathbf{r})]\cdot[\mathbf{A}(\mathbf{r}) - \mathbf{A}'(\mathbf{r})]d\mathbf{r}$$

$$-\int [\mathbf{m}'(\mathbf{r}) - \mathbf{m}(\mathbf{r})]\cdot[\mathcal{B}(\mathbf{r}) - \mathcal{B}'(\mathbf{r})]d\mathbf{r}, \tag{7.214}$$

with $\mathbf{m}(\mathbf{r})$ the magnetization density being the expectation

$$\mathbf{m}(\mathbf{r}) = \langle\Psi|\hat{\mathbf{m}}(\mathbf{r})|\Psi\rangle. \tag{7.215}$$

The inequality is once again a general result.

The third term on the right hand side of (7.214) vanishes if, as in the previous section, a uniform magnetic field is assumed, and the constraint that the orbital angular momentum of the unprimed and primed systems are the same is imposed. Hence next consider the last term of (7.214). With $\mathcal{B}(\mathbf{r}) = B\hat{\mathbf{i}}_z$, the term

$$\int \mathbf{m}(\mathbf{r})\cdot\mathcal{B}(\mathbf{r})d\mathbf{r} = B\int m_z(\mathbf{r})d\mathbf{r}, \tag{7.216}$$

where, with $\mathbf{s}_k \cdot \hat{\mathbf{i}}_z = s_{z,k}$,

$$m_z(\mathbf{r}) = -\frac{1}{c}\sum_\sigma \int \sum_k s_{z,k}\delta(\mathbf{r}_k - \mathbf{r})$$

$$\times\Psi^\star(\mathbf{r}_k\sigma, \mathbf{X}^{N-1})\Psi(\mathbf{r}_k\sigma, \mathbf{X}^{N-1})d\mathbf{X}^{N-1}d\mathbf{r}_k \tag{7.217}$$

$$= -\frac{1}{c}\sum_\sigma \int \sum_k s_{z,k}\int \Psi^\star(\mathbf{r}\sigma, \mathbf{X}^{N-1})$$

$$\times\Psi(\mathbf{r}\sigma, \mathbf{X}^{N-1})d\mathbf{X}^{N-1} \tag{7.218}$$

$$= -\frac{1}{cN}\sum_\sigma S_z\gamma(\mathbf{r}\sigma, \mathbf{r}\sigma), \tag{7.219}$$

where $S_z = \sum_k s_{z,k}$ is the z-component of the total spin $\mathbf{S}$, and $\gamma(\mathbf{xx}') = N \int \Psi^\star(\mathbf{r}\sigma, \mathbf{X}^{N-1})\Psi(\mathbf{r}'\sigma', \mathbf{X}^{N-1})d\mathbf{X}^{N-1}$, the density matrix. Since in the primed system, the spin vectors are different, i.e.. some $\mathbf{s}'_k$, we have

$$\int \big[\mathbf{m}'(\mathbf{r}) - \mathbf{m}(\mathbf{r})\big] \cdot \Delta\mathbf{B}(\mathbf{r})d\mathbf{r} = \Delta B \int \big[m'_z(\mathbf{r}) - m_z(\mathbf{r})\big]d\mathbf{r} \tag{7.220}$$

$$= \frac{\Delta B}{cN} \sum_\sigma \int \big[S'_z\gamma'(\mathbf{r}\sigma, \mathbf{r}\sigma) - S_z\gamma(\mathbf{r}\sigma, \mathbf{r}\sigma)\big]d\mathbf{r}, \tag{7.221}$$

where $\Delta B = B - B'$. Employing the original assumption that the diagonal matrix elements $\gamma(\mathbf{r}\sigma, \mathbf{r}\sigma)$ of the density matrix $\gamma(\mathbf{xx}')$ are the same for the unprimed and primed systems we have the right hand side of (7.221) to be

$$\frac{\Delta B}{cN} \sum_\sigma \int \big[S'_z - S_z\big]\gamma(\mathbf{r}\sigma, \mathbf{r}\sigma) = 0 \tag{7.222}$$

provided $S'_z = S_z$. Hence, the last term of (7.214) vanishes.

Another way of arriving at this conclusion is by rewriting $m_z(\mathbf{r})$ as [8, 9]

$$m_z(\mathbf{r}) = -\frac{1}{2c}[\rho_\alpha(\mathbf{r}) - \rho_\beta(\mathbf{r})], \tag{7.223}$$

with $\rho_\alpha(\mathbf{r})$, $\rho_\beta(\mathbf{r})$ being the spin-up and spin-down spin densities. The last term of the inequality of (7.214) is then

$$\int [\mathbf{m}'(\mathbf{r}) - \mathbf{m}(\mathbf{r})] \cdot \Delta\mathbf{B}(\mathbf{r})d\mathbf{r} = -\frac{1}{2c}\Delta B \int [\{\rho'_\alpha(\mathbf{r}) - \rho'_\beta(\mathbf{r})\} - \{\rho_\alpha(\mathbf{r}) - \rho_\beta(\mathbf{r})\}]d\mathbf{r}. \tag{7.224}$$

If the z-component of the total spin angular momentum S_z for the unprimed and primed systems are the same, the corresponding spin densities are the same, so that the last term of (7.214) vanishes. More generally, the magnetization densities $\mathbf{m}(\mathbf{r})$ and $\mathbf{m}'(\mathbf{r})$ are the same if the total spin angular momentum $\mathbf{S}$ and $\mathbf{S}'$ are the same.

The vanishing of the last two terms of (7.214) once again leads to the contradiction $E + E' < E + E'$. Employing the same reasoning as in the previous section one concludes that the original assumption that Ψ and Ψ' differ is erroneous, and that there can exist a $\{v, \mathbf{A}\}$ and a $\{v', \mathbf{A}'\}$ with the same nondegenerate ground state wave function. With $\Psi = \Psi'$, we have $\rho(\mathbf{r}) = \rho'(\mathbf{r})$, but $\mathbf{j}(\mathbf{r}) \neq \mathbf{j}'(\mathbf{r})$ since $\mathbf{A}(\mathbf{r}) \neq \mathbf{A}'(\mathbf{r})$. This proves that the original assumption that there exists a $\{v', \mathbf{A}'\}$ with the same N, $\mathbf{L}$, and $\mathbf{S}$ as that of $\{v, \mathbf{A}\}$ but leads to the same $\{\rho, \mathbf{j}\}$ to be incorrect. Thus, there can exist only one $\{v, \mathbf{A}\}$ for fixed N, $\mathbf{L}$, and $\mathbf{S}$ that can generate the nondegenerate ground state $\{\rho, \mathbf{j}\}$. The bijective relationship between $\{\rho, \mathbf{j}\}$ and $\{v, \mathbf{A}\}$ for systems defined by the Schrödinger-Pauli Hamiltonian is therefore proved. *Note that the proof explicitly accounts for the many-to-one relationship between the potentials* $\{v, \mathbf{A}\}$ *and the nondegenerate ground state* Ψ. Thus, for electrons with spin, in a uniform magnetic field, the basic variables are $\{\rho(\mathbf{r}), \mathbf{j}(\mathbf{r})\}$.

As was the case for the spinless electrons, one can perform a density and physical current density *preserving* unitary transformation of the Schrödinger-Pauli equation to prove that the wave function is a functional of a gauge function $\alpha(\mathbf{R})$. Hence, the wave function is a functional of $\{\rho, \mathbf{j}, \alpha\}$, i.e. $\Psi(\mathbf{X}) = \Psi[\rho, \mathbf{j}, \alpha]$. As both $\rho(\mathbf{r})$ and $\mathbf{j}(\mathbf{r})$ are gauge invariant, it is the presence of the gauge function that ensures the wave function functional being gauge variant. Again, the choice of vanishing gauge is valid, so that the wave function may be written solely as a functional of $\rho(\mathbf{r})$ and $\mathbf{j}(\mathbf{r})$, i.e. $\Psi(\mathbf{X}) = \Psi[\rho, \mathbf{j}]$.

7.6.3 The Second Hohenberg-Kohn Theorem

We extend here the Second Hohenberg-Kohn Theorem for spinless electrons to the case of the presence of a uniform magnetic field. As proved, the ground state wave function $\Psi(\mathbf{X})$ is a functional of the basic variables $\{\rho(\mathbf{r}), \mathbf{j}(\mathbf{r})\}$ for *fixed electron number N and orbital angular momentum* $\mathbf{L}$, i.e. $\Psi(\mathbf{X}) = \Psi[\rho, \mathbf{j}]$. Thus, the expectation value of the Hamiltonian $\hat{H}_{spinless}$ of (7.11), the ground state energy E is such a functional: $E = E_{v,\mathbf{A}}[\rho, \mathbf{j}]$. Note that in the definition of $\hat{H}_{spinless}$, and therefore that of $E_{v,\mathbf{A}}[\rho, \mathbf{j}]$, *the external potentials* $v(\mathbf{r})$ *and* $\mathbf{A}(\mathbf{r})$ *are fixed* and remain unchanged. The statement of the Second Hohenberg-Kohn Theorem is as follows:

Theorem 4 *The nondegenerate ground state density* $\rho(\mathbf{r})$ *and physical current density* $\mathbf{j}(\mathbf{r})$, *can be determined from the ground state energy functional* $E_{v,\mathbf{A}}[\rho, \mathbf{j}]$ *via the variational principle for the energy by variations only of these densities. The constraints on the corresponding Euler-Lagrange equations are the conservation of electron number, the canonical angular momentum, and the satisfaction of the equation of continuity.*

Proof The ground state energy functional is

$$E_{v,\mathbf{A}}[\rho, \mathbf{j}] = \langle \Psi[\rho, \mathbf{j}] | \hat{H}_{spinless} | \Psi[\rho, \mathbf{j}] \rangle \tag{7.225}$$

$$= F[\rho, \mathbf{j}] + \int \rho(\mathbf{r}) v(\mathbf{r}) d\mathbf{r} + \frac{1}{c} \int \mathbf{j}(\mathbf{r}) \cdot \mathbf{A}(\mathbf{r}) d\mathbf{r} - \frac{1}{2c^2} \int \rho(\mathbf{r}) A^2(\mathbf{r}) d\mathbf{r}, \tag{7.226}$$

where

$$F[\rho, \mathbf{j}] = \langle \Psi[\rho, \mathbf{j}] | \hat{T} + \hat{U} | \Psi[\rho, \mathbf{j}] \rangle, \tag{7.227}$$

is a *universal* energy functional descriptive of all electronic systems in the presence of a magnetic field. As the ground state energy is a functional of the variables $\{\rho(\mathbf{r}), \mathbf{j}(\mathbf{r})\}$, a variational principle exists for arbitrary variations of such $(v, \mathbf{A})$-representable densities. Implicit in such a variational principle, as in all energy variational principles, is that *the external potentials* $\{v(\mathbf{r}), \mathbf{A}(\mathbf{r})\}$ *remain fixed throughout the variation*. According to the variational principle

$$E_{v,\mathbf{A}}[\rho', \mathbf{j}'] > E_{v,\mathbf{A}}[\rho, \mathbf{j}] \text{ for } \{\rho'(\mathbf{r}), \mathbf{j}'(\mathbf{r})\} \neq \{\rho(\mathbf{r}), \mathbf{j}(\mathbf{r})\}, \tag{7.228}$$

$$E_{v,\mathbf{A}}[\rho', \mathbf{j}'] = E_{v,\mathbf{A}}[\rho, \mathbf{j}] \text{ for } \{\rho'(\mathbf{r}), \mathbf{j}'(\mathbf{r})\} = \{\rho(\mathbf{r}), \mathbf{j}(\mathbf{r})\}, \tag{7.229}$$

The corresponding Euler-Lagrange equations that must be solved self-consistently for $\rho(\mathbf{r})$ and $\mathbf{j}(\mathbf{r})$ are

$$\left.\frac{\delta E_{v,\mathbf{A}}[\rho, \mathbf{j}]}{\delta \rho(\mathbf{r})}\right|_{\mathbf{j}(\mathbf{r})} = 0\,; \quad \left.\frac{\delta E_{v,\mathbf{A}}[\rho, \mathbf{j}]}{\delta \mathbf{j}(\mathbf{r})}\right|_{\rho(\mathbf{r})} = 0, \tag{7.230}$$

subject to the constraints

$$\int \rho(\mathbf{r}) d\mathbf{r} = N, \tag{7.231}$$

$$\int \mathbf{r} \times \left[\mathbf{j}(\mathbf{r}) - \frac{1}{c}\rho(\mathbf{r})\mathbf{A}(\mathbf{r})\right] d\mathbf{r} = \mathbf{L}, \tag{7.232}$$

$$\nabla \cdot \mathbf{j}(\mathbf{r}) = 0. \tag{7.233}$$

For electrons with spin as defined by the Hamiltonian $\hat{H}_{spin}$ of (7.12), the basic variables are once again $\{\rho(\mathbf{r}), \mathbf{j}(\mathbf{r})\}$, but with $\mathbf{j}(\mathbf{r})$ now including the magnetization current density component $\mathbf{j}_m(\mathbf{r})$. The added constraint is that of fixed spin angular momentum $\mathbf{S}$. The above proof is readily extended to electrons with spin.

7.6.4 The Percus-Levy-Lieb Constrained-Search Proof

The above proofs of the First and Second Hohenberg-Kohn Theorems in the presence of a uniform magnetic field are valid for the subset of $(v, \mathbf{A})$- representable densities $\{\rho(\mathbf{r}), \mathbf{j}(\mathbf{r})\}$ for fixed $(N, \mathbf{L})$ or fixed $(N, \mathbf{L}, \mathbf{S})$ as the case may be. With the knowledge that the properties $\{\rho(\mathbf{r}), \mathbf{j}(\mathbf{r})\}$ constitute the basic variables, it is then possible to generalize these proofs to the broader subset of N-representable densities for fixed $(N, \mathbf{L})$ or $(N, \mathbf{L}, \mathbf{S})$, and to degenerate states via the Percus-Levy-Lieb (PLL) constrained-search framework. This generalization to spinless electrons defined by the Hamiltonian $\hat{H}_{spinless}$ of (7.11) is described below.

Suppose there exist antisymmetric functions $\Psi_{\rho,\mathbf{j}}(\mathbf{X})$ for fixed $(N, \mathbf{L})$ that lead to the ground state densities $\{\rho(\mathbf{r}), \mathbf{j}(\mathbf{r})\}$. The first question we address is how then does one distinguish these functions from the true ground state wave function $\Psi(\mathbf{X})$? Following PLL, on application of the variational principle to the Hamiltonian $\hat{H}_{spinless}$ we have (*for fixed potentials* $\{v(\mathbf{r}), \mathbf{A}(\mathbf{r})\}$)

$$\langle \Psi_{\rho,\mathbf{j}}(\mathbf{X}) | \hat{H}_{spinless} | \Psi_{\rho,\mathbf{j}}(\mathbf{X}) \rangle \geq \langle \Psi(\mathbf{X}) | \hat{H}_{spinless} | \Psi(\mathbf{X}) \rangle = E_{v,\mathbf{A}}[\rho, \mathbf{j}], \tag{7.234}$$

or equivalently

$$\langle \Psi_{\rho,\mathbf{j}}(\mathbf{X})|\hat{T} + \hat{U}|\Psi_{\rho,\mathbf{j}}(\mathbf{X})\rangle \geq \langle \Psi(\mathbf{X})|\hat{T} + \hat{U}|\Psi(\mathbf{X})\rangle. \tag{7.235}$$

Hence, of all the N-representable $\Psi_{\rho,\mathbf{j}}(\mathbf{X})$ that lead to the ground state $\{\rho(\mathbf{r}), \mathbf{j}(\mathbf{r})\}$, the true wave function $\Psi(\mathbf{X})$ is that which minimizes the expectation $\langle \hat{T} + \hat{U}\rangle$. This constitutes the constrained-search over all $\Psi_{\rho,\mathbf{j}}(\mathbf{X})$ to arrive at $\Psi(\mathbf{X})$.

As in Sect. 7.3.3, it is possible to describe the energy minimization by two nested minimizations. As the wave function $\Psi(\mathbf{X})$ is a functional of $\{\rho(\mathbf{r}), \mathbf{j}(\mathbf{r})\}$, one first defines the universal function $F[\rho, \mathbf{j}]$ of (7.227) in terms of the N-representable $\Psi_{\rho,\mathbf{j}}(\mathbf{X})$ as

$$F[\rho, \mathbf{j}] = \min_{\Psi_{\rho,\mathbf{j}} \to \rho,\mathbf{j}} \langle \Psi_{\rho,\mathbf{j}}(\mathbf{X})|\hat{T} + \hat{U}|\Psi_{\rho,\mathbf{j}}(\mathbf{X})\rangle, \tag{7.236}$$

such that searching over all $\Psi_{\rho,\mathbf{j}}(\mathbf{X})$, the functional $F[\rho, \mathbf{j}]$ delivers the minimum of the expectations $\langle \hat{T} + \hat{U}\rangle$. The functional is equally valid for degenerate states.

(Once again, note that although the functional $F[\rho, \mathbf{j}]$ is defined solely in terms of the operators $\hat{T}$ and $\hat{U}$, the physical system for which the minimization is being performed is known. This is because the constrained search is over all $\Psi_{\rho,\mathbf{j}}(\mathbf{X})$ that generate the ground state $\{\rho, \mathbf{j}\}$. But knowledge of $\{\rho, \mathbf{j}\}$ determines the potentials $\{v, \mathbf{A}\}$ to within a constant and gradient of a scalar function, respectively. Hence, all the operators of the Hamiltonian defining the physical system are known.)

The ground state energy E may then be written as

$$E = \min_{\rho,\mathbf{j}} \left\{ \min_{\Psi_{\rho,\mathbf{j}} \to \rho,\mathbf{j}} \left[\langle \Psi_{\rho,\mathbf{j}}(\mathbf{X})|\hat{H}_{spinless}|\Psi_{\rho,\mathbf{j}}(\mathbf{X})\rangle \right] \right\} \tag{7.237}$$

$$= \min_{\rho,\mathbf{j}} \left\{ F[\rho, \mathbf{j}] + \Big(\int \rho(\mathbf{r}) v(\mathbf{r}) d\mathbf{r} + \frac{1}{c} \int \mathbf{j}(\mathbf{r}) \cdot \mathbf{A}(\mathbf{r}) d\mathbf{r} - \frac{1}{2c^2} \int \rho(\mathbf{r}) A^2(\mathbf{r}) d\mathbf{r} \Big) \right\} \tag{7.238}$$

$$= \min_{\rho,\mathbf{j}} E_{v,\mathbf{A}}[\rho, \mathbf{j}]. \tag{7.239}$$

The variations in (7.239) are over all N-representable $\{\rho(\mathbf{r}), \mathbf{j}(\mathbf{r})\}$ for fixed $(N, \mathbf{L})$. The stringent $(v, \mathbf{A})$-representability requirement, and the restriction to nondegenerate ground states, are thereby overcome. The extension of these arguments to the Schrödinger-Pauli Hamiltonian $\hat{H}_{spin}$ of (7.12) follows similarly.

7.6.5 Remarks on Basic Variables in a Magnetic Field

As noted, the concept of a basic variable in quantum mechanics stems from the First Hohenberg-Kohn Theorem. It is a gauge invariant property, knowledge of which uniquely determines the Hamiltonian and hence the eigenfunctions and eigenvalues of the system. Thus, for a system of spinless electrons of fixed number N, the nondegenerate ground state density $\rho(\mathbf{r})$ constitutes a basic variable. It uniquely determines, (to within an additive constant) the binding potential $v(\mathbf{r})$, and hence the Hamiltonian $\hat{H}$. Solution of the corresponding Schrödinger equation then leads to the wave functions and energies of the system. The key to the proof of this conclusion is the bijective (*one-to-one*) relationship between the potential $v(\mathbf{r})$ and the nondegenerate ground state wave function $\Psi(\mathbf{X})$: the Maps C and C^{-1}.

In the added presence of a magnetic field, the relationship between the corresponding potentials $\{v(\mathbf{r}), \mathbf{A}(\mathbf{r})\}$ and the nondegenerate ground state wave function $\Psi(\mathbf{X})$ can be *many-to-one* and even *infinite-to-one*. There is therefore no equivalent of Maps C and C^{-1}, and a Hohenberg-Kohn-type proof cannot exist. Explicitly accounting for the many-to-one relationship, as proved, the basic variables are shown to be the gauge invariant nondegenerate ground state density $\rho(\mathbf{r})$ and the *physical* current density $\mathbf{j}(\mathbf{r})$. This is the case for both spinless and electrons with spin. These properties determine the potentials $\{v(\mathbf{r}), \mathbf{A}(\mathbf{r})\}$ to within a constant and the gradient of a scalar function, respectively. In the proof, the additional requirement of fixed orbital $\mathbf{L}$ and spin $\mathbf{S}$ angular momentum is revealed.

It is also important to note [15] that the Percus-Levy-Lieb (PLL) constrained-search proofs can be formulated *only following* the Hohenberg-Kohn proofs of what properties constitute the basic variables. Thus, once it is proved that the nondegenerate ground state density $\rho(\mathbf{r})$ is a basic variable, the constrained-search over all functions $\Psi_\rho(\mathbf{X})$ that lead to that $\rho(\mathbf{r})$ becomes meaningful. Similarly, with the knowledge that the basic variables are the ground state $\{\rho(\mathbf{r}), \mathbf{j}(\mathbf{r})\}$, it becomes reasonable to search over all functions $\Psi_{\rho,\mathbf{j}}$, $(\mathbf{X})$ that generate that $\{\rho(\mathbf{r}), \mathbf{j}(\mathbf{r})\}$. (Note that the PLL-type proof in itself does not account for the *many-to-one* relationship between $\{v(\mathbf{r}), \mathbf{A}(\mathbf{r})\}$ and the ground state $\Psi(\mathbf{X})$. This has already been accounted for in the proof of $\{\rho(\mathbf{r}), \mathbf{j}(\mathbf{r})\}$ being the basic variables.)

Many theories in the literature purport other properties to be the basic variables when a magnetic field is present. There are then two types of proofs made to support this claim. The first involves proofs that mimic the original HK proofs for spinless electrons. For example, in spin density functional theory [2, 28, 62], it is assumed that in addition to the density $\rho(\mathbf{r})$, the magnetization current density $\mathbf{m}(\mathbf{r})$ is a basic variable. In current density functional theory [63–65], the added basic variable is the paramagnetic current density $\mathbf{j}_p(\mathbf{r})$. However, both $\mathbf{m}(\mathbf{r})$ and $\mathbf{j}_p(\mathbf{r})$ are gauge variant properties. Furthermore, in the respective proofs, it is *assumed* that the relationship between the potentials $\{v(\mathbf{r}), \mathbf{A}(\mathbf{r})\}$ and the wave function $\Psi(\mathbf{X})$ is *one-to-one*. That is, the Maps C and C^{-1} exist, in contradiction to the physical reality. Having made this assumption, these proofs then rely *solely* on the *one to-one* relationship (Maps D and D^{-1}) between the ground state $\Psi(\mathbf{X})$ and $\mathbf{m}(\mathbf{r})$ or $\mathbf{j}_p(\mathbf{r})$. (A justification by Kohn et al. [66] of the validity of solely Maps (D, D^{-1})-type proofs in fact begin with the assumption of existence of the Maps (C, C^{-1}).) In the solely Maps (D, D^{-1})-type proofs, there is thus no relationship between the properties $\{\rho(\mathbf{r}), \mathbf{m}(\mathbf{r})\}$ or $\{\rho(\mathbf{r}), \mathbf{j}_p(\mathbf{r})\}$ and the potentials $\{v(\mathbf{r}), \mathbf{A}(\mathbf{r})\}$ that is proved. The proofs, that $\mathbf{m}(\mathbf{r})$ and $\mathbf{j}_p(\mathbf{r})$ constitute basic variables, are therefore incorrect. (For further insights into spin density functional theory, see [36].)

Note that the solely Maps (D, D^{-1})-type proofs, (with the intrinsic assumption of existence of the Maps C and C^{-1}), are *always* possible if the operator corresponding to the property of interest appears in the Hamiltonian. However, that property need not constitute a basic variable. For example, for electrons with spin and the corresponding Schrödinger-Pauli Hamiltonian, the basic variables are said to be $\{\rho(\mathbf{r}), \mathbf{j}_p(\mathbf{r}), \mathbf{m}(\mathbf{r})\}$ [67] or $\{\rho(\mathbf{r}), \mathbf{j}_p(\mathbf{r}), \mathbf{m}(\mathbf{r}), \mathbf{j}_{p,m}(\mathbf{r})\}$ [68], where $\mathbf{j}_{p,m}(\mathbf{r})$ are the gauge variant paramagnetic currents of each component of the magnetization density. The properties other than the density $\rho(\mathbf{r})$ thus do not constitute basic variables. What is missing in the solely Maps (D, D^{-1})-type proofs is the intrinsic feature of the many-to-one relationship between the potentials $\{v(\mathbf{r}), \mathbf{A}(\mathbf{r})\}$ and the nondegenerate ground state wave function $\Psi(\mathbf{X})$.

The second category [8, 9, 67–71] of proofs begins with the *assumption* that a specific property is a basic variable, and proceeds directly to a PLL-type proof. Hence, PLL-type proofs assuming that $\{\rho(\mathbf{r}), \mathbf{m}(\mathbf{r})\}$, $\{\rho(\mathbf{r}), \mathbf{j}_p(\mathbf{r})\}$ or $\{\rho(\mathbf{r}), \mathbf{j}_p(\mathbf{r}) + \mathbf{j}_m(\mathbf{r})\}$ are basic variables exist, but with no formal foundational justification. Furthermore, as noted above, PLL-type proofs do not specifically account for the many-to-one relationship between $\{v(\mathbf{r}), \mathbf{A}(\mathbf{r})\}$ and the nondegenerate ground state $\Psi(\mathbf{X})$.

It is interesting to note that although the ground state density $\rho(\mathbf{r})$ does not depend explicitly on the scalar potential $v(\mathbf{r})$, the physical current density $\mathbf{j}(\mathbf{r})$ via its diamagnetic component $\mathbf{j}_d(\mathbf{r}) = (1/c)\rho(\mathbf{r})\mathbf{A}(\mathbf{r})$ depends on the vector potential $\mathbf{A}(\mathbf{r})$. This distinction, however, has no bearing in any application of the variational principle for the energy since the Hamiltonian $\hat{H}$ remains fixed throughout the arbitrary variations of the wave functions. In other words, a rigorous upper bound to the energy can only be obtained if the Hamiltonian $\hat{H}$, and hence the potentials $\{v(\mathbf{r}), \mathbf{A}(\mathbf{r})\}$, remain unchanged throughout the variational procedure. (Of course, according to the HK theorems, knowledge of the ground state $\{\rho(\mathbf{r}), \mathbf{j}(\mathbf{r})\}$ uniquely determines $\{v(\mathbf{r}), \mathbf{A}(\mathbf{r})\}$ to within a constant and the gradient of a scalar function, respectively.)

7.7 Time-Dependent Density Functional Theory

Just as is the case with stationary-state Hohenberg-Kohn-Sham density functional theory, time-dependent theory (TD-DFT) has evolved into a sub-field with all its subtleties. Hence, we present here the basic tenets underlying the theory without proofs together with aspects not stressed in the traditional literature. Additionally, insights into the theory with regard to the electron correlations are arrived at via TD quantal density functional theory (TD-QDFT) (see Chap. 6).

There are two fundamental theorems in TD-DFT. The first, due to Runge and Gross (RG) [4], extends the First Hohenberg-Kohn theorem to the TD case. The First RG Theorem is then generalized by a *density preserving* unitary transformation [43]. This generalization demonstrates the hierarchy in terms of gauge functions that exists within the fundamental theorems of density functional theory, both time-dependent and time-independent. A Corollary to the First RG Theorem is derived [44].

The second fundamental theorem, due to van Leeuwen (vL) [5], is based on the 'Quantal Newtonian' Second Law (see Sect. 2.4). The vL theorem is more general in that the RG theorem constitutes a special case. Further, it provides a justification for the existence of a Kohn-Sham system of noninteracting fermions possessing the same TD density $\rho(\mathbf{y})$ as that of the interacting system defined by the TD Schrödinger equation. There is a brief discussion of the parallelism with the Second Hohenberg-Kohn Theorem proposed by Runge-Gross in terms of Action functionals of the density $\rho(\mathbf{y})$. A rigorous physical interpretation of TD Kohn-Sham theory based on TD-QDFT is described.

(Notation: $\mathbf{Y} = \mathbf{X}t$; $\mathbf{y} = \mathbf{r}t$; $\mathbf{y}' = \mathbf{r}'t$; $\mathbf{X} = \mathbf{x}_1, \ldots, \mathbf{x}_N$; $\mathbf{x} = \mathbf{r}\sigma$; $\mathbf{r}, \sigma$ the spatial and spin coordinates, and $\mathbf{R} = \mathbf{r}_1, \ldots, \mathbf{r}_N$.)

7.7.1 The First Runge-Gross Theorem

The physical system considered is that of N electrons in an external TD electric field $\mathcal{F}^{ext}(\mathbf{y}) = -\nabla v(\mathbf{y})/e$ as defined by the Schrödinger equation (2.1) with wave function $\Psi(\mathbf{Y})$ and Hamiltonian $\hat{H}(t)$ of (2.2). The sole restriction on the scalar potential $v(\mathbf{y})$ is that it is Taylor expandable about some initial time t_0 which is finite.

The statement of the RG Theorem is as follows:

The First Runge-Gross Theorem *Two densities* $\rho(\mathbf{y})$ and $\rho'(\mathbf{y})$ (*or two current densities* $\mathbf{j}(\mathbf{y})$ and $\mathbf{j}'(\mathbf{y})$) *evolving from the same initial state* $\Psi(t_0) = \Psi_0$ *generated by two external potentials* $v(\mathbf{y})$ *and* $v'(\mathbf{y})$ *that are Taylor expandable about* t_0 *are always different provided the potentials differ by more than a purely time-dependent function* $C(t)$, i.e.

$$v(\mathbf{y}) \neq v'(\mathbf{y}) + C(t). \tag{7.240}$$

Note 1 In the proof of the Theorem, it is first proved that potentials $v(\mathbf{y})$ and $v'(\mathbf{y})$ that differ by more than a function $C(t)$ lead to different current densities $\mathbf{j}(\mathbf{y})$ and $\mathbf{j}'(\mathbf{y})$. In other words, there is a one-to-one relationship between the potentials $v(\mathbf{y})$ and current densities $\mathbf{j}(\mathbf{y})$, i.e. $v(\mathbf{y}) \leftrightarrow \mathbf{j}(\mathbf{y})$. Thus, since the kinetic $\hat{T}$ and electron-interaction potential $\hat{U}$ operators are known, knowledge of $\mathbf{j}(\mathbf{y})$ determines $v(\mathbf{y})$ to within a function $C(t)$, and thereby the Hamiltonian $\hat{H}(t)$. Solution of the TD Schrödinger equation (2.1) with the known initial condition then leads to the wave function $\Psi(\mathbf{Y})$. This proves that the current density $\mathbf{j}(\mathbf{y})$ constitutes a basic variable.

Employing the fact that the different $v(\mathbf{y})$ and $v'(\mathbf{y})$ lead to different $\mathbf{j}(\mathbf{y})$ and $\mathbf{j}'(\mathbf{y})$, it is then proved that the corresponding densities $\rho(\mathbf{y})$ and $\rho'(\mathbf{y})$ too are different. Thus, there is also a one-to-one relationship between $v(\mathbf{y})$ and $\rho(\mathbf{y})$, i.e. $v(\mathbf{y}) \leftrightarrow \rho(\mathbf{y})$. Knowledge of $\rho(\mathbf{y})$ thus also leads to the Hamiltonian $\hat{H}(t)$ and the wave function $\Psi(\mathbf{Y})$. Hence, the density $\rho(\mathbf{y})$ too is a basic variable.

In a manner similar to the stationary-state Hohenberg-Kohn path (Sect. 7.1), there is the corresponding TD *RG path* from either basic variable to the Hamiltonian $\hat{H}(t)$:

$$[\rho(\mathbf{y}) \text{ or } \mathbf{j}(\mathbf{y})] \rightarrow v(\mathbf{y}) \rightarrow \hat{H}(t). \tag{7.241}$$

Note 2 In TD-DFT, the fact that the current density $\mathbf{j}(\mathbf{y})$ is a basic variable is not considered any further. This has consequences in terms of electron correlations that must be accounted for when the mapping to a system of noninteracting fermions is considered (see Sect. 7.7.5). Thus, in TD-DFT, the focus is solely on the density $\rho(\mathbf{y})$.

A consequence of the fact that the density $\rho(\mathbf{y})$ is a basic variable is that the wave function $\Psi(\mathbf{Y}) = \Psi[\Psi_0](t)$ which depends on the initial state Ψ_0, is a functional of the density $\rho(\mathbf{y})$ unique to within an arbitrary TD phase factor $\alpha(t)$:

$$\Psi(\mathbf{Y}) = \exp[-\alpha(t)]\tilde{\Psi}[\rho; \Psi_0](t). \tag{7.242}$$

This means that with $\alpha(t_0) = 0$ but otherwise arbitrary, the wave function functional $\tilde{\Psi}[\rho; \Psi_0]$ will have the same density $\rho(\mathbf{y})$ and have the same initial state $\tilde{\Psi}(t_0) = \Psi_0$. The expectation value of any operator $\hat{O}(t)$ (other than one that does not contain time derivatives) is thus a *unique* functional of the density $\rho(\mathbf{y})$:

$$\langle \hat{O}(t)\rangle = \langle \tilde{\Psi}[\rho; \Psi_0](t)|\hat{O}(t)|\tilde{\Psi}[\rho; \Psi_0](t)\rangle, \tag{7.243}$$

with the phase factors canceling out as was the case for the density. The explicit dependence of the wave function on the density is unknown. Hence, the unique functionals of the expectation values are unknown. (These remarks are equally valid for the basic variable $\mathbf{j}(\mathbf{y})$.)

7.7.2 *Generalization of the First Runge-Gross Theorem Via Density Preserving Unitary Transformation*

For the system defined by the Schrödinger equation (2.1) with Hamiltonian $\hat{H}(\mathbf{R}t)$ of (2.2), the First RG Theorem can be generalized to arbitrary gauge functions $\alpha(\mathbf{R}t)$ by a *density preserving* unitary transformation.

Employing the TD unitary operator

$$\hat{U} = e^{j\alpha(\mathbf{R}t)}, \tag{7.244}$$

the transformed wave function $\Psi'(\mathbf{Y})$ and Hamiltonian $\hat{H}'(\mathbf{R}t)$ are respectively

$$\Psi'(\mathbf{Y}) = \hat{U}^{\dagger}\Psi(\mathbf{Y}), \tag{7.245}$$

$$\hat{H}'(\mathbf{R}t) = \hat{H}(\mathbf{R}t) + \frac{d\alpha(\mathbf{R}t)}{dt} + \frac{1}{2}\sum_i \left(\hat{\mathbf{p}}_i \cdot \mathbf{A}_i + \hat{\mathbf{A}}_i \cdot \hat{p}_i + \hat{\mathbf{A}}_i^2\right), \tag{7.246}$$

where $\hat{\mathbf{p}}_i = -j\nabla_i$ and the vector potential operator is defined as $\mathbf{A}_i = \nabla_i\alpha(\mathbf{R}t)$. (Note that $\nabla \times \mathbf{A}_i = 0$.) The transformed Hamiltonian $\hat{H}'(\mathbf{R}t)$ may also be written as

$$\hat{H}'(\mathbf{R}t) = \frac{1}{2}\sum_i \left(\hat{\mathbf{p}}_i + \hat{\mathbf{A}}_i\right)^2 + \hat{W} + \hat{V}', \tag{7.247}$$

where

$$\hat{V}' = \hat{V} + \frac{d\alpha(\mathbf{R}t)}{dt}. \tag{7.248}$$

(The electron-interaction operator is designated as $\hat{W}$.)

The RG Theorem is derived for the case when no magnetic field is present. The Hamiltonian $\hat{H}'(\mathbf{R}t)$ of (7.246), (7.247) is the most general form of the Hamiltonian for which the RG Theorem is valid in the absence of a magnetic field. In addition to the operators of the Hamiltonian $\hat{H}(\mathbf{R}t)$ of (2.2), the transformed Hamiltonian now includes the canonical momentum operator $\hat{\mathbf{p}}_i$, a TD function $C(\mathbf{R}t) = d\alpha(\mathbf{R}t)/dt$, and a TD curl-free vector potential operator $\mathbf{A}_i = \nabla_i\alpha(\mathbf{R}t)$. The one-to-one relationship of the RG Theorem is then $\rho(\mathbf{y}) \leftrightarrow \hat{H}'_k(\mathbf{R}t)$ with $\hat{H}'_k(\mathbf{R}t)$ corresponding to the different gauge functions $\alpha_k(\mathbf{R}t)$.

The First RG Theorem is recovered from the above generalization when the gauge function $\alpha(\mathbf{R}t) = \alpha(t)$ with the function $C(t) = d\alpha(t)/dt$. With the choice of gauge function $\alpha(\mathbf{R}t) = \alpha(\mathbf{R})$, one obtains the generalized Hamiltonians $\hat{H}'(\mathbf{R})$ of (7.49), (7.51) for which the first Hohenberg-Kohn Theorem is valid. Finally, for $\alpha(\mathbf{R}t) = \alpha =$ constant, the case of the original Hohenberg-Kohn Theorem for the Hamiltonian $\hat{H}(\mathbf{R})$ of (7.2) is recovered. The hierarchy of the various theorems in terms of the gauge functions is summarized in the Table below.

Gauge Function	**Theorem**
$\alpha(\mathbf{R}t)$	Generalized Runge-Gross
$\alpha(t)$	Runge-Gross
$\alpha(\mathbf{R})$	Generalized Hohenberg-Kohn
$\alpha =$ constant	Hohenberg-Kohn

Note, however, that unlike the 'Quantal Newtonian' First and Second laws, the stationary-state Hohenberg-Kohn Theorem does not constitute a special case of the TD Runge-Gross Theorem.

As a consequence of the unitary transformation of (7.244), we conclude that the wave function $\Psi(\mathbf{Y})$ must also be a functional of the gauge function, i.e. $\Psi(\mathbf{Y}) = \Psi[\rho(\mathbf{y}); \alpha(\mathbf{R}t); \Psi_0]$. This ensures that the wave function $\Psi(\mathbf{Y})$ when written as a functional is gauge variant. However, as the physical system is *independent* of the choice of gauge function, the choice of $\alpha(\mathbf{R}t) = 0$ is equally valid. This justifies the conclusion of the RG Theorem that the wave function $\Psi(\mathbf{Y})$ can be written solely as a functional of the density $\rho(\mathbf{y})$.

To reiterate, according to the First RG Theorem, there is a one-to-one relationship between the density $\rho(\mathbf{y})$ and the *infinite* number of Hamiltonians $\hat{H}(\mathbf{R}t) + C(\mathbf{R}t)$ representative of the *same physical system*. Thus, knowledge of the density $\rho(\mathbf{y})$ *uniquely* determines the physical system to within a TD function $C(\mathbf{R}t)$. However, as shown in the following section, it is possible to construct an *infinite* set of degenerate

Hamiltonians $\{\hat{H}(t)\}$ that differ by an *intrinsic* function $C(t)$, represent *different physical systems*, but yet possess the same density $\rho(\mathbf{y})$. In this case, the density $\rho(\mathbf{y})$ cannot distinguish between the different physical systems. For such systems, the RG Theorem is not valid.

7.7.3 Corollary to the First Runge-Gross Theorem

According to the First RG Theorem, the Hamilton $\hat{H}(t)$ determines the density $\rho(\mathbf{y})$ via solution of the TD Schrödinger equation (2.1), and knowledge of the density $\rho(\mathbf{y})$ in turn determines the Hamiltonian $\hat{H}(t)$ to within a TD function $C(t)$:

$$\hat{H}(t) + C(t) \leftrightarrow \rho(\mathbf{y}), \tag{7.249}$$

with the two-headed arrow indicative of the one-to-one relationship.

In this section we construct a set of degenerate TD Hamiltonians $\{\hat{H}(t)\}$ which differ by a function $C(t)$ representing different physical systems but which *all* possess the *same* density $\rho(\mathbf{y})$. In this case, the density $\rho(\mathbf{y})$ cannot distinguish between the different physical systems. In equation form then

$$\{\hat{H}(t)\} \rightarrow \rho(\mathbf{y}), \tag{7.250}$$

with the right-directed arrow indicating the lack of invertibility.

The statement of the Corollary to the First RG Theorem is as follows:

Corollary *Degenerate time-dependent Hamiltonians $\{\hat{H}(t)\}$ that represent different physical systems, but which differ by a purely time-dependent function $C(t)$, and which all yield the same density $\rho(\mathbf{r}t)$, cannot be distinguished on the basis of the First Runge-Gross Theorem.*

Consider again the Hooke's species of Sect. 7.2.4, but in this case let us assume the positions of the nuclei are time-dependent, i.e. $\mathbf{R}_j = \mathbf{R}_j(t)$. This could represent the non-Born-Oppenheimer motion of the nuclei. For simplicity, we consider the spring constant to be the same (k') for interaction with all the nuclei.

The external potential energy $v_{\mathcal{N}}(\mathbf{r}t)$ for an arbitrary member of the species which now is

$$v_{\mathcal{N}}(\mathbf{r}t) = \frac{1}{2}k' \sum_{j=1}^{\mathcal{N}} (\mathbf{r} - \mathbf{R}_j(t))^2 \,, \tag{7.251}$$

may then be rewritten as

$$v_{\mathcal{N}}(\mathbf{r}t) = \frac{1}{2}\mathcal{N}k'r^2 - k' \sum_{j=1}^{\mathcal{N}} \mathbf{R}_j(t) \cdot \mathbf{r} + \frac{1}{2}k' \sum_{j=1}^{\mathcal{N}} \mathbf{R}_j^2(t) \,, \tag{7.252}$$

where at some initial time t_0, we have $\mathbf{R}_j(t_0) = \mathbf{R}_{j,0}$. (Note that a spatially uniform time-dependent field $\mathbf{F}(t)$ interacting only with the electrons could be further incorporated by adding a term $\mathbf{F}(t) \cdot \mathbf{r}$ to the external potential energy expression.) The Hamiltonian of an element of the species governed by the number of nuclei $\mathcal{N}$ is then

$$\hat{H}_{\mathcal{N}}(\mathbf{r}_1\mathbf{r}_2 t) = \hat{H}_{\mathcal{N},0} - k' \sum_{j=1}^{\mathcal{N}} [\mathbf{R}_j(t) - \mathbf{R}_{j,0}] \cdot (\mathbf{r}_1 + \mathbf{r}_2) + C(k', \mathcal{N}, t) \,, \quad (7.253)$$

where $\hat{H}_{\mathcal{N},0}$ is the time-independent Hooke's species Hamiltonian (7.73):

$$\hat{H}_{\mathcal{N},0} = \hat{H}_{\mathcal{N}}(k') \,, \quad (7.254)$$

and the time-dependent function

$$C(k', \mathcal{N}, t) = k' \sum_{j=1}^{\mathcal{N}} [\mathbf{R}_j^2(t) - \mathbf{R}_{j,0}^2] \,. \quad (7.255)$$

Note that the function $C(k', \mathcal{N}, t)$ contains physical information about the system: in this case, about the motion of the nuclei about their equilibrium positions. It also differentiates between the different elements of the species. It is intrinsic to the Hamiltonian $\hat{H}_{\mathcal{N}}(\mathbf{r}_1\mathbf{r}_2 t)$.

The solution of the time-dependent Schrödinger equation $\hat{H}_{\mathcal{N}}(t)\Psi(t) = i\partial\Psi(t)/\partial t)$ employing the Harmonic Potential Theorem, (which constitutes a special case of the Generalized Kohn Theorem derived in Sect. 9.2) is

$$\Psi(\mathbf{r}_1\mathbf{r}_2 t) = \exp\{-i\phi(t)\}\exp\left[-i\left\{E_{\mathcal{N},0}t - 2S(t) - 2\frac{d\mathbf{z}}{dt} \cdot \mathbf{y}\right\}\right] \Psi_0(\overline{\mathbf{r}_1}\ \overline{\mathbf{r}_2}) \,, \quad (7.256)$$

where $\overline{\mathbf{r}_i} = \mathbf{r}_i - \mathbf{z}(t), \mathbf{y} = (\mathbf{r}_1 + \mathbf{r}_2)/2$,

$$S(t) = \int_{t_0}^{t} \left[\frac{1}{2}\dot{\mathbf{z}}(t')^2 - \frac{1}{2}k\mathbf{z}(t')^2\right] dt' \,, \quad (7.257)$$

the shift $\mathbf{z}(t)$ satisfies the classical harmonic oscillator equation

$$\ddot{\mathbf{z}}(t) + k\mathbf{z}(t) - k' \sum_{j=1}^{\mathcal{N}} [\mathbf{R}_j(t) - \mathbf{R}_{j,0}] = 0 \,, \quad (7.258)$$

where the additional phase factor $\phi(t)$ is due to the function $C(k', \mathcal{N}, t)$,

$$\phi(t) = \int_{t_0}^{t} C(k', \mathcal{N}, t')dt' , \tag{7.259}$$

and where at the initial time $\Psi(\mathbf{r}_1\mathbf{r}_2t_0) = \Psi_0$ which satisfies $\hat{H}_{\mathcal{N},0}\Psi_0 = E_{\mathcal{N},0}\Psi_0$. Thus, the wave function $\Psi(\mathbf{r}_1\mathbf{r}_2t)$ is the time-independent solution shifted by a time-dependent function $\mathbf{z}(t)$, and multiplied by a phase factor. The explicit contribution of the function $C(k', \mathcal{N}, t)$ to this phase has been separated out. The phase factor cancels out in the determination of the density $\rho(t) = \langle\Psi(t)|\hat{\rho}|\Psi(t)\rangle = \rho(\mathbf{r} - \mathbf{z}(t))$ which is the initial time-independent density $\rho(\mathbf{r}t_0) = \rho_0(\mathbf{r})$ displaced by $\mathbf{z}(t)$.

As in the time-independent case, the 'degenerate Hamiltonians' $\hat{H}_{\mathcal{N}}(\mathbf{r}_1\mathbf{r}_2t)$ of the time-dependent Hooke's species can each be made to generate the *same* density $\rho(\mathbf{r}t)$ by adjusting the spring constant k' such that $\mathcal{N}k' = k$, and provided the density at the initial time t_0 is the *same*. The latter is readily achieved as it constitutes the time-independent Hooke's species case discussed previously.

Thus, we have a set of Hamiltonians describing different physical systems but which can be made to generate the same density $\rho(\mathbf{r}t)$. These Hamiltonians differ by the function $C(k', \mathcal{N}, t)$ that contains information which differentiates between them. In such a case, the density $\rho(\mathbf{r}t)$ cannot distinguish between the different Hamiltonians.

We note that the TD degenerate Hamiltonians $\{\hat{H}(t)\}$ of the Hooke's species, and hence the Corollary, *does not constitute a counter example to the First RG Theorem.* This is because the proof of the RG Theorem is *independent* of whether the function of time $C(t)$ is *additive* or *intrinsic*. In the RG case, the functions $C(t)$ are additive; in the case of the Hooke's species, the functions C(t) are intrinsic. A counter example would be one in which Hamiltonians that differ by more than a function $C(t)$ have the same density $\rho(\mathbf{r}t)$.

7.7.4 The van Leeuwen Theorem

We next address the issue of the mapping of the interacting system as defined by the Schrödinger equation (2.1) with density $\rho(\mathbf{y})$ for time greater than an initial time t_0 and initial condition Ψ_0 to one of *noninteracting* fermions that possess the same TD density. In stationary-state density functional theory, the existence of such a model system is an assumption. In TD-DFT, there is the van Leeuwen Theorem [5] that purports to prove the existence of such a model system. (There has been a critique of this existence theorem, and a response to the critique [43, 72–76].)

The van Leeuwen Theorem is derived for potentials $v(\mathbf{y})$ which are Taylor expandable about an initial time t_0, and employs the 'Quantal Newtonian' Second Law of Sect. 2.4 [77–79]. The statement of the Theorem is the following:

The van Leeuwen Theorem *For a system of electrons with an arbitrary electron-interaction potential U in a time-dependent scalar potential $v(\mathbf{y})$ with density $\rho(\mathbf{y})$ and in an initial state Ψ_0, there exist other systems with a different electron-interaction* *potential U' and scalar potential $v'(\mathbf{y})$ (unique to within a function $C(t)$) which possess the same density $\rho(\mathbf{y})$ provided the corresponding initial state Ψ_0' of this system yields the same density and density time-derivative at the initial time.*

That the RG Theorem constitutes a special case is readily seen if $U' = U$, and $\Psi_0' = \Psi_0$, as then there exists a potential $v'(\mathbf{y})$ (unique to within a function $C(t)$) that generates the same $\rho(\mathbf{y})$.

The noninteracting fermion case corresponds to $U' = 0$. The further assumption is that the initial state of the model system $\Phi_0 = \Psi_0$, and that the initial density and its time-derivative are the same as that of the interacting system. Then, it follows from the van Leeuwen Theorem, that a unique local potential $v_s(\mathbf{y})$ exists (to within a time function $C(t)$) that reproduces the density $\rho(\mathbf{y})$.

As noted previously, although both the TD density $\rho(\mathbf{y})$ and current density $\mathbf{j}(\mathbf{y})$ constitute basic variables, TD-DFT is concerned only with the density $\rho(\mathbf{y})$. Hence, the model Kohn-Sham system of noninteracting fermions is constructed so as to reproduce *solely* the density $\rho(\mathbf{y})$. (Recall that in TD Q-DFT (see Sect. 6.3), the mapping is to a model system of noninteracting fermions which possess the same density $\rho(\mathbf{y})$ and current density $\mathbf{j}(\mathbf{y})$ as that of the interacting system. As such the correlations that must be accounted for in TD-DFT differ and are more complex (see Sect. 7.7.5)).

The local effective potential $v_s(\mathbf{y})$ of TD-DFT is written in the same manner as in stationary-state Kohn-Sham theory (see Sect. 7.4):

$$v_s(\mathbf{y}) = v(\mathbf{y}) + v_{ee}(\mathbf{y}), \tag{7.260}$$

where all the many-body effects are incorporated in the electron-interaction potential $v_{ee}(\mathbf{y})$. As is customary, this potential is written as the sum of the Hartree potential $v_{\mathrm{H}}(\mathbf{y})$ and an 'exchange-correlation' component $v_{xc}(\mathbf{y})$, with the latter being further decomposed into an 'exchange' $v_x(\mathbf{y})$ and a 'correlation' $v_c(\mathbf{y})$ part:

$$v_{ee}(\mathbf{y}) = v_{\mathrm{H}}(\mathbf{y}) + v_{xc}(\mathbf{y}) \tag{7.261}$$

$$= v_{\mathrm{H}}(\mathbf{y}) + v_x(\mathbf{y}) + v_c(\mathbf{y}), \tag{7.262}$$

where

$$v_{\mathrm{H}}(\mathbf{y}) = \int \frac{\rho(\mathbf{y}')}{|\mathbf{r} - \mathbf{r}'|} d\mathbf{r}'. \tag{7.263}$$

The TD Kohn-Sham differential equation for the orbitals $\phi_k(\mathbf{y})$ of the Slater determinant $\Phi\{\phi_k\}$ that generates the density $\rho(\mathbf{y})$ is (with the appropriate initial condition)

$$\left[-\frac{1}{2}\nabla^2 + v_s(\mathbf{y})\right]\phi_k(\mathbf{y}) = i\frac{\partial \phi_k(\mathbf{y})}{\partial t}, \tag{7.264}$$

with

$$\rho(\mathbf{y}) = \sum_k |\phi_k(\mathbf{y})|^2. \tag{7.265}$$

The above equations, when solved self-consistently, then *define* the TD Kohn-Sham 'exchange' $v_x(\mathbf{y})$, 'correlation' $v_c(\mathbf{y})$, and 'exchange-correlation' $v_{xc}(\mathbf{y})$ potentials. (Within TD-QDFT, the local effective potential $v_{ee}(\mathbf{y})$ is such that it reproduces both the density $\rho(\mathbf{y})$ and current density $\mathbf{j}(\mathbf{y})$, and it is *explicitly defined* (see (6.149)–(6.152).)

It is worth noting that in their original work, Runge and Gross had also attempted a parallelism with the Second Hohenberg-Kohn Theorem by constructing an action functional of the density $A[\rho; \Psi_0]$. The basis for the construction of the action functional $A[\rho; \Psi_0]$ is the quantum-mechanical action integral

$$A[\Psi] = \int_{t_0}^{t_1} \langle \Psi(t)|i\frac{\partial}{\partial t} - \hat{H}(t)|\Psi(t)\rangle dt. \tag{7.266}$$

For arbitrary variations $\delta\Psi(t)$ together with the constraints $\delta\Psi(t_0) = \delta\Psi(t_1) = 0$, the action is stationary:

$$\delta A[\Psi] = 0. \tag{7.267}$$

The action functional proposed by Runge-Gross was

$$A[\rho; \Psi_0] = \int_{t_0}^{t_1} \langle \Psi[\rho; \Psi_0](t)|i\frac{\partial}{\partial t} - \hat{H}(t)|\Psi[\rho; \Psi_0](t)\rangle dt. \tag{7.268}$$

Unfortunately, the variation of this action functional does not vanish, i.e.

$$\delta A[\rho; \Psi_0] \neq 0. \tag{7.269}$$

As such there is no Euler-Lagrange equation similar to that of the stationary-state case. Consequently, in the mapping to a system of noninteracting fermions, potentials cannot be defined as functional derivatives of this action functional taken with respect to the density $\rho(\mathbf{y})$.

Over the years, there have been many attempts [80–83] at creating an appropriate action functional $A[\rho; \Psi_0]$ that vanishes for arbitrary variations $\delta\rho(\mathbf{y})$ of the density $\rho(\mathbf{y})$. Such an action functional is then rewritten for the model system of noninteracting fermions thereby defining a Hartree $A_{\mathrm{H}}[\rho]$, an electron-interaction $A_{ee}[\rho]$, and 'exchange-correlation' $A_{xc}[\rho]$, 'exchange' $A_x[\rho]$, and 'correlation' $A_c[\rho]$ action functionals. The functional derivatives of these functionals with respect to the density $\rho(\mathbf{y})$ then provide definitions for the corresponding potentials: $v_{\mathrm{H}}(\mathbf{y}) = \delta A_{\mathrm{H}}[\rho]/\delta\rho(\mathbf{y})$, $v_{ee}(\mathbf{y}) = \delta A_{ee}[\rho]/\delta\rho(\mathbf{y})$, $v_{xc}(\mathbf{y}) = \delta A_{xc}[\rho]/\delta\rho(\mathbf{y})$, $v_x(\mathbf{y}) = \delta A_x[\rho]/\delta\rho(\mathbf{y})$, $v_c(\mathbf{y}) = \delta A_c[\rho]/\delta\rho(\mathbf{y})$.

Hence, within TD-DFT, as described above, there are two ways of defining the local effective potential $v_s(\mathbf{y})$ and its components. All the many-body effects are incorporated in the electron-interaction potential $v_{ee}(\mathbf{y})$ but it is not explained how this is accomplished. A description [79] of the correlations associated with $v_{ee}(\mathbf{y})$ and its various components as determined via TD Quantal Density Functional Theory (TD-QDFT) is provided in the following section.

7.7.5 Physical Interpretation of Time-Dependent Kohn-Sham Theory

In TD-DFT, the interacting system with density $\rho(\mathbf{y})$ and current density $\mathbf{j}(\mathbf{y})$, is mapped to one of noninteracting fermions possessing the same density $\rho(\mathbf{y})$. It is assumed that the current density $\mathbf{j}_s(\mathbf{y})$ of the model system is also equivalent to $\mathbf{j}(\mathbf{y})$. In general, this is not the case. This can be seen from the 'Quantal Newtonian' Second Laws for the interacting (2.78) and model systems (6.140) [79]. The respective current-density fields $\mathcal{J}(\mathbf{y}) = (1/\rho(\mathbf{y}))\partial \mathbf{j}(\mathbf{y})/\partial t$ and $\mathcal{J}_s(\mathbf{y}) = (1/\rho(\mathbf{y}))\partial \mathbf{j}_s(\mathbf{y})/\partial t$ are inequivalent. Thus, there exists a Correlation-Current-Density field $\mathcal{J}_c(\mathbf{y}) = \mathcal{J}_s(\mathbf{y}) - \mathcal{J}(\mathbf{y})$ representative of this difference. (There is also a Correlation- Kinetic field $\mathcal{Z}_{t_c}(\mathbf{y})$ due to the difference in kinetic energies.) The current densities $\mathbf{j}_s(\mathbf{y})$ and $\mathbf{j}(\mathbf{y})$ are equivalent only when both the divergence and curl of the Correlation-Current-Density field $\mathcal{J}_c(\mathbf{y})$ vanishes. That $\nabla \cdot \mathcal{J}_c(\mathbf{y}) = 0$ follows directly from the equation of continuity (2.81) as the densities of the two systems are the same. However, in general, $\nabla \times \mathcal{J}_c(\mathbf{y}) \neq 0$. Thus, the TD-DFT mapping to the model system must account for electron correlations due to the Pauli principle, Coulomb repulsion, Correlation-Kinetic, and Correlation-Current-Density effects. These are the correlations that the TD-DFT electron-interaction potential $v_{ee}(\mathbf{y})$ is representative of [79].

The two definitions of the potential $v_{ee}(\mathbf{y})$ described in Sect. 7.7.4 are strictly *mathematical*: the first definition is via a set of self-consistently solved equations; the second definition is as a functional derivative of an unknown action integral functional. There is, however, a third definition derived via TD-QDFT [77–79] which provides a *rigorous physical interpretation* of this potential. In TD-QDFT, as explained in Sect. 6.3, one maps to a model system possessing both the same density $\rho(\mathbf{y})$ and current density $\mathbf{j}(\mathbf{y})$ as that of the interacting system. There are, therefore, no Correlation-Current-Density effects. However, it is also possible to map to a model system of noninteracting fermions possessing *solely* the same density $\rho(\mathbf{y})$. The model systems in TD-DFT and TD-QDFT are then equivalent. The TD-QDFT definition [79] of the potential $v_{ee}(\mathbf{y})$ is that at each instant of time, it is the work done by the model fermion in a conservative effective field $\mathcal{F}^{\text{eff}}(\mathbf{y})$ representative of electron correlations due to the Pauli principle, Coulomb repulsion, Correlation-Kinetic and Correlation-Current-Density effects.

With the potential $v_{ee}(\mathbf{y})$ decomposed into its Hartree $v_{\mathrm{H}}(\mathbf{y})$ and 'exchange-correlation' $v_{xc}(\mathbf{y})$ components, it then becomes evident that $v_{xc}(\mathbf{y})$ is representative of all the above noted correlations. Once again, at each instant of time, the potential $v_{xc}(\mathbf{y})$ is the work done in a conservative field $(\mathcal{F}^{\text{eff}}(\mathbf{y}) - \mathcal{E}_{\mathrm{H}}(\mathbf{y}))$, where $\mathcal{E}_H(\mathbf{y})$ is the Hartree field.

On application of adiabatic coupling constant (λ) perturbation theory, a *rigorous physical interpretation* of the corresponding 'exchange' $v_x(\mathbf{y})$ and 'correlation' $v_c(\mathbf{y})$ potentials in terms of the various electron correlations can be derived. Thus, at each instant of time, the 'exchange' potential $v_x(\mathbf{y})$ is the work done in a conservative field representative of Pauli correlations and lowest-order $O(\lambda)$ Correlation-Kinetic and Correlation-Current-Density effects. The 'correlation' potential $v_c(\mathbf{y})$ commences in second order $O(\lambda^2)$. At each order, the 'correlation' potential is the work done in a conservative field representative of Coulomb correlations, Correlation-Kinetic effects, and Correlation-Current-Density effects of appropriate order. For details of the proofs, the reader is referred to [79].

Thus, *TD Kohn-Sham 'exchange'* is not comprised solely of Pauli correlations. Additionally, there exist lowest-order $O(\lambda)$ Correlation-Kinetic and Correlation-Current-Density effects. And TD Kohn-Sham 'correlation' represents Coulomb correlations and second $O(\lambda^2)$ and higher-order Correlation-Kinetic and Correlation-Current-Density effects.

References

1. P. Hohenberg, W. Kohn, Phys. Rev. **136**, B864 (1964)
2. W. Kohn, L.J. Sham, Phys. Rev. **140**, A1133 (1965) (It was in their *Note added in proof* that Kohn and Sham recognized they had discovered an *exact* local effective potential theory leading to the true ground state density.)
3. X.-Y. Pan, V. Sahni, J. Chem. Phys. **143**, 174105 (2015)
4. E. Runge, E.K.U. Gross, Phys. Rev. Lett. **52**, 997 (1984)
5. R. van Leeuwen, Phys. Rev. Lett. **82**, 3863 (1999)
6. V. Sahni, *Quantal Density Functional Theory*, 2nd edn. (Springer, Berlin, Heidelberg, 2016) (Referred to as ***QDFT***)
7. V. Sahni, *Quantal Density Functional Theory II: Approximation Methods and Applications* (Springer, Berlin, Heidelberg, 2010) (Referred to as ***QDFT2***)
8. R.G. Parr, W. Yang, *Density Functional Theory of Atoms and Molecules, (Oxford University Press, New York, 1989)*
9. R.M. Dreizler, E.K.U. Gross, *Density Functional Theory* (Springer, Berlin, 1990)
10. N.H. March, *Electron Density Theory of Atoms and Molecules* (Academic, London, 1992)
11. E. Engel, R.M. Dreizler, *Density Functional Theory* (Springer, Heidelberg, Berlin, 2011)
12. A. Ullrich, *Time-dependent Density Functional Theory* (Oxford University Press, Oxford, 2012)
13. J.C. Slater, Phys. Rev. **81**, 385 (1951)
14. J.C. Slater, T.M. Wilson, J.H. Wood, Phys. Rev. **179**, 28 (1969)
15. V. Sahni, X.-Y. Pan, Phys. Rev. A **85**, 052502 (2012)
16. The earliest attribution to Wilson of this argument appears in P.O. Lowdin, Int. J. Quant. Chem. S **19**, 19 (1986). The author thanks Sam Trickey for this information. To the best of the author's knowledge, there is no paper by Wilson in which this argument is stated. The reasoning does not appear in E.B. Wilson, in *Structural Chemistry and Biology*, Eds. A. Rich and N. Davidson, (W.H. Freeman, San Francisco, CA (1968))
17. X.-Y. Pan, V. Sahni, J. Chem. Phys. **119**, 7083 (2003)
18. T. Kato, Commum. Pure Appl. Math. **10**, 151 (1957)
19. E. Steiner, J. Chem. Phys. **39**, 2365 (1963)
20. W.A. Bingel, Z. Naturforsch. **18a**, 1249 (1963)
21. R.T. Pack, W.B. Brown, J. Chem. Phys. **45**, 556 (1966)
22. W.A. Bingel, Theoret. Chim. Acta. (Berl) **8**, 54 (1967)
23. J. Percus, Int. J. Quantum Chem. **13**, 89 (1978)
24. M. Levy, Proc. Natl. Acad. Sci. USA **76**, 6062 (1979)
25. E. Lieb, Int. J. Quantum Chem. **24**, 243 (1983)
26. M. Levy, Int. J. Quantum Chem. **110**, 3140 (2010)

27. B.L. Moiseiwitsch, *Variational Principles* (Wiley, London, 1966)
28. U. von Barth, L. Hedin, J. Phys. C **5**, 1629 (1972)
29. K. Capelle, G. Vignale, Phys. Rev. Lett. **86**, 5546 (2001)
30. H. Eschrig, W.E. Pickett, Solid State Commun. **118**, 123 (2001)
31. A. Laestadius, M. Benedicks, Int. J. Quantum Chem. **114**, 782 (2014)
32. M. Taut, J. Phys. A **27**, 1045 (1994); Corrigenda J. Phys. A **27**, 4723 (1994)
33. M. Taut, H. Eschrig, Z. Phys. Chem. **224**, 631 (2010)
34. V. Sahni, X.-Y. Pan, T. Yang, Computation **4**, 30 (2016). https://doi.org/10.3390/computation4030030
35. V. Sahni, J. Comput. Chem. **39**, 1083 (2018). https://doi.org/10.1002/jcc.24888
36. V. Sahni, Int. J. Quantum Chem. **121**, e26556 (2021)
37. O. Gunnarrsson, B. Lundqvist, Phys. Rev. B **13**, 4274 (1976)
38. Y.-Q. Li, X.-Y. Pan, B. Li, V. Sahni, Phys. Rev. A **85**, 032517 (2012)
39. T.L. Gilbert, Phys. Rev. B **12**, 2111 (1975)
40. J.E. Harriman, Phys. Rev. A **24**, 680 (1981)
41. J. Cioslowski, Phys. Rev. Lett. **60**, 2141 (1988)
42. M. Slamet, X.-Y. Pan, V. Sahni, Phys. Rev. A **84**, 052504 (2011)
43. X.-Y. Pan, V. Sahni, Int. J. Quantum Chem. **108**, 2756 (2008)
44. X.-Y. Pan, V. Sahni, Int. J. Quantum Chem. **95**, 381 (2003)
45. X.-Y. Pan, V. Sahni, J. Chem. Phys. **132**, 164116 (2010)
46. M. Levy, J.P. Perdew, Phys. Rev. A **32**, 2010 (1985)
47. M. Levy, N.H. March, Phys. Rev. A **55**, 1885 (1997)
48. Z. Qian, V. Sahni, Int. J. Quantum Chem. **80**, 555 (2000)
49. A. Görling, M. Levy, Phys. Rev. B **47**, 13105 (1993)
50. V. Sahni, J. Gruenebaum, J.P. Perdew, Phys. Rev. B **26**, 4371 (1982)
51. V. Sahni, M. Levy, Phys. Rev. B **33**, 3869 (1986)
52. R.T. Sharp, G.K. Horton, Phys. Rev. **30**, 317 (1953)
53. J.D. Talman, W.F. Shadwick, Phys. Rev. A **14**, 36 (1976)
54. M. Slamet, V. Sahni, Phys. Rev. B **44**, 10921 (1991)
55. M.K. Harbola, V. Sahni, Phys. Rev. Lett. **62**, 489 (1989)
56. V. Sahni, M.K. Harbola, Int. J. Quantum Chem. **24**, 569 (1990)
57. V. Sahni in *Density Functional Theory*, NATO Advanced Study Institute, vol. 337 Series B: Physics, ed. by E.K.U. Gross, R.M. Driezler (Plenum, New York, 1995)
58. V. Sahni in *Recent Advances in Density Functional Methods, Part I*, ed. by D.P. Chong (World Scientific, 1995)
59. V. Sahni, M. Slamet, Phys. Rev. B **48**, 1910 (1993)
60. M. Slamet, V. Sahni, Phys. Rev. B **45**, 4013 (1992)
61. E.I. Tellgren et al., Phys. Rev. A **86**, 062506 (2012)
62. M.M. Pant, A.K. Rajagopal, Solid State Commun. **10**, 1157 (1972)
63. G. Vignale, M. Rasolt, Phys. Rev. Lett. **59**, 2360 (1987)
64. G. Vignale, M. Rasolt, Phys. Rev. B **37**, 10685 (1988)
65. G. Vignale, M. Rasolt, D.J.W. Geldart, Adv. Quantum Chem. **21**, 235 (1990)
66. W. Kohn, A. Savin, C.A. Ullrich, Int. J. Quantum Chem. **100**, 20 (2004)
67. K. Capelle, E.K.U. Gross, Phys. Rev. Lett. **78**, 1872 (1997)
68. S. Rohra, A. Gorling, Phys. Rev. Lett. **97**, 013005 (2006)
69. J.P. Perdew, A. Zunger, Phys. Rev. B **23**, 5048 (1981)
70. A. Laestadius, M. Benedicks, Phys. Rev. A **91**, 032508 (2015)
71. E.I. Tellgren et al. (2017). arXiv: 1709.10400v1 [phys.chem-ph]. Accessed 29 Sep 2017
72. J. Schirmer, A. Dreuw, Phys. Rev. A **75**, 022513 (2007)
73. N.T. Maitra, R. van Leeuwen, K. Burke, Phys. Rev. A **78**, 056501 (2008)
74. J. Schirmer, A. Dreuw, Phys. Rev. A **78**, 056502 (2008)
75. B.-X. Xu, A.K. Rajagopal, Phys. Rev. A **31**, 2682 (1985)
76. A.K. Dhara, S.K. Ghosh, Phys. Rev. A **35**, 442 (1987)
77. Z. Qian, V. Sahni, Phys. Lett. A **247**, 303 (1998)

78. Z. Qian, V. Sahni, Int. J. Quantum Chem. **78**, 341 (2000)
79. Z. Qian, V. Sahni, Phys. Rev. A **63**, 042508 (2001)
80. A.K. Rajagopal, in *Electron Density Functional Theory*, ed. by J.F. Dobson, G. Vignale, M.P. Das (Plenum, New York, 1998)
81. R. van Leeuwen, Phys. Rev. Lett. **80**, 1280 (1998)
82. R. van Leeuven, Int. J. Mod. Phys. B **15**, 1969 (2001)
83. G. Vignale, Phys. Rev. A **77**, 062511 (2008)

Chapter 8
Wave Function Properties

Abstract For a system of electrons in a binding electrostatic field $\mathcal{E}(\mathbf{r}) = -\nabla v(\mathbf{r})/e$ and a magnetic field $\mathcal{B}(\mathbf{r}) = \nabla \times \mathbf{A}(\mathbf{r})$, with $\{v(\mathbf{r}), \mathbf{A}(\mathbf{r})\}$ the scalar and vector potentials, the following properties of the wave function are described. The derivation of the analytical expressions for the electron-electron and electron-nucleus coalescence constraints on the wave function, (both cusp and node), in dimensions $D \geq 2$ valid for arbitrary state is provided. These expressions lead to the original Kato spherically averaged differential form of these coalescence constraints. The differential form of the electron-nucleus coalescence constraint in terms of the spherically averaged electron density about the nucleus, and the nuclear charge, is derived. The constraint on the pair correlation function as two identical particles coalesce, and the corresponding differential form, are also derived. How the electron-nucleus constraint on the wave function governs the structure of the effective electron-interaction potential $v_{ee}(\mathbf{r})$ of Quantal and Kohn-Sham density functional theories (Q-DFT, KS-DFT) at and near the nucleus is described. The contributions of the individual electron correlations due to the Pauli principle, Coulomb repulsion, and Correlation-Kinetic effects to this structure, is explained. The analytical asymptotic structure of the wave function of finite systems in the classically forbidden region for Coulombic binding potentials $v(\mathbf{r})$ and its relationship to the ionization potential, valid for arbitrary state, is derived. A sum rule for the spherical average of the asymptotic density is obtained. The asymptotic structure of the wave function leads via Q-DFT to the analytical asymptotic structure of the potential $v_{ee}(\mathbf{r})$ of local potential theories in terms of the individual electron correlations that contribute to it. It also provides a rigorous physical interpretation of the highest occupied eigenvalue of the Q-DFT and KS-DFT differential equations as being the negative of the ionization potential. A new symmetry operation of the wave function of 2-electron systems with arbitrary and symmetrical binding potential $v(\mathbf{r})$, and arbitrary electron-interaction $w(|\mathbf{r} - \mathbf{r}'|)$, valid for arbitrary state, is described. The symmetry operation is such that the transformed wave function is equal to the wave function, and referred to as the *Wave Function Identity*. The operation involves a switching of the electrons whilst keeping the spin moment unchanged, followed by an inversion. It is proved that the exact wave function satisfies the identity. The application of the Pauli principle to the identity then proves that the parity of all singlet states is even and that of all triplet states is odd. It is concluded that at electron-electron coalescence, all

V. Sahni, *Schrödinger Theory of Electrons: Complementary Perspectives*,
Springer Tracts in Modern Physics 285,
https://doi.org/10.1007/978-3-030-97409-1_8

singlet states satisfy the cusp coalescence constraint, and all triplet states the node condition. It is further concluded that the parity at each point of electron-electron coalescence is even for singlet states and odd for triplet states. These properties, as well as the Pauli principle, are elucidated by application to both the singlet 2^1S and triplet 2^3S states of a $D = 2$, 2-electron semiconductor quantum dot in a magnetic field.

Introduction

A system of N-electrons possessing a spin moment in the presence of a binding electrostatic field $\mathcal{E}(\mathbf{r}) = -\nabla v(\mathbf{r})/e$ and a magnetostatic field $\mathcal{B}(\mathbf{r}) = \nabla \times \mathbf{A}(\mathbf{r})$, with $\{v(\mathbf{r}), \mathbf{A}(\mathbf{r})\}$ the scalar and vector potentials, is described by the stationary-state Schrödinger-Pauli equation. The special case of spinless electrons is the Schrödinger equation. The corresponding eigenfunctions of these eigenvalue differential equations possess certain properties. As closed-form analytical solutions are generally not derivable, any approximation to the wave function must be constrained to ensure the satisfaction of these properties. These constraints on the wave functions are the following: **(a)** must be single valued, smooth, and bounded; **(b)** satisfy the Pauli principle; **(c)** be normalized with probability density ≥ 0; **(d)** satisfy either the *cusp* or *node* electron-electron coalescence condition; **(e)** satisfy the electron-nucleus constraint for binding potentials $v(\mathbf{r})$ that are singular at the nucleus; **(f)** possess the appropriate number of nodes and the correct asymptotic structure in the classically forbidden region; **(g)** have the correct parity.

In this chapter, we derive properties of the wave function that are obtained in analytical form. The significance of these properties to the understanding they bring to local effective potential theories such as Quantal density functional theory (Q-DFT) (Chap. 6) and Kohn-Sham density functional theory (KS-DFT) (Chap. 7) is explained. Finally we describe a new recently discovered symmetry property of bound 2-electron systems which involves an interchange of electrons whilst keeping the spin coordinate unchanged, followed by an inversion.

We begin with the derivation [1] of the analytical expressions for the electron-electron and electron nucleus coalescence constraints on the wave function in dimensions $D \geq 2$. These expressions are more general than the originally derived differential form due to Kato [2] because they retain the angular dependence of the wave function at coalescence. The expressions are also valid for arbitrary state. The differential form of the electron-nucleus coalescence constraint in terms of the spherically averaged electron density [3] about the nucleus, and the nuclear charge, is derived. The constraint on the pair function $P(\mathbf{r}\mathbf{r}')$ (see (2.27)) as two identical particles coalesce, and the corresponding differential form of the constraint, is also derived for $D \geq 2$. The extension to dimensions other than $D = 3$ [2–6] is done because there is interest now in both the lower dimensional $D = 2$ systems [7–13] as well as in their higher dimensional ($D \geq 4$) generalizations [14, 15]. How the electron-nucleus coalescence constraint then governs the near and at the nucleus structure of the local electron-interaction potential $v_{ee}(\mathbf{r})$ of local effective potential theories, and the individual electron correlations that contribute to this structure, is described [16–19].

In the second component to the chapter, we derive the analytical asymptotic structure of the wave function for finite systems in the classically forbidden region for Coulombic binding potentials $v(\mathbf{r})$ [20–23]. The significance of this asymptotic structure of the wave function, and that of the corresponding density, lies in the fact that it is directly related to the ionization potential. This structure is valid for arbitrary state. A sum rule involving the spherical average of the asymptotic density is derived. Hence, knowledge of the asymptotic density identifies the state of excitation of the system. The asymptotic structure of the wave function also provides a rigorous physical interpretation of the highest occupied eigenvalue of the Q-DFT and KS-DFT differential equations as being the removal energy. The asymptotic wave function structure then leads via Q-DFT to the analytical asymptotic structure of the potential $v_{ee}(\mathbf{r})$ of the local potential theories in terms of the various electron correlations that contribute to it (see ***QDFT2***).

Finally, we describe [24] a new symmetry operation O_{sym} about the center of symmetry of 2-electron systems as described by the Schrödinger-Pauli theory equation with *arbitrary* and *even* binding potential $v(\mathbf{r})$, and arbitrary electron-electron interaction of the form $w(|\mathbf{r} - \mathbf{r}'|)$. The symmetry operation is such that the transformed wave function is equal to the wave function. The equality of the wave function to the transformed wave function is referred to as the *Wave Function Identity*. It is proved that the exact wave function of 2-electron systems defined by the Schrödinger-Pauli equation satisfies this identity. The application of the permutation operation P for fermions to the transformed wave function then proves that the parity of all singlet states is *even*, and that of triplet states is *odd*. Thus, the product of the permutation P and symmetry O_{sym} operations is the inversion or parity operation Π. This in turn leads to the conclusion that at electron-electron coalescence, the singlet states satisfy the *cusp* coalescence constraint of the wave function, whereas triplet states satisfy the *node* coalescence condition. Thus, it is concluded that the *parity about each point of electron-electron coalescence* in configuration space is *even* for singlet states and *odd* for triplet states. All these properties, together with the Pauli principle, are then elucidated by application to the singlet 2^1S and triplet 2^3S states of a $D = 2$, 2-electron 'artificial atom' or semiconductor quantum dot in a magnetic field. For these states of the 'artificial atom', the *exact* closed-form analytical solutions of the corresponding Schrödinger-Pauli equations have been obtained [25–27]. (See Chap. 9 for the derivations). As such the wave function properties are exhibited *exactly*.

8.1 Coalescence Constraints in Dimensions $D \geq 2$

Consider a system of N spinless electrons of mass m and charge $(-e)$ in an external electrostatic field $\mathcal{E}(\mathbf{r}) = -\nabla v(\mathbf{r})/e$. The Hamiltonian $\hat{H}$ of the system is then

$$\hat{H} = \hat{T} + \hat{U} + \hat{V}, \tag{8.1}$$

where the kinetic energy operator $\hat{T}$ is

$$\hat{T} = -\frac{\hbar^2}{2m} \sum_i \nabla_i^2, \tag{8.2}$$

the Coulomb electron-interaction potential operator $\hat{U}$ is

$$\hat{U} = \frac{1}{2} \sum_{i,j}{}' u(|\mathbf{r}_i - \mathbf{r}_j|); \quad u(|\mathbf{r}_i - \mathbf{r}_j|) = -e^2/|\mathbf{r}_i - \mathbf{r}_j|, \tag{8.3}$$

and the external potential operator $\hat{V}$ is

$$\hat{V} = \sum_i v(\mathbf{r}_i). \tag{8.4}$$

The corresponding Schrödinger equation is

$$\hat{H}\psi(\mathbf{R}) = E\psi(\mathbf{R}), \tag{8.5}$$

where $\{\psi(\mathbf{R}), E\}$ are the eigenfunctions and eigenvalues, and $\mathbf{R} = \mathbf{r}_1, \ldots, \mathbf{r}_N$, the coordinates of the electrons.

When two electrons coalesce, i.e. when $\mathbf{r}_{ij} \to 0$ with $r_{ij} = |\mathbf{r}_i - \mathbf{r}_j|$, the electron-interaction operator $\hat{U}$ is singular. If the external field $\mathcal{E}(\mathbf{r})$ is due to a nucleus of charge Ze at the origin, then the potential energy of an electron in the field of the nucleus $v(\mathbf{r}) = -Ze^2/r$. When an electron coalesces with the nucleus, i.e. when $r \to 0$, the external potential operator $\hat{V}$ is singular. These singularities are, respectively, *the electron-electron and electron-nucleus singularities*. For the wave function $\psi(\mathbf{R})$ to be bounded and remain finite at these singularities, it must satisfy coalescence constraints. There are two types of coalescence constraints. One is when the wave function $\psi(\mathbf{R})$ satisfies a *cusp coalescence condition*. This provides the structure of the wave function in the limit as two particles coalesce. The other is when the wave function vanishes at the singularity which is referred to as the *node coalescence condition*.

We next derive [1] the general expression for the wave function $\psi(\mathbf{R})$ at the coalescence of two particles of arbitrary charge and mass thereby encompassing both the above types of singularities. Hence, consider a system of N charged particles in dimensions $D \geq 2$ space with the Hamiltonian $\hat{H}(\hbar = m = 1)$:

$$\hat{H} = -\sum_{i=1}^{N} \frac{1}{2m_i} \nabla_i^2 + \sum_{j>i=1}^{N} \frac{Z_i Z_j}{\mathbf{r}_{ij}}, \tag{8.6}$$

where m_i and Z_i are the mass and charge of the ith particle, and $r_{ij} = |\mathbf{r}_i - \mathbf{r}_j|$.

In D-dimension space, $\mathbf{r} = (x_1, x_2, \ldots, x_D)$, $\mathbf{r} = \sqrt{\sum_{k=1}^{D} x_k x_k}$ and $\nabla^2 = \sum_{k=1}^{D} (\partial/\partial x_k)(\partial/\partial x_k)$. Due to the Coulomb potential energy term, the Hamiltonian is singular when two particles i and j coalesce ($r_{ij} \to 0$). For the wave function $\psi(\mathbf{r}_1, \mathbf{r}_2, \ldots, \mathbf{r}_N)$ which satisfies the Schrödinger equation

$$\hat{H}\psi(\mathbf{r}_1, \mathbf{r}_2, \ldots, \mathbf{r}_N) = E\psi(\mathbf{r}_1, \mathbf{r}_2, \ldots, \mathbf{r}_N) \tag{8.7}$$

to be bounded and remain finite at the singularities, it must satisfy either a cusp or node coalescence constraint. (Here we suppress the spin index.)

We are interested in the form of the wave function when two particles approach each other, i.e., when r_{ij} is very small. Following Pack and Byers Brown [5] we focus our attention on two particles 1 and 2, and transform their coordinates $\mathbf{r}_1$ and $\mathbf{r}_2$ to the center of mass $\mathbf{R}_{12}$ and relative coordinates $\mathbf{r}_{12}$ as

$$\mathbf{R}_{12} = \frac{m_1 \mathbf{r}_1 + m_2 \mathbf{r}_2}{m_1 + m_2}, \tag{8.8}$$

$$\mathbf{r}_{12} = \mathbf{r}_1 - \mathbf{r}_2. \tag{8.9}$$

The Hamiltonian of (8.6) may then be rewritten as

$$\hat{H} = -\frac{1}{2\mu_{12}} \nabla_{\mathbf{r}_{12}}^2 + \frac{Z_1 Z_2}{r_{12}} - \frac{1}{2(m_1 + m_2)} \nabla_{\mathbf{R}_{12}}^2 + \sum_{i=3}^{N} \left\{ Z_i \left(\frac{Z_1}{r_{1i}} + \frac{Z_2}{r_{2i}} \right) \right\} + \sum_{i=3}^{N} \frac{1}{2m_i} \nabla_{\mathbf{r}_i}^2 + \sum_{j>i=3}^{N} \frac{Z_i Z_j}{r_{ij}}, \tag{8.10}$$

where $\mu_{12} = m_1 m_2/(m_1 + m_2)$ is the reduced mass of particles 1 and 2. When particles 1 and 2 are within a small distance of each other ($0 < r_{12} < \epsilon$), and all other particles are well separated, there is only one singularity in the Hamiltonian. Retaining only terms of lower order in r_{12}, (8.7) reduces to

$$\left[-\frac{1}{2\mu_{12}} \nabla_{\mathbf{r}_{12}}^2 + \frac{Z_1 Z_2}{r_{12}} + O(\epsilon^0) \right] \psi\,(\mathbf{r}_1, \mathbf{r}_2, \ldots, \mathbf{r}_N) = 0, \tag{8.11}$$

where $O(\epsilon^0)$ refers to terms of order zero (constant) and higher order in r_{12} and the vector components $(\mathbf{r}_{12})_k$, $k = 1, 2, \ldots, D$.

Equation (8.11) is a one-electron-atom equation in D-dimension space. Furthermore, it is not an eigenvalue equation. We are interested in the solution of the equation that is finite and continuous at the singularities. Thus, in the limit as $\mathbf{r}_1 \to \mathbf{r}_2$, we could write the wave function in two parts as

$$\psi(\mathbf{r}_1, \mathbf{r}_2, \ldots, \mathbf{r}_N) = \psi(\mathbf{r}_2, \mathbf{r}_2, \mathbf{r}_3, \ldots, \mathbf{r}_N) \\ +\delta\psi(\mathbf{r}_1, \mathbf{r}_2, \ldots, \mathbf{r}_N), \tag{8.12}$$

where the term $\delta\psi(\mathbf{r}_1, \mathbf{r}_2, \ldots, \mathbf{r}_N)$ must vanish at the singularity $\mathbf{r}_1 = \mathbf{r}_2$. From the above it follows that we need consider only terms of first order in r_{12} in $\delta\psi$ when $\mathbf{r}_1$ is near $\mathbf{r}_2$. Dropping the subscript 12 and writing r_{12} as r, we see that $\delta\psi$ could have terms of the form $r_D B(\mathbf{r}_2, \mathbf{r}_3, \ldots, \mathbf{r}_N)$, or $\sum_l r_l B_\ell(\mathbf{r}_2, \ldots, \mathbf{r}_N)$ where l denotes the 2 to $(D-1)$-dimensional subspace and r_ℓ the distance in the subspace, or $\mathbf{r} \cdot \mathbf{C}(\mathbf{r}_2, \ldots, \mathbf{r}_N)$. Here, $r_D = r$ is just the conventional distance in D dimensional space. As an example of the second kind of terms, consider the case of $D = 4$. There are four three-dimensional subspaces, and six two-dimensional subspaces. The distances in the three-dimensional subspaces constitute the terms $\sqrt{x_1^2 + x_2^2 + x_3^2}$, $\sqrt{x_1^2 + x_2^2 + x_4^2}$, $\sqrt{x_2^2 + x_3^2 + x_4^2}$, and $\sqrt{x_1^2 + x_4^2 + x_3^2}$. The distances in the two-dimensional subspaces are $\sqrt{x_i^2 + x_j^2}$, where $i \neq j$, and $i, j = 1, 2, 3, 4$. In the limit $r \to 0$, such terms do not effect the $O(D)$ symmetry of the first two terms on the left-hand side of (8.11). However, the Laplacian of these terms is singular since $\nabla^2 r_d = (d-1)/r$, where r_d is the distance in the d-dimensional subspace. These singularities cannot be canceled by the Coulomb potential energy singularity term of (8.11). Thus, their coefficients must vanish. Therefore (recovering the subscript 12), (8.12) must be written as

$$\psi(\mathbf{r}_1, \mathbf{r}_2, \ldots, \mathbf{r}_N) = \psi(\mathbf{r}_2, \mathbf{r}_2, \mathbf{r}_3, \ldots, \mathbf{r}_N) \\ + r_{12} B(\mathbf{r}_2, \mathbf{r}_3, \ldots, \mathbf{r}_N) \\ + \mathbf{r}_{12} \cdot \mathbf{C}(\mathbf{r}_2, \mathbf{r}_3, \ldots, \mathbf{r}_N) + O(\epsilon^2). \tag{8.13}$$

Next substitute (8.13) for $\psi(\mathbf{r}_1, \mathbf{r}_2, \ldots, \mathbf{r}_N)$ into (8.11) and employ $\nabla^2 \mathbf{r} = 0$. For the Coulomb potential energy singularity to be canceled we have

$$-\frac{1}{2\mu_{12}} \frac{D-1}{r_{12}} B(\mathbf{r}_2, \mathbf{r}_3, \ldots, \mathbf{r}_N) \\ + \frac{Z_1 Z_2}{r_{12}} \psi(\mathbf{r}_2, \mathbf{r}_2, \mathbf{r}_3, \ldots, \mathbf{r}_N) = 0. \tag{8.14}$$

At the point of coalescence, the wave function is $\psi(\mathbf{r}_2, \mathbf{r}_2, \mathbf{r}_3, \ldots, \mathbf{r}_N)$. Since we require the wave function to be finite at this point we have

$$B(\mathbf{r}_2, \mathbf{r}_3, \ldots, \mathbf{r}_N) = \frac{2 Z_1 Z_2 \mu_{12}}{D-1} \psi(\mathbf{r}_2, \mathbf{r}_2, \mathbf{r}_3, \ldots, \mathbf{r}_N). \tag{8.15}$$

Thus, in the limit as $r_{12} \to 0$, we may write the wave function as

$$\psi(\mathbf{r}_1, \mathbf{r}_2, \ldots, \mathbf{r}_N) = \psi(\mathbf{r}_2, \mathbf{r}_2, \mathbf{r}_3, \ldots, \mathbf{r}_N)\left(1 + \frac{2Z_1 Z_2 \mu_{12}}{D-1} r_{12}\right) + \mathbf{r}_{12} \cdot \mathbf{C}(\mathbf{r}_2, \mathbf{r}_3, \ldots, \mathbf{r}_N). \tag{8.16}$$

This is the general form of the wave function in D-dimensional space as the two charged particles coalesce. When the wave function at the point of coalescence i.e. $\psi(\mathbf{r}_2, \mathbf{r}_2, \mathbf{r}_3, \ldots, \mathbf{r}_N)$ is finite, then the expression (8.16) is referred to as the *cusp coalescence condition*. The expression is equally valid if the wave function at the point of coalescence vanishes, i.e. $\psi(\mathbf{r}_2, \mathbf{r}_2, \mathbf{r}_3, \ldots, \mathbf{r}_N) = 0$. It is then referred to as the *node coalescence condition*. (See Sect. 8.3 for examples.)

The coalescence constraint of (8.16) is also applicable to electrons with spin in the presence of a magnetic field $\mathcal{B}(\mathbf{r})$.

For the case of electron-nucleus coalescence, $Z_1 = -e$, $Z_2 = Ze$ the nuclear charge, $\mu_{12} \approx m_e$ the mass of the electron. For electron-electron coalescence, $Z_1 = -e$, $Z_2 = -e$, and $\mu_{12} = \frac{1}{2} m_e$.

By taking the spherical average of the wave function of (8.16) about the point of coalescence in D-dimension space, one obtains the *differential form* of the coalescence condition:

$$\left.\frac{\partial \psi_{sp.avg}}{\partial r_{12}}\right|_{r_{12} \to 0} = \frac{2Z_1 Z_2 \mu_{12}}{D-1} \psi(r_{12} = 0), \tag{8.17}$$

where

$$\psi_{sp.avg}(r_{12}) = \frac{1}{4\pi} \int \psi d\Omega_{r_{12}}. \tag{8.18}$$

The differential form was originally derived for $D = 3$ by Kato [2]. It too is valid for both electron-electron and electron-nucleus coalescence.

The differential form of the coalescence condition (8.17) is employed as a required constraint on any approximate wave function [28]. It is also employed [29] when the *electron-nucleus* coalescence condition is expressed in terms of the *electron density* $\rho(\mathbf{r})$ as derived in the following Sect. 8.2. (The electron-electron coalescence condition cannot be so written in terms of the density in differential form.)

The coalescence conditions as expressed by the expressions for the wave function $\psi(\mathbf{r}_1, \ldots, \mathbf{r}_N)$ of (8.16) is more general, and therefore more significant, because it retains the angular dependence of the wave function at coalescence. Further, many properties of a physical system can be derived from this form. As an example, consider two bound electrons with a spin moment, but with no spin-orbit coupling. The corresponding wave function is then of the form

$$\Psi(\mathbf{x}_1, \mathbf{x}_2) = \Psi(\mathbf{r}_1\sigma_1, \mathbf{r}_2\sigma_2) = \psi(\mathbf{r}_1, \mathbf{r}_2)\chi(\sigma_1, \sigma_2), \tag{8.19}$$

$\mathbf{x} = \mathbf{r}\sigma$, the spatial and spin coordinates and $\chi(\sigma_1, \sigma_2)$ the spin component of the wave function. For electron-electron coalescence, the wave function in dimensions $D \geq 2$ is from (8.16)

$$\psi(\mathbf{r}_1, \mathbf{r}_2) = \psi(\mathbf{r}_2, \mathbf{r}_2)[1 + (e^2 m_e/D - 1)r_{12}] + \mathbf{r}_{12} \cdot \mathbf{C}(\mathbf{r}_2). \tag{8.20}$$

From this expression, it is possible to understand why it is that all singlet states satisfy a cusp coalescence condition whereas all triplet states satisfy a node coalescence condition. For a singlet state (with a wave function of the form of (8.19)), the two electrons have opposite spin. Hence, there is a finite (positive-definite) probability of the two electrons being at the same spatial position. That is, at coalescence, $\psi(\mathbf{r}_2, \mathbf{r}_2)$ on the right hand side of (8.20) is *finite*. On the other hand, for the triplet state, the electron spins are parallel. As a consequence of the Pauli principle, the probability of two electrons of parallel spin being at the same physical position is zero. Thus, $\psi(\mathbf{r}_2, \mathbf{r}_2) = 0$ in (8.20), and the wave function vanishes at coalescence. (These conclusions will be seen to also be a consequence of the symmetry of the two-electron wave function as shown in Sect. 8.3.)

8.1.1 Differential Form of Electron-Nucleus Coalescence Condition in Terms of the Density

The differential form of the electron-nucleus coalescence condition as written in terms of the spherically averaged wave function of (8.17) may be expressed in terms of the derivative of the density and the density at the nucleus [3]. In this case with the nuclear positions fixed, the wave function does not depend upon the nuclear coordinates. Then when the electron at $\mathbf{r}$ approaches a nucleus, say fixed at the origin, the integral in the definition of the density which is over all the remaining $(N - 1)$ electrons can be performed.

The density $\rho(\mathbf{r})$ is defined as (2.13)

$$\rho(\mathbf{r}) = N \sum_{\sigma} \int \psi^{\star}(\mathbf{r}\sigma, \mathbf{X}^{N-1}) \psi(\mathbf{r}\sigma, \mathbf{X}^{N-1}) d\mathbf{X}^{N-1}. \tag{8.21}$$

On taking the derivative in the limit of the electron-nucleus coalescence we have

$$\begin{aligned} \lim_{r\to 0} \frac{d\rho(\mathbf{r})}{dr} = N \sum_{\sigma} \int \Bigg\{ & \frac{d\psi^{*}(\mathbf{r}\sigma, \mathbf{X}^{N-1})}{dr}\Big|_{r\to 0} \psi\left(\mathbf{r} = 0, \sigma; \mathbf{X}^{N-1}\right) \\ & + \psi^{*}(\mathbf{r} = 0, \sigma; \mathbf{X}^{N-1}) \frac{d\psi(\mathbf{r}\sigma, \mathbf{X}^{N-1})}{dr}\Big|_{r\to 0} \Bigg\} d\mathbf{X}^{N-1}. \end{aligned} \tag{8.22}$$

Integrating the previous equation over the angular variables of the coalescing electron we obtain

$$\lim_{r\to 0} \frac{d\rho(r)}{dr} = N \sum_{\sigma} \int \Bigg\{ \frac{d\psi^{*}_{sp.av}(r\sigma, \mathbf{X}^{N-1})}{dr}\Big|_{r\to 0} \psi(r = 0, \sigma; \mathbf{X}^{N-1})$$

$$+\psi^*(r=0,\sigma;\mathbf{X}^{N-1}\frac{d\psi_{\text{sp.av}}(r\sigma,\mathbf{X}^{N-1})}{dr}\Big|_{r\to 0}\Big\}d\mathbf{X}^{N-1}. \tag{8.23}$$

On substituting the differential form of the coalescence condition of (8.17) we have

$$\lim_{r\to 0}\frac{d\rho(\mathbf{r})}{dr}=-\frac{4Ze^2m_e}{D-1}N\sum_{\sigma}\int\psi^\star(\mathbf{r}=0,\sigma;\mathbf{X}^{N-1})\psi(\mathbf{r}=0,\sigma;\mathbf{X}^{N-1})d\mathbf{X}^{N-1}. \tag{8.24}$$

The electron-nucleus coalescence condition in terms of the density is then

$$\lim_{r\to 0}\frac{d\rho(\mathbf{r})}{dr}=-\frac{4Ze^2m_e}{D-1}\rho(r=0). \tag{8.25}$$

Thus, the densities in atoms. molecules, and solids exhibit a cusp at the nuclei.

It is evident from the coalescence condition (8.16) and the above definition of the density (8.21), that *there can be no such differential form of the electron-electron coalescence condition in terms of the density.*

8.1.2 Coalescence Constraints for the Pair-Correlation Function

It is also possible to derive the coalescence condition for the pair-correlation function of (2.27) for electron-electron coalescence. The pair function $P(\mathbf{r}_1\mathbf{r}_2)$ is defined as

$$P(\mathbf{r}_1\mathbf{r}_2)=N(N-1)\int\psi^\star(\mathbf{r}_1,\mathbf{r}_2,\ldots,\mathbf{r}_N)$$
$$\times\psi(\mathbf{r}_1,\mathbf{r}_2,\ldots,\mathbf{r}_N)d\mathbf{r}_3,\ldots,d\mathbf{r}_N. \tag{8.26}$$

In the limit as $\mathbf{r}_1\to\mathbf{r}_2$ and employing the coalescence expression of (8.16) we have that to first order in r_{12},

$$P(\mathbf{r}_1\mathbf{r}_2)=P(\mathbf{r}_2\mathbf{r}_2)\left(1+\frac{2Z_1Z_2\mu_{12}}{D-1}r_{12}\right)^2$$
$$+N(N-1)\int(\mathbf{r}_{12}\cdot\mathbf{C}(\mathbf{r}_2,\ldots,\mathbf{r}_N))^2d\mathbf{r}_3\ldots d\mathbf{r}_N$$
$$+2N(N-1)\left(1+\frac{2Z_1Z_2\mu_{12}}{D-1}r_{12}\right)$$
$$\times\int(\mathbf{r}_{12}\cdot\mathbf{C}(\mathbf{r}_2,\ldots,\mathbf{r}_N))\psi(\mathbf{r}_2,\mathbf{r}_2,\ldots,\mathbf{r}_N)$$
$$\times d\mathbf{r}_3\ldots d\mathbf{r}_N+O(\epsilon^3). \tag{8.27}$$

The cusps and node coalescence condition of the pair function $P(\mathbf{r}_1\mathbf{r}_2)$ are reflected in the pair-correlation density $g(\mathbf{r}_1\mathbf{r}_2) = P(\mathbf{r}_1\mathbf{r}_2)/\rho(\mathbf{r}_1)$. (For examples of the cusps in $g(\mathbf{r}_1\mathbf{r}_2)$ for various positions of the electron at $\mathbf{r}_1$, see the example of the singlet 2^1S state of a 2-electron quantum dot in a magnetic field described in Chap. 6. For examples of the node coalescence condition exhibited by $g(\mathbf{r}_1\mathbf{r}_2)$, see the example of the triplet 2^3S state of a 2-electron quantum dot in a magnetic field as described in Chap. 5.)

The differential form of the pair-function $P(\mathbf{r}_1\mathbf{r}_2)$ is

$$\left.\frac{dP_{sp.avg}(\mathbf{r}_1\mathbf{r}_2)}{dr_{12}}\right|_{r_{12}\to 0} = \frac{4Z_1Z_2\mu_{12}}{D-1}P_{sp.avg}(\mathbf{r}_1\mathbf{r}_2), \tag{8.28}$$

where $P_{sp.avg}(\mathbf{r}_1\mathbf{r}_2)$ is the spherical average of the pair function $P(\mathbf{r}_1\mathbf{r}_2)$. The differential coalescence conditions derived for the *uniform electron gas* in dimensions $D = 2, 3$ [30–34] constitute a special case of the general form of (8.28).

8.1.3 Significance of Electron-Nucleus Coalescence Constraint to Local Effective Potential Theories

The electron-nucleus condition of (8.16) plays a significant role [16–19] in local effective potential theories such as Quantal density functional theory (Q-DFT) (Chap. 6) and Kohn-Sham density functional theory (KS-DFT) (Chap. 7). For the description of spinless electrons as defined by the Hamiltonian $\hat{H}$ of (8.1), the basic idea underlying these theories is the mapping from the interacting system described by the Schrödinger equation of (8.5) to one of *noninteracting fermions* (the S system) possessing the *same* density $\rho(\mathbf{r})$, with the density defined as in (8.21). (In KS-DFT, one is restricted to ground state densities.) The corresponding Schrödinger eigenvalue equation for the model fermions is

$$\big[-\frac{1}{2}\nabla^2 + v(\mathbf{r}) + v_{ee}(\mathbf{r})\big]\phi_i(\mathbf{r}) = \epsilon_i\phi_i(\mathbf{r})\ ; \quad i = 1, \ldots, N, \tag{8.29}$$

with $\{\phi_i(\mathbf{r}), \epsilon_i\}$ the orbitals and eigenvalues. Here $v(\mathbf{r}) = \sum_\alpha Z/|\mathbf{r} - \mathbf{R}_\alpha|$ is the potential energy of the electron ($e = 1$) due to the external charge Z of the nuclei at $\mathbf{R}_\alpha$. All the many-body effects are incorporated in the local electron-interaction potential $v_{ee}(\mathbf{r})$. Within Q-DFT, the potential $v_{ee}(\mathbf{r})$ is the work done in a conservative effective field $\mathcal{F}^{\text{eff}}(\mathbf{r})$:

$$v_{ee}(\mathbf{r}) = -\int_\infty^{\mathbf{r}} \mathcal{F}^{\text{eff}}(\mathbf{r}')\cdot d\boldsymbol{\ell}', \tag{8.30}$$

where $\mathcal{F}^{\text{eff}}(\mathbf{r})$ is the sum of the electron-interaction $\mathcal{E}_{ee}(\mathbf{r})$ and Correlation-Kinetic $\mathcal{Z}_{t_c}(\mathbf{r})$ fields (see Chap. 6 for the definitions of the fields):

$$\mathcal{F}^{\mathrm{eff}}(\mathbf{r}) = \mathcal{E}_{ee}(\mathbf{r}) + \mathcal{Z}_{t_c}(\mathbf{r}). \tag{8.31}$$

(Note that the expression for $v_{ee}(\mathbf{r})$ of (8.30) is valid for arbitrary state of the interacting system. In Q-DFT, the state of the model system too is arbitrary.) In KS-DFT, the potential is defined as the functional derivative of the KS electron-interaction $E_{ee}^{\mathrm{KS}}[\rho]$ functional of the density $\rho(\mathbf{r})$ (see Chap. 7 for the definition of the functional):

$$v_{ee}(\mathbf{r}) = \delta E_{ee}^{\mathrm{KS}}[\rho]/\delta\rho(\mathbf{r}). \tag{8.32}$$

The electron density $\rho(\mathbf{r})$ is then given in terms of the orbitals $\phi_i(\mathbf{r})$ as the expectation value of the density operator $\hat{\rho}(\mathbf{r})$ of (2.12):

$$\rho(\mathbf{r}) = \sum_i |\phi_i(\mathbf{r})|^2. \tag{8.33}$$

The relevance of the electron-nucleus constraint to local effective potential theory is that it helps prove the value of the electron-interaction potential $v_{ee}(\mathbf{r})$ at the nucleus to be finite. This result is valid for arbitrary system whether it be atomic, molecular, or solid state, and for arbitrary state and symmetry.

(Historically, there was controversy as to the value of the potential $v_{ee}(\mathbf{r})$ at the nucleus of atoms [35]. In certain work it was shown to be finite. In others it was singular. The structure of $v_{ee}(\mathbf{r})$ at and near the nucleus was resolved by the work of Pan and Sahni [16] and Qian and Sahni [17, 18]. Here we present the Pan-Sahni proof that the potential $v_{ee}(\mathbf{r})$ is finite at the nucleus.)

Proof Let us consider the case of dimension $D = 3$. For 3-dimensional space, the electron-nucleus coalescence constraint of (8.16) reduces to $(e = 1)$

$$\begin{aligned}\psi(\mathbf{r}, \mathbf{r}_2, \ldots, \mathbf{r}_N) = \psi(0, \mathbf{r}_2, \ldots, \mathbf{r}_N)(1 - Zr) \\ + \mathbf{r} \cdot \mathbf{a}(\mathbf{r}_2, \ldots, \mathbf{r}_N) + O(r^2),\end{aligned} \tag{8.34}$$

where the vectors $\mathbf{r}_i$ now represent the position of the electrons from a particular nucleus. The vector $\mathbf{a}(\mathbf{r}_2, \ldots, \mathbf{r}_N)$ is undetermined. The derivation of the cusp condition does not involve any boundary conditions far from the nucleus. Hence, it is valid at the nuclei of atoms, molecules, and periodic, and aperiodic solids.

Substitution of (8.34) into the definition (2.12) or (8.21) for the density leads to

$$\begin{aligned}\rho(\mathbf{r}) = \rho(0)\Bigg[(1 - Zr)^2 + 2(1 - Zr)\sum_{k=1}^{3} B_k r_k \\ + \sum_{l=1,m=1}^{3} r_l r_m A_{lm} + \cdots\Bigg],\end{aligned} \tag{8.35}$$

where $B_k = \int a_k \psi(0, \mathbf{r}_2, \ldots, \mathbf{r}_N)^{-1} d\mathbf{r}_2 \ldots d\mathbf{r}_N$ and $A_{lm} = \int a_l a_m \psi(0, \mathbf{r}_2, \ldots, \mathbf{r}_N)^{-2} d\mathbf{r}_2 \ldots d\mathbf{r}_N$ are constants, and r_k, a_k, etc., components of the vector $\mathbf{r}$ and $\mathbf{a}$. It follows then that

$$\sqrt{\rho(\mathbf{r})} = \sqrt{\rho(0)}\Bigg[(1 - Zr)^2 + 2(1 - Zr)\sum_{k=1}^{3} B_k r_k + \sum_{l=1,m=1}^{3} r_l r_m A_{lm} + \cdots\Bigg]^{\frac{1}{2}}$$
$$= \sqrt{\rho(0)}\big[1 - Zr + \boldsymbol{B}\cdot\mathbf{r} + O(r^2)\big], \tag{8.36}$$

where in the second step we have retained only the terms of $O(r)$.

Inventing the differential equation (8.29), we obtain for any occupied orbital $\phi_i(\mathbf{r})$ the expression for the electron-interaction potential $v_{ee}(\mathbf{r})$ as

$$v_{\text{ee}}(\mathbf{r}) = \epsilon_i + \frac{\nabla^2\phi_i(\mathbf{r})}{2\phi_i(\mathbf{r})} - v(\mathbf{r}). \tag{8.37}$$

Next, we rewrite the orbitals $\phi_i(\mathbf{r})$ as

$$\phi_i(\mathbf{r}) = \sqrt{\rho(\mathbf{r})}c_i(\mathbf{r}), \quad i = 1, 2, \ldots, N, \tag{8.38}$$

where the coefficients $c_i(\mathbf{r})$ satisfy

$$\sum_{i=1}^{N} c_i^2(\mathbf{r}) = 1. \tag{8.39}$$

This definition of the $\phi_i(\mathbf{r})$ is consistent with the density $\rho(\mathbf{r})$ written in terms of the orbitals (see(8.33)). Expanding the coefficient $c_i(\mathbf{r})$ about the nucleus we obtain

$$c_i(\mathbf{r}) = c_i(0) + \nabla c_i(0)\cdot\mathbf{r} + O(r^2) = c_i(0)\left[1 + \mathbf{D}\cdot\mathbf{r} + O(r^2)\right], \tag{8.40}$$

where $\mathbf{D} = \nabla c_i(0)/c_i(0)$ is some constant vector. Inserting (8.36) and (8.40) into (8.38), we obtain the expression for the orbitals $\phi_i(\mathbf{r})$ near the nucleus as

$$\phi_i(\mathbf{r}) = \sqrt{\rho(0)}c_i(0)\left[1 - Zr + (\mathbf{B} + \mathbf{D})\cdot\mathbf{r} + O(r^2)\right]. \tag{8.41}$$

Now the expression for the external potential energy near the nucleus is

$$v(\mathbf{r}) = -\frac{Z}{r} - \sum_{\alpha}{}' \frac{Z}{|\mathbf{r} - \mathbf{R}_\alpha|}, \tag{8.42}$$

where the sum is over all the other nuclei. At the nucleus, the term $-(Z/r)$ is singular, whereas the other terms are constants. From (8.37) it is evident that this singularity must be canceled by the $[\nabla^2\phi_i(\mathbf{r})/2\phi_i(\mathbf{r})]$ term.

Consider the term $[\nabla^2\phi_i(\mathbf{r})/2\phi_i(\mathbf{r})]$ near the nucleus with $\phi_i(\mathbf{r})$ given by (8.41). We have $\nabla^2 r = 2/r$ and $\nabla^2(\mathbf{B}+\mathbf{D})\cdot\mathbf{r} = 0$. After acting by ∇^2 and taking the limit as $r \to 0$, terms of $O(r^2)$ lead to constants while higher-order terms vanish. Thus, near the nucleus, the term $[\nabla^2\phi_i(\mathbf{r})/2\phi_i(\mathbf{r})]$ is $-(Z/r)$ plus some constant, and therefore in this limit the singularity of the external potential energy is canceled. Therefore $v_{\mathrm{ee}}(0)$ is finite.

In the Notes below, we summarize two additional examples of the significance of the electron-nucleus coalescence condition to local effective potential theory.

Note 1 Finiteness at Nucleus of Electron-Interaction Potential $v^B_{ee}(\mathbf{r})$ for Model Bosonic System

As noted in Chap. 6 on Q-DFT, it is possible to map an interacting system of electrons as defined by the Hamiltonian $\hat{H}$ of (8.1) to one of *noninteracting bosons* possessing the *same* density $\rho(\mathbf{r})$ (see Sect. 6.2.5). The corresponding model B system differential equation is for the density amplitude $\sqrt{\rho(\mathbf{r})}$, and is

$$\left[-\frac{1}{2}\nabla^2 + v(\mathbf{r}) + v^B_{ee}(\mathbf{r})\right]\sqrt{\rho(\mathbf{r})} = \mu\sqrt{\rho(\mathbf{r})}, \tag{8.43}$$

where $v^B_{ee}(\mathbf{r})$ is the effective electron-interaction potential energy in which all the many-body effects are incorporated. Within Q-DFT, as was the case for the S system, the potential $v^B_{ee}(\mathbf{r})$ is the work done in a conservative effective field $\mathcal{F}^{\mathrm{eff}}_B(\mathbf{r})$. In KS-DFT the potential is defined as the functional derivative of the energy functional $E^B_{ee}[\rho]$. The single eigenvalue μ is the chemical potential or the negative of the ionization energy. Inverting the B system differential equation, we have the potential energy $v^B_{ee}(\mathbf{r})$ to be

$$v^B_{ee}(\mathbf{r}) = \mu + \frac{\nabla^2\sqrt{\rho(\mathbf{r})}}{2\sqrt{\rho(\mathbf{r})}} - v(\mathbf{r}). \tag{8.44}$$

The proof of the finiteness of the B system potential $v^B_{ee}(\mathbf{r})$ at the nucleus may be derived in two separate ways. The first is to substitute (8.36) into the expression for $v^B_{ee}(\mathbf{r})$ of (8.44) which then leads to the result that $v^B_{ee}(0)$ is finite. Equivalently, this result may also be arrived at as a special case of the S system proof above for which $c_i = 1/\sqrt{N}$.

Note 2 Asymptotic Structure of Electron-Interaction Potential $v_{ee}(\mathbf{r})$ in Terms of Electron Correlations

Employing the electron-nucleus coalescence constraint, it can be shown [17] that for spherically symmetric atoms or sphericalized systems as in the central field approximation, the structure of the potential $v_{ee}(\mathbf{r})$ of (8.29) in the asymptotic $r \to 0$ limit is

$$v_{ee}(\mathbf{r}) = \alpha + \beta r + \gamma r^2, \tag{8.45}$$

where α, β, and γ are constants.

By determining the various fields within Q-DFT in the asymptotic $r \to 0$ region, the following can be shown [18]:

(a) The correlations due to the Pauli principle, Coulomb repulsion and Correlation-Kinetic effects *all* contribute to the coefficient $\alpha = v_{ee}(0)$.

(b) The Pauli and Coulomb correlations do not contribute to the linear term coefficient β.

(c) The linear structure is solely due to Correlation-Kinetic effects, and this contribution is obtained analytically to be

$$\beta = \left(\frac{4Z}{3\rho(0)}\right)[t(0) - t_s(0)], \tag{8.46}$$

where $t(0)$ and $t_s(0)$ are the kinetic energy densities (see (2.122)) of the interacting and noninteracting S systems, respectively.

(d) The Pauli and Coulomb correlations do contribute to the coefficient γ of the quadratic term.

(e) All the above conclusions are equally valid for systems of arbitrary symmetry, provided spherical averages of the properties are employed.

For spherically symmetric systems, the Hartree potential $W_H(\mathbf{r})$ (see (6.83)) has zero slope at the nucleus because the electron density $\rho(\mathbf{r})$ is spherically symmetric about the nucleus. As such on employing (8.45), (8.46), the structure of the KS-DFT 'exchange-correlation' potential $v_{xc}(\mathbf{r})$ near the nucleus is

$$v_{xc}(\mathbf{r}) = v_{xc}(0) + \left(\frac{4Z}{3\rho(0)}\right)[t(0) - t_s(0)] + O(r^2). \tag{8.47}$$

The linear term can thus be identified as being due to Correlation-Kinetic effects. The expressions for $v_{xc}(\mathbf{r})$ can also be derived [19] without employing the Q-DFT definition for $v_{ee}(\mathbf{r})$. However, the advantage of working within the framework of Q-DFT is that the near nucleus structure of the potential $v_{ee}(\mathbf{r})$ can be understood as a function of the separate electron correlations that contribute to it.

8.2 Asymptotic Structure of the Wave Function in the Classically Forbidden Region

An important property of the wave function is its asymptotic structure in the classically forbidden region. The reason is because this structure is related to the ionization potential. Further, this structure is of relevance to both ground and excited state local effective potential theories.

To derive the asymptotic structure of the eigenfunctions $\psi_n(\mathbf{X})$ of the N-electron Hamiltonian $\hat{H}$ of (8.1), we rewrite it as [21, 23]

$$\hat{H} = -\frac{1}{2}\nabla^2 + v(\mathbf{r}) + \sum_{i=2}^{N} \frac{1}{|\mathbf{r} - \mathbf{r}_i|} + \hat{H}^{N-1}, \tag{8.48}$$

where the $(N-1)$ electron Hamiltonian $\hat{H}^{N-1}$ is

$$\hat{H}^{(N-1)} = -\frac{1}{2}\sum_{i=2}^{N} \nabla_i^2 + \sum_{i=2}^{N} v(\mathbf{r}_i) + \frac{1}{2}\sum_{i\neq j\neq 1}^{N} \frac{1}{|\mathbf{r}_i - \mathbf{r}_j|}. \tag{8.49}$$

Now the complete set of eigenfunctions and eigenenergies of the $(N-1)$-electron system are defined by the equation

$$\hat{H}^{N-1}\psi_s^{(N-1)}(\mathbf{X}^{N-1}) = E_s^{(N-1)}\psi_s^{(N-1)}(\mathbf{X}^{N-1}). \tag{8.50}$$

Next expand the N-electron wavefunction $\psi_n(\mathbf{X})$ in terms of the eigenfunctions $\psi_s^{(N-1)}$:

$$\psi_n(\mathbf{r}\sigma, \mathbf{X}^{N-1}) = \sum_s C_{s\sigma}(\mathbf{r})\psi_s^{(N-1)}(\mathbf{X}^{N-1}), \tag{8.51}$$

and rewrite the Schrödinger equation (8.5) as

$$\left(-\frac{1}{2}\nabla^2 + v(\mathbf{r}) + \sum_{i=2}^{N} \frac{1}{|\mathbf{r} - \mathbf{r}_i|} + \hat{H}^{(N-1)}\right)\sum_s C_{s\sigma}(\mathbf{r})\psi_s^{(N-1)}(\mathbf{X}^{N-1})$$
$$= E_n \sum_s C_{s\sigma}(\mathbf{r})\psi_s^{(N-1)}(\mathbf{X}^{N-1}). \tag{8.52}$$

For asymptotic positions of the electron we have by Taylor expansion

$$\frac{1}{|\mathbf{r} - \mathbf{r}_i|} = \frac{1}{r} + \frac{\mathbf{r}_i \cdot \mathbf{r}}{r^3} + \frac{1}{2}\sum_{\alpha,\beta} r_{i\alpha} r_{i\beta} \frac{\partial^2}{\partial r_\alpha \partial r_\beta}\frac{1}{r} + \ldots, \tag{8.53}$$

so that on retaining just the leading term, (8.52) becomes

$$\left[-\frac{1}{2}\nabla^2 + v(\mathbf{r}) + \frac{N-1}{r}\right]\sum_s C_{s\sigma}(\mathbf{r})\psi_s^{(N-1)}(\mathbf{X}^{N-1})$$
$$= \sum_s \left[E_n - E_s^{(N-1)}\right] C_{s\sigma}(\mathbf{r})\psi_s^{(N-1)}(\mathbf{X}^{N-1}). \tag{8.54}$$

Multiplying (8.54) by $\psi_s^{(N-1)*}(\mathbf{X}^{N-1})$ from the left, integrating over $\int d\mathbf{X}^{N-1}$, and employing the orthonormality condition

$$\left\langle \psi_{s'}^{(N-1)} | \psi_{s}^{(N-1)} \right\rangle = \delta_{ss'}, \tag{8.55}$$

we have

$$\left(-\frac{1}{2}\nabla^2 + v(\mathbf{r}) + \frac{N-1}{r} + I_{s,n} \right) C_{s\sigma}(\mathbf{r}) = 0, \tag{8.56}$$

where the ionization potential $I_{s,n}$ is

$$I_{s,n} = E_s^{(N-1)} - E_n. \tag{8.57}$$

The $I_{s,n}$ are the ionization potentials from the N-electron state with energy E_n into various states of the $(N-1)$-electron ion. It is assumed that $I_{s,n} < I_{s+1,n}$ etc.

The derivation thus far is for arbitrary external potential $v(\mathbf{r})$. The expressions arrived at for the remainder of this section and following subsection are restricted to *Coulombic potentials* $v(\mathbf{r})$.

For atomic systems, $v(\mathbf{r}) = -Z/r$. For molecules in the far asymptotic region, $v(\mathbf{r}) = -Q/r$, where Q is the total nuclear charge. Thus, the Schrödinger equation in the asymptotic region is of the form

$$\left[-\frac{1}{2}\nabla^2 - \frac{(Z-N+1)}{r} + I_{s,n} \right] C_{s\sigma}(\mathbf{r}) = 0, \tag{8.58}$$

and the asymptotic solution is

$$C_{s\sigma}(\mathbf{r})_{r \rightarrow \infty} \; r^{\beta_s} e^{-\alpha_s r} \chi(\sigma), \tag{8.59}$$

where $(1+\beta_s) = (Z-N+1)/\alpha_s$, and $\alpha_s = \sqrt{2I_{s,n}}$, with $\chi(\sigma)$ the spin component. The satisfaction of the differential equation (8.58) with this solution occurs on neglecting the $0(1/r^2)$ term of $\nabla^2 C_{s\sigma}(\mathbf{r})$. Hence, the asymptotic structure of the wave function $\psi_n(\mathbf{X})$ in the classically forbidden region is

$$\psi_n(\mathbf{X})_{r \rightarrow \infty} e^{-\sqrt{2I_{s,n}}r} \chi(\sigma), \tag{8.60}$$

and is directly related to the corresponding ionization potential.

8.2.1 Asymptotic Structure of the Density and Sum Rule

For the form of the wave function $\psi_n(\mathbf{X})$ as defined by (8.51), the density $\rho_n(\mathbf{r})$ from its definition of (8.21) is

$$\rho_n(\mathbf{r}) = N \sum_{\sigma} \sum_{s} |C_{s\sigma}(\mathbf{r})|^2. \tag{8.61}$$

The asymptotic structure of the density in the classically forbidden region employing (8.59) is then

$$\rho_n(\mathbf{r})_{r \to \infty} \exp(-2\sqrt{2I_{s,n}}r). \quad (8.62)$$

Thus, the asymptotic structure of the density for arbitrary state too is directly related to the corresponding first ionization potential.

With the density $\rho_n(\mathbf{r})$ spherically averaged:

$$\rho_{n,sp.av}(\mathbf{r}) = \frac{1}{4\pi} \int \rho_n(\mathbf{r}) d\Omega, \quad (8.63)$$

it is evident on employing (8.62), that the asymptotic structure of $\rho_{n,sp.av}(\mathbf{r})$ as $r \to \infty$ is

$$\rho_{n,sp.av}(\mathbf{r}) \bigg|_{r \to \infty} = e^{-2\sqrt{2I_{s,n}}r}. \quad (8.64)$$

Therefore,

$$\frac{d\rho_{n,sp.av}(\mathbf{r})}{dr} \bigg|_{r \to \infty} = -2\sqrt{2I_{s,n}} \rho_{n,sp.av}(\mathbf{r}) \bigg|_{r \to \infty}. \quad (8.65)$$

Equivalently, the sum rule on the asymptotic structure of the spherically averaged density may be written as

$$\frac{d \, ln \rho_{n,sp.av}(\mathbf{r})}{dr} \bigg|_{r \to \infty} = -2\sqrt{2I_{s,n}}. \quad (8.66)$$

(As the asymptotic structure of the wave function $\psi_n(\mathbf{X})$ is known (see (8.59), (8.60)), the above sum rule may also be derived via the procedure of Sect. 8.1.1 but taking the limit as $r \to \infty$ instead.)

8.2.2 *Significance of Asymptotic Structure of Wave Function and Density to Local Effective Potential Theories*

The asymptotic structure of the density $\rho(\mathbf{r})$ in the classically forbidden region has considerable significance for local effective potential theories. These are described in the Notes below.

Note 1 Physical Interpretation of Highest Occupied Eigenvalue ϵ_m

In local effective potential theories such as Q-DFT and KS-DFT, the potential energy of each model fermion $v_s(\mathbf{r}) = v(\mathbf{r}) + v_{ee}(\mathbf{r})$ (see (8.29)) of the S system is the same. Hence, the asymptotic decay of each orbital $\phi_i(\mathbf{r})$ in the classically forbidden region depends on the corresponding occupied eigenvalue ϵ_i. The orbital that decays the farthest thus depends on the highest occupied eigenvalue ϵ_m. The asymptotic structure of the density $\rho_n(\mathbf{r})$ for the system in state n is then

$$\lim_{r\to\infty} \rho_n(\mathbf{r}) = |\phi_m(\mathbf{r})|^2 \sim \exp(-2\sqrt{-2\epsilon_m r}). \tag{8.67}$$

A comparison of the asymptotic structure of the density $\rho_n(\mathbf{r})$ of (8.62) as derived by solution of the Schrödinger equation with that for the local effective potential theory of (8.67) then shows that the highest occupied eigenvalue ϵ_m of the S system is the negative of the first ionization potential:

$$\epsilon_m = -I_{s,n}. \tag{8.68}$$

In Q-DFT, the state of the model system is *arbitrary*, irrespective of the state of the interacting system as described by the Schrödinger equation. Thus, it is possible to map a *ground state* of the interacting system to a model system either in its *ground state* or in an *excited state* possessing the same density. It is also possible to map an *excited state* of the interacting system to a model system either in an arbitrary *excited state* or one in its *ground state*. (For such examples, see [27, 36–41]). In KS-DFT, a ground state theory, the mapping is always from a *ground state* of the interacting system to a model system also in its *ground state*. In all these cases, irrespective of the state of the model system, the *highest occupied eigenvalue* ϵ_m *thus has the rigorous physical interpretation of being the negative of the first ionization potential.* (This interpretation of ϵ_m, for the specific case of the mapping from a ground state of the interacting system to a ground state of the model fermionic system, has also been arrived at in [42–44].)

Note 2 Asymptotic Structure of the Electron-Interaction Potential $v_{ee}(\mathbf{r})$ ***in Terms of Electron Correlations***

Within Q-DFT, it is possible to determine the asymptotic structure of the electron-interaction potential $v_{ee}(\mathbf{r})$ (see (8.29)) in the classically forbidden region in terms of the separate electron correlations that contribute to it [20–22]. Consider the case when both the N-electron atom and the $(N-1)$-electron ion are nondegenerate. In such a case, both the atom and ion are spherically symmetric. By including higher-order terms in the expansion of the wave function beyond that given in Sect. 8,2, and thereby determining the resulting structure of the single particle density matrix $\gamma(\mathbf{rr}')$, density $\rho(\mathbf{r})$ and pair-correlation density $g(\mathbf{rr}')$, it can be derived that that the asymptotic structure of $v_{ee}(\mathbf{r})$ as $\mathbf{r} \to \infty$ is

$$v_{ee}(\mathbf{r})(\mathbf{r} \to \infty) = \frac{N}{r} - \frac{1}{r} - \frac{\alpha}{2r^4} + \frac{8\kappa_0\chi}{5r^5}. \tag{8.69}$$

This structure can be explained in terms of electron correlations as follows:

(a) The first term N/r is the Hartree potential energy. It is the work done $W_H(\mathbf{r})$ in the field due to the quantal source of the *static* electron density $\rho(\mathbf{r})$.
(b) The second term $(-1/r)$ of the potential $v_{ee}(\mathbf{r})$ is *solely* due to correlations arising from the Pauli principle. It is the work done $W_x(\mathbf{r})$ in the field of the *dynamic* Fermi hole charge $\rho_x(\mathbf{rr}')$ quantal source.
(c) The third term $(-\alpha/2r^4)$ of the potential $v_{ee}(\mathbf{r})$, where α is the ground state polarizability of the $(N-1)$-electron ion, is *solely* due to Coulomb correlations

beyond those of the Hartree term. It is the work done $W_c(\mathbf{r})$ in the field of the dynamic Coulomb hole distribution $\rho_c(\mathbf{rr}')$ quantal source.

(d) It is shown that there is no $O(1/r^5)$ term due to Coulomb correlations.

(e) The final term $(8\kappa_0\chi/5r^5)$ of the potential $v_{ee}(\mathbf{r})$, where $\kappa_0^2/2$ is the ionization potential energy and χ an expectation of the $(N-1)$-electron ion, is *solely* due to Correlation-Kinetic effects. It is the work done $W_{t_c}(\mathbf{r})$ in the field due to the corresponding density matrix quantal source $\gamma_{t_c}(\mathbf{rr}') = [\gamma(\mathbf{rr}') - \gamma_s(\mathbf{rr}')]$. Here $\gamma(\mathbf{rr}')$ and $\gamma_s(\mathbf{rr}')$ are the single particle and Dirac density matrices, respectively.

(f) The far asymptotic structure of $v_{ee}(\mathbf{r})$ is then $(N-1)/r$, and governed *solely* by the Hartree electrostatic term and correlations due to the Pauli principle.

Note 3 Asymptotic Density Sum Rule and Excitation State

The expression for the asymptotic structure of the density (8.62), or equivalently the sum rule of (8.66), involve the ionization potential $I_{s,n} = E_s^{N-1} - E_n$. The N-electron system with energy E_n could be an excited state. Thus, it is evident from these expressions that knowledge of the asymptotic density uniquely defines the explicit excited state E_n.

8.3 A New Symmetry for 2-Electron Systems in an Electromagnetic Field

In this section we describe [24] a new symmetry property of the wave function for 2-electron systems in an electromagnetic field, and the subsequent additional properties of the system that emanate as a consequence of this symmetry. The symmetry and the other properties, together with the Pauli principle are then elucidated by example.

We consider the 2 electrons to possess a spin moment, and to have an interaction of the *general* form $w(|\mathbf{r} - \mathbf{r}'|)$. The electrons are in an *arbitrary* binding electrostatic field $\mathcal{E}(\mathbf{r}) = -\nabla v(\mathbf{r})/e$, where $v(\mathbf{r})$ is a *symmetrical* scalar potential, and a magnetostatic field $\mathcal{B}(\mathbf{r}) = \nabla \times \mathbf{A}(\mathbf{r})$, with $\mathbf{A}(\mathbf{r})$ the vector potential. The Schrödinger-Pauli theory [45, 46] eigenvalue equation for the bound 2-electron system is

$$\hat{H}\Psi(\mathbf{X}) = E\Psi(\mathbf{X}), \tag{8.70}$$

with $\{\Psi(\mathbf{X}), E\}$ the eigenfunctions and eigenvalues; $\mathbf{X} = \mathbf{x}_1, \ldots, \mathbf{x}_N$; $\mathbf{x} = \mathbf{r}\sigma$; $\mathbf{r}$ and σ the spatial and spin coordinates. The Hamiltonian $\hat{H}$ for spin $\frac{1}{2}$ particles is comprised of the sum of the Feynman kinetic $\hat{T}_F$, the *general* electron-interaction potential $\hat{W}$, and electrostatic binding potential $\hat{V}$ operators. In atomic units (charge of electron $-e$; $e = \hbar = m = 1$), and with all summations $k = 1$ to 2,

$$\hat{H} = \hat{T}_F + \hat{W} + \hat{V}, \tag{8.71}$$

where

$$\hat{T}_F = \frac{1}{2}\sum_k \left(\boldsymbol{\sigma}\cdot\hat{\mathbf{p}}_{k,\mathrm{phys}}\right)\left(\boldsymbol{\sigma}\cdot\hat{\mathbf{p}}_{k,\mathrm{phys}}\right) \tag{8.72}$$

$$= \frac{1}{2}\sum_k \left(\hat{\mathbf{p}}_k + \frac{1}{c}\mathbf{A}(\mathbf{r}_k)\right)^2 + \frac{1}{c}\sum_k \boldsymbol{\mathcal{B}}(\mathbf{r}_k)\cdot\mathbf{s}_k, \tag{8.73}$$

$$\hat{W} = \frac{1}{2}\sum_{k,\ell}{}' w(|\mathbf{r}_k - \mathbf{r}_\ell|), \tag{8.74}$$

$$\hat{V} = \sum_k v(\mathbf{r}_k). \tag{8.75}$$

Here the physical momentum operator $\hat{\mathbf{p}}_{\mathrm{phys}} = \left(\hat{\mathbf{p}} + \frac{1}{c}\mathbf{A}(\mathbf{r})\right)$, with $\hat{\mathbf{p}} = -i\boldsymbol{\nabla}$ the canonical momentum operator; $\boldsymbol{\sigma}$ is the Pauli spin matrix, $\mathbf{s} = \frac{1}{2}\boldsymbol{\sigma}$, and $\mathbf{s}$ the electron spin angular momentum vector operator. The general electron-interaction function $w(|\mathbf{r} - \mathbf{r}'|)$ could be Coulombic, harmonic, screened-Coulomb, etc. The binding scalar electrostatic potential is $v(\mathbf{r})$. For natural atoms and molecules this potential is Coulombic, whereas for 'artificial atoms' and 'artificial molecules', it is harmonic. The use of the Feynman kinetic energy operator leads to the correct gyromagnetic ratio of $g = 2$. The eigenfunctions $\Psi(\mathbf{X})$ are of the form

$$\Psi(\mathbf{x}_1, \mathbf{x}_2) \equiv \Psi(\mathbf{r}_1\sigma_1, \mathbf{r}_2\sigma_2) = \psi(\mathbf{r}_1, \mathbf{r}_2)\chi(\sigma_1, \sigma_2), \tag{8.76}$$

where $\psi(\mathbf{r}_1, \mathbf{r}_2)$, $\chi(\sigma_1, \sigma_2)$ are, respectively, the spatial and spin components. That these components are separable is a consequence of the lack of any spin-orbit interaction term in the Hamiltonian.

The physical system employed to exhibit the various properties of the wave function described below is the 2D 2-electron 'artificial atom' or semiconductor quantum dot in a magnetic field [10–12]. The motion of the electrons is confined to two dimensions in a quantum well within a thin layer of semiconductor such as GaAs which is sandwiched between two much thicker layers of another semiconductor AlGaAs. The lateral confinement is achieved by placing an electrostatic gate on this system. The electrons are further constrained by application of a magnetic field perpendicular to the plane of motion. For the 'artificial atom' the free electron mass m is replaced by the semiconductor band effective mass $m^\star$, and the Coulomb electron-interaction modified by the dielectric constant ϵ.

In contrast to natural atoms, the electrons in the 'artificial atom' are bound by a harmonic potential $v(\mathbf{r}) = \frac{1}{2}m^\star\omega_0^2 r^2$, with ω_0 the harmonic frequency. The Schrödinger-Pauli equation (8.70) is then

$$\left[\frac{1}{2m^\star}\sum_{k=1}^{2}\left(\hat{\mathbf{p}}_k + \frac{e}{c}\mathbf{A}(\mathbf{r}_k)\right)^2 + (g^\star\mu_B/\hbar)\sum_{k=1}^{2}\mathcal{B}(\mathbf{r}_k)\cdot\mathbf{s}_k + \frac{1}{\epsilon}\frac{e^2}{|\mathbf{r}_1-\mathbf{r}_2|}\right.$$
$$\left. + \frac{1}{2}m^\star\omega_0^2\sum_{k=1}^{2}r_k^2\right]\Psi(\mathbf{x}_1,\mathbf{x}_2) = E\Psi(\mathbf{x}_1,\mathbf{x}_2), \quad (8.77)$$

where $\mu_B = e\hbar/2mc$ is the Bohr magneton; $g^\star$ the gyromagnetic ratio; $\mathbf{s} = \frac{\hbar}{2}\boldsymbol{\sigma}$; $\mathbf{x} = \mathbf{r}\sigma$; $\mathbf{r} = (r\theta)$; $\mathbf{r}\sigma$ the spatial and spin coordinates.

In the symmetric gauge $\mathbf{A}(\mathbf{r}) = \frac{1}{2}\mathcal{B}(\mathbf{r})\times\mathbf{r}$ with the magnetic field in the z-direction $\mathcal{B}(\mathbf{r}) = \mathcal{B}\mathbf{i}_z$, the Schrödinger-Pauli equation can be solved [25–27] in closed analytical form by the method of Taut [47] for the first excited singlet 2^1S and triplet 2^3S states for a denumerably infinite set of frequencies ω_0 and ω_L such that the effective force constant $k_{\text{eff}} = \omega_0^2 + \omega_L^2$, where $\omega_L = \mathcal{B}/2c$ is the Larmor frequency. Effective atomic units are employed: $e^2/\epsilon = \hbar = m^\star = c = 1$. The effective Bohr radius is $a_0^\star = a_0(m\epsilon/m^\star)$, the effective energy unit is $(a.u.)^\star = (a.u.)(m^\star/mc^2)$. The wave functions are of the form $\Psi(\mathbf{x}_1,\mathbf{x}_2)$ of (8.76). The expressions for the spatial components $\psi(\mathbf{r}_1,\mathbf{r}_2)$ of the singlet and triplet states and their respective energies E are given below. (See Chap. 9 for the derivation.)

Singlet 2^1S ***State***

$$\begin{aligned}
\psi(\mathbf{r}_1,\mathbf{r}_2) &= Ne^{-\sqrt{k_{\text{eff}}}(r_1^2+r_2^2)/2}[1 + |\mathbf{r}_1-\mathbf{r}_2| + c_2|\mathbf{r}_1-\mathbf{r}_2|^2 + c_3|\mathbf{r}_1-\mathbf{r}_2|^3]\\
N &= 0.108563 \quad \text{(Normalization constant)}\\
k_{\text{eff}} &= 0.471716 \quad \text{(Effective force constant)}\\
c_2 &= -0.265111; \quad c_3 = -0.182082 \quad \text{(Coefficients of expansion)}\\
E &= 3.434066\ (a.u.)^\star \quad \text{(Energy)}. \qquad (8.78)
\end{aligned}$$

Observe that since the spin component $\chi(\sigma_1,\sigma_2)$ is antisymmetric in an interchange of the spin coordinates σ_1 and σ_2, the spatial component $\psi(\mathbf{r}_1,\mathbf{r}_2)$ is symmetric in an interchange of the coordinates $\mathbf{r}_1$ and $\mathbf{r}_2$.

Triplet 2^3S ***State***

$$\begin{aligned}
\psi(\mathbf{r}_1,\mathbf{r}_2) &= Ne^{im\theta-\sqrt{k_{\text{eff}}}(r_1^2+r_2^2)/2}[|\mathbf{r}_1-\mathbf{r}_2| + c_2|\mathbf{r}_1-\mathbf{r}_2|^2 + c_3|\mathbf{r}_1-\mathbf{r}_2|^3\\
&\quad + c_4|\mathbf{r}_1-\mathbf{r}_2|^4]\\
N &= 0.022466 \text{ (Normalization constant)}\\
m &= 1 \text{ (Angular momentum quantum number)}\\
k_{\text{eff}} &= 0.072217 \text{ (Effective force constant)}\\
c_2 &= \frac{1}{3}; \quad c_3 = -0.059108; \quad c_4 = -0.015884 \text{ (Coefficients of expansion)}\\
E &= 1.612392\ (a.u.)^\star \text{ (Energy)}. \qquad (8.79)
\end{aligned}$$

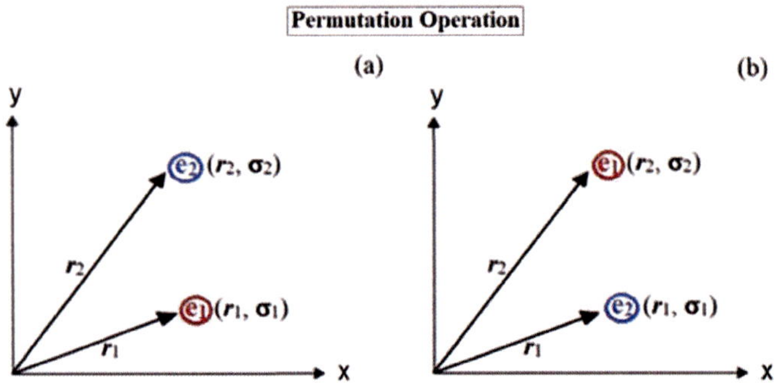

Fig. 8.1 Vector description of the permutation operation of the electronic coordinates. (**a**) The coordinates before switching are spatial $\mathbf{r}_1$ and spin σ_1 coordinates for electron 1, and spatial $\mathbf{r}_2$ and spin σ_2 coordinates for electron 2; thus $e_1(\mathbf{r}_1, \sigma_1)$ and $e_2(\mathbf{r}_2, \sigma_2)$. (**b**) As a result of switching both the spatial and spin coordinates for each electron the new coordinates are spatial $\mathbf{r}_2$ and spin σ_2 coordinates for electron 1, and spatial $\mathbf{r}_1$ and spin σ_1 coordinates for electron 2; thus $e_1(\mathbf{r}_2, \sigma_2)$ and $e_2(\mathbf{r}_1, \sigma_1)$

Note that since the spin component $\chi(\sigma_1, \sigma_2)$ is symmetric in an interchange of the coordinates σ_1, σ_2, the spatial component $\psi(\mathbf{r}_1, \mathbf{r}_2)$ is antisymmetric in an interchange of $\mathbf{r}_1$ and $\mathbf{r}_2$. That this is the case results from the presence of the phase factor $e^{im\theta}$. When $\mathbf{r}_1$ and $\mathbf{r}_2$ are interchanged, the magnitude of the relative vector $\mathbf{s} = \mathbf{r}_2 - \mathbf{r}_1$ does not change, but its angle θ (which points from the tip of $\mathbf{r}_1$ to the tip of $\mathbf{r}_2$) changes to $\theta + \pi$. This changes the sign of the phase factor $e^{im\theta}$.

The properties to be described are equally valid for the eigenfunctions $\Psi(\mathbf{X})$ of bound 2-electron systems in the presence of a magnetic field as described by the Schrödinger theory of spinless electrons [45].

As the ground 1^1S state of the quantum dot is a singlet state, it possesses the same characteristics as that of the first excited singlet 2^1S state described below. The sole difference is that being a ground state, there are no nodes to the wave function.

To contrast the switching of electrons in the new symmetry operation O_{sym} with that of the Pauli principle [48–50], we begin with a brief discussion of the permutation operation P_{12} that leads to the Pauli principle.

8.3.1 Permutation Operation and the Pauli Principle

The permutation operation P_{12} permutes the coordinates $\mathbf{x}_1, \mathbf{x}_2$ of the electrons 1 and 2. Figure 8.1 is a 2D vector description of the switching of the electronic coordinates. Figure 8.1(a) corresponds to the initial coordinates of the two electrons, and Fig. 8.1(b) to the switched coordinates.

The corresponding permutation operator $\hat{P}_{12}$ commutes with the Hamiltonian $\hat{H}$:

$$[\hat{P}_{12}, \hat{H}] = 0, \tag{8.80}$$

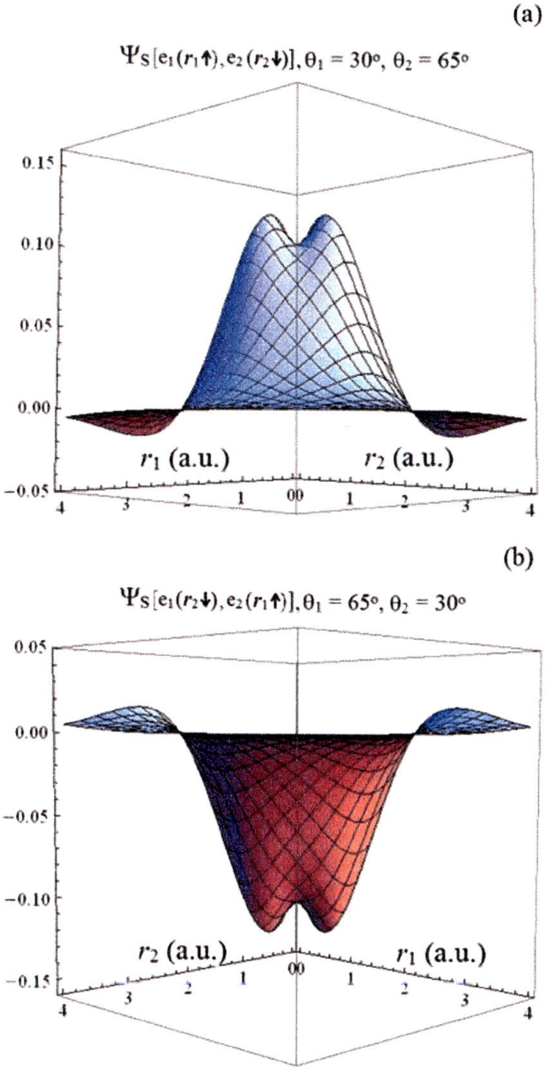

Fig. 8.2 Satisfaction of the Pauli principle for the singlet 2^1S state wave function of the 'artificial atom': (**a**) $\Psi_S[e_1(\mathbf{r}_1\uparrow), e_2(\mathbf{r}_2\downarrow)](\theta_1 = 30^\circ, \theta_2 = 65^\circ)$; (**b**) $\Psi_S[e_1(\mathbf{r}_2\downarrow), e_2(\mathbf{r}_1\uparrow)](\theta_1 = 65^\circ, \theta_2 = 30^\circ)$ (Note the switch of the coordinate axes labels in Fig. 8.2(b))

so that the eigenfunctions of $\hat{H}$ with eigenvalue E (see (8.70)) are also eigenfunctions of the operator $\hat{P}_{12}$. The action of the operator $\hat{P}_{12}$ on $\Psi(\mathbf{x}_1, \mathbf{x}_2)$ is thus

$$\hat{P}_{12}\Psi(\mathbf{x}_1, \mathbf{x}_2) = \Psi(\mathbf{x}_2, \mathbf{x}_1). \tag{8.81}$$

Operating with $\hat{P}_{12}$ on (8.81) one obtains

$$\hat{P}_{12}\hat{P}_{12}\Psi(\mathbf{x}_1, \mathbf{x}_2) = \Psi(\mathbf{x}_1, \mathbf{x}_2), \tag{8.82}$$

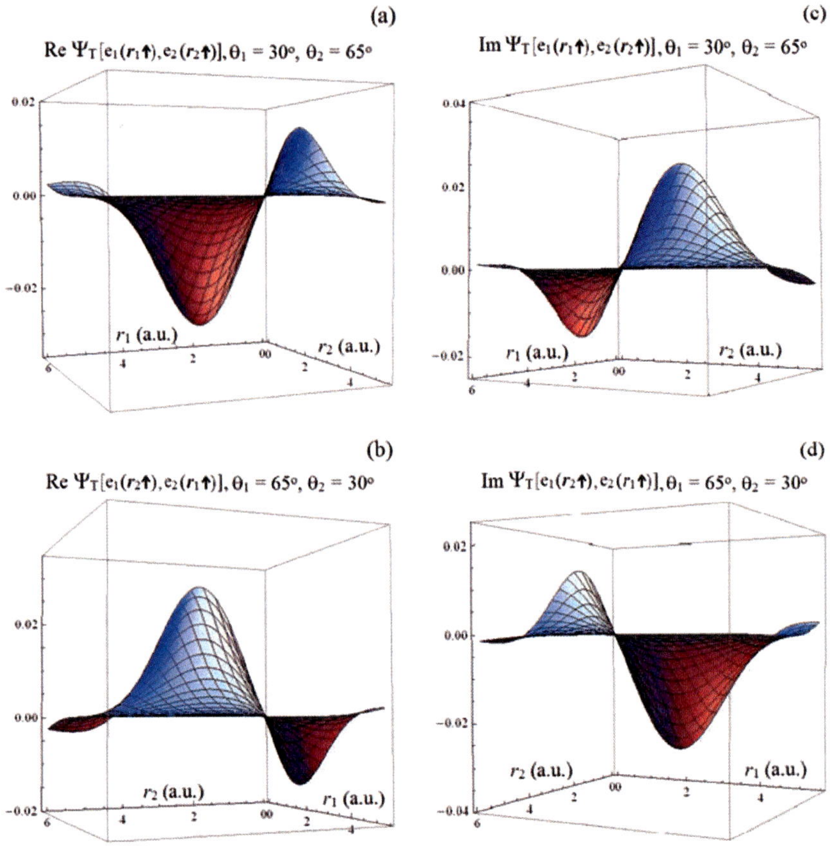

Fig. 8.3 Satisfaction of the Pauli principle for the Real and Imaginary parts of the triplet 2^3S state wave function for the 'artificial atom': **(a)** $\Re\Psi_T[e_1(\mathbf{r}_1\uparrow), e_2(\mathbf{r}_2\uparrow)](\theta_1 = 30°, \theta_2 = 65°)$; **(b)** $\Re\Psi_T[e_1(\mathbf{r}_2\uparrow), e_2(\mathbf{r}_1\uparrow)]$ $(\theta_1 = 65°, \theta_2 = 30°)$. (Note the switching of the coordinate axes labels In Fig. 8.3(b).) **(c)** $\Im\Psi_T[e_1(\mathbf{r}_1\uparrow), e_2(\mathbf{r}_2\uparrow)]$ $(\theta_1 = 30°, \theta_2 = 65°)$; **(d)** $\Im\Psi_T[e_1(\mathbf{r}_2\uparrow), e_2(\mathbf{r}_1\uparrow)]$ $(\theta_1 = 65°, \theta_2 = 30°)$ (Note the switching of the coordinate axes labels in Fig. 8.3(d))

so that

$$\hat{P}_{12}^2 = \hat{I}, \tag{8.83}$$

with $\hat{I}$ the unit operator. Therefore the eigenvalues of the operator $\hat{P}_{12}$ are $\epsilon = \pm 1$. The eigenfunctions that correspond to the eigenvalue $\epsilon = -1$ are such that

$$\hat{P}_{12}\Psi(\mathbf{x}_1, \mathbf{x}_2) = \Psi(\mathbf{x}_2, \mathbf{x}_1) = (-1)\Psi(\mathbf{x}_1, \mathbf{x}_2) \tag{8.84}$$

are antisymmetric under the permutation P_{12}. This is the statement of the Pauli principle:

$$\Psi\left[e_1(\mathbf{r}_1\sigma_1), e_2(\mathbf{r}_2\sigma_2)\right] = -\Psi\left[e_1(\mathbf{r}_2\sigma_2), e_2(\mathbf{r}_1\sigma_1)\right]. \tag{8.85}$$

This statement is independent of the analytical structure or symmetry of the binding potential. It is independent of dimensionality. It is valid for arbitrary state. For wave functions of the form of (8.76), the statement of the Pauli principle is

$$\psi(\mathbf{r}_1, \mathbf{r}_2)\chi(\sigma_1, \sigma_2) = -\psi(\mathbf{r}_2, \mathbf{r}_1)\chi(\sigma_2, \sigma_1). \tag{8.86}$$

A pictorial description of the Pauli principle for the excited singlet 2^1S and triplet 2^3S states of a quantum dot follows:

Singlet 2^1S ***state***: Employing the exact analytical expression for the excited singlet state of the 'artificial atom' or quantum dot (8.78), the wave function $\Psi_S[e_1(\mathbf{r}_1\uparrow), e_2(\mathbf{r}_2\downarrow)](\theta_1 = 30°, \theta_2 = 65°)$ is plotted in Fig. 8.2(a). (This corresponds to the case of Fig. 8.1(a).) The plot for the wave function when the coordinates are switched, (corresponding to Fig. 8.1(b)), which is $\Psi_S[e_1(\mathbf{r}_2\downarrow), e_2(\mathbf{r}_1\uparrow)](\theta_1 = 65°, \theta_2 = 30°)$ is given in Fig. 8.2(b). (Note the switching of the coordinate axes labels in Fig. 8.2(b).) The satisfaction of the Pauli principle of (8.85) is evident.

Triplet 2^3S ***state***: Employing the exact analytical expression for the triplet state wave function for the 'artificial atom' given (8.79), the Real part of the wave function $\Re\Psi_T[e_1(\mathbf{r}_1\uparrow), e_2(\mathbf{r}_2\uparrow)](\theta_1 = 30°, \theta_2 = 65°)$ is plotted in Fig. 8.3(a). The Real part of the wave function with the switched coordinates $\Re\Psi_T[e_1(\mathbf{r}_2\uparrow), e_2(\mathbf{r}_1\uparrow)](\theta_1 = 65°, \theta_2 = 30°)$ is plotted in Fig. 8.3(b). In Fig. 8.3(c, d), the Imaginary parts of the wave function $\Im\Psi_T[e_1(\mathbf{r}_1\uparrow), e_2(\mathbf{r}_2\uparrow)](\theta_1 = 30°, \theta_2 = 65°)$ and $\Im\Psi_T[e_1(\mathbf{r}_2\uparrow), e_2(\mathbf{r}_1\uparrow)](\theta_1 = 65°, \theta_2 = 30°)$, respectively, are plotted. (Note the switching of the coordinate axes labels in Fig. 8.3(b, d), respectively.) The satisfaction of the Pauli principle of (8.85) by the Real and Imaginary parts of the wave function is evident.

8.3.2 New Symmetry Operation and Wave Function Identity

With the initial coordinates of the electrons as exhibited in Fig. 8.1(a), the new symmetry operation O_{sym} is a two-step process on the electron coordinates as explained in Fig. 8.4. In Step 1 the spatial coordinates of the electrons are switched ($\mathbf{r}_1 \leftrightarrow \mathbf{r}_2$) while keeping the spin coordinates associated with each electron unchanged (see quadrant 1). Step 2 is an inversion through the center of symmetry ($\mathbf{r}_1 \rightarrow -\mathbf{r}_1$ and $\mathbf{r}_2 \rightarrow -\mathbf{r}_2$) while the spin coordinates remain unchanged (see quadrant 3).

The operator $\hat{O}_{sym}$ corresponding to the symmetry operation O_{sym} commutes with the Hamiltonian $\hat{H}$:

$$\left[\hat{O}_{sym}, \hat{H}\right] = 0, \tag{8.87}$$

so that the eigenfunctions of $\hat{O}_{sym}$ are the same as those of the Hamiltonian $\hat{H}$. Operating with the operator $\hat{O}_{sym}$ on the wave function $\Psi(\mathbf{x}_1, \mathbf{x}_2)$ one obtains

$$\hat{O}_{sym}\Psi\big[e_1(\mathbf{r}_1\sigma_1), e_2(\mathbf{r}_2\sigma_2)\big] = \Psi\big[e_1(-\mathbf{r}_2\sigma_1), e_2(-\mathbf{r}_1\sigma_2)\big]. \tag{8.88}$$

Operating with $\hat{O}_{sym}$ a second time leads to

$$\hat{O}_{sym}\hat{O}_{sym}\Psi\big[e_1(\mathbf{r}_1\sigma_1), e_2(\mathbf{r}_2\sigma_2)\big] = \Psi\big[e_1(\mathbf{r}_1\sigma_1), e_2(\mathbf{r}_2\sigma_2)\big], \tag{8.89}$$

so that

$$\hat{O}^2_{sym} = \hat{I}. \tag{8.90}$$

Thus, the eigenvalues of the operator $\hat{O}_{sym}$ are $\lambda = \pm 1$. The wave functions that correspond to the eigenvalue $\lambda = +1$ are such that

$$\hat{O}_{sym}\Psi\big[e_1(\mathbf{r}_1\sigma_1), e_2(\mathbf{r}_2\sigma_2)\big] = (+1)\Psi\big[e_1(\mathbf{r}_1\sigma_1), e_2(\mathbf{r}_2\sigma_2)\big]. \tag{8.91}$$

The equality of the symmetry transformed wave function to the wave function is referred to as the *Wave Function Identity:*

$$\Psi\big[e_1(\mathbf{r}_1\sigma_1), e_2(\mathbf{r}_2\sigma_2)\big] = \Psi\big[e_1(-\mathbf{r}_2\sigma_1), e_2(-\mathbf{r}_1\sigma_2)\big]. \tag{8.92}$$

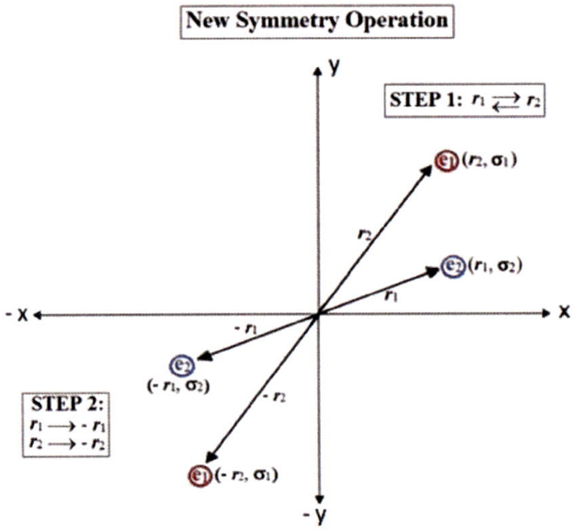

Fig. 8.4 Vector description to illustrate the new symmetry operation. The original electronic coordinates are those described in Fig. 8.1(a). The symmetry operation is a two-step process. STEP 1 (quadrant 1): The spatial coordinates are interchanged while the spin coordinates remain unchanged which results in spatial $\mathbf{r}_2$ and spin σ_1 coordinates for electron 1, and spatial $\mathbf{r}_1$ and spin σ_2 coordinates for electron 2; thus $e_1(\mathbf{r}_2, \sigma_1)$ and $e_2(\mathbf{r}_1, \sigma_2)$. STEP 2 (quadrant 3) is an inversion about the origin. Thus the final result is spatial $-\mathbf{r}_2$ and spin σ_1 coordinates for electron 1, and spatial $-\mathbf{r}_1$ and spin σ_2 coordinates for electron 2; hence $e_1(-\mathbf{r}_2, \sigma_1)$ and $e_2(-\mathbf{r}_1, \sigma_2)$

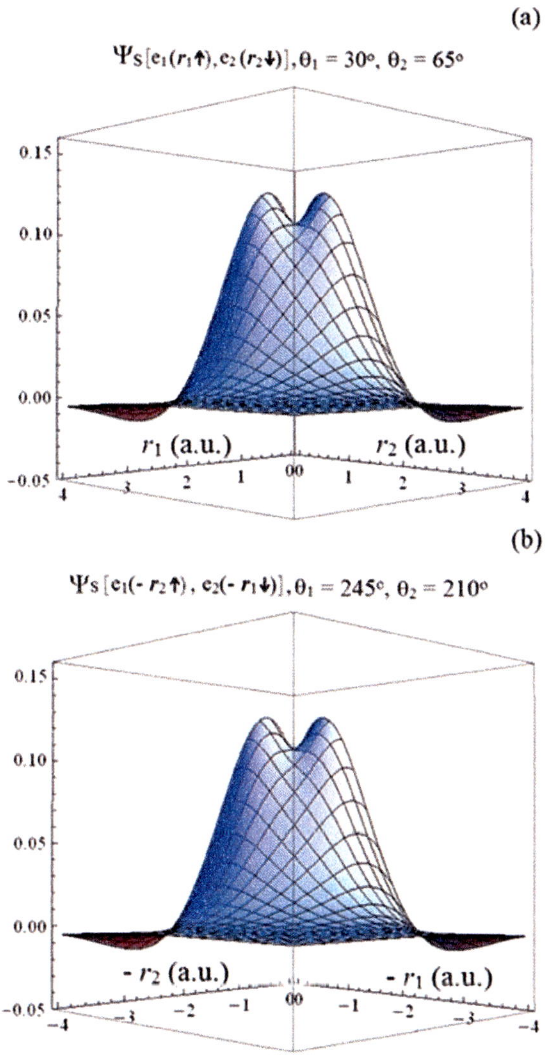

Fig. 8.5 Satisfaction of the Wave Function Identity for the singlet 2^1S state of the 'artificial atom': (**a**) $\Psi_S[e_1(\mathbf{r}_1\uparrow), e_2(\mathbf{r}_2\downarrow)]$ $(\theta_1 = 30°, \theta_2 = 65°)$ and (**b**) $\Psi_S[e_1(-\mathbf{r}_2\uparrow), e_2(-\mathbf{r}_1\downarrow)]$ $(\theta_1 = 245°, \theta_2 = 210°)$

In terms of the wave function $\Psi(\mathbf{x}_1, \mathbf{x}_2)$ of (8.76), the Wave Function Identity statement is

$$\psi(\mathbf{r}_1, \mathbf{r}_2)\chi(\sigma_1, \sigma_2) = \psi(-\mathbf{r}_2, -\mathbf{r}_1)\chi(\sigma_1, \sigma_2). \quad (8.93)$$

The Wave Function Identity is valid only for binding potentials $v(\mathbf{r})$ that are even. In common with the Pauli principle, it is valid for binding potentials of arbitrary analytical structure, and arbitrary dimensionality. It is also valid for both singlet and triplet states.

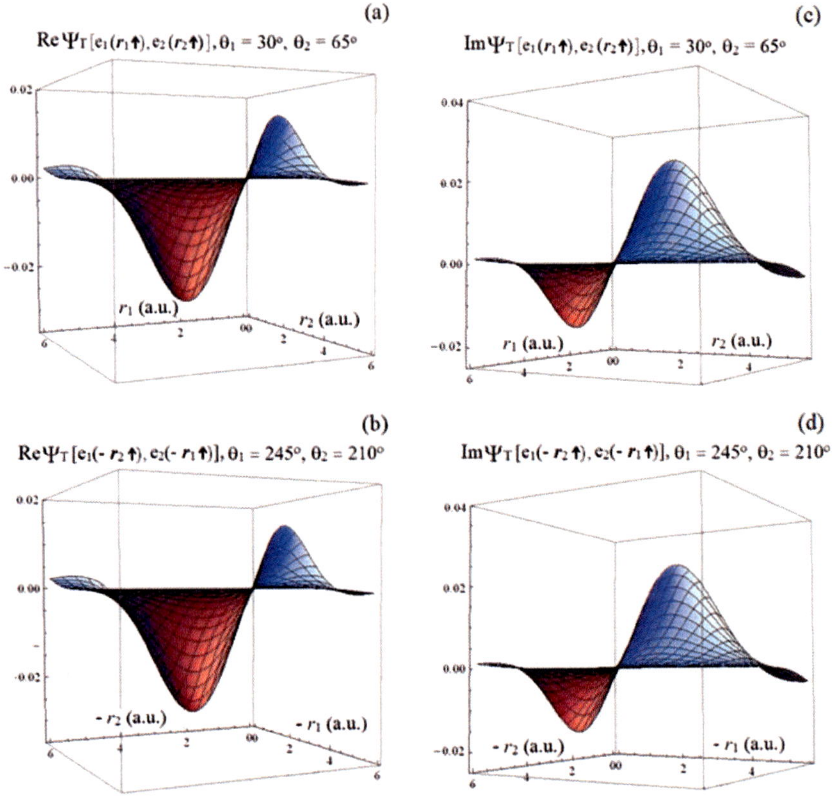

Fig. 8.6 Satisfaction of the Wave Function Identity for the Real and Imaginary parts of the triplet 2^3S state of the 'artificial atom': **(a)** $\Re\Psi_T[e_1(\mathbf{r}_1\uparrow), e_2(\mathbf{r}_2\uparrow)]$ $(\theta_1 = 30°, \theta_2 = 65°)$ and **(b)** $\Re\Psi_T[e_1(-\mathbf{r}_2\uparrow), e_2(-\mathbf{r}_1\uparrow)]$ $(\theta_1 = 245°, \theta_2 = 210°)$ and **(c)** $\Im\Psi_T[e_1(\mathbf{r}_1\uparrow), e_2(\mathbf{r}_2\uparrow)]$ $(\theta_1 = 30°, \theta_2 = 65°)$ and **(d)** $\Im\Psi_T[e_1(-\mathbf{r}_2\uparrow), e_2(-\mathbf{r}_1\uparrow)]$ $(\theta_1 = 245°, \theta_2 = 210°)$

A pictorial representation of the Wave Function Identity for the singlet and triplet states of a quantum dot is provided in Figs. 8.5 and 8.6.

Singlet 2^1S ***state***: The Wave Function Identity of (8.92) is exhibited in Fig. 8.5. In Fig. 8.5(a, b), respectively, the wave functions $\Psi_S[e_1(\mathbf{r}_1\uparrow), e_2(\mathbf{r}_2\downarrow)](\theta_1 = 30°, \theta_2 = 65°)$ and $\Psi_S[e_1(-\mathbf{r}_2\uparrow), e_2(-\mathbf{r}_1\downarrow)](\theta_1 = 245°, \theta_2 = 210°)$ for the 'artificial atom' are plotted. (Note the change in the labels of the coordinate axes in Fig. 8.5(b).) The satisfaction of the Wave Function Identity is evident.

Triplet 2^3S ***state***: In Fig. 8.6(a, b), respectively, the Real part of the triplet state wave functions for the 'artificial atom' $\Re\Psi_T[e_1(\mathbf{r}_1\uparrow), e_2(\mathbf{r}_2\uparrow)](\theta_1 = 30°, \theta_2 = 65°)$ and $\Re\Psi_T[e_1(-\mathbf{r}_2\uparrow), e_2(-\mathbf{r}_1\uparrow)](\theta_1 = 245°, \theta_2 = 210°)$ are plotted. In Fig. 8.6(c, d), the Imaginary part of the wave functions $\Im\Psi_T[e_1(\mathbf{r}_1\uparrow), e_2(\mathbf{r}_2\uparrow)](\theta_1 = 30°, \theta_2 = 65°)$ and $\Im\Psi_T[e_1(-\mathbf{r}_2\uparrow), e_2(-\mathbf{r}_1\uparrow)](\theta_1 = 245°, \theta_2 = 210°)$ are plotted. (Note the change in the labels of the coordinate axes in Fig. 8.6(b, d).)These figures demonstrate the satisfaction of the Wave Function Identity for the Real and Imaginary parts of the wave function for the triplet state.

8.3.3 Parity of Singlet and Triplet State Wave Functions

We next show by application of the permutation operator $\hat{P}_{12}$ to the Wave Function Identity that the parity of the singlet states is even, and that of the triplet states odd. To do so, consider the form of the wave function $\Psi(\mathbf{x}_1, \mathbf{x}_2)$ to be that of (8.76). We know from the previous section that

$$\hat{O}_{sym}\big[\psi(\mathbf{r}_1, \mathbf{r}_2)\chi(\sigma_1, \sigma_2)\big] = \psi(-\mathbf{r}_2, -\mathbf{r}_1)\chi(\sigma_1, \sigma_2). \tag{8.94}$$

On application of the permutation operator $\hat{P}_{12}$ to (8.94), we have

$$\hat{P}_{12}\hat{O}_{sym}\big[\psi(\mathbf{r}_1, \mathbf{r}_2)\chi(\sigma_1, \sigma_2)\big] = -\psi(-\mathbf{r}_1, -\mathbf{r}_2)\chi(\sigma_2, \sigma_1). \tag{8.95}$$

(Note that the negative sign on the right is a consequence of the Pauli principle.) Let us consider the singlet and triplet states separately.

Singlet 2^1S ***state***: For the singlet state $\Psi_S(\mathbf{x}_1, \mathbf{x}_2)$, the spin component of the wave function $\chi_S(\sigma_1, \sigma_2)$ is antisymmetric in an interchange of the spin coordinates. Hence, the spatial component $\psi_S(\mathbf{r}_1, \mathbf{r}_2)$ is symmetric in an interchange of the spatial coordinates. Employing these constraints, we then have from (8.95)

$$\begin{aligned}
\hat{P}_{12}\hat{O}_{sym}\big[\psi_S(\mathbf{r}_1, \mathbf{r}_2)\chi_S(\sigma_1, \sigma_2)\big] &= -\psi_S(-\mathbf{r}_1, -\mathbf{r}_2)\chi_S(\sigma_2, \sigma_1) &(8.96)\\
&= \big[-\psi_S(-\mathbf{r}_1, -\mathbf{r}_2)\big]\big[-\chi_S(\sigma_1, \sigma_2)\big] &(8.97)\\
&= \psi_S(-\mathbf{r}_1, -\mathbf{r}_2)\chi_S(\sigma_1, \sigma_2) &(8.98)\\
&= \psi_S(-\mathbf{r}_2, -\mathbf{r}_1)\chi_S(\sigma_1, \sigma_2). &(8.99)
\end{aligned}$$

From a comparison of the Wave Function Identity of (8.93) to that of (8.99), together with (8.98), it follows that for the singlet state

$$\psi_S(\mathbf{r}_1, \mathbf{r}_2)\chi(\sigma_1, \sigma_2) = \psi_S(-\mathbf{r}_1, -\mathbf{r}_2)\chi_S(\sigma_1, \sigma_2) \tag{8.100}$$

or more generally

$$\Psi_S(\mathbf{x}_1, \mathbf{x}_2) = \Psi_S(-\mathbf{x}_1, -\mathbf{x}_2), \tag{8.101}$$

where $-\mathbf{x} = -\mathbf{r}\sigma$. The right hand side of (8.100) corresponds to an inversion about the center of symmetry, and thus the singlet state has even parity. In Fig. 8.7 we demonstrate the even parity of the singlet state wave function for the 'artificial atom' by plotting $\Psi_S[e_1(\mathbf{r}_1\uparrow), e_2(\mathbf{r}_2\downarrow)](\theta_1 = 30°; \theta_2 = 65°)$ and $\Psi_S[e_1(-\mathbf{r}_1\uparrow), e_2(-\mathbf{r}_2\downarrow)](\theta_1 = 210°; \theta_2 = 245°)$.

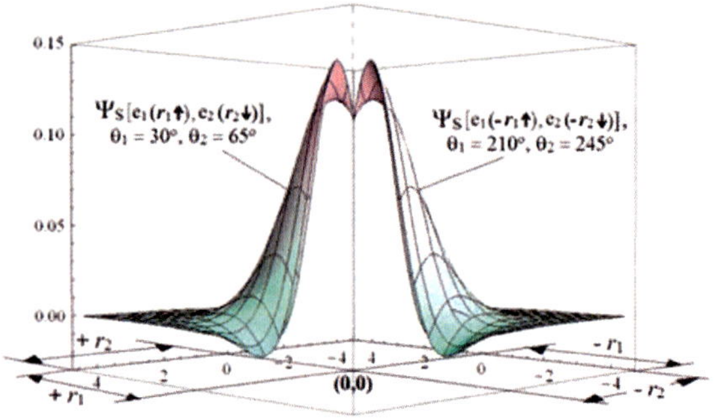

Fig. 8.7 Even Parity of the singlet 2^1S state wave function of the 'artificial atom'. The functions $\Psi_S[e_1(\mathbf{r}_1\uparrow), e_2(\mathbf{r}_2\downarrow)]$ $(\theta_1 = 30°, \theta_2 = 65°)$ and $\Psi_S[e_1(-\mathbf{r}_1\uparrow), e_2(-\mathbf{r}_2\downarrow)]$ $(\theta_1 = 210°, \theta_2 = 245°)$ are plotted

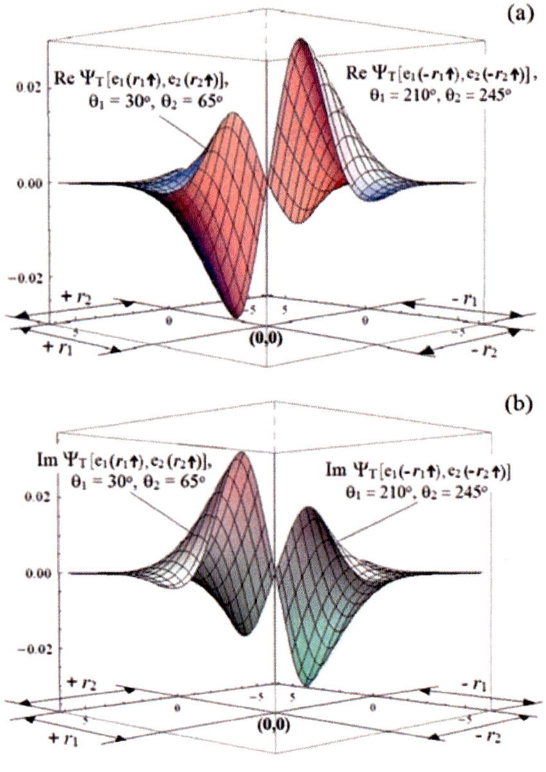

Fig. 8.8 Odd Parity of the Real and Imaginary parts of the triplet 2^3S wave function for the 'artificial atom'. The following functions are plotted **(a)** $\Re\Psi_T[e_1(\mathbf{r}_1\uparrow), e_2(\mathbf{r}_2\uparrow)](\theta_1 = 30°, \theta_2 = 65°)$ and $\Re\Psi_T[e_1(-\mathbf{r}_1\uparrow), e_2(-\mathbf{r}_2\uparrow)](\theta_1 = 210°, \theta_2 = 245°)$; **(b)** $\Im\Psi_T[e_1(\mathbf{r}_1\uparrow), e_2(\mathbf{r}_2\uparrow)](\theta_1 = 30°, \theta_2 = 65°)$ and $\Im\Psi_T[e_1(-\mathbf{r}_1\uparrow), e_2(-\mathbf{r}_2\uparrow)](\theta_1 = 210°, \theta_2 = 245°)$

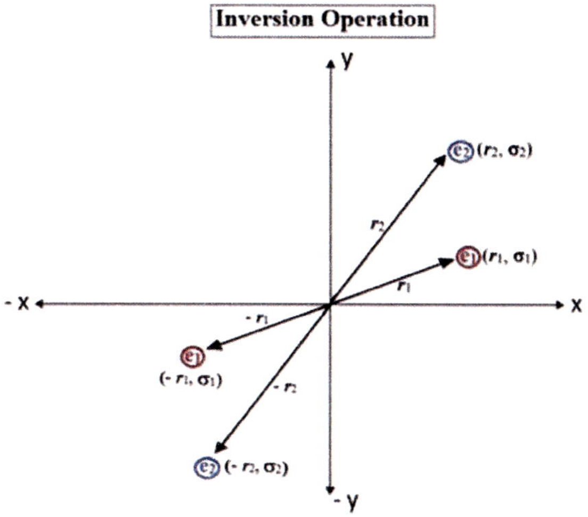

Fig. 8.9 Vector description of the inversion operation. The figure is valid for both the singlet and triplet states. Quadrant 1 is the same as Fig. 8.1(a). Quadrant 3: Electron 1 with spatial $-\mathbf{r}_1$ and spin σ_1 coordinates: $e_1(-\mathbf{r}_1, \sigma_1)$; Electron 2 with spatial $-\mathbf{r}_2$ and spin σ_2 coordinates: $e_2(-\mathbf{r}_2, \sigma_2)$

Triplet 2^3S ***state***: For the triplet state, the spin component of the wave function $\chi_T(\sigma_1, \sigma_2)$ is symmetric in an interchange of the spin coordinates, and therefore the spatial component $\psi_T(\mathbf{r}_1, \mathbf{r}_2)$ is antisymmetric in an interchange of the spatial coordinates. Employing these constraints, we then have from (8.95)

$$\hat{P}_{12}\hat{O}_{sym}\big[\psi_T(\mathbf{r}_1, \mathbf{r}_2)\chi_T(\sigma_1, \sigma_2)\big] = -\psi_T(-\mathbf{r}_1, -\mathbf{r}_2)\chi_T(\sigma_2, \sigma_1) \tag{8.102}$$

$$= -\psi_T(-\mathbf{r}_1, -\mathbf{r}_2)\chi_T(\sigma_1, \sigma_2) \tag{8.103}$$

$$= -[-\psi_T(-\mathbf{r}_2, -\mathbf{r}_1)]\chi_T(\sigma_1, \sigma_2) \tag{8.104}$$

$$= \psi_T(-\mathbf{r}_2, -\mathbf{r}_1)\chi_T(\sigma_1, \sigma_2). \tag{8.105}$$

From a comparison of the Wave Function Identity of (8.93) to that of (8.105), together with (8.103), it follows that for the triplet state

$$\psi_T(\mathbf{r}_1, \mathbf{r}_2)\chi(\sigma_1, \sigma_2) = -\psi_T(-\mathbf{r}_1, -\mathbf{r}_2)\chi_T(\sigma_1, \sigma_2) \tag{8.106}$$

or more generally

$$\Psi_T(\mathbf{x}_1, \mathbf{x}_2) = -\Psi_T(-\mathbf{x}_1, -\mathbf{x}_2). \tag{8.107}$$

Since on inversion, the wave function changes sign, the parity of the triplet states is odd.

To exhibit the odd parity of the triplet state wave function, we plot in Fig. 8.8(a) $\Re\Psi_T[e_1(\mathbf{r}_1\uparrow), e_2(\mathbf{r}_2\uparrow)](\theta_1 = 30°, \theta_2 = 65°)$ and $\Re\Psi_T[e_1(-\mathbf{r}_1\uparrow), e_2(-\mathbf{r}_2\uparrow)](\theta_1 = 210°, \theta_2 = 245°)$. In Fig. 8.8(b) we plot $\Im\Psi_T[e_1(\mathbf{r}_1\uparrow), e_2(\mathbf{r}_2\uparrow)](\theta_1 = 30°, \theta_2 = 65°)$ and $\Im\Psi_T[e_1(-\mathbf{r}_1\uparrow), e_2(-\mathbf{r}_2\uparrow)](\theta_1 = 210°, \theta_2 = 245°)$.

It is evident from the above that the symmetry operation O_{sym} followed by a permutation P_{12} is equivalent to an inversion (see Fig. 8.9). In operator form,

$$\hat{P}_{12}\hat{O}_{sym} = \hat{\Pi}, \tag{8.108}$$

where $\hat{\Pi}$ is the parity operator. The properties of the parity operator are the following: $[\hat{\Pi}, \hat{H}] = 0$; $\hat{\Pi}^2 = \hat{I}$; the eigenvalues of $\hat{\Pi}$ are $\alpha = \pm 1$.

As a consequence of (8.101), (8.107), (8.108), we conclude that the eigenfunctions of the parity operator for eigenvalue $\alpha = 1$ are singlet states, and those of the eigenvalue $\alpha = -1$ are triplet states. Thus,

$$\hat{P}_{12}\hat{O}_{sym}\Psi_S(\mathbf{r}_1\sigma_1, \mathbf{r}_2\sigma_2) = \hat{\Pi}\Psi_S(\mathbf{r}_1\sigma_1, \mathbf{r}_2\sigma_2) = \Psi_S(-\mathbf{r}_1\sigma_1, -\mathbf{r}_2\sigma_2), \tag{8.109}$$

and

$$\hat{P}_{12}\hat{O}_{sym}\Psi_T(\mathbf{r}_1\sigma_1, \mathbf{r}_2\sigma_2) = \hat{\Pi}\Psi_T(\mathbf{r}_1\sigma_1, \mathbf{r}_2\sigma_2) = -\Psi_T(-\mathbf{r}_1\sigma_1, -\mathbf{r}_2\sigma_2). \tag{8.110}$$

Note that *a priori* it is not known that the positive eigenvalue of the parity operator $\hat{\Pi}$ corresponds to the singlet state, whereas the negative eigenvalue that of the triplet state. It is only *a posteriori*, that is following the proof that the parity of singlet states is even and that of triplet states odd, that one can associate the positive eigenvalue of the parity operator $\hat{\Pi}$ with singlet states and the negative eigenvalue with the triplet state.

8.3.4 Parity About All Points of Electron-Electron Coalescence

As explained in Sect. 8.1, one may conclude from the electron-electron coalescence condition of (8.20), that the singlet states satisfy the cusp coalescence constraint whereas the triplet states satisfy the node coalescence condition. In other words, the singlet state wave functions exhibit a cusp at *all* points of electron-electron coalescence, and triplet states exhibit a node at these points. We have also proved that the singlet states have even parity and the triplet states odd parity. The parity of a wave function is with respect to the center (origin) of the system as defined by the symmetrical binding potential. Now the origin of the system also constitutes a point of electron-electron coalescence. Thus for singlet states, the cusp at the origin must be such that the slope of the wave function as the origin is approached from the right must be the same but the negative of the slope when approached from the left. The slopes have a discontinuity at the origin. From the perspective of the coalescence of electrons, the origin is not a special point of configuration space. There is no reason why the parity about all other points of electron-electron coalescence should not also be the same. Hence, it follows that the parity of the singlet state wave functions *at all points of electron-electron coalescence* is even.

To show that this is the case, we plot in Fig. 8.10 the singlet 2^1S state wave function of the 'artificial atom' as a function of the center of mass $\mathbf{R} = (\mathbf{r}_1 + \mathbf{r}_2)/2$

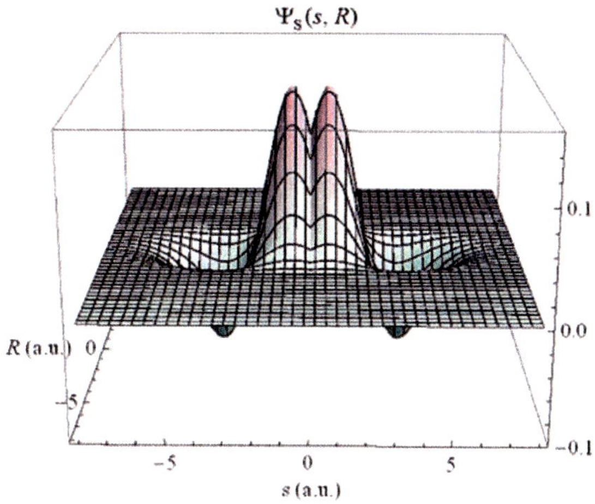

Fig. 8.10 The singlet 2^1S state wave function of the 'artificial atom' plotted as a function of the center of mass **R** and relative **s** coordinates

and relative $\mathbf{s} = \mathbf{r}_2 - \mathbf{r}_1$ coordinates. As may be observed from the figure, the parity of the wave function about the line $\mathbf{s} = 0$ is even.

The triplet state wave functions exhibit a node at the origin, and have odd parity. Hence, the functions are smooth, i.e.. continuous and with continuous first derivatives, about the origin. Again, the origin corresponds to just one point of electron-electron coalescence. Thus, one may conclude that triplet state wave functions have odd parity about *all points of electron-electron coalescence*. In Fig. 8.11(a, b), we plot two views of the Real part of the triplet 2^3S state wave function of the 'artificial atom' as a function of the coordinates (**s**, **R**) for $\alpha = 30°$, where α is the angle of the relative coordinate. Observe that the parity of the wave function about the line $\mathbf{s} = 0$ is odd. In Fig. 8.11(c) and (d), two views of the Imaginary part of the wave function are plotted. Again, the parity about the line $\mathbf{s} = 0$ is odd.

Finally, implicit in the understanding that the singlet state wave functions have even parity is that they satisfy the *cusp* electron-electron coalescence constraint. This too is evident in Fig. 8.10 for all points $\mathbf{s} = 0$. The fact that triplet states have odd parity then implies that these wave functions satisfy the *node* coalescence condition. This is evident for both the Real and Imaginary parts of the triplet state wave function in Fig. 8.11 along the $\mathbf{s} = 0$ line.

8.3.5 Proof of Satisfaction of the Wave Function Identity By the Exact Wave Function

In this section we prove that the exact wave function of any 2-electron system with symmetrical binding potential $v(\mathbf{r})$ must satisfy the Wave Function Identity. The examples of the previous section show that this is the case for the 2D 'artificial atoms'

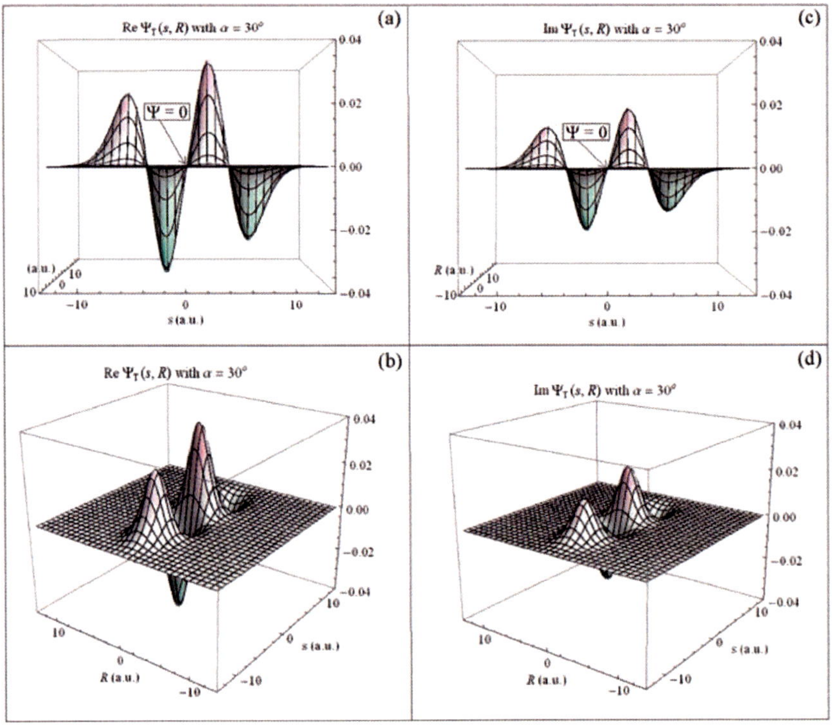

Fig. 8.11 Two views of the Real (**a**), (**b**) and Imaginary (**c**), (**d**) parts of the triplet 2^3S wave function for the 'artificial atom' plotted as a function of the center of mass **R** and relative **s** coordinates for $\alpha = 30°$, where α is the angle of the relative coordinate **s**

with symmetric harmonic binding potential. This is also the case for 3D 'artificial atoms' with the magnetic field absent. This can be seen from the solutions of the corresponding Schrödinger equations which can also be obtained *exactly* in closed analytical form [36, 37, 51–54]. Here we prove the result for *arbitrary* symmetrical potential $v(\mathbf{r})$ and *arbitrary* interaction of the form $w(|\mathbf{r} - \mathbf{r}'|)$.

The exact wave function $\Psi(\mathbf{x}_1, \mathbf{x}_2)$ of (8.76) may be written as

$$\Psi(\mathbf{x}_1, \mathbf{x}_2) = \left[\sum_i c_i \Phi_i(\mathbf{r}_1, \mathbf{r}_2)\right] \chi(\sigma_1, \sigma_2), \tag{8.111}$$

so that the spatial part $\psi(\mathbf{r}_1, \mathbf{r}_2)$ is given by the infinite series expansion

$$\psi(\mathbf{r}_1, \mathbf{r}_2) = \sum_i c_i \Phi_i(\mathbf{r}_1, \mathbf{r}_2), \tag{8.112}$$

where the functions $\Phi_i(\mathbf{r}_1, \mathbf{r}_2)$ form a complete orthonormal set, with the coefficients c_i being suitably chosen constants. If it is ensured that each term $\Phi_i(\mathbf{r}_1, \mathbf{r}_2)$ satisfies the Wave Function Identity, the exact wave function will satisfy the identity.

Triplet States

Since for the triplet state, the spin component of the wave function $\chi(\sigma_1, \sigma_2)$ is symmetric in an interchange of the spin coordinates (σ_1, σ_2), each spatial component function $\Phi_i(\mathbf{r}_1, \mathbf{r}_2)$ must be antisymmetric in an interchange of the spatial coordinates $(\mathbf{r}_1, \mathbf{r}_2)$. Thus, an arbitrary determinant $\Phi(\mathbf{r}_1, \mathbf{r}_2)$ is given as

$$\Phi(\mathbf{r}_1, \mathbf{r}_2) = \phi_1(\mathbf{r}_1)\phi_2(\mathbf{r}_2) - \phi_2(\mathbf{r}_1)\phi_1(\mathbf{r}_2). \tag{8.113}$$

As the binding potential $v(\mathbf{r})$ is symmetrical, it is possible to ensure that the orbital $\phi_1(\mathbf{r})$ has *even* parity and the orbital $\phi_2(\mathbf{r})$ has *odd* parity, i.e..

$$\phi_1(\mathbf{r}) = \phi_1(-\mathbf{r})\,; \quad \phi_2(\mathbf{r}) = -\phi_2(-\mathbf{r}). \tag{8.114}$$

Then, the spatial part of the right hand side of the Wave Function Identity of (8.93) (with $\psi(-\mathbf{r}_2, -\mathbf{r}_1)$ replaced by $\Phi(-\mathbf{r}_2, -\mathbf{r}_1)$) employing (8.113) becomes

$$\Phi(-\mathbf{r}_2, -\mathbf{r}_1) = \phi_1(-\mathbf{r}_2)\phi_2(-\mathbf{r}_1) - \phi_2(-\mathbf{r}_2)\phi_1(-\mathbf{r}_1). \tag{8.115}$$

On substituting (8.114) on the right hand side of (8.115), we have

$$\Phi(-\mathbf{r}_2, -\mathbf{r}_1) = \phi_1(\mathbf{r}_2)[-\phi_2(\mathbf{r}_1)] - [-\phi_2(\mathbf{r}_2)]\phi_1(\mathbf{r}_1) \tag{8.116}$$
$$= \phi_1(\mathbf{r}_1)\phi_2(\mathbf{r}_2) - \phi_2(\mathbf{r}_1)\phi_1(\mathbf{r}_2) \tag{8.117}$$
$$= \Phi(\mathbf{r}_1, \mathbf{r}_2). \tag{8.118}$$

This proves that the Wave Function Identity is satisfied for each determinant $\Phi_i(\mathbf{r}_1, \mathbf{r}_2)$ of $\psi(\mathbf{r}_1, \mathbf{r}_2)$ of (8.112), and thereby by the *exact* triplet state wave function $\Psi(\mathbf{x}_1, \mathbf{x}_2)$ of (8.111).

Singlet States

For the singlet state, the spin component of the wave function $\chi(\sigma_1, \sigma_2)$ is antisymmetric in an interchange of the spin coordinates (σ_1, σ_2). Thus, each spatial component term $\Phi_i(\mathbf{r}_1, \mathbf{r}_2)$ must be symmetric in an interchange of the spatial coordinates $(\mathbf{r}_1, \mathbf{r}_2)$. Hence, the permanent $\Phi_i(\mathbf{r}_1, \mathbf{r}_2)$ is given as

$$\Phi_i(\mathbf{r}_1, \mathbf{r}_2) = \phi_1(\mathbf{r}_1)\phi_2(\mathbf{r}_2) + \phi_2(\mathbf{r}_1)\phi_1(\mathbf{r}_2). \tag{8.119}$$

If one now ensures that *both* orbitals $\phi_1(\mathbf{r})$ and $\phi_2(\mathbf{r})$ have *even* parity, then once again following the procedure above, we have

$$\Phi(-\mathbf{r}_2, -\mathbf{r}_1) = \Phi(\mathbf{r}_1, \mathbf{r}_2), \tag{8.120}$$

Thus, each permanent $\Phi_i(\mathbf{r}_1, \mathbf{r}_2)$, and therefore the *exact* singlet state wave function $\Psi(\mathbf{x}_1, \mathbf{x}_2)$, too satisfies the Wave Function Identity.

What one also learns from the above, is that in the construction of approximate configuration-interaction type wave functions of the form of (8.111), one orbital $\phi_1(\mathbf{r})$ must always have even parity, whereas the other $\phi_2(\mathbf{r})$ must have even parity for singlet states and odd parity for triplet states. This will ensure not only the satisfaction of the Wave Function Identity, but also that the approximate singlet and triplet state wave functions will have even and odd parity, respectively, as they must. The approximate wave functions will then also have the correct parity about each point of electron-electron coalescence.

References

1. X.-Y. Pan, V. Sahni, J. Chem. Phys. **119**, 7083 (2003)
2. T. Kato, Commun. Pure Appl. Math. **10**, 151 (1957)
3. E. Steiner, J. Chem. Phys. **39**, 2365 (1963)
4. W.A. Bingel, Z. Naturforsch. A **18A**, 1249 (1963)
5. R.T. Pack, W.Byers Brown, J. Chem. Phys. **45**, 556 (1966)
6. W.A. Bingel, Theor. Chim. Acta **8**, 54 (1967)
7. T. Ando, A.B. Fowler, F. Stern, Rev. Mod. Phys. **54**, 437 (1982)
8. E. Abrahams, S.V. Kravechenko, M.P. Sarachik, Rev. Mod. Phys. **73**, 251 (2001)
9. R.B. Laughlin, Rev. Mod. Phys. **71**, 863 (1998)
10. R.C. Ashoori, Nature **379**, 413 (1996)
11. S.M. Reimann, M. Manninen, Rev. Mod. Phys. **74**, 1283 (2002)
12. H. Saarikovski et al., Rev. Mod. Phys. **82**, 2785 (2010)
13. W. Zhou, J.J. Coleman, Curr. Opin. Solid State **20**, 352 (2016)
14. S.C. Zhang, J. Hu, Science **294**, 823 (2001)
15. D. Karabali, V.P. Nair, Nucl. Phys. B **641**, 533 (2002)
16. X.-Y. Pan, V. Sahni, Phys. Rev. A **67**, 012501 (2003)
17. Z. Qian, V. Sahni, Int. J. Quantum Chem. **79**, 205 (2000)
18. Z. Qian, V. Sahni, Phys. Rev. A **75**, 1 (2007)
19. Z. Qian, Phys. Rev B **75**, 193104 (2007)
20. Z. Qian, V. Sahni, Phys. Rev. A **57**, 4041 (1998)
21. Z. Qian, V. Sahni, Int. J. Quantum Chem. **70**, 671 (1998)
22. Z. Qian, V. Sahni, Int. J. Quantum Chem. **80**, 555 (2000)
23. J. Katriel, E.R. Davidson, Proc. Natl. Acad. Sci. USA **77**, 4403 (1980)
24. M. Slamet, V. Sahni, Chem. Phys. **556**, 111453 (2022)
25. V. Sahni, Int. J Quantum Chem. **121**, e26556 (2021)
26. M. Slamet, V. Sahni, Chem. Phys. **546**, 111073 (2021)
27. M. Slamet, V. Sahni, Comput. Theor. Chem. **1114**, 125 (2017)
28. A.J. Thakkar, V.J. Smith Jr., Chem. Phys. Lett. **42**, 476 (1976); W. Kutzelnigg, W. Kopper, J. Chem. Phys. **94**, 1985 (1991); H. Cox, S.J. Smith, B.T. Sutcliffe, Phys. Rev. A **49**, 4520 (1994); F.S. Carvalho, J.P. Braga, J. Phys. B **51**, 135001 (2018)
29. A. Nagy, K.D. Sen, Chem. Phys. Lett. **332**, 154 (2000); J. Phys. B **33**, 1745 (2000); J. Chem. Phys. **115**, 6300 (2001)
30. J.C. Kimball, Phys. Rev. A **7**, 1648 (1973)
31. A.K. Rajagopal, J.C. Kimball, Phys. Rev. B **15**, 2819 (1977)
32. A.K. Rajagopal, J.C. Kimball, M. Bannerjee, Phys. Rev. B **18**, 2339 (1978)
33. J.P. Perdew, Y. Wang, Phys. Rev. B **46**, 12947 (1992)

34. N.H. March et al., Phys. Rev. A **63**, 012520 (2000)
35. D.W. Smith, S. Jagannathan, G.S. Handler, Int. J. Quantum Chem. Symp. **13**, 103 (1979); E.R. Davidson, Int. J. Quantum Chem. **37**, 811 (1990); C.J. Umrigar, X. Gonze, Phys. Rev. A **50**, 3827 (1994); C.-O. Almbladh, A.C. Pedroza, Phys. Rev. A **29**, 2322 (1984); A.C. Pedroza, Phys. Rev. A **33**, 804 (1986); Q. Zhao, R.C. Morrison, R.G. Parr, Phys. Rev. A **50**, 2138 (1994); R.C. Morrison, Q. Zhao, Phys. Rev. A **51**, 1980 (1995); C. Fillipi, X. Gonze, C.J. Umrigar, in *Recent Developments and Applications of Density Functional Theory*, ed. By J. Seminario (Elsevier, Ansterdam, 1996); J. Chen, R.O. Esquivel, M.J. Stott, Phil. Mag. B **69**, 1001 (1994); O. V. Gritsenko, R. van Leeuwen, E.J. Baerends, Phys. Rev. A **52**, 1870 (1995); S. Liu, R.G. Parr, A. Nagy, Phys. Rev. A **52**, 2645 (1995); R. Singh, L. Massa, V. Sahni, Phys. Rev. A **60**, 41355 (1999); D.C. Langreth, M.J. Mehl, Phys. Rev. B **28**, 1809 (1983); J.P. Perdew, Y. Wang, Phys. Rev. B **33**, 8800 (1986); J.P. Perdew, in *Electronic Structure of Solids '91*, ed. By P.Ziesche, H. Eschrig (Akademic Verlag, Berlin, 1991); A.D. Becke, Phys. Rev. A **38**, 3098 (1988)
36. Z. Qian, V. Sahni, Phys. Rev. A **57**, 2527 (1998)
37. M. Slamet, V. Sahni, Int. J. Quantum Chem. **85**, 436 (2001)
38. V. Sahni, M. Slamet. Int. J. Quantum Chem. **100**, 858 (2004)
39. V. Sahni, L. Massa, R. Singh, M. Slamet, Phys. Rev. Lett. **87**, 113002 (2001)
40. M. Slamet, R. Singh, L. Massa, V. Sahni Phys. Rev. A **68**, 042504 (2003)
41. T. Yang, X.-Y. Pan, V. Sahni, Phys Rev A **83**, 042518 (2011)
42. J.P. Perdew, R.G. Parr, M. Levy, J.L. Balduz, Phys. Rev. Lett. **49**, 1691 (1982)
43. M. Levy, J.P. Perdew, V. Sahni, Phys. Rev. A **30**, 2745 (1984)
44. C.-O. Almbladh, U. von Barth, Phys. Rev. B **31**, 3231 (1985)
45. E. Schrödinger, Ann. Physik. **79**, 362 (1926); ibid **79**, 489 (1926)
46. W. Pauli, Z. Physik **43**, 601 (1927)
47. M. Taut, J. Phys. A **27**, 1045 (1994); Corrigenda J. Phys. A **27**, 4723 (1994)
48. W. Pauli, Z. Physik **31**, 765 (1925)
49. P.A.M. Dirac, Proc. Roy. Soc. A **112**, 661 (1926)
50. W. Heisenberg, Z. Physik **38**, 411 (1926)
51. V. Sahni, X.-Y. Pan, Phys. Rev. Lett. **90**, 123001 (2003)
52. N.R. Kestner, O. Sinanoglu, Phys. Rev. **128**, 2687 (1962)
53. S. Kais, D.R. Herschbach, R.D. Levine, J. Chem. Phys. **91**, 7791 (1989)
54. M. Taut, Phys. Rev. A **48**, 3561 (1993)

Chapter 9
Wave Functions for Harmonically Bound Electrons in an Electromagnetic Field

Abstract This Chapter is comprised of two derivations of closed-form analytical expressions for wave functions of systems of *interacting* electrons, one for the stationary-state and the other for the temporal case. The expressions then allow for the determination of the exact properties of these systems. In particular, they provide an explicit description of the electron correlations due to the Pauli principle and Coulomb repulsion within the context of both the traditional and complementary descriptions of Schrödinger theory. The stationary-state derivation is for the general analytical form of the wave functions of the Schrödinger-Pauli eigenvalue equation for two harmonically bound electrons in a magnetic field. The Frobenius power series method of solution is employed. For each state, an infinite number of analytical solutions exist. Expressions for the corresponding eigenvalues too are obtained. The analytical expressions for the wave functions and energies for a ground and a first excited triplet 2^3S state are thereby derived. A *first principles* derivation via the 'interaction' representation of time-dependent (TD) Schrödinger theory is provided for the Generalized Kohn Theorem (GKT) wave function. This is the solution to the time-dependent Schrödinger equation of a system of N harmonically bound electrons in a magnetic field perturbed by a spatially uniform time-dependent electric field. The GKT wave function is comprised of a phase factor times the unperturbed wave function in which the coordinates of each electron are translated by a TD value that satisfies the corresponding classical equation of motion. Hence, if the unperturbed wave functions are known, the time-evolution of the wave functions and thereby that of the electron correlations of the system, are then determinable. In particular, the evolution of properties which are expectations of Hermitian operators that do not involve a differential operator, is simply that of the unperturbed system translated by a time-dependent function. As the solutions to the stationary-state system of two harmonically bound electrons in both the presence and absence of a magnetic field are known analytically, the time evolution of all properties of these systems in a TD electric field is determined.

V. Sahni, *Schrödinger Theory of Electrons: Complementary Perspectives*,
Springer Tracts in Modern Physics 285,
https://doi.org/10.1007/978-3-030-97409-1_9

Introduction

In texts on quantum mechanics, detailed derivations are provided for the analytical solutions to the stationary-state Schrödinger equation [1] for the Hydrogen atom [1] employing the power series method. Solutions may also be obtained via an analytical approach employing knowledge of the solution in the asymptotic near nucleus and classically forbidden regions, with a polynomial solution in the intermediate region. Additionally, there is the derivation for the harmonic oscillator [2] employing the power series method. The latter example leads naturally to the solutions for a free electron in a magnetic field, and that of a harmonically bound electron in a magnetic field—the Fock-Darwin states [3, 4]. However, as one moves to the next element of the Periodic Table—the two-electron Helium atom—there are no closed-form analytical solutions to the corresponding eigenvalue equation. As such it is not possible to study the effects of electron correlations, *viz.* those due to the Pauli principle and Coulomb repulsion analytically, or even to study the Pauli principle [5–7]. Hence, one resorts to the variational principle for the energy [8] whereby for an approximate wave function Ψ correct to $O(\delta)$, a rigorous upper bound to the energy is obtained correct to $O(\delta^2)$. Other properties of the system, which are expectation values of Hermitian single and two-particle operators, are then obtained correct only to $O(\delta)$. Variational principles for single particle properties such as the density, diamagnetic susceptibility, nuclear magnetic constant, size of atom, coherent atomic scattering factor (the Fourier transform of the density) have consequently been developed [9]. Thus, starting with a wave function Ψ correct to $O(\delta)$, the corresponding expectation values are obtained correct to $O(\delta^2)$. These are stationary-state rather than minimum principles. Similar variational principles have been developed [10] for properties such as the single-particle density matrix, and hence for the momentum density and properties in momentum space such as the Compton profile. There is then the idea [11] of constructing *constrained-search variational wave functions* Ψ that are functionals of functions χ, *i.e.* $\Psi = \Psi[\chi]$. This then expands the space of variations beyond those of parameterized functions. As the space of variations is large, one constrains the search for the functions χ to those for which $\Psi[\chi]$ reproduces a known property which is the expectation of a single or two-particle Hermitian operator whilst simultaneously obtaining a rigorous upper bound to the energy. Thus, the wave function functionals $\Psi[\chi]$ are not only accurate in the region of the property of interest but also lead to an accurate energy. Nevertheless, all these variational wave functions and wave function functionals are approximate. It is as such not possible to elucidate in an *exact* manner the electron correlation properties of an interacting system, properties such as the Pauli principle, the nonlocal pair-correlation density, the Fermi and Fermi-Coulomb hole charge distributions, the single-particle density matrix, the fields arising from these quantal sources, and the satisfaction and structure of the wave function at electron-electron coalescence.

It has, however, been possible to obtain closed-form analytical solutions to the Schrödinger and Schrödinger-Pauli theory equations for two harmonically bound electrons in the absence and presence of a magnetic field in an arbitrary state [12–17].

As a consequence of advances in technology, such ‘artificial atoms’ or semiconductor quantum dots have been created. The properties of such ‘atoms’ are very similar to those of natural atoms. It is the length scales that are different. The fact that analytical solutions exist thus enables the study of electron correlations in finite systems exactly. Most properties of the system are also obtained in closed analytical form. (For such studies, see Chaps. 5, 6, 8 and corresponding Appendices.) In the first section of this Chapter, we provide a detailed derivation for the general analytical solutions of the Schrödinger-Pauli equation for the 2-dimensional 2-electron quantum dot in a magnetic field. The power series method of Taut [15] is followed. The specific examples of a ground state and the first excited triplet 2^3S state wave functions are then derived.

The next question that arises is whether there exist examples whereby it is possible to describe the time evolution of an interacting system of electrons. In other words, is it possible to solve the corresponding temporal Schrödinger and Schrödinger-Pauli equations, and thereby study the electron correlations and the Pauli principle as they evolve in time? The answer is in the affirmative. In the second section of the chapter, we derive from *first principles* via the ‘interaction’ representation the Generalized Kohn Theorem (GKT) wave function. The GKT wave function $\Psi(\mathbf{r}_1, \mathbf{r}_2, \ldots, \mathbf{r}_N; t)$ is the solution to the time-dependent (TD) Schrödinger equation for a system of N harmonically bound electrons in a magnetostatic field $\mathcal{B}(\mathbf{r}) = \boldsymbol{\nabla} \times \mathbf{A}(\mathbf{r})$, with $\mathbf{A}(\mathbf{r})$ the vector potential, perturbed by a spatially homogeneous TD electric field $\mathbf{E}(t)$. The GKT wave function is comprised of a phase factor times the unperturbed wave function in which the coordinates of each electron are translated by a TD function that satisfies the corresponding classical equation of motion. Hence, if the unperturbed wave function is known, the time evolution of the wave function and properties of the system are then known. Additionally, the evolution of all properties that are expectations of non-differential Hermitian operators such as the density, the pair-correlation density, the Fermi-Coulomb hole charge, the single-particle density matrix, and the fields derived from these quantal sources, are *simply those of the unperturbed system translated by a TD function*. As a consequence of the phase factor, this is not the case for properties such as the physical current density which is an observable involving differential operators. Nonetheless, as the temporal wave function is known, the time evolution of such properties can be determined. In particular, as shown in the first section, the stationary-state wave functions for two harmonically bound electrons in a magnetostatic field can be derived in closed analytical form. Thus the evolution with time of an arbitrary state of such a system and of its properties can be studied. The derivation by Zhu, et al. [18] of the GKT wave function via the interaction ‘representation’ of the Schrödinger equation is provided. With the above closed-form analytical stationary-state and time-dependent solutions, the exact electron correlations and other properties of the quantum system can be studied within both the traditional and complementary perspectives to Schrödinger theory as described in the earlier chapters of the book.

9.1 Wave Functions for the 2D 2-Electron Harmonically Bound 'Artificial Atom' in a Magnetic Field

Consider a 2-dimensional system of 2 electrons of mass m, charge $-e$, and spin angular momentum $\mathbf{s}$ in an external binding field $\mathcal{E}(\mathbf{r}) = -\nabla v(\mathbf{r})/e$ and a magnetic field $\mathcal{B}(\mathbf{r}) = \nabla \times \mathbf{A}(\mathbf{r})$, where $\{v(\mathbf{r}), \mathbf{A}(\mathbf{r})\}$ are their respective scalar and vector potentials. If the binding potential is harmonic with frequency ω_0, the system represents a 2-dimensional 'atom'. The corresponding Schrödinger-Pauli theory Hamiltonian $\hat{H}$ is

$$\hat{H} = \hat{T}_F + \hat{U} + \hat{V}, \tag{9.1}$$

where the Feynman kinetic energy operator $\hat{T}_F$ is

$$\hat{T}_F = \frac{1}{2m}\sum_{k=1}^{2}\Big(\boldsymbol{\sigma}\cdot\hat{\mathbf{p}}_{k,phys}\Big)\Big(\boldsymbol{\sigma}\cdot\hat{\mathbf{p}}_{k,phys}\Big), \tag{9.2}$$

with $\boldsymbol{\sigma}$ the Pauli spin matrix defined in terms of the spin momentum $\mathbf{s} = \frac{\hbar}{2}\boldsymbol{\sigma}$; the physical momentum $\hat{\mathbf{p}}_{phys} = (\hat{\mathbf{p}} + \frac{e}{c}\mathbf{A})$; and the canonical momentum $\hat{\mathbf{p}} = -i\hbar\nabla$. Thus, the kinetic energy operator is

$$\hat{T}_F = \frac{1}{2m}\sum_{k=1}^{2}\Big(\hat{\mathbf{p}}_k + \frac{e}{c}\mathbf{A}(\mathbf{r}_k)\Big)^2 + (g\mu_B/\hbar)\sum_{k=1}^{2}\mathcal{B}(\mathbf{r}_k)\cdot\mathbf{s}_k, \tag{9.3}$$

where the gyromagnetic ratio $g = 2$, and $\mu_B = e\hbar/2mc$ is the Bohr magneton. The electron Coulomb interaction operator $\hat{U}$ is

$$\hat{U} = \frac{1}{|\mathbf{r}_1 - \mathbf{r}_2|}, \tag{9.4}$$

and the binding potential operator $\hat{V}$ is

$$\hat{V} = \frac{1}{2}m\omega_0\sum_{k=1}^{2}r_k^2. \tag{9.5}$$

The Hamiltonian in units where $e = \hbar = m = c = 1$ is then

$$\hat{H} = \frac{1}{2}\sum_{k=1}^{2}\Big(\hat{\mathbf{p}}_k + \mathbf{A}(\mathbf{r}_k\Big)^2 + \sum_{k=1}^{2}\mathcal{B}(\mathbf{r}_k)\cdot\mathbf{s}_k + \frac{1}{|\mathbf{r}_1 - \mathbf{r}_2|} + \frac{1}{2}\omega_0^2\sum_{k=1}^{2}r_k^2, \tag{9.6}$$

with the Schrödinger-Pauli equation being

$$\hat{H}\Psi(\mathbf{x}_1\mathbf{x}_2) = E\Psi(\mathbf{x}_1\mathbf{x}_2), \tag{9.7}$$

with $\mathbf{x} = \mathbf{r}\sigma$, the spatial and spin coordinates, respectively. It turns out that there exist closed-form analytical solutions of the differential equations (9.7) for a denumerably infinite set of effective oscillator frequencies.

$$\Omega = \sqrt{\omega_0^2 + \omega_L^2}, \tag{9.8}$$

where the Larmor frequency $\omega_L = \frac{\omega_c}{2}$, with the cyclotron frequency $\omega_c = \frac{B}{c}$. In the above units $\omega_L = \mathcal{B}/2c$.

In the symmetric gauge $\mathbf{A}(\mathbf{r}) = \frac{1}{2}\boldsymbol{\mathcal{B}}(\mathbf{r}) \times \mathbf{r}$, with the magnetic field assumed uniform and in the z-direction perpendicular to the plane of the atom defined by the polar coordinates (r, θ): $\boldsymbol{\mathcal{B}}(\mathbf{r}) = \mathcal{B}\hat{\mathbf{e}}_z$ and $\mathbf{r} = r\hat{\mathbf{e}}_r$, we have

$$\mathbf{A}(\mathbf{r}) = \omega_L r \hat{\mathbf{e}}_\theta. \tag{9.9}$$

Next, rewrite the Hamiltonian $\hat{H}$ of (9.6) as

$$\hat{H} = \hat{H}_1 + \hat{H}_2 + \hat{H}_3 + \hat{H}_4 + \hat{H}_5 + \hat{H}_{spin}, \tag{9.10}$$

where on expanding the first and fourth terms,

$$\hat{H}_1 = \frac{1}{2}(\hat{p}_1^2 + \hat{p}_2^2) \tag{9.11}$$

$$\hat{H}_2 = \frac{1}{2}[A^2(\mathbf{r}_1) + A^2(\mathbf{r}_2)] \tag{9.12}$$

$$\hat{H}_3 = [\hat{\mathbf{p}}_1 \cdot A(\mathbf{r}_1) + \hat{\mathbf{p}}_2 \cdot A(\mathbf{r}_2)] \tag{9.13}$$

$$\hat{H}_4 = \frac{1}{2}\omega_0^2(r_1^2 + r_2^2) \tag{9.14}$$

$$\hat{H}_5 = 1/|\mathbf{r}_1 - \mathbf{r}_2| \tag{9.15}$$

$$\hat{H}_{spin} = \sum_{k=1}^{2} \mathcal{B}\hat{\mathbf{e}}_z \cdot \mathbf{s}_k. \tag{9.16}$$

9.1.1 *Decoupling of the Hamiltonian*

As a consequence of the binding potential $v(\mathbf{r})$ being harmonic, it is possible to decouple the Hamiltonian $\hat{H}$ of (9.6) into its center of mass $\mathbf{R}$ and relative $\mathbf{s}$ coordinate systems. Thus, with

$$\mathbf{R} = \frac{\mathbf{r}_1 + \mathbf{r}_2}{2} \quad \text{and} \quad \mathbf{s} = \mathbf{r}_2 - \mathbf{r}_1, \tag{9.17}$$

we have

$$R^2 + \frac{1}{4}s^2 = \frac{1}{2}(r_1^2 + r_2^2). \tag{9.18}$$

Define the new frequencies ω_s and ω_R as

$$\omega_s = \frac{\omega_0}{2} \quad \text{and} \quad \omega_R = 2\omega_0, \tag{9.19}$$

and the new momentum operators $\hat{\mathbf{q}}$ and $\hat{\mathbf{Q}}$ as

$$\hat{\mathbf{q}} = \frac{1}{i}\nabla_{\mathbf{s}} = \frac{1}{2}(\hat{\mathbf{p}}_2 - \hat{\mathbf{p}}_1) \quad \text{and} \quad \hat{\mathbf{Q}} = \frac{1}{i}\nabla_{\mathbf{R}} = (\hat{\mathbf{p}}_1 + \hat{\mathbf{p}}_2). \tag{9.20}$$

Therefore, on employing (9.18) and (9.19),

$$\hat{H}_4 = \frac{1}{4}\omega_R^2 R^2 + \omega_s^2 s^2. \tag{9.21}$$

On employing (9.20),

$$\hat{H}_1 = \hat{q}^2 + \frac{1}{4}\hat{Q}^2. \tag{9.22}$$

As the magnetic field $\boldsymbol{\mathcal{B}}(\mathbf{r})$ is uniform, one may define

$$\mathbf{A}(\mathbf{s}) = \mathbf{A}(\mathbf{r}_2) - \mathbf{A}(\mathbf{r}_1) \quad \text{and} \quad \mathbf{A}(\mathbf{R}) = \frac{1}{2}[\mathbf{A}(\mathbf{r}_1) + \mathbf{A}(\mathbf{r}_2)], \tag{9.23}$$

so that

$$\hat{H}_2 = A^2(\mathbf{R}) + \frac{1}{4}A^2(\mathbf{s}). \tag{9.24}$$

Define the new vector potentials $\mathbf{A_R}$ and $\mathbf{A_s}$ as

$$\mathbf{A_R} = 2\mathbf{A}(\mathbf{R}) \quad \text{and} \quad \mathbf{A_s} = \frac{1}{2}\mathbf{A_s}. \tag{9.25}$$

In terms of these potentials, the sum of $\hat{H}_1$ (9.22) and $\hat{H}_2$ (9.24) is

$$\hat{H}_1 + \hat{H}_2 = (\hat{q}^2 + A_{\mathbf{s}}^2) + \frac{1}{4}(\hat{Q}^2 + A_{\mathbf{R}}^2). \tag{9.26}$$

To express $\hat{H}_3$ of (9.13) in terms of the new momentum ($\hat{\mathbf{q}}$, $\hat{\mathbf{Q}}$) and vector potential ($\mathbf{A_s}$, $\mathbf{A_R}$) operators, the procedure is to double the terms, write them as two separate sets of terms, multiply by $\frac{1}{2}$, add and subtract the cross-terms $(\hat{\mathbf{p}}_1 \cdot \mathbf{A}(\mathbf{r}_2))$, $(\hat{\mathbf{p}}_2 \cdot \mathbf{A}(\mathbf{r}_1))$. Thus,

$$
\begin{aligned}
H_3 &= \frac{1}{2}[\hat{\mathbf{p}}_1 \cdot \mathbf{A}(\mathbf{r}_1) + \hat{\mathbf{p}}_2 \cdot \mathbf{A}(\mathbf{r}_2)] + \frac{1}{2}[\hat{\mathbf{p}}_1 \cdot \mathbf{A}(\mathbf{r}_1) + \hat{\mathbf{p}}_2 \cdot \mathbf{A}(\mathbf{r}_2)] \\
&= \frac{1}{2}[\hat{\mathbf{p}}_1 \cdot \mathbf{A}(\mathbf{r}_1) - \hat{\mathbf{p}}_1 \cdot \mathbf{A}(\mathbf{r}_2) + \hat{\mathbf{p}}_2 \cdot \mathbf{A}(\mathbf{r}_2) - \hat{\mathbf{p}}_2 \cdot \mathbf{A}(\mathbf{r}_1)] \\
&+ \frac{1}{2}[\hat{\mathbf{p}}_1 \cdot \mathbf{A}(\mathbf{r}_1) + \hat{\mathbf{p}}_1 \cdot \mathbf{A}(\mathbf{r}_2) + \hat{\mathbf{p}}_2 \cdot \mathbf{A}(\mathbf{r}_2) + \hat{\mathbf{p}}_2 \cdot \mathbf{A}(\mathbf{r}_1)] \\
&= \frac{1}{2}\big\{(\hat{\mathbf{p}}_2 - \hat{\mathbf{p}}_1) \cdot [\mathbf{A}(\mathbf{r}_2) - \mathbf{A}(\mathbf{r}_1)]\big\} + \frac{1}{2}\big\{(\hat{\mathbf{p}}_2 + \hat{\mathbf{p}}_1) \cdot [\mathbf{A}(\mathbf{r}_2) + \mathbf{A}(\mathbf{r}_1)]\big\} \\
&= \hat{\mathbf{q}} \cdot 2\mathbf{A}_s + \frac{1}{2}\hat{\mathbf{Q}} \cdot \mathbf{A}_{\mathbf{R}}.
\end{aligned}
\tag{9.27}
$$

Therefore, (see (9.26) and (9.27)),

$$
\hat{H}_1 + \hat{H}_2 + \hat{H}_3 = (\hat{\mathbf{q}} + \mathbf{A}_s)^2 + \frac{1}{4}(\hat{\mathbf{Q}} + \mathbf{A}_{\mathbf{R}})^2. \tag{9.28}
$$

The Hamiltonian $\hat{H}$ of (9.6) may then be written as

$$
\hat{H} = 2\hat{H}_s + \frac{1}{2}\hat{H}_{\mathbf{R}} + \hat{H}_{spin}, \tag{9.29}
$$

where

$$
\hat{H}_s = \frac{1}{2}(\hat{\mathbf{q}} + \mathbf{A}_s)^2 + \frac{1}{2}\omega_s^2 s^2 + \frac{1}{2s}, \tag{9.30}
$$

$$
\hat{H}_{\mathbf{R}} = \frac{1}{2}(\hat{\mathbf{Q}} + \mathbf{A}_{\mathbf{R}})^2 + \frac{1}{2}\omega_R^2 R^2, \tag{9.31}
$$

and $\hat{H}_{spin}$ is defined by (9.16). With the Hamiltonian thus separated into its relative, center of mass and spin components, the solution to the Schrödinger-Pauli equation (9.7) may then be expressed as

$$
\Psi(\mathbf{x}_1\mathbf{x}_2) = \phi(\mathbf{s})\Phi(\mathbf{R})\chi(\sigma_1\sigma_2), \tag{9.32}
$$

where $\phi(\mathbf{s})$ is the solution to

$$
\hat{H}_s\phi(\mathbf{s}) = \epsilon_s\phi(\mathbf{s}), \tag{9.33}
$$

and $\Phi(\mathbf{R})$ to

$$
\hat{H}_{\mathbf{R}}\Phi(\mathbf{R}) = \eta_R\Phi(\mathbf{R}), \tag{9.34}
$$

with $\chi(\sigma_1\sigma_2)$ the spin component of the wave function. The energy eigenvalue E of (9.7) is then

$$
E = 2\epsilon_s + \frac{1}{2}\eta_R + E_{spin}. \tag{9.35}
$$

Note that the eigenvalue equation (9.34) for the center of mass coordinate $\mathbf{R}$ is that of a 2-dimensional harmonic oscillator in a magnetic field. Further, the electron interaction potential term appears only in the relative coordinate component of the Hamiltonian.

The center of mass coordinate wave function component $\Phi(\mathbf{R})$ is symmetric in an interchange of the coordinates $\mathbf{r}_1$ and $\mathbf{r}_2$. Since for *singlet* states, the spin function $\chi(\sigma_1\sigma_2)$ is antisymmetric in an interchange of the spin coordinates σ_1 and σ_2, the relative coordinate component $\phi(\mathbf{s})$ must be symmetric in an interchange of $\mathbf{r}_1$ and $\mathbf{r}_2$, *i.e.* $\phi(-\mathbf{s}) = \phi(\mathbf{s})$. For *triplet* states, the spin function is symmetric in an interchange of the coordinates σ_1 and σ_2. Hence, $\phi(\mathbf{s})$ for these states must be antisymmetric in an interchange of $\mathbf{r}_1$ and $\mathbf{r}_2$, *i.e.* $\phi(-\mathbf{s}) = -\phi(\mathbf{s})$.

9.1.2 The Relative Coordinate Wave Function Component $\phi(\mathbf{s})$

We next determine the differential equation corresponding to the relative coordinate Hamiltonian $\hat{H}_s$ of (9.30) for the choice of vector potential $\mathbf{A}(\mathbf{r})$ gauge of (9.9). From (9.9),

$$\mathbf{A}(\mathbf{s}) = \omega_L s\hat{\mathbf{e}}_\theta, \tag{9.36}$$

and employing (9.25) we have

$$\mathbf{A}(\mathbf{s}) = \frac{\omega_L}{2} s\hat{\mathbf{e}}_\theta. \tag{9.37}$$

With $\mathbf{q} = \frac{1}{i}\nabla_{\mathbf{s}}$, the relative coordinate equation (9.33) is

$$\left\{\frac{1}{2}\Big[\frac{1}{i}\nabla_{\mathbf{s}} + \frac{\omega_L}{2}s\hat{\mathbf{e}}_\theta\Big]^2 + \frac{1}{2}\omega_s^2 s^2 + \frac{1}{2s}\right\}\phi(\mathbf{s}) = \epsilon_s\phi(\mathbf{s}). \tag{9.38}$$

Expanding, the first term of (9.38) on the left is

$$\begin{aligned}
&\frac{1}{2}\left[\frac{1}{i}\nabla_{\mathbf{s}} + \frac{\omega_L}{2}s\hat{\mathbf{e}}_\theta\right]\cdot\left[\frac{1}{i}\nabla_{\mathbf{s}} + \frac{\omega_L}{2}s\hat{\mathbf{e}}_\theta\right]\phi(\mathbf{s}) \\
&= \frac{1}{2}\Big[-\nabla_{\mathbf{s}}^2\phi(\mathbf{s}) + \frac{1}{i}\nabla_{\mathbf{s}}\cdot(\frac{\omega_L}{2}s)\phi(\mathbf{s})\hat{\mathbf{e}}_\theta + (\frac{\omega_L}{2}s)\hat{\mathbf{e}}_\theta\cdot\frac{1}{i}\nabla_{\mathbf{s}}\phi(\mathbf{s}) \\
&\quad + (\frac{\omega_L}{2}s)^2\phi(\mathbf{s})\hat{\mathbf{e}}_\theta\cdot\hat{\mathbf{e}}_\theta\Big].
\end{aligned} \tag{9.39}$$

As the system has cylindrical symmetry, we have the vector relations:

$$\nabla F(\mathbf{r}) = \frac{\partial F}{\partial r}\hat{\mathbf{e}}_r + \frac{1}{r}\frac{\partial F}{\partial \theta}\hat{\mathbf{e}}_\theta + \frac{\partial F}{\partial z}\hat{\mathbf{e}}_z, \tag{9.40}$$

$$\nabla \cdot F(\mathbf{r}) = \nabla \cdot [F_r\hat{\mathbf{e}}_r + F_\theta\hat{\mathbf{e}}_\theta + F_z\hat{\mathbf{e}}_z] \tag{9.41}$$

$$= \frac{1}{r}\frac{\partial}{\partial r}(rF_r) + \frac{1}{r}\frac{\partial F_\theta}{\partial \theta} + \frac{\partial F_z}{\partial z}, \tag{9.42}$$

$$\nabla^2 F(\mathbf{r}) = \frac{1}{r}\frac{\partial}{\partial r}\left(r\frac{\partial F}{\partial r}\right) + \frac{1}{r^2}\frac{\partial^2 F}{\partial \theta^2} + \frac{\partial^2 F}{\partial z^2}. \tag{9.43}$$

However, the electrons are confined to the $x - y$ plane. Thus, the divergence term of (9.39) employing (9.42) is

$$\begin{aligned} \frac{1}{i}\nabla_{\mathbf{s}} \cdot (\frac{\omega_L}{2}s)\phi(\mathbf{s})\hat{\mathbf{e}}_\theta &= \frac{1}{i}\frac{1}{s}\frac{\partial}{\partial \theta}(\frac{\omega_L}{2}s\phi(\mathbf{s})) \\ &= \frac{1}{i}\frac{\omega_L}{2}\frac{\partial \phi(\mathbf{s})}{\partial \theta}. \end{aligned} \tag{9.44}$$

The gradient term of (9.39) employing (9.40) is

$$\begin{aligned} (\frac{\omega_L}{2}s)\hat{\mathbf{e}}_\theta \cdot \frac{1}{i}\nabla_{\mathbf{s}}\phi(\mathbf{s}) &= \frac{1}{i}\frac{\omega_L}{2}s\frac{1}{s}\frac{\partial \phi(\mathbf{s})}{\partial \theta} \\ &= \frac{1}{i}\frac{\omega_L}{2}\frac{\partial \phi(\mathbf{s})}{\partial \theta}, \end{aligned} \tag{9.45}$$

which is identical to (9.44). The Laplacian term of (9.39) employing (9.43) is

$$-\nabla_{\mathbf{s}}^2\phi(\mathbf{s}) = -\frac{1}{s}\frac{\partial}{\partial s}(s\frac{\partial \phi(\mathbf{s})}{\partial s}) - \frac{1}{s^2}\frac{\partial^2 \phi(\mathbf{s})}{\partial \theta^2}. \tag{9.46}$$

Substituting (9.44)–(9.46) into (9.39) then leads to the first term of the relative coordinate equation (9.38) to be

$$\frac{1}{2}\left[-\frac{1}{s}\frac{\partial}{\partial s}(s\frac{\partial \phi(\mathbf{s})}{\partial s}) - \frac{1}{s^2}\frac{\partial^2 \phi(\mathbf{s})}{\partial \theta^2} + 2\frac{1}{i}\frac{\omega_L}{2}\frac{\partial \phi(\mathbf{s})}{\partial \theta} + \frac{\omega_L^2}{4}s^2\phi(\mathbf{s})\right]. \tag{9.47}$$

On employing (9.47), the relative coordinate differential equation (9.38) becomes

$$\begin{aligned} -\frac{1}{2s}\frac{\partial}{\partial s}\left(s\frac{\partial \phi(\mathbf{s})}{\partial s}\right) - \frac{1}{2s^2}\frac{\partial^2 \phi(\mathbf{s})}{\partial \theta^2} - i\frac{\omega_L}{2}\frac{\partial \phi(\mathbf{s})}{\partial \theta} + \frac{\omega_L^2}{8}s^2\phi(\mathbf{s}) \\ + \frac{\omega_s^2}{2}s^2\phi(\mathbf{s}) + \frac{1}{2s}\phi(\mathbf{s}) = \epsilon_s\phi(\mathbf{s}). \end{aligned} \tag{9.48}$$

In order to reduce the above equation to be solely in terms of the relative coordinate s, let us assume that

$$\phi(\mathbf{s}) = \frac{e^{im\theta}}{\sqrt{2\pi}} \frac{u(s)}{\sqrt{s}} \, ; \quad m = 0, \pm 1, \pm 2, \ldots \tag{9.49}$$

where m is the angular momentum quantum number. (This will eliminate terms involving $\partial/\partial s$, $\partial/\partial\theta$, and $\partial^2/\partial\theta^2$.) With this ansatz, we next determine each term on the left hand side of (9.48).

$$\begin{aligned}
\textit{1st Term:} \quad & -\frac{e^{im\theta}}{\sqrt{2\pi}} \frac{1}{2s} \frac{\partial}{\partial s} \left[s \frac{1}{\sqrt{s}} \frac{\partial u}{\partial s} + s(-\frac{1}{2}) \frac{1}{s^{3/2}} u \right] \\
&= -\frac{e^{im\theta}}{\sqrt{2\pi}} \frac{1}{2s} \frac{\partial}{\partial s} \left[\sqrt{s} \frac{\partial u}{\partial s} - \frac{1}{2\sqrt{s}} u \right] \\
&= -\frac{e^{im\theta}}{\sqrt{2\pi}} \frac{1}{2s} \left[\frac{1}{2\sqrt{s}} \frac{\partial u}{\partial s} + \sqrt{s} \frac{\partial^2 u}{\partial s^2} + \frac{1}{4} \frac{1}{s^{3/2}} u - \frac{1}{2\sqrt{s}} \frac{\partial u}{\partial s} \right] \\
&= \frac{e^{im\theta}}{\sqrt{2\pi}} \frac{1}{\sqrt{s}} \left[-\frac{1}{2} \frac{\partial^2 u}{\partial s^2} - \frac{1}{8 s^2} u \right].
\end{aligned} \tag{9.50}$$

Since

$$\frac{\partial}{\partial\theta} e^{im\theta} = im e^{im\theta} \, ; \quad \frac{\partial^2}{\partial\theta^2} = -m^2 e^{im\theta}, \tag{9.51}$$

we have for the

$$\begin{aligned}
\textit{2nd Term:} \quad & -\frac{1}{2s^2} \frac{\partial^2}{\partial\theta^2} \left[\frac{e^{im\theta}}{\sqrt{2\pi}} \frac{u(s)}{\sqrt{s}} \right] = \frac{m^2}{2 s^{5/2}} \frac{e^{im\theta}}{\sqrt{2\pi}} u \\
&= \frac{e^{im\theta}}{\sqrt{2\pi}} \frac{1}{\sqrt{s}} \left(\frac{1}{s^2} \frac{m^2}{2} u \right),
\end{aligned} \tag{9.52}$$

and the

$$\begin{aligned}
\textit{3rd Term:} \quad & -i \frac{\omega_L}{2} \frac{\partial}{\partial\theta} \left(\frac{e^{im\theta}}{\sqrt{2\pi}} \frac{u}{\sqrt{s}} \right) \\
&= \frac{e^{im\theta}}{\sqrt{2\pi}} \frac{1}{\sqrt{s}} \left(\frac{m\omega_L}{2} u \right).
\end{aligned} \tag{9.53}$$

Employing (9.19) and (9.8), and defining

$$\Omega_s = \frac{\Omega}{2}, \tag{9.54}$$

we have

$$\frac{\omega_L^2}{8} + \frac{\omega_s^2}{2} = \frac{1}{8}(\omega_L^2 + \omega_0^2) = \frac{1}{8}\Omega^2 = \frac{1}{2}\Omega_s^2. \tag{9.55}$$

Thus, the

$$\begin{aligned}\textit{4 and 5th Terms:}\quad &(\frac{\omega_L^2}{8}+\frac{\omega_s^2}{2})s^2\phi\\ &=\frac{e^{im\theta}}{\sqrt{2\pi}}\frac{1}{\sqrt{s}}(\frac{\Omega_s^2}{2}s^2u).\end{aligned}\tag{9.56}$$

$$\begin{aligned}\textit{6th Term:}\quad \frac{1}{2s}\phi(u) &= \frac{e^{im\theta}}{\sqrt{2\pi}}\frac{1}{2s^{3/2}}u\\ &=\frac{e^{im\theta}}{\sqrt{2\pi}}\frac{1}{\sqrt{s}}(\frac{1}{2s}u).\end{aligned}\tag{9.57}$$

The right hand side of (9.48) is

$$\frac{e^{im\theta}}{\sqrt{2\pi}}\frac{1}{\sqrt{s}}(\epsilon_s u).\tag{9.58}$$

On substitution of (9.50), (9.52), (9.53), (9.56), (9.57), (9.58) into (9.48) and canceling the $e^{im\theta}/\sqrt{2\pi s}$ in each term, one obtains the differential equation for $u(s)$ as

$$\left[-\frac{1}{2}\frac{d^2}{ds^2}+\frac{1}{2}(m^2-\frac{1}{4})\frac{1}{s^2}+\frac{1}{2}\Omega_s^2 s^2+\frac{1}{2s}\right]u(s)=(\epsilon_s-\frac{1}{2}m\omega_L)u(s).\tag{9.59}$$

The next step is to transform the relative coordinate s by the linear transformation

$$s=\frac{1}{\sqrt{\Omega_s}}\xi.\tag{9.60}$$

(The reason for this transformation is explained below.) On applying the transformation, the differential equation (9.59) becomes

$$\begin{aligned}&\left\{-\frac{1}{2}\Omega_s\frac{d^2}{d\xi^2}+\frac{\Omega_s}{2}(m^2-\frac{1}{4})\frac{1}{\xi^2}+\frac{1}{2}\Omega_s\xi^2\right.\\ &\qquad\left.+\frac{\sqrt{\Omega_s}}{2\xi}\right\}u(s)=(\epsilon_s-\frac{1}{2}m\omega_L)u(s).\end{aligned}\tag{9.61}$$

To simplify the equation further, multiply it by $(-2/\Omega_s)$ to obtain

$$\begin{aligned}\frac{d^2u(\xi)}{d\xi^2}&+\left[-(m^2-\frac{1}{4})\frac{1}{\xi^2}-\xi^2-\frac{1}{\sqrt{\Omega_s}\xi}\right]u(\xi)\\ &=-\frac{2}{\Omega_s}\left[\epsilon_s-\frac{1}{2}m\omega_L\right]u(\xi)\qquad(9.62)\\ &=-\beta u(\xi),\qquad(9.63)\end{aligned}$$

where the eigenvalue β is defined as

$$\beta = \frac{2}{\Omega_s}[\epsilon_s - \frac{1}{2}m\omega_L]. \tag{9.64}$$

In the asymptotic limit as $\xi \to \infty$, the differential equation (9.63) reduces to

$$\frac{d^2u(\xi)}{d\xi^2} - \xi^2 u(\xi) = 0. \tag{9.65}$$

The physically meaningful solutions to this equation must then be of the form

$$u(\xi) = e^{-\xi^2/2}t(\xi), \tag{9.66}$$

where $t(\xi)$ is a function to be determined and one that is independent of the above asymptotic solution. To prove that (9.66) is the solution to (9.65) determine

$$\frac{du}{d\xi} = e^{-\xi^2/2}(\frac{dt}{d\xi} - \xi t), \tag{9.67}$$

$$\frac{d^2u}{d\xi^2} = e^{-\xi^2/2}(-2\xi\frac{dt}{d\xi} + \frac{d^2t}{d\xi^2} - t + \xi^2 t). \tag{9.68}$$

Thus, in the limit as $\xi \to \infty$, the first three terms on the right vanish, with the last term being the dominant one, so that

$$\lim_{\xi\to\infty} \frac{d^2u}{d\xi} = e^{-\xi^2/2}\xi^2 t, \tag{9.69}$$

and (9.65) is proved. The reason for the linear transformation of (9.60) is to obtain the simplified asymptotic structure (9.65) of the differential equation (9.63) for which the solution form is known.

Substitution of the solution $u(\xi)$ of (9.66) into (9.63) employing (9.68) leads to

$$e^{-\xi^2/2}\left[t'' - 2\xi t' + \xi^2 t - t - \left(m^2 - \frac{1}{4}\right)\frac{t}{\xi^2} - \xi^2 t - \frac{t}{\sqrt{\Omega_s}}\frac{1}{\xi} + \beta t\right] = 0. \tag{9.70}$$

On canceling the $\xi^2 t$ terms, noting that $e^{-\xi^2/2}$ is finite, one then obtains the differential equation for $t(\xi)$ to be

$$t'' - 2\xi t' + \left[\beta - 1 - \frac{1}{\sqrt{\Omega_s}}\frac{1}{\xi} - (m^2 - \frac{1}{4})\frac{1}{\xi^2}\right]t = 0. \tag{9.71}$$

This is a second-order linear homogeneous differential equation which has a *regular* singular point at $\xi = 0$. *At least one* solution of the equation is of the form

$$t(\xi) = \xi^c \sum_{j=0}^{\infty} a_j \xi^j = \sum_{j=0}^{\infty} a_j \xi^{j+c}, \tag{9.72}$$

where the exponent c is a number that may be real or complex and where $a_0 \neq 0$. Such a series is called a generalized power series or *Frobenius series* [19]. Solution of (9.71) then requires the determination of the number c and the expansion coefficients $\{a_j\}$. To do so, substitute the expansion (9.72) into the differential equation (9.71). Differentiating (9.72) one obtains

$$t' = \sum_{j=0}^{\infty} a_j (j + c) \xi^{j+c-1}, \tag{9.73}$$

so that the various terms of (9.71) are the following:

1st Term: $$t'' = \sum_{j=0}^{\infty} a_j (j + c)(j + c - 1) \xi^{j+c-2} \tag{9.74}$$

2st Term: $$-2\xi t' = -2 \sum_{j=0}^{\infty} a_j (j + c) \xi^{j+c} \tag{9.75}$$

3st Term: $$(\beta - 1)t = (\beta - 1) \sum_{j=0}^{\infty} a_j \xi^{j+c} \tag{9.76}$$

4st Term: $$-\frac{1}{\sqrt{\Omega_s}} \frac{1}{\xi} t = \frac{1}{\sqrt{\Omega_s}} \sum_{j=0}^{\infty} a_j \xi^{j+c-1} \tag{9.77}$$

5th Term: $$-\left(m^2 - \frac{1}{4}\right) \frac{1}{\xi^2} t = -\left(m^2 - \frac{1}{4}\right) \sum_{j=0}^{\infty} a_j \xi^{j+c-2}. \tag{9.78}$$

Next, multiply each term by ξ^{2-c} so that the terms reduce to the following:

$$\textit{1st Term:} \quad \sum_{j=0}^{\infty} a_j (j+c)(j+c-1)\xi^j \tag{9.79}$$

$$\textit{2nd Term:} \quad -\sum_{j=0}^{\infty} a_j (j+c)\xi^{j+2} \tag{9.80}$$

$$\textit{3rd Term:} \quad (\beta-1)\sum_{j=0}^{\infty} a_j \xi^{j+2} \tag{9.81}$$

$$\textit{4st Term:} \quad -\frac{1}{\sqrt{\Omega_s}}\sum_{j=0}^{\infty} a_j \xi^{j+1} \tag{9.82}$$

$$\textit{5th Term:} \quad -\left(m^2-\frac{1}{4}\right)\sum_{j=0}^{\infty} a_j \xi^{j}. \tag{9.83}$$

Next introduce a new dummy index n such that in *each term* ξ is raised to the same power n. Thus,

$$\textit{1st Term:}\text{set } \ j=n: \sum_{n=0}^{\infty} a_n (n+c)(n+c-1)\xi^n \tag{9.84}$$

$$\textit{2nd Term:}\text{set } \ j+2=n: -2\sum_{n=2}^{\infty} a_{n-2}(n-2+c)\xi^n \tag{9.85}$$

$$\textit{3rd Term:}\text{set } \ j+2=n: (\beta-1)\sum_{n=2}^{\infty} a_{n-2}\xi^n \tag{9.86}$$

$$\textit{4th Term:}\text{set } \ j+1=n: -\frac{1}{\sqrt{\Omega_s}}\sum_{n=1}^{\infty} a_{n-1}\xi^n \tag{9.87}$$

$$\textit{5th Term:}\text{set } \ j=n: -\left(m^2-\frac{1}{4}\right)\sum_{n=0}^{\infty} a_n \xi^n. \tag{9.88}$$

Next, expand each of the above terms so that the summation begins at $n = 2$. Thus,

$$\textit{1st Term:} \quad a_0 c(c-1) + a_1(1+c)c\xi + \sum_{n=2}^{\infty} a_n(n+c)(n+c-1)\xi^n \tag{9.89}$$

$$\textit{2nd Term:} \quad -2\sum_{n=2}^{\infty} a_{n-2}(n-2+c)\xi^n \tag{9.90}$$

$$\textit{3rd Term:} \quad (\beta-1)\sum_{n=2}^{\infty} a_{n-2}\xi^n \tag{9.91}$$

$$\textit{4th Term:} \quad -\frac{1}{\sqrt{\Omega_s}}a_0\xi - \frac{1}{\sqrt{\Omega_s}}\sum_{n=2}^{\infty} a_{n-1}\xi^n \tag{9.92}$$

$$\textit{5th Term:} \quad -\left(m^2-\frac{1}{4}\right)a_0 - \left(m^2-\frac{1}{4}\right)a_1\xi - \left(m^2-\frac{1}{4}\right)\sum_{n=2}^{\infty} a_n\xi^n. \tag{9.93}$$

Recall that $a_0 \neq 0$. (It will be determined by normalizing $\phi(\mathbf{s})$.) In order to determine a_1 in terms of a_0, and to derive the general recursion expression for a_n, group terms of equal power of ξ, and then set the corresponding coefficient to zero. This is because the right hand side of (9.71) is zero. Employing (9.89) to (9.93) we have

$$\textit{Terms of } O(\xi^0): \quad a_0 c(c-1) - \left(m^2-\frac{1}{4}\right)a_0 = 0, \tag{9.94}$$

$$\text{or} \quad a_0\left[c^2 - c - \left(m^2-\frac{1}{4}\right)\right] = 0, \tag{9.95}$$

which on solving the quadratic equation leads to

$$c = \frac{1}{2} \pm |m|. \tag{9.96}$$

The value of $c = \frac{1}{2} - m$ is excluded because it makes $\phi(\mathbf{s})$ singular when $s \to 0$. This is readily seen to be the case since on employing (9.49), (9.60), (9.66), (9.72), we have

$$\phi(\mathbf{s}) \propto \frac{e^{-\Omega_r s^2/2}}{s^{1/2}} s^{(\frac{1}{2}-|m|)}(a_0 + a_1' s + a_2' s^2 + \cdots), \tag{9.97}$$

so that

$$\lim_{s\to 0}\phi(\mathbf{s}) \propto \frac{1}{s^{|m|}}a_0 = \infty. \tag{9.98}$$

$$\textit{Terms of } O(\xi): \quad a_1(1+c)c - \left(m^2-\frac{1}{4}\right)a_1 - \frac{1}{\sqrt{\Omega_s}}a_0 = 0. \tag{9.99}$$

On substituting for c from (9.96), we have

$$a_1\left[\left(1+\frac{1}{2}+|m|\right)(1+|m|)-\left(m^2-\frac{1}{4}\right)\right]-\frac{1}{\sqrt{\Omega_s}}a_0=0 \tag{9.100}$$

$$\text{or}\quad a_1(2|m|+1)=\frac{1}{\sqrt{\Omega_s}}a_0, \tag{9.101}$$

so that

$$a_1=\frac{a_0}{\sqrt{\Omega_s}(2|m|+1)}. \tag{9.102}$$

$$\textit{Terms of } O(\xi^n):\quad a_n\left[(n+c)(n+c-1)-\left(m^2-\frac{1}{4}\right)\right]$$
$$-\frac{1}{\sqrt{\Omega_s}}a_{n-1}+a_{n-2}[-2(n-2+c)+\beta-1]=0. \tag{9.103}$$

On substituting for c from (9.96) and simplifying, we have

$$a_n n(n+2|m|)-\frac{1}{\sqrt{\Omega_s}}a_{n-1}-a_{n-2}[2(n+|m|-1)-\beta]=0. \tag{9.104}$$

Thus,

$$a_n=\frac{1}{n(n+2|m|)}\left\{\frac{a_{n-1}}{\sqrt{\Omega_s}}+a_{n-2}\big[2(n+|m|-1)-\beta\big]\right\}\quad \text{for } n\geq 2. \tag{9.105}$$

From (9.105), it is evident that in order to determine the $\{a_n\}$, one must obtain the parameters Ω_s, (and hence $\Omega=2\Omega_s$), and β. In order to do so, note that the normalization of $\phi(\mathbf{s})$ is possible only if the coefficients $\{a_n\}$ in the Frobenius series solution of (9.72) vanish for some $j=n, n\geq 2$. That is, one chooses the number of terms n that are to be non-zero: $a_0\neq 0, a_1\neq 0,\ldots, a_{n-1}\neq 0$, but $a_n=a_{n+1}=\cdots=0$. Thus, the polynomial for $t(\xi)$ of (9.72) is

$$t(\xi)=\sum_{j=0}^{n-1}a_j\xi^{j+c}. \tag{9.106}$$

To determine the parameter β, we then set the coefficient $a_{n+1}=0$ in the recursion formula (9.105) because the polynomial is terminated at n, *i.e.* $a_n=0$. Hence, from (9.105) we have

$$a_{n+1}=\frac{(a_n/\sqrt{\Omega_s})+a_{n-1}[2(n+|m|)-\beta]}{(n+1)(n+1+2|m|)}=0 \tag{9.107}$$

so that since $a_n = 0$, $a_{n-1} \neq 0$,

$$\beta = 2(n + |m|). \tag{9.108}$$

On substituting for β into the expression for a_n of (9.105), and recursively obtaining the a_{n-1} and a_{n-2} in terms of a_0, the equation for a_n is then of the form

$$a_n = F(n, |m|, \Omega_s, \beta)a_0. \tag{9.109}$$

Since $a_n = 0$ but $a_0 \neq 0$, it follows that

$$F(n, |m|, \Omega_s, \beta) = 0. \tag{9.110}$$

For the choice of the number n of the terms in the power series expansion, and the angular momentum quantum number m, which together determine the parameter β from (9.108), equation (9.110) is then an equation for the determination of the parameter Ω_s. That is, the Ω_s constitute the roots of the function $F(n, |m|, \Omega_s, \beta)$. Then the corresponding effective frequency Ω of (9.8) is $\Omega = 2\Omega_s$ (see (9.54)). The solutions of (9.110) with the smaller Ω_s has zero nodes and corresponds to a ground state, that with the second largest Ω_s has one node corresponding to the first excited state, and so on.

Thus, beginning with the definition of $\phi(\mathbf{s})$ of (9.49), and employing (9.54), (9.60), (9.66), (9.96) and (9.106), we have

$$\phi(\mathbf{s}) = \frac{1}{\sqrt{2\pi}}\left(\sqrt{\frac{\Omega}{2}}\right)^{|m|+\frac{1}{2}} \frac{e^{im\theta}}{s^{\frac{1}{2}}} s^{|m|+\frac{1}{2}} e^{-\Omega s^2/4} \sum_{j=0}^{n-1} a_j \left(\sqrt{\frac{\Omega}{2}}s\right)^j \tag{9.111}$$

or

$$\begin{aligned}\phi(\mathbf{s}) = \; & N_s' e^{im\theta} e^{-\Omega s^2/4} s^{|m|} \left[a_0 + a_1 \sqrt{\frac{\Omega}{2}} s \right. \\ & \left. + \, a_2 \left(\frac{\Omega}{2}\right) s^2 + a_3 \left(\frac{\Omega}{2}\right)^{3/2} s^3 + \cdots + a_{n-1} \left(\frac{\Omega}{2}\right)^{(n-1)/2} s^{n-1} \right],\end{aligned} \tag{9.112}$$

where the constant N_s' is

$$N_s' = \frac{1}{\sqrt{2\pi}}\left(\sqrt{\frac{\Omega}{2}}\right)^{|m|+\frac{1}{2}}. \tag{9.113}$$

The coefficient a_0 is determined by the normalization condition

$$\int \phi^2(\mathbf{s}) d\mathbf{s} = 1. \tag{9.114}$$

9.1.3 The Center of Mass Coordinate Wave Function Component $\Phi(\mathbf{R})$

The center of mass coordinate Hamiltonian $\hat{H}_{\mathbf{R}}$ of (9.31) and the corresponding Schrödinger equation for the eigenfunctions $\Phi(\mathbf{R})$ and eigenvalues η_R of (9.34) is that of a two-dimensional harmonic oscillator in a uniform magnetic field. Hence, the center of mass coordinate wave functions $\Phi(\mathbf{R})$ for arbitrary state are known. However, in order to simplify the system wave function $\Psi(\mathbf{x}_1\mathbf{x}_2)$ of (9.32), one can simply employ the *ground state* $\Phi_0(\mathbf{R})$ as the center of mass component. The state of the system is then governed by the relative coordinate component $\phi(\mathbf{s})$.

The ground state center of mass component $\Phi_0(\mathbf{R})$ solution of (9.34) is

$$\Phi_0(\mathbf{R}) = N_R e^{-m\Omega_R R^2/2\hbar} \tag{9.115}$$

where $\Omega_R = 2\Omega = 2\sqrt{\omega_0^2 + \omega_L^2}$ is the effective frequency. In units of $m = \hbar = 1$, then

$$\Phi_0(\mathbf{R}) = N_R e^{-\Omega R^2}, \tag{9.116}$$

with N_R obtained by the normalization condition

$$\int \Phi_0^2(\mathbf{R}) d\mathbf{R} = 1 \tag{9.117}$$

to be

$$N_R = \sqrt{\frac{2\Omega}{\pi}}. \tag{9.118}$$

9.1.4 The General Wave Function $\Psi(\mathbf{x}_1\mathbf{x}_2)$ and Total Energy E

The general form of the wave function $\Psi(\mathbf{x}_1\mathbf{x}_2)$ of the two-dimensional 'atom' in a magnetic field is then (see (9.32))

$$\Psi(\mathbf{x}_1\mathbf{x}_2) = \psi(\mathbf{R}\mathbf{s})\chi(\sigma_1\sigma_2) = \Phi_0(\mathbf{R})\phi(\mathbf{s})\chi(\sigma_1\sigma_2), \tag{9.119}$$

where

$$\Phi_0(\mathbf{R}) = \sqrt{\frac{2\Omega}{\pi}} e^{-\Omega R^2}, \tag{9.120}$$

$$\phi(\mathbf{s}) = N_s' e^{im\theta} e^{-\Omega s^2/4} s^{|m|} [a_0 + a_1 \sqrt{\frac{\Omega}{2}} s$$
$$+ a_2 (\frac{\Omega}{2}) s^2 + a_3 (\frac{\Omega}{2})^{3/2} + \cdots + a_{n-1} (\frac{\Omega}{2})^{(n-1)/2} s^{n-1}], \tag{9.121}$$

$$N_s' = \frac{1}{\sqrt{2\pi}} \left(\sqrt{\frac{\Omega}{2}} \right)^{|m|+\frac{1}{2}}. \tag{9.122}$$

Note that the components $\Phi_0(\mathbf{R})$, $\phi(\mathbf{s})$ and $\chi(\sigma_1\sigma_2)$ are each normalized to unity.

The procedure for the determination of the wave function $\Psi(\mathbf{x}_1\mathbf{x}_2)$ and energy E for a particular state is the following:

Step 1: Select the value of the angular momentum quantum number m.

Step 2: Select the number n in the polynomial of $\phi(\mathbf{s})$. The first coefficient is a_0. Thus, if $n = 4$ is chosen, then there are 4 non-zero terms: a_0, a_1, a_2, a_3 and higher-order coefficients $a_{n=4} = a_{n+1=5} = a_{n+2=6} = 0$.

Step 3: Hence, one knows the value of $\beta = 2(n + |m|)$ (see (9.108)).

Step 4: Determine the coefficient a_1 in terms of a_0 from (9.102) which is repeated below:

$$a_1 = \frac{a_0}{\sqrt{\Omega_s}(2|m| + 1)}. \tag{9.123}$$

Step 5: Determine the coefficients $a_2, a_3, \ldots$ from (9.105) which is also repeated below:

$$a_n = \frac{1}{n(n + 2|m|)} \left\{ \frac{a_{n-1}}{\sqrt{\Omega_s}} + a_{n-2}\big[2(n + |m| - 1) - \beta\big] \right\} \quad \text{for } n \geq 2. \tag{9.124}$$

Each term $a_2, a_3, \ldots$ will be in terms of a_0.

Step 6: Set $a_n = 0$. This is equivalent to setting the function $F(n, |m|, \Omega_s, \beta) = 0$ (see (9.109), (9.110)). Then determine the roots Ω_s of $F(n, |m|, \Omega_s, \beta)$. If there is more than one root, then the lowest value root corresponds to a ground state (zero node), the next higher value root to the first excited state (one node), etc. The coefficients $a_1, a_2, \ldots, a_{n-1}$ which depend on Ω_s are thus known in terms of a_0. The coefficient a_0 is obtained by normalization.

Step 7: The effective frequency $\Omega = 2\Omega_s$ (see (9.54)).

Step 8: With the effective frequency Ω, and the coefficients $a_0, a_1, \ldots, a_{n-1}$ known, the wave function is determined.

Step 9: Since β is known from Step 3, and Ω_s from Step 6, we then have from (9.64)

$$\left[\epsilon_s - \frac{1}{2} m\omega_L\right] = \frac{\beta\Omega_s}{2}. \tag{9.125}$$

For a choice of Larmor frequency ω_L value, the eigenvalue ϵ_s of the relative coordinate differential equation (9.33) is then obtained. However, since β and Ω_s are determined, so is the term $(\epsilon_s - \frac{1}{2}m\omega_L)$.

Step 10: The eigenvalue η_R of the center of mass coordinate differential equation (9.34) for the two-dimensional atom is then

$$\begin{aligned} \eta_R &= \Omega_R(\nu + 1) \\ &= \Omega_R \\ &= 2\Omega, \end{aligned} \tag{9.126}$$

where $\nu = 0$ as the oscillator is assumed to be in its ground state, and $\Omega_R = 2\Omega$ (see below (9.115)).

Step 11: The total energy E from (9.35) is then

$$\begin{aligned} E &= 2\epsilon_s + \frac{1}{2}\eta_R + E_{spin} \\ &= 2\left[\epsilon_s - \frac{1}{2}m\omega_L\right] + \frac{\eta_R}{2}. \end{aligned} \tag{9.127}$$

From the above procedure, with the choice of state via the angular momentum quantum number m, and that of the number n of terms in the power series expansion, one obtains (via the coefficients $\{a_j\}$ with $j = 0, 1, \ldots, n-1$), the effective frequency Ω, and thereby a closed-form analytical expression for the wave function $\Psi(\mathbf{x}_1\mathbf{x}_2)$ from (9.119). It is evident then that since the number of terms n of the power series is arbitrary, that there exist an infinite number of such analytical solutions for a particular state.

9.1.5 *Wave Function for a Ground and First Excited Triplet State*

We next provide the derivations for the ground (singlet) state [20], and the first excited triplet 2^3S [21, 22] state wave functions and energies of the 2-electron 'atom' in a magnetic field. These wave functions are employed in the book in Chaps. 5, 6 and 8.

Example 1: **The Ground State**

This is a singlet states with the electrons having opposite spins. Thus, the spatial part of the wave function is symmetric in an interchange of the spatial coordinates.

Step 1:

$$\text{Choose } \; m = 0 \tag{9.128}$$

Step 2:

$$\text{Choose } \; n = 2 \tag{9.129}$$

Step 3: From (9.108),

$$\beta = 2(n + |m|) = 4 \tag{9.130}$$

Step 4: From (9.102),

$$a_1 = \frac{a_0}{\sqrt{\Omega_s}(2|m| + 1)} = \frac{a_0}{\sqrt{\Omega_s}}. \tag{9.131}$$

Step 5: From (9.124),

$$\begin{aligned} a_2 &= \frac{1}{4}\left\{\frac{1}{\sqrt{\Omega_s}} a_1 - 2a_0\right\} \\ &= \frac{1}{4}\left\{\frac{1}{\sqrt{\Omega_s}}\frac{a_0}{\sqrt{\Omega_s}} - 2a_0\right\} = \frac{a_0}{4}\left\{\frac{1}{\Omega_s} - 2\right\}. \end{aligned} \tag{9.132}$$

Step 6: Since $a_n = 0$, we put

$$a_2 \equiv F(2, 0, \Omega_s, 4) = 0 \tag{9.133}$$

and since $a_0 \neq 0$, we have from (9.132)

$$\Omega_s = \frac{1}{2}. \tag{9.134}$$

Note that there is only one root of (9.133).

Step 7: From (9.54), the effective frequency

$$\Omega = 2\Omega_s = 1. \tag{9.135}$$

Step 8: Therefore, from (9.120) and (9.121), respectively,

$$\Phi(\mathbf{R}) = \sqrt{\frac{2}{\pi}} e^{-R^2} \tag{9.136}$$

$$\begin{aligned} \phi(\mathbf{s}) &= N_s' e^{-s^2/4}\left[a_0 + \frac{a_0}{\sqrt{\frac{1}{2}}}\sqrt{\frac{1}{2}} s\right] \\ &= N_s e^{-s^2/4}[1 + s] \end{aligned} \tag{9.137}$$

where $N_s = N_s' a_0$ is the normalization constant obtained from

$$N_s^2 \int_0^\infty \phi^2(s) 2\pi s ds = 1. \tag{9.138}$$

Thus, the spatial part of the ground state wave function $\psi(\mathbf{Rs})$ is

$$\psi(\mathbf{R}\mathbf{s}) = Ne^{-(R^2+\frac{s^2}{4})}[1+s], \tag{9.139}$$

with $N = 1/\pi\sqrt{3+\sqrt{2\pi}} = 0.135646$.

Step 9: From (9.125),

$$\epsilon_s = 1 \tag{9.140}$$

Step 10: From (9.126),

$$\eta_R = 2. \tag{9.141}$$

Step 11: From (9.127), the ground state energy E is

$$E = 3 \text{ (a.u.)}. \tag{9.142}$$

Note that the value of the energy E obtained as the eigenvalue of Schrödinger-Pauli Theory equation (9.7) for the ground state is the same as that obtained via Quantal Density Functional Theory (see Table 6.1).

Example 2: **The First Excited Triplet State**

In the triplet state, the electron spins are parallel, and hence the spatial part of the wave function must be antisymmetric in an interchange of the spatial coordinates.

Step 1:

$$\text{Choose } m = 1. \tag{9.143}$$

Step 2:

$$\text{Choose } n = 4. \tag{9.144}$$

Step 3: From (9.108),

$$\beta = 2(n+|m|) = 10. \tag{9.145}$$

Step 4: From (9.102),

$$a_1 = \frac{a_0}{\sqrt{\Omega_s}(2|m|+1)} = \frac{a_0}{3\sqrt{\Omega_s}}. \tag{9.146}$$

Step 5: From (9.124),

$$a_2 = \frac{1}{8}\left\{\frac{a_1}{\sqrt{\Omega_s}} + (-6)a_0\right\} = \frac{a_0}{8}\left(\frac{1}{3\Omega_s} - 6\right) \tag{9.147}$$

$$a_3 = \frac{1}{15}\left\{\frac{a_2}{\sqrt{\Omega_s}} + (-4)a_1\right\} = \frac{a_0}{180\sqrt{\Omega_s}}\left(\frac{1}{2\Omega_s} - 25\right). \tag{9.148}$$

Step 6: Since $a_n = 0$, we first determine a_4 and then put

$$a_4 = F(4, 1, \Omega_s, 10) = 0. \tag{9.149}$$

Thus, from (9.124)

$$a_4 = \frac{1}{24}\left\{\frac{a_3}{\sqrt{\Omega_s}} + (-2)a_2\right\} \tag{9.150}$$

$$= \frac{1}{90\Omega_s^2} - \frac{25}{45\Omega_s} - \frac{1}{3\Omega_s} + 6 = 0. \tag{9.151}$$

This is a quadratic equation:

$$540\Omega_s^2 - 80\Omega_s + 1 = 0 \tag{9.152}$$

whose roots are

$$\Omega_s = 0.0137821, 0.1343660. \tag{9.153}$$

We must employ the root of higher value as this corresponds to a single node type wave function.

Step 7: From (9.54), the effective frequency

$$\Omega = 2\Omega_s = 0.2687320 \tag{9.154}$$

Step 8: Therefore, from (9.120),

$$\Phi_0(\mathbf{R}) = \sqrt{\frac{2\Omega}{\pi}}e^{-\Omega R^2} \tag{9.155}$$

with the value of Ω from (9.154). From (9.121)

$$\phi(\mathbf{s}) = N_s' e^{i\theta} e^{-\Omega s^2/4} s\left[a_0 + a_1\sqrt{\frac{\Omega}{2}}s + a_2\left(\frac{\Omega}{2}\right)s^2 + a_3\left(\frac{\Omega}{2}\right)^{3/2}s^3\right]$$

$$= N_s' a_0 e^{i\theta} e^{-\Omega s^2/4} s\left[1 + \frac{a_1}{a_0}\sqrt{\frac{\Omega}{2}}s + \frac{a_2}{a_0}\left(\frac{\Omega}{2}\right)s^2 + \frac{a_3}{a_0}\left(\frac{\Omega}{2}\right)^{3/2}s^3\right] \tag{9.156}$$

$$= N_s e^{i\theta} e^{-\Omega s^2/4}[s + c_2 s^2 + c_3 s^3 + c_4 s^4], \tag{9.157}$$

where $N_s = N_s' a_0$ and

$$c_2 = \frac{a_1}{a_0}\sqrt{\frac{\Omega}{2}} = \frac{1}{3} \tag{9.158}$$

$$c_3 = \frac{a_2}{a_0}\left(\frac{\Omega}{2}\right) = -0.0591078 \tag{9.159}$$

$$c_4 = \frac{a_3}{a_0}\left(\frac{\Omega}{2}\right)^{3/2} = -0.0158842. \tag{9.160}$$

Normalizing the relative coordinate wave function as in (9.138), the spatial part of the triplet state wave function is

$$\psi(\mathbf{Rs}) = N e^{i\theta} e^{-\Omega(R^2+\frac{s^2}{4})}[s + c_2 s^2 + c_3 s^3 + c_4 s^4], \tag{9.161}$$

where $N = 0.0224663$.

Step 9: From (9.125),

$$\epsilon_s - \frac{1}{2}\omega_L = 0.671830. \tag{9.162}$$

Step 10: From (9.126),

$$\eta_R = 2\Omega. \tag{9.163}$$

Step 11: The total energy E of (9.127):

$$E = 2\left[\epsilon_s - \frac{1}{2}m\omega_L\right] + \frac{\eta_R}{2}, \tag{9.164}$$

on employing (9.154), (9.162), (9.163) is then

$$\begin{aligned} E &= 2(0.671830) + 0.2687320 \\ &= 1.612392 \text{ a.u.} \end{aligned} \tag{9.165}$$

Observe that the spatial part of the wave function $\psi(\mathbf{Rs})$ of (9.161) is antisymmetric in an interchange of the spatial coordinates. This is ensured as a result of the phase factor $e^{i\theta}$. When the coordinates $\mathbf{r}_1$ and $\mathbf{r}_2$ are interchanged, the magnitude of the relative coordinate $\mathbf{s}$ remains unchanged. However, the angle θ changes to $\theta + \pi$.

Note that the energy E of (9.165) obtained as the eigenvalue of the Schrödinger-Pauli Theory equation (9.7) for this triplet 2^3S state is the same as given in Table 5.1 in which the value is determined via the various fields of the complementary perspective to Schrödinger-Pauli theory.

9.2 The Generalized Kohn Theorem Wave Function

The statement [18] of the Generalized Kohn Theorem (GKT) is the following:

Consider a system of N harmonically confined electrons in a magnetostatic field $\mathcal{B}(\mathbf{r}) = \nabla \times \mathbf{A}(\mathbf{r})$. On application of a spatially homogeneous time-dependent electric field $\mathbf{E}(t)$, the Hamiltonian is

$$\hat{H}(t) = \hat{H}(0) + e\mathbf{E}(t) \cdot \sum_i \mathbf{r}_i, \tag{9.166}$$

where the unperturbed Hamiltonian is

$$\hat{H}(0) = \sum_i \left[\frac{1}{2m} \left(\hat{\mathbf{p}}_i + \frac{e}{c} \mathbf{A}(\mathbf{r}_i) \right)^2 + \frac{1}{2} \mathbf{r}_i \cdot \mathbb{K} \cdot \mathbf{r}_i \right] + \sum_{(i,j)} u(\mathbf{r}_i - \mathbf{r}_j), \tag{9.167}$$

with $i = 1, \ldots, N$ denotes the electrons; (i, j) denotes all pairs; $\mathbf{r}_i = (x_i, y_i, z_i)$; $u(\mathbf{r})$ the electron interaction operator; and $\mathbb{K}$ the force constant matrix

$$\mathbb{K} = m \begin{pmatrix} \omega_x^2 & 0 & 0 \\ 0 & \omega_y^2 & 0 \\ 0 & 0 & \omega_z^2 \end{pmatrix}, \tag{9.168}$$

and ω_x, w_y, w_z the directional harmonic frequencies. The corresponding time-dependent Schrödinger equation is

$$\left(i\hbar \frac{\partial}{\partial t} - \hat{H}(t) \right) \Psi(t) = 0. \tag{9.169}$$

Note that by modifying the force constant matrix $\mathbb{K}$, the Hamiltonian of (9.167) can represent a quantum well ($\omega_x = \omega_y = 0, \omega_z \neq 0$); quantum dot ($\omega_x \neq \omega_y \neq 0, \omega_z = 0$; $\omega_x = \omega_y \neq \omega_z$); or a two-dimensional electron gas ($\omega_x = \omega_y = \omega_z = 0$). The absence of the magnetic field $\mathcal{B}$ in turn also constitutes special cases. In the symmetric gauge $\mathbf{A}(\mathbf{r}) = \frac{1}{2}\mathcal{B}(\mathbf{r}) \times \mathbf{r}$, the solution to (9.169) is the following:

$$\begin{aligned} \Psi(\mathbf{r}_1, \ldots, \mathbf{r}_N; t) &= e^{\frac{i}{\hbar}(E_n t + S_0[\mathbf{R}_m(t)] - \mathbf{P}_{\mathbf{R}_m(t)} \cdot \mathbf{R})} \\ &\times \Psi_n(\mathbf{r}_1 - \mathbf{R}_m(t), \mathbf{r}_2 - \mathbf{R}_m(t), \ldots, \mathbf{r}_N - \mathbf{R}_m(t)), \end{aligned} \tag{9.170}$$

where E_n, Ψ_n are the eigenenergies and eigenfunctions of the unperturbed Hamiltonian:

$$\hat{H}(0)\Psi_n(\mathbf{r}_1, \ldots, \mathbf{r}_N) = E_n \Psi_n(\mathbf{r}_1, \ldots, \mathbf{r}_N); \tag{9.171}$$

$S_0[\mathbf{R}_m(t)]$ the total classical action:

$$S_0[\mathbf{R}_m(t)] = S_0^{XY}(t) + S_0^Z(t), \tag{9.172}$$

$$S_0^{XY}[\mathbf{R}_{||}(t)] = \frac{1}{2} M \int_0^t (\dot{\mathbf{R}}_{||}^2 - \tilde{\mathbf{R}}_{||} \Omega^2 \mathbf{R}_{||} + \omega_c \tilde{R}_{||} J \dot{\mathbf{R}}_{||}) dt', \tag{9.173}$$

$$S_0^Z[Z(t)] = \frac{1}{2}M\int_0^t (\dot{Z}^2 - \omega_z^2 Z^2)dt'. \tag{9.174}$$

In (9.170), $\mathbf{R}_m(t)$ is the three dimensional vector $\begin{pmatrix} X_m(t) \\ Y_m(t) \\ Z_m(t) \end{pmatrix}$, and $\mathbf{P}_{R_m(t)}$ the corresponding canonical momentum (see (9.261) below). The vector $\mathbf{R}_m(t)$ satisfies the classical equation of motion

$$m\frac{d^2\mathbf{R}_m(t)}{dt^2} = -\mathbb{K}\cdot\mathbf{R}_m(t) - eE(t) - \frac{e}{c}\frac{d\mathbf{R}_m(t)}{dt}\times\mathcal{B}. \tag{9.175}$$

In (9.173), $\mathbf{R}_{||}(t)$ is the two-dimensional vector $\begin{pmatrix} X(t) \\ Y(t) \end{pmatrix}$; $\tilde{\mathbf{R}}_{||}$ the transpose of the vector $\mathbf{R}_{||}$; $\Omega^2 = \begin{pmatrix} \omega_x^2 & 0 \\ 0 & \omega_y^2 \end{pmatrix}$; and $J = \begin{pmatrix} 0 & -1 \\ 1 & 0 \end{pmatrix}$. Finally, the vector in (9.170) $\mathbf{R} = \frac{1}{N}\sum_i \mathbf{r}_i = \begin{pmatrix} X \\ Y \\ Z \end{pmatrix}$, $\omega_c = \frac{NeB}{Mc} = \frac{eB}{mc}$ the cyclotron frequency, and $M = Nm$.

The origin of the GKT lies in the original work of Kohn et al. [23–27]. For the unperturbed Hamiltonian of (9.167) representing a *uniform electron gas, i.e.* without the harmonic term, as well as when the harmonic term is included, it was shown that the response of the system to the TD electric field was identical to that of a single particle of mass Nm, charge Ne, and thus independent of the number of electrons and the electron-interaction. The GKT wave function of (9.170) for the case of the uniform electron gas is referred to as the *Kohn Theorem* (KT) wave function. In the absence of the magnetic field in the Hamiltonian (9.167), the GKT wave function reduces to that of the *Harmonic Potential Theorem* [28] (HPT). For a *first principles* derivation of the HPT wave function via the Feynman Path Integral method, and the operator method, see [29]. An alternative first principles proof of the HPT via the interaction 'representation' method is given in [30]. For a *first principles* Feynman path-integral method derivation of the HPT for *time-dependent* harmonically confined electrons, see [31].

The GKT wave function of (9.170) is comprised of a phase factor times the unperturbed wave function in which the coordinates of each electron $\mathbf{r}_i$ are translated by a time-dependent function that satisfies the corresponding classical equation of motion (9.175). Thus, the time evolution of properties $O(\mathbf{r})$ or $O(\mathbf{rr}')$ that are expectation values of Hermitian operators $\hat{O}(\mathbf{r})$, $\hat{O}(\mathbf{rr}')$ are simply those of the unperturbed system translated by the function $\mathbf{R}_m(t)$, *i.e.* $O(\mathbf{r} - \mathbf{R}_m(t))$, $O(\mathbf{r} - \mathbf{R}_m(t), \mathbf{r}')$. Hence, properties of the system such as the density, the pair-correlation density, the single-particle density matrix, and the corresponding Hartree, electron-interaction, and kinetic fields as defined in Chap. 2 all obey this translational property. Observables that involve expectations of differential operators such as the physical current density do not possess this translational property because of the phase factor.

9.2.1 Decoupling of the Hamiltonian

The key to the derivation of the GKT wave function is that the Hamiltonian of (9.166) is separable into its center of mass and relative coordinate components. The center of mass component is that of a single particle of mass Nm and charge Ne confined harmonically by N times the potential of a single particle. *The effect of the TD electric field also appears only in this component.* The electron interaction potential term appears only in the relative coordinate component of the Hamiltonian. Furthermore, the in-plane center of mass motion component of the Hamiltonian is separable from the motion in the plane perpendicular to it which is the assumed direction of the magnetic field.

To see this, define the center of mass and relative coordinates and momentum as

$$\mathbf{R} \equiv \mathbf{R}^{(1)} = \frac{1}{N}\sum_i \mathbf{r}_i; \qquad \Pi \equiv \Pi^{(1)} = \sum_i \hat{\pi}_i; \qquad \hat{\pi}_i = -i\hbar\nabla_i + \frac{e}{c}\mathbf{A}(\mathbf{r}_i), \tag{9.176}$$

and

$$\begin{aligned} X^{(2)} &= x_1 - x_2, \\ X^{(3)} &= x_1 + x_2 - 2x_3, \ldots, \\ X^{(N)} &= x_1 + x_2 + \cdots + x_{N-1} - (N-1)x_N, \end{aligned} \tag{9.177}$$

and similarly for $Y^{(2)}, \ldots Y^{(N)}, Z^{(2)}, \ldots Z^{(N)}$, and $\Pi^{(2)}, \ldots \Pi^{(N)}$. The Hamiltonian can then be rewritten as

$$\hat{H} = \hat{H}_{cm} + \hat{H}_{rel}. \tag{9.178}$$

where the center of mass Hamiltonian $\hat{H}_{cm}(t)$ is

$$\hat{H}_{cm}(t) = \frac{\hat{\Pi}^2}{2M} + \frac{M}{2}(\omega_x^2 X^2 + \omega_y^2 Y^2 + \omega_z^2 Z^2) - Ne\mathbf{E}(t)\cdot\mathbf{R}, \tag{9.179}$$

with

$$\hat{\Pi} = -i\hbar\nabla_{\mathbf{R}} + \frac{Ne}{c}\mathbf{A}(\mathbf{R}), \tag{9.180}$$

and $M = Nm$. $\hat{H}_{rel}$ is the Hamiltonian of the relative coordinates and contains the effects of the interaction. It can be readily shown that $[\hat{H}_{cm}, \hat{H}_{rel}] = 0$ so that the center-of-mass motion and the relative motion are separable. Therefore, the eigenstates of the Hamiltonian are the product of the eigenstates of the center-of-mass motion $\Phi(\mathbf{R}, t)$ and the relative motion $\varphi_{rel}(\mathbf{R}^{(2)}, \ldots, \mathbf{R}^{(N)})$:

$$\Psi(\mathbf{r}_1, \mathbf{r}_2, \ldots \mathbf{r}_N, t) = \Phi(\mathbf{R}, t)\varphi_{rel}(\mathbf{R}^{(2)}, \ldots, \mathbf{R}^{(N)}). \tag{9.181}$$

The relative motion wave function $\varphi_{rel}(\mathbf{R}^{(2)}, \ldots, \mathbf{R}^{(N)})$ satisfies

$$\hat{H}_{rel}\varphi_{rel}(\mathbf{R}^{(2)}, \ldots, \mathbf{R}^{(N)}) = E_{rel}\varphi_{rel}(\mathbf{R}^{(2)}, \ldots, \mathbf{R}^{(N)}), \quad (9.182)$$

where E_{rel} is the corresponding eigenvalue. The derivation of $\hat{H}_{rel}$ is given in Sect. 9.2.5.

We next focus on the $\hat{H}_{cm}(t)$ since the effect of the external electrical field appears only in the center-of-mass motion. For simplicity, choosing the direction of the magnetic field as the z direction, then in the symmetric gauge $\mathbf{A}(\mathbf{R}) = (-\frac{BY}{2}, \frac{BX}{2}, 0)$, the $\hat{H}_{cm}(t)$ can be decomposed into two parts

$$\hat{H}_{cm}(t) = \hat{H}_{XY}(t) + \hat{H}_Z(t), \quad (9.183)$$

where $\hat{H}_{XY}(t)$ describes the *in-plane* motion, and

$$\hat{H}_Z(t) = \frac{1}{2M}\hat{P}_Z^2 + \frac{1}{2}M{\omega_z}^2 Z^2 + NeE_z(t)Z, \quad (9.184)$$

describes a trivial motion in the Z-direction without the influence of the magnetic field. This motion is independent of the in-plane motion, and as such the center-of-mass wave function $\Phi(\mathbf{R}, t)$ is just the product of the in-plane motion wave function and the Z-direction wave function. The latter is simply the one-dimensional HPT wave function.

9.2.2 In-plane Center-of-Mass Motion

We next focus on the *in-plane* motion described by $\hat{H}_{XY}(t)$, which may be further separated as

$$\hat{H}_{XY}(t) = \hat{H}_0 + \hat{H}_1(t), \quad (9.185)$$

where the time-independent part is

$$\begin{aligned}\hat{H}_0 &= \frac{1}{2M}(\hat{\Pi}_X^2 + \hat{\Pi}_Y^2) + \frac{1}{2}M({\omega_x}^2X^2 + {\omega_y}^2Y^2) \\ &= \frac{1}{2M}(\hat{P}_X^2 + \hat{P}_Y^2) + \frac{1}{2}M({\omega_1}^2X^2 + {\omega_2}^2Y^2) + \frac{\omega_c}{2}\hat{L}_Z,\end{aligned} \quad (9.186)$$

with ${\omega_1}^2 = {\omega_x}^2 + \frac{{\omega_c}^2}{4}, {\omega_2}^2 = {\omega_y}^2 + \frac{{\omega_c}^2}{4}$, $\omega_c = \frac{NeB}{Mc} = \frac{eB}{mc}$ is the cyclotron frequency, and $\hat{L}_Z = X\hat{P}_Y - Y\hat{P}_X$ the angular momentum component along the Z axis. The time dependent part is

$$\hat{H}_1(t) = Ne[E_x(t)X + E_y(t)Y]. \quad (9.187)$$

In the interaction 'representation' the evolution of state $|\Phi\rangle$ at time t denoted by $|\Phi_t\rangle$ is obtained as

$$|\Phi_t\rangle = \exp\left\{-\frac{i}{\hbar}\hat{H}_0 t\right\} T\exp\left\{-\frac{i}{\hbar}\int_0^t du\hat{H}_1^{int}(u)\right\}|\Phi_0\rangle, \tag{9.188}$$

where T is the time-ordering operator, and where

$$\hat{H}_1^{int}(t) = e^{i\hat{H}_0 t/\hbar}\hat{H}_1(t)e^{-i\hat{H}_0 t/\hbar} = Ne[E_x(t)X(t) + E_y(t)Y(t)], \tag{9.189}$$

with

$$\begin{aligned} X(t) &= e^{i\hat{H}_0 t/\hbar} X e^{-i\hat{H}_0 t/\hbar} \\ &= a_1(t)X + a_2(t)Y + a_3(t)\hat{P}_X + a_4(t)\hat{P}_Y, \end{aligned} \tag{9.190}$$

with

$$a_1(t) = \gamma_1 \cos\sqrt{\lambda_1}t + \gamma_2\cos\sqrt{\lambda_2}t, \tag{9.191}$$

$$a_2(t) = \gamma_3\left(\frac{\sin\sqrt{\lambda_1}t}{\sqrt{\lambda_1}}\right) + \gamma_4\left(\frac{\sin\sqrt{\lambda_2}t}{\sqrt{\lambda_2}}\right), \tag{9.192}$$

$$a_3(t) = \gamma_5\left(\frac{\sin\sqrt{\lambda_1}t}{\sqrt{\lambda_1}}\right) + \gamma_6\left(\frac{\sin\sqrt{\lambda_2}t}{\sqrt{\lambda_2}}\right), \tag{9.193}$$

$$a_4(t) = \gamma_7 \cos\sqrt{\lambda_1}t + \gamma_8\cos\sqrt{\lambda_2}t, \tag{9.194}$$

where the coefficients

$$\gamma_1 = \left(\frac{\omega_2^2 - \omega_1^2}{2\sqrt{\Delta}} + \frac{1}{2}\right), \qquad \gamma_2 = \left(\frac{\omega_1^2 - \omega_2^2}{2\sqrt{\Delta}} + \frac{1}{2}\right), \tag{9.195}$$

$$\gamma_3 = \frac{\omega_L}{2\sqrt{\Delta}}\left(\frac{3\omega_2^2 + \omega_1^2 - \sqrt{\Delta}}{2}\right), \qquad \gamma_4 = \frac{\omega_L}{2\sqrt{\Delta}}\left(\frac{-\sqrt{\Delta} - 3\omega_2^2 - \omega_1^2}{2}\right), \tag{9.196}$$

$$\gamma_5 = \frac{1}{2M\sqrt{\Delta}}\left(\sqrt{\Delta} - \omega_1^2 + \omega_2^2 - \omega_c^2\right), \quad \gamma_6 = \frac{1}{2M\sqrt{\Delta}}\left(\sqrt{\Delta} + \omega_1^2 - \omega_2^2 + \omega_c^2\right), \tag{9.197}$$

$$\gamma_7 = -\frac{\omega_c}{M\sqrt{\Delta}}, \qquad \gamma_8 = \frac{\omega_c}{M\sqrt{\Delta}}, \tag{9.198}$$

Similarly,

$$\begin{aligned} Y(t) &= e^{i\hat{H}_0 t/\hbar} Y e^{-i\hat{H}_0 t/\hbar} \\ &= b_1(t)X + b_2(t)Y + b_3(t)\hat{P}_x + b_4(t)\hat{P}_y, \end{aligned} \tag{9.199}$$

with

$$b_1(t) = \eta_1\left(\frac{\sin\sqrt{\lambda_1}t}{\sqrt{\lambda_1}}\right) + \eta_2\left(\frac{\sin\sqrt{\lambda_2}t}{\sqrt{\lambda_2}}\right), \tag{9.200}$$

$$b_2(t) = \eta_3 \cos\sqrt{\lambda_1}t + \eta_4 \cos\sqrt{\lambda_2}t, \tag{9.201}$$

$$b_3(t) = \eta_5 \cos\sqrt{\lambda_1}t + \eta_6 \cos\sqrt{\lambda_2}t, \tag{9.202}$$

$$b_4(t) = \eta_7\left(\frac{\sin\sqrt{\lambda_1}t}{\sqrt{\lambda_1}}\right) + \eta_8\left(\frac{\sin\sqrt{\lambda_2}t}{\sqrt{\lambda_2}}\right), \tag{9.203}$$

where the coefficients

$$\eta_1 = \frac{\omega_c}{2\sqrt{\Delta}}\left(\frac{-3\omega_1^2 - \omega_2^2 + \sqrt{\Delta}}{2}\right), \quad \eta_2 = \frac{\omega_c}{2\sqrt{\Delta}}\left(\frac{\sqrt{\Delta} + 3\omega_1^2 + \omega_2^2}{2}\right), \tag{9.204}$$

$$\eta_3 = \left(\frac{\omega_1^2 - \omega_2^2}{2\sqrt{\Delta}} + \frac{1}{2}\right), \quad \eta_4 = \left(\frac{\omega_2^2 - \omega_1^2}{2\sqrt{\Delta}} + \frac{1}{2}\right), \tag{9.205}$$

$$\eta_5 = \frac{\omega_c}{M\sqrt{\Delta}}, \quad \eta_6 = -\frac{\omega_c}{M\sqrt{\Delta}}, \tag{9.206}$$

$$\eta_7 = \frac{1}{2M\sqrt{\Delta}}\left(\sqrt{\Delta} + \omega_1^2 - \omega_2^2 - \omega_c^2\right), \quad \eta_8 = \frac{1}{2M\sqrt{\Delta}}\left(\sqrt{\Delta} - \omega_1^2 + \omega_2^2 + \omega_c^2\right). \tag{9.207}$$

Notice that in above equations

$$\Delta = (\omega_1^2 - \omega_2^2)^2 + 2\omega_c^2(\omega_1^2 + \omega_2^2), \tag{9.208}$$

$$\lambda_{1,2} = \frac{1}{2}\left(\frac{\omega_c^2}{2} + \omega_1^2 + \omega_2^2 \mp \sqrt{(\omega_1^2 - \omega_2^2)^2 + 2\omega_c^2(\omega_1^2 + \omega_2^2)}\right). \tag{9.209}$$

The details of how to calculate the coefficients $a_i(t)$ and $b_i(t), i = 1, 2, 3, 4$ of $X(t)$ (9.190) and $Y(t)$ (9.199) are provided at the end of this subsection.

Proceeding with the substitution of (9.190), (9.199) into (9.189) yields

$$\hat{H}_1^{int}(t) = c_1(t)X + c_2(t)Y + c_3(t)\hat{P}_X + c_4(t)\hat{P}_Y, \tag{9.210}$$

with

$$c_i(t) = Ne[E_x(t)a_i(t) + E_y(t)b_i(t)], i = 1, 2, 3, 4. \tag{9.211}$$

Note that the commutator of $\hat{H}_1^{int}(u)$ at different times is a c number,

$$[\hat{H}_1^{int}(u), \hat{H}_1^{int}(v)] = i\hbar \cdot g(u, v), \tag{9.212}$$

where

$$g(u, v) = c_1(u)c_3(v) + c_2(u)c_4(v) - c_1(v)c_3(u) - c_2(v)c_4(u). \tag{9.213}$$

One then obtains

$$\begin{aligned} T\exp\left\{-\frac{i}{\hbar}\int_0^t du\hat{H}_1^{int}(u)\right\} &= \exp\left\{-\frac{i}{\hbar}\int_0^t du\hat{H}_1^{int}(u)\right\} \\ &\quad \exp\left\{-\frac{1}{2\hbar^2}\int_0^t du\int_0^u dv[\hat{H}_1^{int}(u), \hat{H}_1^{int}(v]\right\} \\ &= \exp\left\{-\frac{i}{\hbar}\int_0^t du\hat{H}_1^{int}(u)\right\}\exp\left\{\frac{-i}{\hbar}\alpha(t)\right\}, \end{aligned} \tag{9.214}$$

where

$$\alpha(t) = \frac{1}{2}\int_0^t du\int_0^u dvg(u, v). \tag{9.215}$$

Substituting (9.214) into (9.188), one obtains the in-plane center of mass wave function component to be

$$|\Phi_t\rangle = \exp\left\{-\frac{i}{\hbar}\hat{H}_0 t\right\}\exp\left\{-\frac{i}{\hbar}\int_0^t du\hat{H}_1^{int}(u)\right\}\exp\left\{-\frac{i}{\hbar}\alpha(t)\right\}|\Phi_0\rangle. \tag{9.216}$$

Derivation of $X(t)$ and $Y(t)$.

Employing

$$e^L A e^{-L} = A + [L, A] + \frac{1}{2!}[L, [L, A]] + \frac{1}{3!}[L, [L, [L, A]]] + \cdots \tag{9.217}$$

we have

$$\begin{aligned} X(t) &= e^{i\hat{H}_0 t/\hbar} X e^{-i\hat{H}_0 t/\hbar} = \hat{X} + \frac{it}{\hbar}[\hat{H}_0, \hat{X}] + \frac{1}{2!}\left(\frac{it}{\hbar}\right)^2[\hat{H}_0, [\hat{H}_0, \hat{X}]] + \\ &\frac{1}{3!}\left(\frac{it}{\hbar}\right)^3[\hat{H}_0, [\hat{H}_0, [\hat{H}_0, \hat{X}]]] + \cdots, \end{aligned} \tag{9.218}$$

Employing the expression for $\hat{H}_0$ of (9.186), we obtain the following equations

$$[\hat{H}_0, X] = -\alpha_1 \hat{P}_X + \alpha_2 Y, \tag{9.219}$$

$$[\hat{H}_0, Y] = -\alpha_1 \hat{P}_Y - \alpha_2 X, \tag{9.220}$$

$$[\hat{H}_0, \hat{P}_X] = \alpha_2 \hat{P}_Y + M^2\omega_1^2\alpha_1 X, \tag{9.221}$$

$$[\hat{H}_0, \hat{P}_Y] = -\alpha_2 \hat{P}_X + M^2\omega_2^2\alpha_1 Y, \tag{9.222}$$

where

$$\alpha_1 = \frac{i\hbar}{M}, \quad \alpha_2 = \frac{i\hbar\omega_c}{2}. \tag{9.223}$$

Denote

$$A_0 = X = c_0 \hat{P}_Y + d_0 X, \tag{9.224}$$
$$A_1 = [\hat{H}_0, X] = c_1 \hat{P}_X + d_1 Y, \tag{9.225}$$
$$A_2 = [\hat{H}_0, [\hat{H}_0, X]] = c_2 \hat{P}_Y + d_2 X, \tag{9.226}$$
$$A_3 = [\hat{H}_0, [\hat{H}_0, [\hat{H}_0, X]]] = c_3 \hat{P}_X + d_3 Y, \tag{9.227}$$
$$A_4 = [\hat{H}_0, [\hat{H}_0, [\hat{H}_0, [\hat{H}_0, X]]]] = c_4 V + d_4 X, \tag{9.228}$$
$$\dots,$$

where $c_0 = 0$, $d_0 = 1$, $c_1 = -\alpha_1$, $d_1 = \alpha_2$. Then $X(t)$ of (9.218) can be written as

$$\begin{aligned} X(t) &= A_0 + \left(\frac{it}{\hbar}\right)A_1 + \frac{1}{2!}\left(\frac{it}{\hbar}\right)^2 A_2 + \frac{1}{3!}\left(\frac{it}{\hbar}\right)^3 A_3 + \frac{1}{4!}\left(\frac{it}{\hbar}\right)^4 A_4 + \cdots \\ &= (c_0 P_Y + d_0 X) + \left(\frac{it}{\hbar}\right)(c_1 P_X + d_1 Y) \\ &\quad + \frac{1}{2!}\left(\frac{it}{\hbar}\right)^2 (c_2 P_Y + d_2 X) + \frac{1}{3!}\left(\frac{it}{\hbar}\right)^3 (c_3 P_X + d_3 Y) + \cdots, \end{aligned} \tag{9.229}$$

The above equation can be rewritten as

$$X(t) = \sum_{n=0}^{\infty}\left\{\frac{1}{(2n)!}\left(\frac{it}{\hbar}\right)^{2n}(c_{2n}P_Y + d_{2n}X) + \frac{1}{(2n+1)!}\left(\frac{it}{\hbar}\right)^{2n+1}(c_{2n+1}P_X + d_{2n+1}Y)\right\}, \tag{9.230}$$

Note that from

$$A_{2n+1} = [\hat{H}_0, A_{2n}], \tag{9.231}$$

we have

$$\begin{pmatrix} c_{2n+1} \\ d_{2n+1} \end{pmatrix} = \begin{pmatrix} -\alpha_2 & -\alpha_1 \\ \alpha_1 M^2 \omega_2^2 & \alpha_2 \end{pmatrix} \begin{pmatrix} c_{2n} \\ d_{2n} \end{pmatrix}, \tag{9.232}$$

and from

$$A_{2n} = [\hat{H}_0, A_{2n-1}], \tag{9.233}$$

we obtain

$$\begin{pmatrix} c_{2n} \\ d_{2n} \end{pmatrix} = \begin{pmatrix} \alpha_2 & -\alpha_1 \\ \alpha_1 M^2 \omega_1^2 & -\alpha_2 \end{pmatrix} \begin{pmatrix} c_{2n-1} \\ d_{2n-1} \end{pmatrix}. \tag{9.234}$$

From (9.232) and (9.234), we obtain

$$\begin{pmatrix} c_{2n+1} \\ d_{2n+1} \end{pmatrix} = M_1 \begin{pmatrix} c_{2n-1} \\ d_{2n-1} \end{pmatrix} = \cdots = M_1^n \begin{pmatrix} c_1 \\ d_1 \end{pmatrix} = M_1^n \begin{pmatrix} -\alpha_1 \\ \alpha_2 \end{pmatrix}, \tag{9.235}$$

$$\begin{pmatrix} c_{2n} \\ d_{2n} \end{pmatrix} = M_2 \begin{pmatrix} c_{2n-2} \\ d_{2n-2} \end{pmatrix} = \cdots = M_2^n \begin{pmatrix} c_0 \\ d_0 \end{pmatrix} = M_1^n \begin{pmatrix} 0 \\ 1 \end{pmatrix}, \tag{9.236}$$

where

$$M_1 = \hbar^2 \begin{pmatrix} \frac{\omega_c^2}{4} + \omega_1^2 & -\frac{\omega_c}{M} \\ -\frac{M\omega_c}{2}(\omega_1^2 + \omega_2^2) & \frac{\omega_c^2}{4} + \omega_2^2 \end{pmatrix}, \quad M_2 = \hbar^2 \begin{pmatrix} \frac{\omega_c^2}{4} + \omega_2^2 & \frac{\omega_c}{M} \\ \frac{M\omega_c}{2}(\omega_1^2 + \omega_2^2) & \frac{\omega_c^2}{4} + \omega_1^2 \end{pmatrix}. \tag{9.237}$$

Combining (9.231), (9.236) and (9.237), we have

$$X(t) = (P_Y \quad X) \sum_{n=0}^{\infty} \frac{1}{(2n)!} \left(\frac{it}{\hbar}\right)^{2n} M_2^n \begin{pmatrix} 0 \\ 1 \end{pmatrix} + (P_X \quad Y) \sum_{n=0}^{\infty} \frac{1}{(2n+1)!} \left(\frac{it}{\hbar}\right)^{2n+1} M_1^n \begin{pmatrix} -\alpha_1 \\ \alpha_2 \end{pmatrix}. \tag{9.238}$$

Note that M_1, M_2 can be diagonalized as

$$M_1 = \hbar^2 U_1 \begin{pmatrix} \lambda_1 & 0 \\ 0 & \lambda_2 \end{pmatrix} U_1^{-1}, \quad M_2 = \hbar^2 U_2 \begin{pmatrix} \lambda_3 & 0 \\ 0 & \lambda_4 \end{pmatrix} U_2^{-1}, \tag{9.239}$$

with

$$U_1 = \begin{pmatrix} \frac{\omega_1^2-\omega_2^2-\sqrt{\Delta}}{-M\omega_c(\omega_1^2+\omega_2^2)} & \frac{\omega_1^2-\omega_2^2+\sqrt{\Delta}}{-M\omega_c(\omega_1^2+\omega_2^2)} \\ 1 & 1 \end{pmatrix}, \quad U_1^{-1} = \begin{pmatrix} \frac{M\omega_c(\omega_1^2+\omega_2^2)}{2\sqrt{\Delta}} & \frac{\omega_1^2-\omega_2^2+\sqrt{\Delta}}{2\sqrt{\Delta}} \\ \frac{-M\omega_c(\omega_1^2+\omega_2^2)}{2\sqrt{\Delta}} & \frac{\omega_2^2-\omega_1^2+\sqrt{\Delta}}{2\sqrt{\Delta}} \end{pmatrix}, \tag{9.240}$$

and

$$U_2 = \begin{pmatrix} \frac{\omega_2^2-\omega_1^2-\sqrt{\Delta}}{M\omega_c(\omega_1^2+\omega_2^2)} & \frac{\omega_2^2-\omega_1^2+\sqrt{\Delta}}{M\omega_c(\omega_1^2+\omega_2^2)} \\ 1 & 1 \end{pmatrix}, \quad U_2^{-1} = \begin{pmatrix} \frac{-M\omega_c(\omega_1^2+\omega_2^2)}{2\sqrt{\Delta}} & \frac{\omega_2^2-\omega_1^2+\sqrt{\Delta}}{2\sqrt{\Delta}} \\ \frac{M\omega_c(\omega_1^2+\omega_2^2)}{2\sqrt{\Delta}} & \frac{\omega_1^2-\omega_2^2+\sqrt{\Delta}}{2\sqrt{\Delta}} \end{pmatrix}, \tag{9.241}$$

where

$$\Delta = (\omega_1^2 - \omega_2^2)^2 + 2\omega_c^2(\omega_1^2 + \omega_2^2), \tag{9.242}$$

$$\lambda_{1,2} = \lambda_{3,4} = \frac{1}{2}\left(\frac{\omega_c^2}{2} + \omega_1^2 + \omega_2^2 \mp \sqrt{(\omega_1^2 - \omega_2^2)^2 + 2\omega_c^2(\omega_1^2 + \omega_2^2)}\right). \tag{9.243}$$

Hence, we have

$$(P_Y \;\; X)\sum_{n=0}^{\infty} \frac{1}{(2n)!}\left(\frac{it}{\hbar}\right)^{2n} M_2^n \begin{pmatrix} 0 \\ 1 \end{pmatrix} = (P_Y \;\; X)U_2 \begin{pmatrix} \cos\sqrt{\lambda_3}t & \\ & \cos\sqrt{\lambda_4}t \end{pmatrix} U_2^{-1} \begin{pmatrix} 0 \\ 1 \end{pmatrix}, \tag{9.244}$$

and

$$(P_X \;\; Y)\sum_{n=0}^{\infty} \frac{1}{(2n+1)!}\left(\frac{it}{\hbar}\right)^{2n+1} M_1^n \begin{pmatrix} -\alpha_1 \\ \alpha_2 \end{pmatrix} = (P_X \;\; Y)U_1 \frac{i}{\hbar} \begin{pmatrix} \frac{\sin\sqrt{\lambda_1}t}{\sqrt{\lambda_1}} & \\ & \frac{\sin\sqrt{\lambda_2}t}{\sqrt{\lambda_2}} \end{pmatrix} U_1^{-1} \begin{pmatrix} -\alpha_1 \\ \alpha_2 \end{pmatrix}, \tag{9.245}$$

Substituting (9.239), (9.240), (9.241) into (9.238), we finally obtain $X(t)$ of (9.190):

Similarly, we can obtain for $Y(t)$ of (9.199):

$$Y(t) = e^{i\hat{H}_0 t/\hbar} Y e^{-i\hat{H}_0 t/\hbar} = \hat{Y} + \frac{it}{\hbar}[\hat{H}_0, \hat{Y}] + \frac{1}{2!}\left(\frac{it}{\hbar}\right)^2 [\hat{H}_0, [\hat{H}_0, \hat{Y}]] + \frac{1}{3!}\left(\frac{it}{\hbar}\right)^3 [\hat{H}_0, [\hat{H}_0, [\hat{H}_0, \hat{Y}]]] + \cdots$$
$$= (P_X \;\; Y)\sum_{n=0}^{\infty} \frac{1}{(2n)!}\left(\frac{it}{\hbar}\right)^{2n} M_1^n \begin{pmatrix} 0 \\ 1 \end{pmatrix} + (P_Y \;\; X)\sum_{n=0}^{\infty} \frac{1}{(2n+1)!}\left(\frac{it}{\hbar}\right)^{2n+1} M_2^n \begin{pmatrix} -\alpha_1 \\ -\alpha_2 \end{pmatrix}. \tag{9.246}$$

9.2.3 *Evolution of the Eigenstates of the In-plane Motion*

Since the eigenstates $|n\rangle$ of $\hat{H}_0$ of (9.186) form a complete set, any state can be expanded in terms of them. We next calculate the evolution of the eigenstates of the in-plane motion, i.e., with $|\Phi_0\rangle = |n\rangle$, and $\hat{H}_0|n\rangle = E_n^{XY}|n\rangle$ we then have from (9.216)

$$|\Phi_t\rangle = \exp\left\{-\frac{i}{\hbar}\hat{H}_0 t\right\}\exp\left\{-\frac{i}{\hbar}\int_0^t du\hat{H}_1^{int}(u)\right\}\exp\left\{-\frac{i}{\hbar}\alpha(t)\right\}|n\rangle$$
$$= e^{-\frac{i}{\hbar}\alpha(t)}e^{-i\hat{H}_0 t/\hbar}e^{-\frac{i}{\hbar}\int_0^t du\hat{H}_1^{int}(u)}e^{i\hat{H}_0 t/\hbar}e^{-iE_n^{XY}t/\hbar}|n\rangle. \tag{9.247}$$

Using the identity

$$e^A e^B e^{-A} = e^{\{e^A B e^{-A}\}}, \tag{9.248}$$

and that of (9.217), the equation (9.247) can be rewritten as

$$|\Phi_t\rangle = e^{-\frac{i}{\hbar}[\alpha(t)+E_n^{XY}t]}\exp\left\{e^{-i\hat{H}_0 t/\hbar}\left(-\frac{i}{\hbar}\int_0^t du\hat{H}_1^{int}(u)\right)e^{i\hat{H}_0 t/\hbar}\right\}|n\rangle$$
$$= e^{-\frac{i}{\hbar}[\alpha(t)+E_n^{XY}t]}$$
$$\times\exp\left\{-\frac{i}{\hbar}\int_0^t du\left[e^{i\hat{H}_0(u-t)/\hbar}\hat{H}_1(u)e^{-i\hat{H}_0(u-t)/\hbar}\right]\right\}|n\rangle. \tag{9.249}$$

Note that

$$\hat{H}_1(u) = Ne[E_x(u)X + E_y(u)Y], \tag{9.250}$$

so that

$$e^{i\hat{H}_0(u-t)/\hbar}\hat{H}_1(u)e^{-i\hat{H}_0(u-t)/\hbar} = Ne[E_x(u)X(u-t) + E_y(u)Y(u-t)]$$
$$= c_1(u,t)X + c_2(u,t)Y + c_3(u,t)\hat{P}_X + c_4(u,t)\hat{P}_Y, \tag{9.251}$$

with

$$c_i(u,t) = Ne[E_x(u)a_i(u-t) + E_y(u)b_i(u-t)], i = 1, 2, 3, 4. \tag{9.252}$$

Inserting (9.251) into (9.249), we obtain

$$|\Phi_t\rangle = |e^{-\frac{i}{\hbar}\left(\alpha_1(t)X+\alpha_2(t)\hat{P}_X+\beta_1(t)Y+\beta_2(t)\hat{P}_Y\right)}|n\rangle e^{\frac{-i}{\hbar}[E_n^{XY}t+\alpha(t)]} \tag{9.253}$$

where

$$\alpha_1(t) = \int_0^t c_1(u,t)du, \quad \alpha_2(t) = \int_0^t c_3(u,t)du, \tag{9.254}$$

$$\beta_1(t) = \int_0^t c_2(u,t)du, \quad \beta_2(t) = \int_0^t c_4(u,t)du. \tag{9.255}$$

Then, the evolution $|\Phi_t\rangle$ in the coordinate representation is

$$\langle X, Y|\Phi_t\rangle = \langle X, Y|\exp\left\{-\frac{i}{\hbar}[\alpha_1(t)X + \alpha_2(t)\hat{P}_X + \beta_1(t)Y + \beta_2(t)\hat{P}_Y]\right\}|n\rangle \\ \times \exp\left\{-\frac{i}{\hbar}[E_n^{XY}t + \alpha(t)]\right\}. \tag{9.256}$$

Note that for operators A, B if their commutator $[A, B]$ is a c number, then

$$e^{A+B} = e^A e^B e^{-\frac{1}{2}[A,B]}. \tag{9.257}$$

Hence, (9.256) can be rewritten as

$$\langle X, Y|\Phi_t\rangle = < X, Y|e^{-\frac{i}{\hbar}\left(\alpha_1(t)X+\alpha_2(t)\hat{P}_X\right)}e^{-\frac{i}{\hbar}\left(\beta_1(t)Y+\beta_2(t)\hat{P}_Y\right)}|n\rangle e^{\frac{-i}{\hbar}\left(E_n^{XY}t+\alpha(t)\right)} \\ = \exp\left\{-\frac{i}{\hbar}[\alpha_1(t)X + \beta_1(t)Y]\right\}\exp\left\{\frac{i}{2\hbar}(\alpha_1(t)\alpha_2(t) + \beta_1(t)\beta_2(t))\right\} \\ \langle X - \alpha_2(t), Y - \beta_2(t)|n\rangle \exp\left\{-\frac{i}{\hbar}[E_n^{XY}t + \alpha(t)]\right\}. \tag{9.258}$$

From the above equation, we can immediately see that the wave function is shifted from the original wave function, and the phase angle is changed. To see the physical meaning of the shift, we next investigate the properties of the translation functions $\alpha_2(t), \beta_2(t)$.

9.2.4 Total Wave Function and Classical Equation of Motion

The corresponding Lagrangian for the in-plane Hamiltonian of (9.185) is,

$$L_{XY} = \frac{1}{2}M(\dot{\mathbf{R}}_{||}^2 - \tilde{\mathbf{R}}_{||}\Omega^2\mathbf{R}_{||} + \omega_c\tilde{\mathbf{R}}_{||}J\dot{\mathbf{R}}_{||}) \\ - NeE_x(t)X - NeE_y(t)Y, \tag{9.259}$$

where

$$\mathbf{R}_{||}(t) = \begin{pmatrix} X(t) \\ Y(t) \end{pmatrix}, \tag{9.260}$$

denotes the mass center position vector perpendicular to the magnetic field $\mathbf{B}$, $\tilde{\mathbf{R}}_{||}$ the transpose of vector $\mathbf{R}_{||}$, $J = \begin{pmatrix} 0 & -1 \\ 1 & 0 \end{pmatrix}$, and $\Omega^2 = \begin{pmatrix} \omega_x^2 & 0 \\ 0 & \omega_y^2 \end{pmatrix}$. From the Lagrangian one can obtain the canonical momentum as

$$P_X = M\dot{X} + \frac{M\omega_c Y}{2}, \quad P_Y = M\dot{Y} - \frac{M\omega_c X}{2}, \tag{9.261}$$

and the equations of motion

$$\ddot{X} + \omega_c \dot{Y} + \omega_x^2 X + \frac{Ne}{M} E_x(t) = 0,$$
$$\ddot{Y} - \omega_c \dot{X} + \omega_y^2 Y + \frac{Ne}{M} E_y(t) = 0. \tag{9.262}$$

After some algebraic manipulations, one can verify that the position vector

$$\begin{pmatrix} \alpha_2(t) \\ \beta_2(t) \end{pmatrix} = \mathbf{R}_{||,m}(t) \equiv \begin{pmatrix} X_m(t) \\ Y_m(t) \end{pmatrix}, \tag{9.263}$$

satisfies the equation of motion of (9.262) with the initial condition $\alpha_2(0) = 0, \beta_2(0) = 0$. Moreover,

$$\alpha_1(t) = -P_{X_m}, \quad \beta_1(t) = -P_{Y_m}, \tag{9.264}$$

the corresponding canonical momentum for $\mathbf{R}_{||,m}$.

The classical action between the points $(0, \mathbf{R}_{||,0})$ and $(t, \mathbf{R}_{||,t})$ without the external electric field term is

$$\begin{aligned} S_0^{XY}[\mathbf{R}_{||}(t)] &= \frac{1}{2} M \int_0^t du (\dot{\mathbf{R}}_{||}^2 - \tilde{\mathbf{R}}_{||} \Omega^2 \mathbf{R}_{||} + \omega_L \tilde{\mathbf{R}}_{||} J \dot{\mathbf{R}}_{||}) \\ &= \frac{M}{2} \{\tilde{\mathbf{R}}_{||,t} \dot{\mathbf{R}}_{||,t} - \tilde{\mathbf{R}}_{||,0} \dot{\mathbf{R}}_{||,0}\} \\ &+ \frac{Ne}{2} \int_0^t du \tilde{\mathbf{R}}_{||}(u) \cdot \mathbf{E}(u). \end{aligned} \tag{9.265}$$

Thus,

$$\begin{aligned} S_0^{XY}[\mathbf{R}_{||,m}] &= \frac{M}{2} \{\tilde{\mathbf{R}}_{||,m} \dot{\mathbf{R}}_{||,m}\} + \frac{Ne}{2} \int_0^t du \tilde{\mathbf{R}}_{||,m}(u) \cdot \mathbf{E}(u) \\ &= -\frac{1}{2} [\alpha_1(t) \alpha_2(t) + \beta_1(t) \beta_2(t)] \\ &+ \frac{Ne}{2} \int_0^t du [E_x(u) X_m(u) + E_y(u) Y_m(u)]. \end{aligned} \tag{9.266}$$

We next show

$$\alpha(t) = \frac{Ne}{2} \int_0^t du [E_x(u) X_m(u) + E_y(u) Y_m(u)]. \tag{9.267}$$

Using (9.252), (9.254), and (9.255), the equation (9.267) can be rewritten as

$$\alpha(t) = \frac{(Ne)^2}{2}\int_0^t du\Big\{E_x(u)\int_0^u dv[E_x(v)a_3(v-u) + E_y(v)b_3(v-u)]$$
$$+E_y(u)\int_0^u dv[E_x(v)a_4(v-u) + E_y(v)b_4(v-u)]\Big\}. \quad (9.268)$$

On the other hand, from the definition of $\alpha(t)$ (9.215), combining (9.211) and (9.213), we only need to show

$$\alpha(t) = \frac{(Ne)}{2}\int_0^t du\{E_x(u)\int_0^u dv[a_1(u)c_3(v) + a_2(u)c_4(v) - a_3(u)c_1(v) - a_4(u)c_2(v)]$$
$$+E_y(u)\int_0^u dv[b_1(u)c_3(v) + b_2(u)c_4(v) - b_3(u)c_1(v) - b_4(u)c_2(v)]\}. \quad (9.269)$$

Using the following identities,

$$[a_1(u)a_3(v) + a_2(u)a_4(v) - a_3(u)a_1(v) - a_4(u)a_2(v)] = a_3(v-u), \quad (9.270)$$

$$[a_1(u)b_3(v) + a_2(u)b_4(v) - a_3(u)b_1(v) - a_4(u)b_2(v)] = b_3(v-u), \quad (9.271)$$

$$[b_1(u)a_3(v) + b_2(u)a_4(v) - b_3(u)a_1(v) - b_4(u)a_2(v)] = a_4(v-u), \quad (9.272)$$

$$[b_1(u)b_3(v) + b_2(u)b_4(v) - b_3(u)b_1(v) - b_4(u)b_2(v)] = b_4(v-u), \quad (9.273)$$

it is not difficult to verify that (9.269) is true.

Finally, we have the wave function of the in-plane motion to be

$$\langle X, Y|\Phi_t\rangle = \exp\{\frac{i}{\hbar}[P_{X_m}X + P_{Y_m}Y]\}\langle X - X_m(t), Y - Y_m(t)|n\rangle$$
$$\times \exp\{-\frac{i}{\hbar}(E_n^{XY}t + S_0^{XY}[\mathbf{R}_{||,m}])\}. \quad (9.274)$$

Hence, the total wave function including the in-plane, Z-component and relative motions is then given by (9.170).

9.2.5 The Relative Coordinate Hamiltonian

With the center of mass and relative coordinates and momentum defined by (9.176) and (9.177), the transformed coordinates may be expressed as

$$\begin{pmatrix} \mathbf{R}^{(1)} \\ \mathbf{R}^{(2)} \\ \mathbf{R}^{(3)} \\ \vdots \\ \mathbf{R}^{(N)} \end{pmatrix} = T \begin{pmatrix} \mathbf{r}_1 \\ \mathbf{r}_2 \\ \mathbf{r}_3 \\ \vdots \\ \mathbf{r}_N \end{pmatrix}, \tag{9.275}$$

where the transformation matrix T is

$$\begin{pmatrix} \frac{1}{N} & \frac{1}{N} & \frac{1}{N} \cdots & \frac{1}{N} \\ 1 & -1 & 0 \ldots & 0 \\ 1 & 1 & -2 \ldots & 0 \\ \vdots & \vdots & \vdots \ddots & \vdots \\ 1 & 1 & 1 \ldots & -(N-1) \end{pmatrix}. \tag{9.276}$$

From (9.275)

$$\begin{pmatrix} \mathbf{r}_1 \\ \mathbf{r}_2 \\ \vdots \\ \mathbf{r}_N \end{pmatrix} = T^{-1} \begin{pmatrix} \mathbf{R}^{(1)} \\ \mathbf{R}^{(2)} \\ \vdots \\ \mathbf{R}^{(N)} \end{pmatrix} \quad \text{or} \quad \mathbf{r}_i = \sum_{\ell=1}^{N} (T^{-1})_{i\ell} \mathbf{R}^{(\ell)}, \tag{9.277}$$

where the inverse matrix is

$$T^{-1} = \begin{pmatrix} 1 & \frac{1}{2} & \frac{1}{6} & \frac{1}{12} & \cdots & \frac{1}{N(N-1)} \\ 1 & -\frac{1}{2} & \frac{1}{6} & \frac{1}{12} & \cdots & \frac{1}{N(N-1)} \\ 1 & 0 & \frac{1}{3} & \frac{1}{12} & \cdots & \frac{1}{N(N-1)} \\ 1 & 0 & 0 & -\frac{1}{4} & \cdots & \frac{1}{N(N-1)} \\ \vdots & \vdots & \vdots & \vdots & \ddots & \vdots \\ 1 & 0 & 0 & 0 & \ldots & -\frac{1}{N} \end{pmatrix}. \tag{9.278}$$

Since

$$\boldsymbol{\nabla}_{\mathbf{r}_i} = \frac{\partial \mathbf{R}^{(1)}}{\partial \mathbf{r}_i} \boldsymbol{\nabla}_{\mathbf{R}^{(1)}} + \frac{\partial \mathbf{R}^{(2)}}{\partial \mathbf{r}_i} \boldsymbol{\nabla}_{\mathbf{R}^{(2)}} + \cdots + \frac{\partial \mathbf{R}^{(N)}}{\partial \mathbf{r}_i} \boldsymbol{\nabla}_{\mathbf{R}^{(N)}}, \tag{9.279}$$

we have

$$\begin{pmatrix} \boldsymbol{\nabla}_{\mathbf{r}_1} \\ \boldsymbol{\nabla}_{\mathbf{r}_2} \\ \boldsymbol{\nabla}_{\mathbf{r}_3} \\ \vdots \\ \boldsymbol{\nabla}_{\mathbf{r}_N} \end{pmatrix} = \tilde{T} \begin{pmatrix} \boldsymbol{\nabla}_{\mathbf{R}^{(1)}} \\ \boldsymbol{\nabla}_{\mathbf{R}^{(2)}} \\ \boldsymbol{\nabla}_{\mathbf{R}^{(3)}} \\ \vdots \\ \boldsymbol{\nabla}_{\mathbf{R}^{(N)}} \end{pmatrix}, \tag{9.280}$$

where $\tilde{T}$ is the transpose of T. Thus,

$$\frac{1}{2m}\sum_i \hat{p}_i^2 = -\frac{\hbar^2}{2m}\sum_i \nabla_{\mathbf{r}_i}^2 = -\frac{\hbar^2}{2m}(\nabla_{\mathbf{r}_1}\nabla_{\mathbf{r}_2}\ldots\nabla_{\mathbf{r}_N})\begin{pmatrix}\nabla_{\mathbf{r}_1}\\ \nabla_{\mathbf{r}_2}\\ \vdots\\ \nabla_{\mathbf{r}_N}\end{pmatrix}$$

$$= -\frac{\hbar^2}{2m}(\nabla_{\mathbf{R}^{(1)}}\nabla_{\mathbf{R}^{(2)}}\ldots\nabla_{\mathbf{R}^{(N)}})\ T\tilde{T}\begin{pmatrix}\nabla_{\mathbf{R}^{(1)}}\\ \nabla_{\mathbf{R}^{(2)}}\\ \vdots\\ \nabla_{\mathbf{R}^{(N)}}\end{pmatrix}. \qquad (9.281)$$

The product of the matrices

$$T\tilde{T} = \begin{pmatrix}\frac{1}{N} & 0 & 0 & 0 & \ldots & 0\\ 0 & 2 & 0 & 0 & \ldots & 0\\ 0 & 0 & 6 & 0 & \ldots & 0\\ 0 & 0 & 0 & 12 & \ldots & 0\\ \vdots & \vdots & \vdots & \vdots & \ddots & \vdots\\ 0 & 0 & 0 & 0 & \ldots & N(N-1)\end{pmatrix} \qquad (9.282)$$

is a diagonal matrix. This means that for each transformed coordinate $\mathbf{R}^{(i)}$ there corresponds an effective mass m_i. Thus, $m_1 = Nm = M$, $m_2 = m/2$, $m_3 = m/6, \ldots, m_N = m/N(N-1)$.

The kinetic energy operator $\hat{T}$ of the Hamiltonian $\hat{H}(t)$ of (9.167) is

$$\hat{T} = \frac{1}{2m}\sum_{i=1}^{N}[\hat{p}_i^2 + (\frac{e}{c})^2 A^2(\mathbf{r}_i) + 2\frac{e}{c}\mathbf{A}(\mathbf{r}_i)\cdot\hat{\mathbf{p}}_i], \qquad (9.283)$$

where we have employed the Coulomb gauge $\nabla\cdot\mathbf{A} = 0$.

The first term $\hat{T}_1$ of (9.283) is from (9.281), (9.282):

$$\hat{T}_1 = \frac{\hbar^2}{2m}\sum_{i=1}^{N}\nabla_{\mathbf{r}_i}^2 = -\sum_{i=1}^{N}\frac{\hbar^2}{2m_i}\nabla_{\mathbf{R}^{(i)}}^2$$

$$= \frac{P^2}{2M} - \sum_{i=2}^{N}\frac{\hbar^2}{2m_i}\nabla_{\mathbf{R}^{(i)}}^2. \qquad (9.284)$$

The second term $\hat{T}_2$ of (9.283) in the symmetric gauge $\mathbf{A} = \frac{1}{2}\boldsymbol{\mathcal{B}}\times\mathbf{r}$ is

$$\hat{T}_2 = \frac{1}{2m}\sum_{i=1}^{N}\frac{e^2}{c^2}A^2(\mathbf{r}_i) = \frac{1}{2m}\frac{e^2B^2}{4c^2}\sum_{i=1}^{N} r_i^2$$
$$= \frac{m}{2}\frac{\omega_c^2}{4}\sum_{r=1}^{N} r_i^2$$
$$= \frac{\omega_c^2}{4}\sum_{i=1}^{N}\frac{m_i}{2}\tilde{\mathbf{R}}^{(i)}\cdot\mathbf{R}^{(i)}. \tag{9.285}$$

The last step in the previous equation is the same as that for the external harmonic potential operator $\hat{V}$ derived below, and will be incorporated into it.

The third term $\hat{T}_3$ of (9.283) can be written as

$$\hat{T}_3 = \frac{e(-i\hbar)}{mc}\sum_{i=1}^{N}\mathbf{A}(\mathbf{r}_i)\cdot\hat{\nabla}_{\mathbf{r}_i} = \frac{\omega_c(-i\hbar)}{2}\sum_{i=1}^{N}\tilde{\mathbf{r}}_i\cdot J\hat{\nabla}_{\mathbf{r}_i}, \tag{9.286}$$

Employing the transformation equations (9.275), (9.277), (9.280), (9.282), we obtain

$$\hat{T}_3 = \frac{\omega_c}{2}(\mathbf{R}^{(1)}\mathbf{R}^{(2)}\cdots\mathbf{R}^{(N)})\tilde{T}^{-1}\tilde{T}J\begin{pmatrix}\hat{\nabla}_{\mathbf{R}^{(1)}}\\ \hat{\nabla}_{\mathbf{R}^{(2)}}\\ \vdots\\ \hat{\nabla}_{\mathbf{R}^{(N)}}\end{pmatrix}$$
$$= \frac{\omega_c(-i\hbar)}{2}\sum_{i=1}^{N}\tilde{\mathbf{R}}^{(i)}\cdot J\hat{\nabla}_{\mathbf{R}^{(i)}} = \frac{\omega_c}{2}\sum_{i=1}^{N}\hat{L}_{Z^{(i)}}, \tag{9.287}$$

where $\hat{L}_{Z^{(i)}} = X^{(i)}\hat{P}_{Y^{(i)}} - Y^{(i)}\hat{P}_{X^{(i)}}$ being the Z-component of the angular momentum for the coordinates defined by (9.176).

Employing the transformation (9.277), the external harmonic potential energy operator $\hat{V}$ which is

$$\hat{V} = \frac{1}{2}m\sum_{i=1}^{N}\tilde{\mathbf{r}}_i\cdot\mathbb{K}\cdot\mathbf{r}_i \tag{9.288}$$

may be written as

$$\hat{V} = \frac{1}{2}m(\mathbf{R}^{(1)}\mathbf{R}^{(2)}\cdots\mathbf{R}^{(N)})\tilde{T}^{-1}\mathbb{K}(t)T^{-1}\begin{pmatrix}\mathbf{R}^{(1)}\\ \mathbf{R}^{(2)}\\ \vdots\\ \mathbf{R}^{(N)}\end{pmatrix}$$
$$= \frac{1}{2}m(\mathbf{R}^{(1)}\mathbf{R}^{(2)}\cdots\mathbf{R}^{(N)})\begin{pmatrix} N & 0 & 0 & \dots & 0\\ 0 & \frac{1}{2} & 0 & \dots & 0\\ 0 & 0 & \frac{1}{6} & \dots & 0\\ \vdots & \vdots & \vdots & \ddots & \vdots\\ 0 & 0 & 0 & \dots & \frac{1}{N(N-1)}\end{pmatrix}\mathbb{K}(t)\begin{pmatrix}\mathbf{R}^{(1)}\\ \mathbf{R}^{(2)}\\ \vdots\\ \mathbf{R}^{(N)}\end{pmatrix}$$
$$= \sum_{i=1}^{N}\frac{1}{2}m_i\tilde{\mathbf{R}}^{(i)}\cdot\mathbb{K}(t)\cdot\mathbf{R}^{(i)}. \tag{9.289}$$

Observe that the transformed operator $\hat{T}$ and $\hat{V}$ are of the same form as in the original Hamiltonian, but with effective masses m_i replacing the electron mass m. From (9.277),

$$\mathbf{r}_i - \mathbf{r}_j = \sum_{\ell=1}^{N}[(T^{-1})_{i\ell} - (T^{-1})_{j\ell}]\mathbf{R}^{(\ell)}$$
$$= \sum_{\ell=2}^{N}[(T^{-1})_{i\ell} - (T^{-1})_{j\ell}]\mathbf{R}^{(\ell)}, \tag{9.290}$$

since $(T^{-1})_{i1} = 1$. Thus, the transformed electron-interaction operator is

$$\hat{U} = \sum_{i\neq j}u\left(\sum_{\ell=2}^{N}[(T^{-1})_{i\ell} - (T^{-1})_{j\ell}]\mathbf{R}^{(\ell)}\right). \tag{9.291}$$

Observe that the transformed operator depends only on the relative coordinates. The relative coordinates Hamiltonian is then

$$\hat{H}_{rel} = \sum_{i=2}^{N}\left[-\frac{\hbar^2}{2m_i}\nabla^2_{\mathbf{R}^{(i)}} + \frac{\omega_c}{2}\hat{L}_{Z^{(i)}} + \frac{m_i}{2}\tilde{\mathbf{R}}^{(i)}\cdot\Xi^2\mathbf{R}^{(i)}\right]$$
$$+\sum_{i\neq j}u\left(\sum_{\ell=2}^{N}[(T^{-1})_{i\ell} - (T^{-1})_{j\ell}]\mathbf{R}^{(\ell)}\right), \tag{9.292}$$

where

$$\Xi^2 = \begin{pmatrix}\omega_1^2 & 0\\ 0 & \omega_2^2\end{pmatrix}. \tag{9.293}$$

References

1. E. Schrödinger, *Collected Papers on Wave Mechanics* (American Mathematical Society Chelsea Publishing, Providence, Rhode Island, Reprinted, 2014)
2. A. Sommerfeld, *Wave Mechanics* (Dutton, New York, 1929)
3. V. Fock, Z. Physik **47**, 446 (1928)
4. C.G. Darwin, Cambridge Phil. Soc. **27**, 86 (1930)
5. W. Pauli, Z. Physik **31**, 765 (1925)
6. P.A.M. Dirac, Proc. Roy. Soc. **A112**, 661 (1926)
7. W. Heisenberg, Z. Physik **38**, 411 (1926)
8. B.L. Moiseiwitsch, *Variational Principles* (Wiley, London, 1996)
9. V. Sahni, J.B. Krieger, Int. J. Quantum Chem. Symp. **5**, 47 (1971); V. Sahni, J.B. Krieger, Int. J. Quantum Chem. Symp. **6**, 103 (1972); J.B. Krieger, V. Sahni, Phys. Rev. A **6**,919 (1972); V. Sahni, J.B. Krieger, Phys. Rev. A **6**, 928 (1972); V. Sahni, J.B. Krieger, Phys. Rev. A **8**, 65 (1973)
10. V. Sahni, J.B. Krieger, Phys. Rev. A **11**, 409 (1975); V. Sahni, J.B. Krieger, J. Gruenebaum, Phys. Rev. A **12**, 768 (1975)
11. X.-Y. Pan, V. Sahni, L. Massa, Phys. Rev. Lett. **93**, 13401 (2004); X.-Y. Pan, V. Sahni, L. Massa, Phys. Rev. A **72**, 032505 (2005); X.-Y. Pan, V. Sahni, L. Massa, Phil. Mag. **86**, 2673 (2006); X.-Y. Pan, M. Slamet, V. Sahni, Phys. Rev. A **81**, 042524 (2010); M. Slamet, X.-Y. Pan, V. Sahni, Phys. Rev. A **84**, 052504 (2011)
12. N.R. Kestner, O. Sinanoglu, Phys. Rev. **128**, 2687 (1962)
13. S. Kais, D.R. Herschbach, R.D. Levine, J. Chem. Phys. **91**, 7791 (1989)
14. M. Taut, Phys. Rev. A **48**, 3561 (1993)
15. M. Taut, J. Phys. A **27**, 1045 (1994); Corrigenda J. Phys. A **27**, 4723 (1994)
16. M. Taut, J. Phys. Condens. Matter **12**, 3689 (2000)
17. M. Taut, H. Eschrig, Z. Phys, Chem. **224**, 631 (2010)
18. H.-M, Zhu, J.-W. Chen, X.-Y. Pan, V. Sahni, J. Chem. Phys. **140**, 024318 (2014)
19. K.F. Riley, M.P. Hobson, S.J. Pence, *Mathematical Methods for Physics and Engineering*, 3rd edn. (Cambridge University Press, 2006)
20. T. Yang, X.-Y. Pan, V. Sahni, Phys. Rev. A **83**, 042518 (2011)
21. V. Sahni, Int. J. Quantum Chem. **121**, e26556 (2021)
22. M. Slamet, V. Sahni, Chem. Phys. **546**, 111073 (2021)
23. W. Kohn, Phys. Rev. **123**, 1242 (1961)
24. L. Brey, N.F. Johnson, B.I. Halperin, Phys. Rev. B **40**, 10647 (1989)
25. S.K. Yip, Phys. Rev. B **43**, 1707 (1991)
26. P.A. Maksym, T. Chakraborty, Phys. Rev. Lett. **65**, 108 (1990)
27. F.M. Peters, Phys. Rev. **42**, 1486 (1990)
28. J. Dobson, Phys. Rev. Lett. **73**, 2244 (1994)
29. V. Sahni, *Quantal Density Functional Theory*, 2nd edn. (Springer, Berlin, Heidelberg, 2016) (Referred to as ***QDFT***)
30. J.-W. Chen, T. Yang, X.-Y. Pan, Chin. Phys. Lett. **30**, 020303 (2013)
31. Y.-Q. Li, X.-Y. Pan, V. Sahni, J. Chem. Phys. **139**, 114301 (2013)

Chapter 10
Epilogue

Abstract The Epilogue constitutes a brief review of the complementary perspectives to Schrödinger (and Schrödinger-Pauli) theory of electrons in an electromagnetic field as described in the book. The principal new idea is the description of the *individual* electron in the sea of electrons via its equations of motion – the 'Quantal Newtonian' First and Second Laws – descriptive of stationary-state and time-dependent theory, respectively. These Laws are in terms of 'classical' fields whose sources are quantum-mechanical expectation values of Hermitian operators taken with respect to the wave function. New physical and mathematical insights into Schrödinger theory are thereby revealed. The local effective potential theory complement to both stationary-state and temporal Schrödinger theory – Quantal Density Functional Theory – is described. This description too is in terms of 'classical' fields and quantal sources. Finally, the tenets of Modern Density Functional Theory, both the stationary ground state and temporal local effective potential theory complements to Schrödinger theory are explained. New understandings of the theory, and a rigorous physical interpretation of its mathematical entities, are provided. Various properties of the many-electron stationary-state wave function are derived, and a new symmetry for a two-electron system described. The analytical wave functions for harmonically bound electrons in time-independent and dependent electromagnetic fields are also derived. These wave functions are then employed to explicate the complementary perspectives.

This book provides a complementary perspective to the Schrödinger theory of electrons. The perspective is ensconced within the Born interpretation of the wave function Ψ as a probability amplitude. It provides physical and mathematical insights into the theory. The solution to the stationary-state Schrödinger equation is the *many-electron* wave function Ψ. The wave function is descriptive of the *entire electronic* system. The solution to the temporal equation $\Psi(t)$ then describes the evolution of this *system of electrons* with time. The perspective presented in the book, on the other hand, is that of the *individual* electron in the sea of electrons. This perspective

V. Sahni, *Schrödinger Theory of Electrons: Complementary Perspectives*,
Springer Tracts in Modern Physics 285,
https://doi.org/10.1007/978-3-030-97409-1_10

is described via the equations of motion—the 'Quantal Newtonian' First and Second Laws—for *each* electron. These Laws are derived from the Schrödinger equation.

The 'Quantal Newtonian' Laws are written in terms of 'classical' fields that pervade all space. The fields obey a classical equation of motion similar to that of the Coulomb-Lorentz Law. There is, however, a fundamental difference between the fields of classical physics and those of the 'Quantal Newtonian' Laws. The sources of the fields of the 'Quantal Newtonian' Laws are quantum-mechanical in that they are expectation values of Hermitian operators taken with respect to the wave function. Thus, the individual-electron-perspective hews to the Born interpretation. However, once the quantal sources are defined, the properties of these fields, being similar to those of classical physics, imply their determinism.

The statement of the 'Quantal Newtonian' First law is similar to that of Newton's First Law for the *individual* classical charged particle amongst all the other such particles in an electromagnetic field. Each electron experiences the external field $\mathcal{F}^{\mathrm{ext}}(\mathbf{r})$ and an internal field $\mathcal{F}^{\mathrm{int}}(\mathbf{r})$, the sum of which vanishes. As in classical physics, each electron experiences the external electrostatic field, but in addition there now is a quantum-mechanical Lorentz field component. (In quantum mechanics, the external electrostatic binding field is not necessarily a Coulomb field, it could be harmonic, or any other field defined by a scalar potential.) The internal field $\mathcal{F}^{\mathrm{int}}(\mathbf{r})$ experienced by each electron is comprised of components representative of properties of the system, *viz.* the correlations due to the Pauli principle and Coulomb repulsion, kinetic effects, the electron density, and an internal magnetic component. That each electron experiences all these different fields is a new understanding of the physics of electrons. The total energy and its components can be expressed in terms of these fields in integral virial form. In classical physics, one assumes that any internal forces between the particles obey Newton's Third Law. Hence, on summing over all the particles, the contribution of the internal forces vanish, leading to Newton's First Law. In a similar manner, on summing over all the electrons, the contributions of the internal field $\mathcal{F}^{\mathrm{int}}(\mathbf{r})$ vanishes, leading thereby to Ehrenfest's First Law Theorem. This then is a more insightful derivation of Ehrenfest's Theorem.

The 'Quantal Newtonian' Second Law for each electron parallels Newton's Second Law for each classical particle. The latter Law states that the response of each particle—the time rate of change of its momentum—equals the sum of the external and internal forces acting on it. In summing over all the particles, the contribution of the internal forces vanishes, leading to Newton's Second Law. In a similar manner, the statement of the 'Quantal Newtonian' Second Law is that the sum of the time-dependent external $\mathcal{F}^{\mathrm{ext}}(\mathbf{r}t)$ and internal $\mathcal{F}^{\mathrm{int}}(\mathbf{r}t)$ fields causes a response by *each electron* represented by a Current Density $\mathcal{J}(\mathbf{r}t)$ field involving the time rate of change of the current density. The fields experienced by each electron are the temporal versions of those of the 'Quantal Newtonian' First Law. On summing over all the electrons, the contributions of the internal field $\mathcal{F}^{\mathrm{int}}(\mathbf{r}t)$ vanish, leading to a more insightful derivation of Ehrenfest's Second Law Theorem.

The individual-electron-perspective via the 'Quantal Newtonian' Laws further provides mathematical insights into the Schrödinger equation itself. In the traditional description of Schrödinger theory, the binding scalar potential $v(\mathbf{r})$ of each electron is considered an external known input into the Hamiltonian $\hat{H}$ of the system. According to the 'Quantal Newtonian' First Law, the potential $v(\mathbf{r})$ has the rigorous physical interpretation of being the work done in a *conservative* field $\mathcal{F}(\mathbf{r})$. (Thus, $v(\mathbf{r})$ is a potential energy in the classical sense.) The field $\mathcal{F}(\mathbf{r})$ is comprised of the Lorentz field and all the components of the internal field $\mathcal{F}^{\mathrm{int}}(\mathbf{r})$. Thus, the potential $v(\mathbf{r})$ is inherently related to the physical system itself via the internal field $\mathcal{F}^{\mathrm{int}}(\mathbf{r})$ components which each represent a different property. This then is the quantum-mechanical definition of the binding potential $v(\mathbf{r})$. It can no longer be considered an extrinsic input to the electronic system. The fact that the binding potential $v(\mathbf{r})$ is related to various properties of the physical system may be understood in a more fundamental way. As the quantal sources of all the individual fields are expectation values of Hermitian operators taken with respect to the wave function Ψ, the conservative field $\mathcal{F}(\mathbf{r})$, and therefore the potential $v(\mathbf{r})$ is a functional of Ψ, i.e. $v(\mathbf{r}) = v[\Psi](\mathbf{r})$. Since the individual fields are explicitly defined, this functional is *exactly known* and *universal*. This in turn means that the Hamiltonian $\hat{H}$ is an exactly known universal functional of Ψ, i.e. $\hat{H} = \hat{H}[\Psi]$. The Schrödinger eigenvalue equation may then be written as $\hat{H}[\Psi]\Psi = E[\Psi]\Psi$. This equation can be solved self-consistently for the $\{\Psi, E\}$. In doing so, the potential $v[\Psi](\mathbf{r})$ is simultaneously determined. Thus, the potential $v[\Psi](\mathbf{r})$ is self-consistently related to the properties of the system via field $\mathcal{F}(\mathbf{r})$. What one learns in addition, however, is that in common with theories such as those of Hartree and Hartree-Fock, stationary-state Schrödinger theory too is intrinsically self- consistent. In a similar manner, it follows from the 'Quantal Newtonian' Second Law, that the time-dependent Hamiltonian $\hat{H}(t)$ is an *exactly known universal* functional of the solution $\Psi(t)$ to the temporal Schrödinger equation, i.e. $\hat{H}(t) = \hat{H}[\Psi](t)$. Thus, the temporal Schrödinger equation may be written as $\{\hat{H}[\Psi](t)\}\Psi(t) = i\hbar\partial\Psi(t)/\partial t$. Once again, this makes evident that time-dependent Schrödinger theory is intrinsically self-consistent.

A further complementary perspective to Schrödinger theory is provided by Quantal Density Functional Theory (Q-DFT). Stationary-state Q-DFT constitutes the mapping from an interacting system of electrons in a *ground* or *excited* state to one of noninteracting fermions possessing the same gauge invariant properties of the density $\rho(\mathbf{r})$ and physical current density $\mathbf{j}(\mathbf{r})$. The state of the model system is *arbitrary*. Q-DFT is thus a local effective potential theory. (By this is meant that the potential energy of each model fermion is the same, and that the orbitals of these fermions are obtained self-consistently.) In essence, Q-DFT provides the understanding of how the (*nonlocal*) electron correlations due to the Pauli principle and Coulomb repulsion are incorporated into a *local* (multiplicative) effective electron-interaction potential $v_{ee}(\mathbf{r})$. This description is once again in terms of fields experienced by each model fermion. It is proved that the potential $v_{ee}(\mathbf{r})$ has the *rigorous physical interpretation* of being the work done in a conservative effective field $\mathcal{F}^{\mathrm{eff}}(\mathbf{r})$. The field $\mathcal{F}^{\mathrm{eff}}(\mathbf{r})$ accounts not only for the correlations due to the Pauli principle and Coulomb repulsion, but also of the contributions of these correlations to kinetic properties—the Correlation-Kinetic effects. Within Q-DFT, it is further possible to separate the correlations due to the Pauli principle and Coulomb repulsion. Thus, the *separate* Pauli, Coulomb, and Correlation-Kinetic contributions to both the potential $v_{ee}(\mathbf{r})$ as well as to the corresponding energy components—the Pauli (exchange), Coulomb, and Correlation-Kinetic energy—are determined. The total energy is expressed in terms of the properties of the model system. Additionally, irrespective of the state of the model system, the highest occupied eigenvalue of the corresponding differential equation is the negative of the Ionization Potential. Therefore, by the Q-DFT mapping to the model of noninteracting fermions, additional properties of the interacting electronic system are thereby obtained. Temporal Q-DFT constitutes the local effective potential complement to time-dependent Schrödinger theory which ensures the equivalence of the electronic $\rho(\mathbf{r}t)$ and physical current $\mathbf{j}(\mathbf{r}t)$ densities.

Yet another different and complementary perspective on Schrödinger theory is that of Hohenberg and Kohn (HK) density functional theory (DFT). For a system of electrons in an external electrostatic field, HK prove that all the information about the physical system is ensconced in the gauge invariant nondegenerate ground state density $\rho(\mathbf{r})$. This is a fundamental insight into quantum mechanics. What is proved is that knowledge of the density $\rho(\mathbf{r})$ determines the system Hamiltonian $\hat{H}$ and therefore, via the Schrödinger equation, all its eigenfunctions and eigenvalues. All properties of the system are functionals of the density $\rho(\mathbf{r})$. The density $\rho(\mathbf{r})$ is referred to as a *basic variable* of quantum mechanics. In the added presence of a uniform magnetostatic field, Pan and Sahni have proved that the corresponding *basic variables* are the gauge invariant nondegenerate ground state density $\rho(\mathbf{r})$ and physical current density $\mathbf{j}(\mathbf{r})$. Once again, knowledge of $\{\rho(\mathbf{r}), \mathbf{j}(\mathbf{r})\}$ determines the corresponding Hamiltonian, eigenfunctions, and eigenvalues. Therefore, all the properties of the system are functionals of $\{\rho(\mathbf{r}), \mathbf{j}(\mathbf{r})\}$. In the temporal case, Runge and Gross have proved that the basic variables are the density $\rho(\mathbf{r}t)$ and physical current density $\mathbf{j}(\mathbf{r}t)$. Knowledge of either $\rho(\mathbf{r}t)$ or $\mathbf{j}(\mathbf{r}t)$ determines the time-dependent Hamiltonian $\hat{H}(t)$ and wave function $\Psi(t)$. And hence, properties are functionals of the density $\rho(\mathbf{r}t)$ or current density $\mathbf{j}(\mathbf{r}t)$. The focus of the chapter on DFT is thus on these fundamental theorems, and of our more recent further understanding of them via density preserving unitary transformations and various Corollaries.

The application of these theorems on DFT is via the corresponding Kohn and Sham (KS) density functional theories. Stationary-state KS DFT—a ground state theory—and its temporal counterpart, are local effective potential theories. (The origins of local effective potential theory lie in the work of Slater (see ***QDFT2***) where approximate local potentials are constructed. Stationary-state KS-DFT provides one definition of *ground-state* local effective potential theory. The broader perspective of local effective potential theory is provided by Q-DFT.) In KS-DFT, the energy and action are described in terms of unknown functionals of the density, with the potentials then defined via the Euler-Lagrange equation as their functional derivatives. Based on the physical description of local effective potential theory via Q-DFT, a rigorous explanation of these mathematical entities in terms of the various electron correlations which contribute to them is provided.

As the reader will have observed, there has been a consistent emphasis in the book on electron correlations. These correlations are all embedded in the many-electron wave function. In the wave function, spin correlations which arise as a result of the Pauli Exclusion Principle are explicitly represented by the Dirac-Heisenberg requirement that it be antisymmetric in an interchange of the coordinates of any two electrons. For natural systems, i.e. with a Coulombic binding potential, the manner in which correlations due to Coulomb repulsion are incorporated in the wave function is yet unknown. It is, however, possible to determine the structure of the wave function at points where the Schrödinger equation is singular, i.e. at electron-electron and electron-nucleus coalescence. The electron-electron coalescence constraint is, of course, valid for arbitrary binding potential. The electron-nucleus coalescence constraint is valid for all potentials that are singular at the nucleus. For Coulombically bound electrons, the structure of the wave function in the asymptotic classically forbidden region can be determined. A new symmetry property for arbitrary-potential bound two-electron systems is also described. This then leads to new insights into the structure and the parity of the wave functions for such systems.

For natural systems then, the study of electron correlations is based on accurate but approximate wave functions. With the advent of the creation of semiconductor quantum dots or 'artificial atoms', it has become possible to study electron correlations in an exact manner. In such 'artificial atoms', the binding potential of the electrons is harmonic. For two harmonically bound electrons, in the absence and presence of a magnetic field, it is possible to solve *analytically* the Schrödinger and Schrödinger-Pauli equations, respectively. Hence, closed-form analytical solutions for *arbitrary state* of such 'artificial atoms' may be derived. These wave functions, of course, obey the Pauli principle. But significantly, they inform about how Coulomb correlations are accounted for in these wave functions. Hence, all the stationary-state properties of such 'artificial atoms', including in particular the various electron correlations, can be described in an *exact* manner. The detailed derivation of the general wave function of two harmonically bound electrons in a magnetic field is provided. The *exact* time evolution of such systems, when perturbed by a time-dependent electric field, is then determined via the Generalized Kohn Theorem. The detailed derivation of the Generalized Kohn Theorem is also provided. Thus, all the complementary perspectives to Schrödinger theory described in the book are elucidated in an *exact* manner by application to such 'artificial atoms'.

I hope you have enjoyed the book, and the physics of the many-electron problem revealed by the new complimentary perspectives to Schrödinger theory described therein.

Appendix A

Derivation of the 'Quantal Newtonian' Second and First Laws

Part A: Interacting Electrons

'Quantal Newtonian' Second Law

The derivation of the 'Quantal Newtonian' Second Law in this Appendix is for the general case [1] of a system of N electrons in a time-dependent electric field $\mathcal{E}(\mathbf{y}) = -\nabla v(\mathbf{y})$ in the added presence of an electromagnetic field $\mathcal{B}(\mathbf{y}) = \nabla \times \boldsymbol{A}(\mathbf{y}), \mathbf{E}(\mathbf{y}) = -\nabla \Phi(\mathbf{y}) - \partial \boldsymbol{A}(\mathbf{y})/\partial t$ with $\{v(\mathbf{y}), \Phi(\mathbf{y})\}$ scalar and $\boldsymbol{A}(\mathbf{y})$ vector potentials. (The scalar potentials $v(\mathbf{y})$ and $\Phi(\mathbf{y})$ could be combined but are kept separate because of the difference in their origins. For example, $v(\mathbf{y})$ includes the binding potential of the electrons.) The time-independent 'Quantal Newtonian' First Law then follows as a special case.

The units employed are the following: charge of electron $-e$, $|e| = \hbar = m = 1$, together with the assumption $c = 1$. In the expressions replace $\boldsymbol{A}(\mathbf{y})$ by $\boldsymbol{A}(\mathbf{y})/c$ for them to be in atomic units.

Notation: $\mathbf{Y} = \mathbf{X}t$; $\mathbf{X} = \mathbf{x}_1, \ldots, \mathbf{x}_N$; $\mathbf{x} = \mathbf{r}\sigma$; $(\mathbf{r}\sigma)$ the spatial and spin coordinates; $\mathbf{y} = \mathbf{r}t$; $\mathbf{y}' = \mathbf{r}'t$; $\mathbf{y}_k = \mathbf{r}_k t$; $\mathbf{X}^{N-1} = \mathbf{x}_2, \mathbf{x}_3, \ldots, \mathbf{x}_N$; $d\mathbf{X}^{N-1} = d\mathbf{x}_2, \ldots, d\mathbf{x}_N$; $\int d\mathbf{x} = \sum_\sigma \int d\mathbf{r}$.

The corresponding time-dependent Schrödinger equation is

$$\left[\frac{1}{2}\sum_k (\hat{\mathbf{p}}_k + \boldsymbol{A}(\mathbf{y}_k))^2 + \frac{1}{2}\sum_{k,\ell}{}' \frac{1}{|\mathbf{r}_k - \mathbf{r}_\ell|} + \sum_k \left[v(\mathbf{y}_k) - \Phi(\mathbf{y}_k)\right]\right]\Psi(\mathbf{Y}) = i\frac{\partial \Psi(\mathbf{Y})}{\partial t}, \tag{A.1}$$

V. Sahni, *Schrödinger Theory of Electrons: Complementary Perspectives*,
Springer Tracts in Modern Physics 285,
https://doi.org/10.1007/978-3-030-97409-1

where the terms of the Hamiltonian are the physical kinetic $\hat{T}_A$, (with $\hat{\mathbf{p}}_k = -i\nabla_{\mathbf{r}_k}$ the 'canonical' momentum operator), electron-interacting potential $\hat{U}$, and external scalar potential $\hat{V}$ energy operators. The quantal sources of the electron density $\rho(\mathbf{y})$, the single-particle density matrix $\gamma(\mathbf{rr}'t)$, the pair-function $\mathbf{P}(\mathbf{rr}'t)$ are defined in Chap. 2. The physical current density $\mathbf{j}(\mathbf{y})$ is the expectation value

$$\mathbf{j}(\mathbf{y}) = \langle \Psi(\mathbf{Y})|\hat{\mathbf{j}}(\mathbf{y})|\Psi(\mathbf{Y})\rangle \tag{A.2}$$

$$= \mathbf{j}_p(\mathbf{y}) + \mathbf{j}_d(\mathbf{y}), \tag{A.3}$$

where $\mathbf{j}_p(\mathbf{r})$ is the paramagnetic and $\mathbf{j}_d(\mathbf{r})$ the diamagnetic current density components. The physical current density operator $\hat{\mathbf{j}}(\mathbf{y})$ is the sum of the paramagnetic $\hat{\mathbf{j}}_p(\mathbf{r})$ and diamagnetic $\hat{\mathbf{j}}_d(\mathbf{y})$ operators:

$$\hat{\mathbf{j}}(\mathbf{y}) = \hat{\mathbf{j}}_p(\mathbf{r}) + \hat{\mathbf{j}}_d(\mathbf{y}) \tag{A.4}$$

with

$$\hat{j}_p(\mathbf{r}) = \frac{1}{2i}\sum_k \Big[\nabla_{\mathbf{r}_k}\delta(\mathbf{r}_k - \mathbf{r}) + \delta(\mathbf{r}_k - \mathbf{r})\nabla_{\mathbf{r}_k}\Big] \tag{A.5}$$

and

$$\hat{\mathbf{j}}_d(\mathbf{y}) = \hat{\rho}(\mathbf{r})\boldsymbol{A}(\mathbf{y}). \tag{A.6}$$

The physical current density $\hat{\mathbf{j}}_d(\mathbf{r})$ may also be written as

$$\mathbf{j}(\mathbf{y}) = N\Re\sum_\sigma \int \Psi^\star(\mathbf{r}\sigma, \mathbf{X}^{N-1}, t)\big(\hat{\mathbf{p}} + \boldsymbol{A}(\mathbf{y})\big)\Psi(\mathbf{r}\sigma, \mathbf{X}^{N-1}, t)d\mathbf{X}^{N-1} \tag{A.7}$$

with

$$\mathbf{j}_p(\mathbf{y}) = N\Re\sum_\sigma \int \Psi^\star(\mathbf{r}\sigma, \mathbf{X}^{N-1}, t)\hat{\mathbf{p}}\Psi(\mathbf{r}\sigma, \mathbf{X}^{N-1}, t)d\mathbf{X}^{N-1}, \tag{A.8}$$

$$= i\frac{N}{2}\sum_\sigma \int \Big[\Psi(\mathbf{Y})\nabla\Psi^\star(\mathbf{Y}) - \Psi^\star(\mathbf{Y})\nabla\Psi(\mathbf{Y})\Big]d\mathbf{X}^{N-1}, \tag{A.9}$$

and

$$\mathbf{j}_d(\mathbf{y}) = \boldsymbol{A}(\mathbf{y})N\sum_\sigma \int \Psi^\star(\mathbf{r}\sigma, \mathbf{X}^{N-1}, t)\Psi(\mathbf{r}\sigma, \mathbf{X}^{N-1}, t)d\mathbf{X}^{N-1} \tag{A.10}$$

$$= \rho(\mathbf{y})\boldsymbol{A}(\mathbf{y}). \tag{A.11}$$

We rewrite the Hamiltonian of (A.1) as

$$\hat{H}(t) = -\frac{1}{2}\sum_k \nabla_k^2 + \frac{1}{2}\sum_{k,\ell}{}' \frac{1}{|\mathbf{r}_k - \mathbf{r}_\ell|} + \sum_k \Big[v(\mathbf{y}_k) - \Phi(\mathbf{y}_k) \Big] + \sum_k \hat{\omega}\Big[\mathbf{y}_k; \mathbf{A}(\mathbf{y}_k) \Big], \tag{A.12}$$

where

$$\hat{\omega}\Big[\mathbf{y}; \mathbf{A}(\mathbf{y}) \Big] = \frac{1}{2}A^2(\mathbf{y}) - i\hat{\Omega}(\mathbf{y}), \tag{A.13}$$

$$\hat{\Omega}(\mathbf{y}) = \frac{1}{2}\Big[\nabla \cdot \mathbf{A}(\mathbf{y}) + 2\mathbf{A}(\mathbf{y}) \cdot \nabla \Big]. \tag{A.14}$$

In general, the wave function may be written as $\Psi(\mathbf{Y}) = \Psi^R(\mathbf{Y}) + i\Psi^I(\mathbf{Y})$, where $\Psi^R(\mathbf{Y}) \equiv \Psi^R(t)$ and $\Psi^I(\mathbf{Y}) \equiv \Psi^I(t)$ are its real and imaginary parts. On substituting the wave function into the Schrödinger equation of (A.12), we have

$$i\frac{\partial \Psi^R(t)}{\partial t} - \frac{\partial \Psi^I(t)}{\partial t} = \Bigg\{ -\frac{1}{2}\sum_k \nabla_k^2 + \sum_k \Big[v(\mathbf{y}_k) - \Phi(\mathbf{y}_k) \Big]$$
$$+\frac{1}{2}\sum_{k,\ell}{}' \frac{1}{|\mathbf{r}_k - \mathbf{r}_\ell|} + \sum_k \hat{\omega}\Big[(\mathbf{y}_k)\mathbf{A}(\mathbf{y}_k) \Big] \Bigg\} \Big(\Psi^R(t) + i\Psi^I(t) \Big). \tag{A.15}$$

Equating the real parts of (A.15) yields

$$\sum_k \Big[v(\mathbf{y}_k) - \Phi(\mathbf{y}_k) \Big] + \frac{1}{2}\sum_{k,\ell}{}' \frac{1}{|\mathbf{r}_k - \mathbf{r}_\ell|} + \frac{1}{2}\sum_k A^2(\mathbf{y}_k)$$
$$+\frac{1}{\Psi^R(t)}\frac{\partial}{\partial t}\Psi^I(t) = \frac{1}{\Psi^R(t)}\Bigg\{ \frac{1}{2}\sum_k \Big[\nabla_k^2 \Psi^R(t) - \hat{\Omega}\Psi^I(t) \Big] \Bigg\}. \tag{A.16}$$

On differentiating (A.16) with respect $r_{1\alpha}$, where $r_{1\alpha}$ is the α coordinate of $\mathbf{r}_1$, and then multiplying both sides by $[\Psi^R(t)]^2$, we obtain

$$\frac{\partial}{\partial r_{1\alpha}}\Bigg[v(\mathbf{y}_k) - \Phi(\mathbf{y}_k) + \frac{1}{2}\sum_{\ell=2} \frac{1}{|\mathbf{r}_k - \mathbf{r}_\ell|} + \frac{1}{2}A^2(\mathbf{y}_k)$$
$$+\frac{1}{\Psi^R(t)}\frac{\partial}{\partial t}\Psi^I(t) \Bigg] [\Psi^R(t)]^2$$
$$= \frac{1}{2}\sum_k \sum_\beta \Bigg[\Psi^R(t)\frac{\partial^3 \Psi^R(t)}{\partial r_{1\alpha}\partial r_{k\beta}^2} - \frac{\partial \Psi^R(t)}{\partial r_{1\alpha}}\frac{\partial^2 \Psi^R(t)}{\partial r_{k\beta}^2} \Bigg]$$
$$-\sum_k \sum_\beta \Bigg[\Psi^R(t)\frac{\partial}{\partial r_{1\alpha}}\Big(A_{k\beta}\frac{\partial \Psi^I(t)}{\partial r_{k\beta}} \Big) - A_{k\beta}\frac{\partial \Psi^I(t)}{\partial r_{k\beta}}\frac{\partial \Psi^R(t)}{\partial r_{1\alpha}} \Bigg]$$
$$-\frac{1}{2}\sum_k \sum_\beta \Bigg[\Psi^R(t)\frac{\partial}{\partial r_{1\alpha}}\Big(\Psi^I(t)\frac{\partial A_{k\beta}}{\partial r_{k\beta}} \Big) - \Psi^I(t)\frac{\partial A_{k\beta}}{\partial r_{k\beta}}\frac{\partial \Psi^R(t)}{\partial r_{1\alpha}} \Bigg]. \tag{A.17}$$

Now since

$$\frac{1}{4}\frac{\partial^3\Psi^R(t)\Psi^R(t)}{\partial r_{k\beta}\partial r_{k\beta}\partial r_{1\alpha}} = \frac{1}{2}\frac{\partial^2\Psi^R(t)}{\partial r_{k\beta}^2}\frac{\partial\Psi^R(t)}{\partial r_{1\alpha}} + \frac{\partial\Psi^R(t)}{\partial r_{k\beta}}\frac{\partial^2\Psi^R(t)}{\partial r_{k\beta}\partial r_{1\alpha}} + \frac{1}{2}\Psi^R(t)\frac{\partial^3\Psi^R(t)}{\partial r_{k\beta}\partial r_{k\beta}\partial r_{1\alpha}}, \tag{A.18}$$

the first term of the right side in (A.17) can be written as

$$\frac{1}{4}\frac{\partial^3\Psi^R(t)\Psi^R(t)}{\partial r_{k\beta}\partial r_{k\beta}\partial r_{1\alpha}} - \frac{\partial}{\partial r_{k\beta}}\left[\frac{\partial\Psi^R(t)}{\partial r_{1\alpha}}\frac{\partial\Psi^R(t)}{\partial r_{k\beta}}\right] = \frac{1}{2}\Psi^R(t)\frac{\partial^3\Psi^R(t)}{\partial r_{k\beta}\partial r_{k\beta}\partial r_{1\alpha}} - \frac{1}{2}\frac{\partial\Psi^R(t)}{\partial r_{1\alpha}}\frac{\partial^2\Psi^R(t)}{\partial r_{k\beta}^2}. \tag{A.19}$$

Thus, (A.17) is

$$\begin{aligned}
&\frac{\partial}{\partial r_{1\alpha}}\left[v(\mathbf{y}_1) - \Phi(\mathbf{y}_1) + \frac{1}{2}A^2(\mathbf{y}_1) + \frac{1}{2}\sum_{\ell=2}\frac{1}{|\mathbf{r}_1 - \mathbf{r}_\ell|}\right.\\
&\left.+\frac{1}{\Psi^R(t)}\frac{\partial}{\partial t}\Psi^I(t)\right][\Psi^R(t)]^2\\
&= \sum_k\sum_\beta\left[\frac{1}{4}\frac{\partial^3\Psi^R(t)\Psi^R(t)}{\partial r_{1\alpha}\partial^2 r_{k\beta}} - \frac{\partial}{\partial r_{k\beta}}\left(\frac{\partial\Psi^R(t)}{\partial r_{1\alpha}}\frac{\partial\Psi^R(t)}{\partial r_{1\alpha}}\right)\right]\\
&-\sum_k\sum_\beta\left[\Psi^R(t)\frac{\partial}{\partial r_{1\alpha}}\left(A_{k\beta}\frac{\partial\Psi^I(t)}{\partial r_{k\beta}}\right) - A_{k\beta}\frac{\partial\Psi^I(t)}{\partial r_{k\beta}}\frac{\partial\Psi^R(t)}{\partial r_{1\alpha}}\right]\\
&-\frac{1}{2}\sum_k\sum_\beta\left[\Psi^R(t)\frac{\partial}{\partial r_{1\alpha}}\left(\Psi^I(t)\frac{\partial A_{k\beta}}{\partial r_{k\beta}}\right) - \Psi^I(t)\frac{\partial A_{k\beta}}{\partial r_{k\beta}}\frac{\partial\Psi^R(t)}{\partial r_{1\alpha}}\right].
\end{aligned} \tag{A.20}$$

Similarly we have the following equation for the imaginary part of the wave function $\Psi^I(t)$:

$$\begin{aligned}
&\frac{\partial}{\partial r_{1\alpha}}\left[v(\mathbf{y}_1) - \Phi(\mathbf{y}_1) + \frac{1}{2}A^2(\mathbf{y}_1) + \frac{1}{2}\sum_{\ell=2}\frac{1}{|\mathbf{r}_1 - \mathbf{r}_\ell|}\right.\\
&\left.+\frac{1}{\Psi^I(t)}\frac{\partial}{\partial t}\Psi^R(t)\right][\Psi^I(t)]^2\\
&= \sum_k\sum_\beta\left[\frac{1}{4}\frac{\partial^3\Psi^I(t)\Psi^I(t)}{\partial r_{1\alpha}\partial^2 r_{k\beta}} - \frac{\partial}{\partial r_{k\beta}}\left(\frac{\partial\Psi^I(t)}{\partial r_{1\alpha}}\frac{\partial\Psi^I(t)}{\partial r_{k\beta}}\right)\right]\\
&-\sum_k\sum_\beta\left[\Psi^I(t)\frac{\partial}{\partial r_{1\alpha}}\left(A_{k\beta}\frac{\partial\Psi^R(t)}{\partial r_{k\beta}}\right) - A_{k\beta}\frac{\partial\Psi^R(t)}{\partial r_{k\beta}}\frac{\partial\Psi^I(t)}{\partial r_{1\alpha}}\right]\\
&-\frac{1}{2}\sum_k\sum_\beta\left[\Psi^I(t)\frac{\partial}{\partial r_{1\alpha}}\left(\Psi^R(t)\frac{\partial A_{k\beta}}{\partial r_{k\beta}}\right) - \Psi^R(t)\frac{\partial A_{k\beta}}{\partial r_{k\beta}}\frac{\partial\Psi^I(t)}{\partial r_{1\alpha}}\right].
\end{aligned} \tag{A.21}$$

Using the fact that

$$\frac{\partial}{\partial r_{1\alpha}}\left(\frac{1}{\Psi^R(t)}\frac{\partial}{\partial t}\Psi^I(t)\right)[\Psi^R(t)]^2 - \frac{\partial}{\partial r_{1\alpha}}\left(\frac{1}{\Psi^I(t)}\frac{\partial}{\partial t}\Psi^R(t)\right)[\Psi^I(t)]^2$$
$$= \frac{\partial}{\partial t}\left(\Psi^R(t)\frac{\partial}{\partial r_{1\alpha}}\Psi^R(t) - \Psi^I(t)\frac{\partial}{\partial r_{1\alpha}}\Psi^R(t)\right), \tag{A.22}$$

then on summing (A.20) and (A.21) and operating by $N\sum_{\sigma_1}\int d\mathbf{X}^{N-1}$ on the resulting equation, we have that

$$N\sum_{\sigma_1}\int d\mathbf{X}^{N-1}\left[\frac{\partial}{\partial r_{1\alpha}}\left(v(\mathbf{y}_1) - \Phi(\mathbf{y}_1)\right.\right.$$
$$\left.\left.+\frac{1}{2}A^2(\mathbf{y}_1) + \frac{1}{2}\sum_{\ell=2}\frac{1}{|\mathbf{r}_1 - \mathbf{r}_\ell|}\right)\right]|\Psi(t)|^2$$
$$+N\sum_{\sigma_1}\int d\mathbf{X}^{N-1}\frac{\partial}{\partial t}\left(\Psi^R(t)\frac{\partial}{\partial r_{1\alpha}}\Psi^I(t) - \Psi^I(t)\frac{\partial}{\partial r_{1\alpha}}\Psi^R(t)\right)$$
$$= N\sum_{\sigma_1}\int d\mathbf{X}^{N-1}\sum_\beta\left[\frac{1}{4}\frac{\partial^3|\Psi(t)|^2}{\partial r_{1\alpha}\partial^2 r_{1\beta}} - \frac{\partial}{\partial r_{1\beta}}\left(\frac{\partial\Psi^R(t)}{\partial r_{1\alpha}}\frac{\partial\Psi^R(t)}{\partial r_{1\beta}} + \frac{\partial\Psi^I(t)}{\partial r_{1\alpha}}\frac{\partial\Psi^I(t)}{\partial r_{1\beta}}\right)\right]$$
$$+N\sum_{\sigma_1}\int d\mathbf{X}^{N-1}\sum_\beta\left[-\Psi^R(t)\frac{\partial}{\partial r_{1\alpha}}\left(A_{1\beta}\frac{\partial\Psi^I(t)}{\partial r_{1\beta}}\right) + A_{1\beta}\frac{\partial\Psi^R(t)}{\partial r_{1\alpha}}\frac{\partial\Psi^I(t)}{\partial r_{1\beta}}\right.$$
$$\left.+\Psi^I(t)\frac{\partial}{\partial r_{1\alpha}}\left(A_{1\beta}\frac{\partial\Psi^R(t)}{\partial r_{1\beta}}\right) - A_{1\beta}\frac{\partial\Psi^I(t)}{\partial r_{1\alpha}}\frac{\partial\Psi^R(t)}{\partial r_{1\beta}}\right]$$
$$+N\sum_{\sigma_1}\int d\mathbf{X}^{N-1}\sum_\beta\left[-\Psi^R(t)\frac{\partial A_{1\beta}}{\partial r_{1\beta}}\frac{\partial\Psi^I(t)}{\partial r_{1\alpha}} + \Psi^I(t)\frac{\partial A_{1\beta}}{\partial r_{1\beta}}\frac{\partial\Psi^R(t)}{\partial r_{1\alpha}}\right]$$
$$+N\sum_{\sigma_1}\int d\mathbf{X}^{N-1}\left\{\sum_{\ell=2}\sum_\beta\left[\frac{1}{4}\frac{\partial^3|\Psi(t)|^2}{\partial r_{1\alpha}\partial^2 r_{\ell\beta}} - \frac{\partial}{\partial r_{\ell\beta}}\left(\frac{\partial\Psi^R(t)}{\partial r_{1\alpha}}\frac{\partial\Psi^R(t)}{\partial r_{\ell\beta}} + \frac{\partial\Psi^I(t)}{\partial r_{1\alpha}}\frac{\partial\Psi^I(t)}{\partial r_{\ell\beta}}\right)\right]\right.$$
$$+\sum_{\ell=2}\sum_\beta\left[-\Psi^R(t)\frac{\partial}{\partial r_{1\alpha}}\left(A_{k\beta}\frac{\partial\Psi^I(t)}{\partial r_{\ell\beta}}\right) + A_{k\beta}\frac{\partial\Psi^R(t)}{\partial r_{1\alpha}}\frac{\partial\Psi^I(t)}{\partial r_{\ell\beta}}\right.$$
$$\left.+\Psi^I(t)\frac{\partial}{\partial r_{1\alpha}}\left(A_{\ell\beta}\frac{\partial\Psi^R(t)}{\partial r_{\ell\beta}}\right) - A_{\ell\beta}\frac{\partial\Psi^I(t)}{\partial r_{1\alpha}}\frac{\partial\Psi^R(t)}{\partial r_{\ell\beta}}\right]$$
$$\left.+\sum_{\ell=2}\sum_\beta\left[-\Psi^R(t)\frac{\partial A_{k\beta}}{\partial r_{\ell\beta}}\frac{\partial\Psi^I(t)}{\partial r_{1\alpha}} + \Psi^I(t)\frac{\partial A_{\ell\beta}}{\partial r_{\ell\beta}}\frac{\partial\Psi^R(t)}{\partial r_{1\alpha}}\right]\right\}. \tag{A.23}$$

As the wave function and its derivative vanish at infinity, the last term of (A.23) vanishes on integration over $dr_{\ell\beta}$. The second and third terms of (A.23) may be rewritten as

$$\sum_\beta \Big[-\Psi^R(t)\frac{\partial}{\partial r_{1\alpha}}\Big(A_{1\beta}\frac{\partial \Psi^I(t)}{\partial r_{1\beta}}\Big) + \Psi^I(t)\frac{\partial}{\partial r_{1\alpha}}\Big(A_{1\beta}\frac{\partial \Psi^R(t)}{\partial r_{1\beta}}\Big)$$

$$+ A_{1\beta}\Big(\frac{\partial \Psi^R(t)}{\partial r_{1\alpha}}\frac{\partial \Psi^I(t)}{\partial r_{1\beta}} - \frac{\partial \Psi^I(t)}{\partial r_{1\alpha}}\frac{\partial \Psi^R(t)}{\partial r_{1\beta}}\Big)$$

$$- \Psi^R(t)\frac{\partial A_{1\beta}}{\partial r_{1\beta}}\frac{\partial \Psi^I(t)}{\partial r_{1\alpha}} + \Psi^I(t)\frac{\partial A_{1\beta}}{\partial r_{1\beta}}\frac{\partial \Psi^R(t)}{\partial r_{1\alpha}}\Big]$$

$$= \sum_\beta \Big\{ \frac{\partial A_{1\beta}}{\partial r_{1\alpha}}\Big[\Psi^I(t)\frac{\partial \Psi^R(t)}{\partial r_{1\beta}} - \Psi^R(t)\frac{\partial \Psi^I(t)}{\partial r_{1\beta}}\Big]$$

$$+ \frac{\partial}{\partial r_{1\beta}}\Big[A_{1\beta}\Big(\Psi^I(t)\frac{\partial \Psi^R(t)}{\partial r_{1\alpha}} - \Psi^R(t)\frac{\partial \Psi^I(t)}{\partial r_{1\alpha}}\Big)\Big]\Big\}. \quad \text{(A.24)}$$

In turn (A.24) can be expressed in terms of the paramagnetic current density $\mathbf{j}_p(\mathbf{y})$ as

$$k_\alpha(\mathbf{y}\mathbf{j}_p\mathbf{A}) = \sum_\beta \big[\{\nabla_\alpha(\mathbf{r})A_\beta(\mathbf{y})\}\mathbf{j}_{p\beta}(\mathbf{y}) + \nabla_\beta\{A_\beta(\mathbf{y})\mathbf{j}_{p\alpha}(\mathbf{y})\}\big], \quad \text{(A.25)}$$

where $\mathbf{j}_p(\mathbf{y})$ is defined in (A.9).

Finally, to address the first term of (A.23), we note that the kinetic 'force' $\mathbf{z}(\mathbf{y})$ is defined by its component as

$$z_\alpha(\mathbf{y}) = 2\sum_\beta \frac{\partial}{\partial r_\beta} t_{\alpha\beta}(\mathbf{y}), \quad \text{(A.26)}$$

where $t_{\alpha\beta}(\mathbf{y})$ is the kinetic energy density tensor defined in terms of the single-particle density matrix $\gamma(\mathbf{r}\mathbf{r}'t)$ as

$$t_{\alpha\beta}(\mathbf{y}) = \frac{1}{4}\Big[\frac{\partial^2}{\partial r'_\alpha \partial r''_\beta} + \frac{\partial^2}{\partial r'_\beta \partial r''_\alpha}\Big]\gamma(\mathbf{r}'\mathbf{r}''t)\big|_{\mathbf{r}'=\mathbf{r}''=\mathbf{r}}. \quad \text{(A.27)}$$

On substituting the complex form of the wave function into the expression for $\gamma(\mathbf{r}\mathbf{r}'t)$, the tensor may be written as

$$t_{\alpha\beta}(\mathbf{y}) = \frac{N}{2}\sum_\sigma \int d\mathbf{X}^{N-1}\Big[\frac{\partial}{\partial r_\alpha}\Psi^R(\mathbf{r}\sigma, \mathbf{X}^{N-1}t)\frac{\partial}{\partial r_\beta}\Psi^R(\mathbf{r}\sigma, \mathbf{X}^{N-1}t)$$

$$+\frac{\partial}{\partial r_\alpha}\Psi^I(\mathbf{r}\sigma, \mathbf{X}^{N-1}t)\frac{\partial}{\partial r_\beta}\Psi^I(\mathbf{r}\sigma, \mathbf{X}^{N-1}t)\Big]. \quad \text{(A.28)}$$

Employing (A.25), (A.26), we obtain (A.23) in terms of the various 'forces' to be

$$\rho(\mathbf{y})\nabla\left[v(\mathbf{y}) + \frac{1}{2}A^2(\mathbf{y}) - \Phi(\mathbf{y})\right] - \mathbf{e}_{\text{ee}}(\mathbf{y}) + \mathbf{z}(\mathbf{y}) + \mathbf{d}(\mathbf{y}) \\ + \mathbf{k}(\mathbf{y}\mathbf{j}_p\mathbf{A}) + \frac{\partial}{\partial t}\mathbf{j}_p(\mathbf{y}) = 0, \tag{A.29}$$

where the 'forces' $\mathbf{e}_{\text{ee}}(\mathbf{y})$, $\mathbf{d}(\mathbf{y})$ are defined in (2.45) and (2.50), respectively.

In terms of the physical current density $\mathbf{j}(\mathbf{y})$ which is defined by (A.3), the terms $k_\alpha(\mathbf{y}\mathbf{j}_p\mathbf{A})$ may be written as

$$k_\alpha(\mathbf{y}\mathbf{j}_p\mathbf{A}) = k_\alpha(\mathbf{y}\mathbf{j}\mathbf{A}) - \sum_\beta \big[\rho(\mathbf{y})A_\beta\nabla_\alpha A_\beta + A_\beta\nabla_\beta(\rho(\mathbf{y})A_\alpha) \\ + \rho(\mathbf{y})A_\alpha\nabla_\beta A_\beta\big]. \tag{A.30}$$

Additionally, since the continuity equation is

$$\nabla\cdot\mathbf{j}(\mathbf{y}) + \frac{\partial\rho(\mathbf{y})}{\partial t} = 0, \tag{A.31}$$

we have

$$\begin{aligned}\frac{\partial}{\partial t}\mathbf{j}_p(\mathbf{y}) &= \frac{\partial}{\partial t}\big[\mathbf{j}(\mathbf{y}) - \rho(\mathbf{y})\mathbf{A}(\mathbf{y})\big] \\ &= \frac{\partial}{\partial t}\mathbf{j}(\mathbf{y}) + \mathbf{A}(\mathbf{y})\nabla\cdot\mathbf{j}(\mathbf{y}) - \rho(\mathbf{y})\frac{\partial\mathbf{A}(\mathbf{y})}{\partial t}.\end{aligned} \tag{A.32}$$

Employing (A.30) and (A.32), (A.29) is

$$\rho(\mathbf{y})\big[\nabla v(\mathbf{y}) + \mathbf{E}(\mathbf{y})\big] - \mathbf{e}_{\text{ee}}(\mathbf{y}) + \mathbf{z}(\mathbf{y}) + \mathbf{d}(\mathbf{y}) + \frac{\partial}{\partial t}\mathbf{j}(\mathbf{y}) \\ + \mathbf{A}(\mathbf{y})\nabla\cdot\mathbf{j}(\mathbf{y}) + \mathbf{k}(\mathbf{y}\mathbf{j}\mathbf{A}) - \sum_\beta \nabla_\beta\big[\rho(\mathbf{y})\mathbf{A}(\mathbf{y})A_\beta(\mathbf{y}) = 0, \tag{A.33}$$

with $\mathbf{E}(\mathbf{y}) = -\nabla\Phi(\mathbf{y}) - \frac{\partial\mathbf{A}(\mathbf{y})}{\partial t}$.

The last three terms of (A.33) contain the vector potential $\mathbf{A}(\mathbf{r})$ and can be afforded a physical interpretation as the sum of the external Lorentz 'force' $\boldsymbol{\ell}(\mathbf{y})$ and a contribution $\mathbf{i}_m(\mathbf{y})$ to the internal 'force'. The Lorentz 'force' is

$$\boldsymbol{\ell}(\mathbf{y}) = \mathbf{j}(\mathbf{y}) \times \boldsymbol{\mathcal{B}}(\mathbf{y}), \tag{A.34}$$

so that with $\boldsymbol{\mathcal{B}}(\mathbf{y}) = \nabla\times\mathbf{A}(\mathbf{y})$, we have

$$\ell_\alpha(\mathbf{y}) = \sum_\beta \big[j_\beta(\mathbf{y})\nabla_\alpha A_\beta(\mathbf{y}) - j_\beta(\mathbf{y})\nabla_\beta A_\alpha(\mathbf{y})\big]. \tag{A.35}$$

The contribution of the magnetic field to the internal 'force' $\boldsymbol{i}_m(\mathbf{y})$ is defined by its components as

$$i_{m,\alpha}(\mathbf{y}) = \sum_{\beta} \nabla_{\beta} I_{\alpha\beta}(\mathbf{y}), \tag{A.36}$$

where

$$I_{\alpha\beta}(\mathbf{y}) = \big[j_{\alpha}(\mathbf{y})A_{\beta}(\mathbf{y}) + j_{\beta}(\mathbf{y})A_{\alpha}(\mathbf{y})\big] - \rho(\mathbf{y})A_{\alpha}(\mathbf{y})A_{\beta}(\mathbf{y}). \tag{A.37}$$

Hence, the sum

$$\begin{aligned} \ell_{\alpha}(\mathbf{y}) + i_{m,\alpha}(\mathbf{y}) = k_{\alpha}(\mathbf{yjA}) - \sum_{\beta} \nabla_{\beta}\big[\rho(\mathbf{y})A_{\alpha}(\mathbf{y})A_{\beta}(\mathbf{y})\big] \\ + \sum_{\beta} A_{\alpha}(\mathbf{y})\nabla_{\beta} j_{\beta}(\mathbf{y}). \end{aligned} \tag{A.38}$$

Thus, (A.33) may be written as the 'Quantal Newtonian' second law in terms of fields as

$$\mathcal{F}^{\text{ext}}(\mathbf{y}) + \mathcal{F}^{\text{int}}(\mathbf{y}) = \mathcal{J}(\mathbf{y}), \tag{A.39}$$

where the external field is

$$\mathcal{F}^{\text{ext}}(\mathbf{y}) = \mathcal{E}(\mathbf{y}) - \mathcal{L}(\mathbf{y}) - \mathbf{E}(\mathbf{y}) \tag{A.40}$$

$$= -\boldsymbol{\nabla} v(\mathbf{y}) - \mathcal{L}(\mathbf{y}) + \boldsymbol{\nabla}\Phi(\mathbf{y}) + \frac{\partial \mathbf{A}(\mathbf{y})}{\partial t}, \tag{A.41}$$

with the Lorentz field defined as

$$\mathcal{L}(\mathbf{y}) = \frac{\boldsymbol{\ell}(\mathbf{y})}{\rho(\mathbf{y})}. \tag{A.42}$$

The internal field is

$$\mathcal{F}^{\text{int}}(\mathbf{y}) = \mathcal{E}_{\text{ee}}(\mathbf{y}) - \mathcal{Z}(\mathbf{y}) - \mathcal{D}(\mathbf{y}) - \mathcal{I}_m(\mathbf{y}), \tag{A.43}$$

where the electron-interaction $\mathcal{E}_{\text{ee}}(\mathbf{y})$, kinetic $\mathcal{Z}(\mathbf{y})$, and differential density $\mathcal{D}(\mathbf{r})$ fields are defined as in (2.44), (2.51), and (2.49), respectively, and where the internal magnetic field contribution is defined as

$$\mathcal{I}_m(\mathbf{y}) = \frac{\boldsymbol{i}_m(\mathbf{y})}{\rho(\mathbf{y})}. \tag{A.44}$$

The response of each electron is described by the current density field

$$\mathcal{J}(\mathbf{y}) = \frac{1}{\rho(\mathbf{y})} \frac{\partial \mathbf{j}(\mathbf{y})}{\partial t}. \tag{A.45}$$

We next consider the following special cases of the external fields. The derivation of the corresponding 'Quantal Newtonian' Second and First Laws follow from that of the general case given above: one simply drops the appropriate terms in the derivation and result.

Special Case A: External field: $\boldsymbol{\mathcal{E}}(\mathbf{y}) = -\boldsymbol{\nabla} v(\mathbf{y})$.

The Schrödinger equation is

$$\left[\frac{1}{2}\sum_k \hat{p}_k^2 + \frac{1}{2}\sum_{k,\ell}{}' \frac{1}{|\mathbf{r}_k - \mathbf{r}_\ell|} + \sum_k v(\mathbf{r}_k)\right]\Psi(\mathbf{Y}) = i\frac{\partial \Psi(\mathbf{Y})}{\partial t}. \tag{A.46}$$

The 'Quantal Newtonian' second law is [2,3]

$$\mathcal{F}^{\mathrm{ext}}(\mathbf{y}) + \mathcal{F}^{\mathrm{int}}(\mathbf{y}) = \mathcal{J}(\mathbf{y}), \tag{A.47}$$

where

$$\mathcal{F}^{\mathrm{ext}}(\mathbf{y}) = \boldsymbol{\mathcal{E}}(\mathbf{y}) - \boldsymbol{\nabla} v(\mathbf{y}), \tag{A.48}$$

$$\mathcal{F}^{\mathrm{int}}(\mathbf{y}) = \boldsymbol{\mathcal{E}}_{\mathrm{ee}}(\mathbf{y}) - \boldsymbol{\mathcal{Z}}(\mathbf{y}) - \boldsymbol{\mathcal{D}}(\mathbf{y}), \tag{A.49}$$

$$\mathcal{J}(\mathbf{y}) = \frac{1}{\rho(\mathbf{y})} \frac{\partial \mathbf{j}(\mathbf{y})}{\partial t}, \tag{A.50}$$

with

$$\mathbf{j}(\mathbf{y}) = \mathbf{j}_p(\mathbf{y}) = \langle \Psi(\mathbf{Y}) | \hat{\mathbf{j}}_p(\mathbf{r}) | \Psi(\mathbf{Y}) \rangle. \tag{A.51}$$

'Quantal Newtonian' First Law

The stationary state 'Quantal Newtonian' first law constitutes a special case of the second law. With the wave function written as

$$\Psi(\mathbf{Y}) = \psi(\mathbf{X})e^{-iEt}, \tag{A.52}$$

one then obtains the time-independent Schrödinger equation and the 'Quantal Newtonian' first law for the following cases.

Special Case B: External fields: $\mathcal{E}(\mathbf{r}) = -\nabla v(\mathbf{r})$; $\mathcal{B}(\mathbf{r}) = \nabla \times \boldsymbol{A}(\mathbf{r})$.

The Schrödinger equation is

$$\left[\frac{1}{2}\sum_k \left(\hat{\mathbf{p}}_k + \boldsymbol{A}(\mathbf{r}_k)\right)^2 + \frac{1}{2}\sum_{k,\ell}{}^{'} \frac{1}{|\mathbf{r}_k - \mathbf{r}_\ell|} + \sum_k v(\mathbf{r}_k)\right]\psi(\mathbf{X}) = E\psi(\mathbf{x}). \quad (A.53)$$

The 'Quantal Newtonian' first law is [4,5]

$$\mathcal{F}^{\text{ext}}(\mathbf{r}) + \mathcal{F}^{\text{int}}(\mathbf{r}) = 0, \quad (A.54)$$

where

$$\mathcal{F}^{\text{ext}}(\mathbf{r}) = \mathcal{E}(\mathbf{r}) - \mathcal{L}(\mathbf{r}) \quad (A.55)$$

$$= -\nabla v(\mathbf{r}) - \mathcal{L}(\mathbf{r}), \quad (A.56)$$

$$\mathcal{F}^{\text{int}}(\mathbf{r}) = \mathcal{E}_{\text{ee}}(\mathbf{r}) - \mathcal{Z}(\mathbf{r}) - \mathcal{D}(\mathbf{r}) - \mathcal{I}_m(\mathbf{r}), \quad (A.57)$$

with the density, the various fields, and the current density $\mathbf{j}(\mathbf{r})$ defined as in the time-dependent case but with the variable $\mathbf{y} = \mathbf{r}t$ replaced by the variable $\mathbf{r}$. The law is derived employing the continuity equation

$$\nabla \cdot \mathbf{j}(\mathbf{r}) = 0, \quad (A.58)$$

where the current density $\mathbf{j}(\mathbf{r})$ is the sum of its paramagnetic $\mathbf{j}_p(\mathbf{r})$ and diamagnetic $\mathbf{j}_d(\mathbf{r})$ components.

Special Case C: External field: $\mathcal{E}(\mathbf{r}) = -\nabla v(\mathbf{r})$.

The Schrödinger equation is

$$\left[\frac{1}{2}\sum_k \hat{p}_k^2 + \frac{1}{2}\sum_{k,\ell}{}^{'} \frac{1}{|\mathbf{r}_k - \mathbf{r}_\ell|} + \sum_k v(\mathbf{r}_k)\right]\psi(\mathbf{X}) = E\psi(\mathbf{x}). \quad (A.59)$$

The 'Quantal Newtonian' first law is [6, 7, 8]

$$\mathcal{F}^{\text{ext}}(\mathbf{r}) + \mathcal{F}^{\text{int}}(\mathbf{r}) = 0, \quad (A.60)$$

where

$$\mathcal{F}^{\text{ext}}(\mathbf{r}) = \mathcal{E}(\mathbf{r}) = -\nabla v(\mathbf{r}), \quad (A.61)$$

$$\mathcal{F}^{\text{int}}(\mathbf{r}) = \mathcal{E}_{\text{ee}}(\mathbf{r}) - \mathcal{Z}(\mathbf{r}) - \mathcal{D}(\mathbf{r}), \quad (A.62)$$

with the various fields defined as in the time-dependent case but with the variable $\mathbf{y} = \mathbf{r}t$ replaced by the variable $\mathbf{r}$.

Part B: Noninteracting Fermions

'Quantal Newtonian' Second Law

With the assumption that the external field $\mathcal{F}^{ext}(\mathbf{y})$ experienced by the noninteracting fermions is the same as for the interacting electrons (see (A.40)), the local effective potential differential equation for the single p[article orbitals $\phi_j(\mathbf{y})$ is

$$\left[\frac{1}{2}(\hat{\mathbf{p}} + \mathbf{A}(y))^2 + v_s(\mathbf{y})\right]\phi_j(\mathbf{y}) = i\frac{\partial \phi_j(\mathbf{y})}{\partial t} \; ; \quad j = 1, \dots, N, \tag{A.63}$$

with

$$v_s(\mathbf{y}) = v(\mathbf{y}) - \Phi(\mathbf{y}) + v_{ee}(\mathbf{y}), \tag{A.64}$$

where all the many-body effects are incorporated into the local effective electron-interaction potential $v_{ee}(\mathbf{y})$. The model system wave function is the Slater determinant $\Phi\{\phi_j\}$ of these orbitals. All quantal sources such as the density $\rho_s(\mathbf{r})$, the Dirac density matrix $\gamma_s(\mathbf{rr}'t)$, and the current density $\mathbf{j}_s(\mathbf{y})$ are expectation values of the corresponding Hermitian operators taken with respect to the determinant $\Phi\{\phi_j\}$.

The derivation of the 'Quantal Newtonian' Second Law for the noninteracting fermions is similar to that for interacting electrons. One writes the orbitals as $\phi_j(\mathbf{y}) = \phi_j^R(\mathbf{y}) + i\phi_j^I(\mathbf{y})$, where $\phi_j^R(\mathbf{y})$ and $\phi_j^I(\mathbf{y})$ are its real and imaginary components. On substitution into the differential equation (A.63) one then obtains the Second Law. With the added assumption that the densities of the noninteracting system $\{\rho_s(\mathbf{y}), \mathbf{j}_s(\mathbf{y})\}$ and interacting system $\{\rho(\mathbf{y}), \mathbf{j}(\mathbf{y})\}$ are the same, the statement of the Law is

$$\mathcal{F}^{ext}(\mathbf{y}) + \mathcal{F}^{int}(\mathbf{y}) = \mathcal{J}_s(\mathbf{y}) = \mathcal{J}(\mathbf{y}), \tag{A.65}$$

where

$$\mathcal{F}^{int}(\mathbf{y}) = -\nabla v_{ee}(\mathbf{y}) - \mathcal{Z}_s(\mathbf{y}) - \mathcal{D}(\mathbf{y}) - \mathcal{I}_m(\mathbf{y}). \tag{A.66}$$

The current density $\mathcal{J}_s(\mathbf{y})$ and kinetic $\mathcal{Z}_s(\mathbf{y})$ fields are defined as for the interacting system but in terms of the quantal sources $\mathbf{j}_s(\mathbf{y})$ and $\gamma_s(\mathbf{rr}'t)$. The differential density $\mathcal{D}(\mathbf{y})$ and internal magnetic $\mathcal{I}_m(\mathbf{y})$ fields are the same as those of the interacting system.

'Quantal Newtonian' First Law

The time-independent First Law constitutes a special case with the external fields being $\mathcal{E}(\mathbf{r}) = -\nabla v(\mathbf{r})$ and $\mathcal{B}(\mathbf{r}) = \nabla \times \mathbf{A}(\mathbf{r})$. The corresponding eigenvalue differential equation for the model fermions is

$$\left[\frac{1}{2}(\hat{\mathbf{p}}+\mathbf{A}(\mathbf{r}))^2+v_s(\mathbf{r})\right]\phi_j(\mathbf{r})=\epsilon_j\phi_j(\mathbf{r})\ ;\quad j=1,\ldots,N, \tag{A.67}$$

with

$$v_s(\mathbf{r})=v(\mathbf{r})+v_{ee}(\mathbf{r}). \tag{A.68}$$

The ‘Quantal Newtonian’ First Law is

$$\mathcal{F}^{ext}(\mathbf{r})+\mathcal{F}_s^{int}(\mathbf{r})=0, \tag{A.69}$$

where

$$\mathcal{F}^{ext}(\mathbf{r})=\mathcal{E}(\mathbf{r})-\mathcal{L}(\mathbf{r}), \tag{A.70}$$

$$\mathcal{F}_s^{int}(\mathbf{r})=-\nabla v_{ee}(\mathbf{r})-\mathcal{Z}_s(\mathbf{r})-\mathcal{D}(\mathbf{r})-\mathcal{I}_m(\mathbf{r}). \tag{A.71}$$

The definitions of the fields are the same as for the temporal case but with the time parameter absent.

The special cases of the ‘Quantal Newtonian’ Second and First Laws without the magnetic field are obtained by setting the corresponding Lorentz and internal magnetic fields components to zero.

References

1. V. Sahni, X.-Y. Pan, T. Yang, Computation **4**, 30 (2016); https://doi.org/10.3390/computation4030030
2. Z. Qian, V. Sahni, Phys. Lett. A **247**, 303 (1998).
3. Z. Qian, V. Sahni, Int. J. Quantum Chem. **78**, 341 (2000).
4. T. Yang, X.-Y. Pan and V. Sahni, Phys. Rev. A **83**, 042518 (2011).
5. A. Holas, N.H. March, Phys. Rev. A **56**, 4595 (1997).
6. V. Sahni, Phys. Rev. A **55**, 1846 (1997).
7. V. Sahni, Top. Curr. Chem. **182**, 1 (1996).
8. A. Holas, N.H. March, Phys. Rev. A **51**, 2040 (1995).

Appendix B

Analytical Expressions for Properties of the Triplet 2^3S State of a Two-Electron Quantum Dot in a Magnetic Field

This Appendix provides [1] the closed-form analytical and semi-analytical expressions for various properties of the 2^3S state of the 2-electron quantum dot in a magnetic field. The properties are determined by employing the solution $\Psi(\mathbf{x}_1\mathbf{x}_2)$ of (5.2) of the corresponding Schrödinger-Pauli theory equation (5.1). In these expressions the zeroth- and first-order modified Bessel functions $I_0(x)$ and $I_1(x)$ appear. The general Bessel function $I_\nu(x)$ is defined as [2]

$$I_\nu(x) = \sum_{n=0}^{\infty} \frac{1}{n!\Gamma(n+\nu+1)} \left(\frac{1}{2}x\right)^{2n+\nu}, \tag{B.1}$$

where the Gamma function $\Gamma(n)$ is [2]

$$\Gamma(n) = \int_0^{\infty} t^{n-1} e^{-t} dt \;\; ; \;\; n > 0. \tag{B.2}$$

Also the parameter Ω is related to the binding frequency ω_0, the Larmor frequency ω_L, and the effective force constant k_{eff} as

$$\Omega = \sqrt{k_{\mathrm{eff}}} = \sqrt{\omega_0^2 + \omega_L^2} = 0.268732 \tag{B.3}$$

Other constants that appear in the expressions are

V. Sahni, *Schrödinger Theory of Electrons: Complementary Perspectives*,
Springer Tracts in Modern Physics 285,
https://doi.org/10.1007/978-3-030-97409-1

$$A = \frac{1}{3}\,;\; B = \frac{1}{8}\left(\frac{1}{3} - 3\Omega\right) = -0.059108$$

$$C = \frac{1}{360}(1 - 25\Omega) = -0.015884. \tag{B.4}$$

Electron density $\rho(\mathbf{r})$

$$\begin{aligned}\rho(r) &= 4\pi N^2 e^{-2\Omega r^2} \int_0^\infty e^{-\Omega x^2} x\left[x + c_2x^2 + c_3x^3 + c_4x^4\right]^2 I_0(2\Omega r x)dx \\ &= \tfrac{N^2}{4\Omega^9}\{8\pi\Omega^4 e^{-\Omega r^2}\left(K_1 + L_1r^2 + M_1r^4 + N_1r^6 + O_1r^8\right) + \\ &\pi^{3/2}\Omega^{9/2}\, e^{-\frac{3}{2}\Omega r^2}\big[\left(K_2 + L_2r^2 + M_2r^4 + N_2r^6 + O_2r^8\right) I_o\left(\Omega r^2/2\right) + \\ &2\Omega\left(L_3r^2 + M_3r^4 + N_3r^6 + O_3r^8\right) I_1\left(\Omega r^2/2\right)\big]\},\end{aligned} \tag{B.5}$$

where

$$\begin{aligned} N &= 0.02246632108, \\ K_1 &= 24\,C^2 + (6B^2 + 12\,AC)\,\Omega + 2\,A^2\,\Omega^2, \\ L_1 &= 96\,C^2\,\Omega + (18B^2 + 36\,AC)\,\Omega^2 + (4\,A^2 + 8B)\,\Omega^3 + \Omega^4, \\ M_1 &= 2\,C^2\,\Omega^2 + (9B^2 + 18\,AC)\,\Omega^3 + (A^2 + 2B)\,\Omega^4, \\ N_1 &= 16\,C^2\,\Omega^3 + (B^2 + 2\,AC)\,\Omega^4, \\ O_1 &= C^2\,\Omega^4, \\ K_2 &= 105BC + 30(AB + C)\,\Omega + 12\,A\,\Omega^2, \\ L_2 &= 420BC\,\Omega + 90(AB + C)\,\Omega^2 + 24A\,\Omega^3, \\ M_2 &= 376BC\,\Omega^2 + 56(AB + C)\,\Omega^3 + 8\,A\,\Omega^4, \\ N_2 &= 104BC\,\Omega^3 + 8(AB + C)\,\Omega^4, \\ O_2 &= 8BC\,\Omega^4, \\ L_3 &= 88BC + 23(AB + C)\,\Omega + 8\,A\,\Omega^2, \\ M_3 &= 142BC\,\Omega + 24(AB + C)\;\Omega^2 + 4\,A\,\Omega^3, \\ N_3 &= 48BC\,\Omega^2 + 4(AB + C)\,\Omega^3, \\ O_3 &= 4BC\,\Omega^3. \end{aligned} \tag{B.6}$$

The asymptotic structure of $\rho(\mathbf{r})$ near the center of the quantum dot, and in the classically forbidden region, respectively, is as follows:

$$\rho(r) \underset{r\to 0}{\sim} 0.0555 - 0.00625\,r^2 - 0.000230\,r^4 + \cdots \tag{B.7}$$

with $\rho(0) = 0.0555377\; a.u.^{\star}$,

$$\rho(r) \underset{r\to\infty}{\sim} e^{-\Omega r^2}\,10^{-5}\,(8.03\,r^4 + 12.6\,r^5 + 9.35\,r^6 + 2.22\,r^7 + 0.298\,r^8 + \cdots) \tag{B.8}$$

Physical current density j(r) and its components

$$\mathbf{j}(\mathbf{r}) = \mathbf{j}_p(\mathbf{r}) + \mathbf{j}_d(\mathbf{r}) + \mathbf{j}_m(\mathbf{r}), \tag{B.9}$$

where $\mathbf{j}_p(\mathbf{r})$, $\mathbf{j}_d(\mathbf{r})$, and $\mathbf{j}_m(\mathbf{r})$ are the paramagnetic, diamagnetic and magnetization current density components, respectively. For definitions of these components, see Chap. 4. The components are

$$\begin{aligned}
\mathbf{j}_p(\mathbf{r}) = j_p(r)\,\hat{\mathbf{i}}_\theta = \{\tfrac{2\pi}{\Omega} N^2 e^{-2\Omega r^2} \\
\tfrac{\partial}{\partial r}\big[\textstyle\int_0^\infty e^{-\Omega x^2}\tfrac{1}{x}(x + c_2 x^2 + c_3 x^3 + c_4 x^4)^2 I_0(2\Omega r x)dx\big]\}\,\hat{\mathbf{i}}_\theta \\
= \tfrac{\pi N^2 e^{-3\Omega r^2/2}}{4\Omega^4}\{8e^{\Omega r^2/2}\big[\big(24\,C^2 + (6B^2 + 12\,AC)\Omega + (2A^2 + 4B)\Omega^2 + \Omega^3\big)r + \\
\big(36\,C^2\Omega + (6B^2 + 12\,AC)\Omega^2 + (A^2 + 2B)\Omega^3\big)r^3 + \big(12C^2\Omega^2 + (B^2 + 2\,AC)\Omega^3\big)r^5 + \\
C^2\Omega^3 r^7 + \sqrt{\pi\Omega}\big[\big(105BC + 30(AB + C)\Omega + 12\,A\Omega^2\big)r + \big(180BC\Omega + 36(AB + \\
C)\Omega^2 + 8A\Omega^3\big)r^3 + \big(76BC\Omega^2 + 8(AB + C)\Omega^3\big)r^5 + 8BC\Omega^3 r^7\big]I_0(\Omega r^2/2) + \\
\sqrt{\pi\Omega}\big[\big(15BC + 6(AB + C)\Omega + 4\,A\Omega^2\big)r + \big(116BC\Omega + 28(AB + C)\Omega^2 + \\
8\,A\Omega^3\big)r^3 + \big(68BC\Omega^2 + 8(AB + C)\Omega^3\big)r^5 + 8BC\Omega^3 r^7\big]I_1(\Omega r^2/2)\}\,\hat{\mathbf{i}}_\theta,
\end{aligned} \tag{B.10}$$

$$\mathbf{j}_d(\mathbf{r}) = j_d(r)\,\hat{\mathbf{i}}_\theta = r\,\omega_L\,\rho(r)\,\hat{\mathbf{i}}_\theta, \tag{B.11}$$

$$\mathbf{j}_m(\mathbf{r}) = j_m(r)\,\hat{\mathbf{i}}_\theta = -\tfrac{1}{2}\,\tfrac{\partial\rho(r)}{\partial r}\,\hat{\mathbf{i}}_\theta. \tag{B.12}$$

The asymptotic structures of $\mathbf{j}(\mathbf{r})$, and its components are

$$\begin{aligned}
\mathbf{j}(\mathbf{r}) &\underset{r\to 0}{\sim} 0.0267\,r - 0.00501\,r^3 - 0.0000511\,r^5, \\
\mathbf{j}(\mathbf{r}) &\underset{r\to\infty}{\sim} e^{-\Omega r^2}10^{-6}\,(1.10\,r^9 + 8.17\,r^8 + 25.6\,r^7), \\
\mathbf{j}_p(\mathbf{r}) &\underset{r\to 0}{\sim} 0.0149\,r - 0.00485\,r^3 + 0.000848\,r^5, \\
\mathbf{j}_p(\mathbf{r}) &\underset{r\to\infty}{\sim} e^{-\Omega r^2}10^{-3}\,(0.00298\,r^7 + 0.0222\,r^6 + 0.049\,r^5 - 0.118\,r^4), \\
\mathbf{j}_d(\mathbf{r}) &\underset{r\to 0}{\sim} 10^{-3}\,(5.55\,r - 0.625\,r^3 - 0.0230\,r^5), \\
\mathbf{j}_d(\mathbf{r}) &\underset{r\to\infty}{\sim} e^{-\Omega r^2}10^{-6}\,(0.298\,r^9 + 2.22\,r^8 + 9.35\,r^7 + 17.0\,r^6), \\
\mathbf{j}_m(\mathbf{r}) &\underset{r\to 0}{\sim} 10^{-3}\,(6.25\,r + 0.461\,r^3 - 0.876\,r^5), \\
\mathbf{j}_m(\mathbf{r}) &\underset{r\to\infty}{\sim} e^{-\Omega r^2}10^{-5}\,(0.0800\,r^9 + 0.596\,r^8 + 1.32\,r^7).
\end{aligned} \tag{B.13}$$

Pair-correlation density $g(\mathbf{rr}')$

$$g(\mathbf{rr}') = \frac{2\,N^2\,e^{-\Omega(r^2+r'^2)}\left[|\mathbf{r}-\mathbf{r}'| + A\,|\mathbf{r}-\mathbf{r}'|^2 + B|\mathbf{r}-\mathbf{r}'|^3 + C|\mathbf{r}-\mathbf{r}'|^4\right]^2}{\rho(\mathbf{r})}. \quad \text{(B.14)}$$

Single-particle density matrix $\gamma(\mathbf{rr}')$

$$\gamma(\mathbf{rr}') = 2\,N^2\,e^{-\Omega(r^2+r'^2)/2}\int e^{-\Omega\,y^2}\left[|\mathbf{y}-\mathbf{r}| + A|\mathbf{y}-\mathbf{r}|^2 + \right.$$
$$\left. B|\mathbf{y}-\mathbf{r}|^3 + C|\mathbf{y}-\mathbf{r}|^4\right]\left[|\mathbf{y}-\mathbf{r}'| + A|\mathbf{y}-\mathbf{r}'|^2 + B|\mathbf{y}-\mathbf{r}'|^3 + C|\mathbf{y}-\mathbf{r}'|^4\right]d\mathbf{y}. \quad \text{(B.15)}$$

Electron-interaction field $\mathcal{E}_{ee}(\mathbf{r})$

$$\mathcal{E}_{ee}(\mathbf{r}) = \frac{4\,\pi\,N^2}{32\,\Omega^{7/2}\rho(\mathbf{r})}\{\sqrt{\pi}\,e^{-3\Omega r^2/2}[I_1(\Omega\,r^2/2)\,([15\,C^2 + (6B^2 + 12\,AC)\Omega +$$
$$(4\,A^2 + 8B)\Omega^2 + 8\Omega^3]r + [116\,C^2\,\Omega + (28B^2 + 56\,AC)\,\Omega^2 + (8\,A^2 +$$
$$16B)\,\Omega^3\,]r^3 + [68\,C^2\,\Omega^2 + (8B^2 + 16\,AC)\,\Omega^3]r^5 + 8\,C^2\,\Omega^3\,r^7) +$$
$$I_0(\Omega\,r^2/2)\,([105\,C^2 + (30B^2 + 60\,AC)\,\Omega + (12\,A^2 + 24\,AB)\,\Omega^2 +$$
$$8\,\Omega^3]r + [180\,C^2\,\Omega + (36B^2 + 72\,AC)\,\Omega^2 + (8\,A^2 + 16B)\,\Omega^3]r^3 +$$
$$[76\,C^2\,\Omega^2 + (8B^2 + 16\,AC)\,\Omega^3]r^5 + 8\,C^2\,\Omega^3 r^7)\,] + 32\sqrt{\Omega}\,e^{-\Omega r^2}[\,(6BC +$$
$$(2\,AB + 2\,C)\,\Omega + A\,\Omega^2)r + (6BC\,\Omega + (AB + C)\,\Omega^2)r^3 + BC\,\Omega^2 r^5]\,\}, \quad \text{(B.16)}$$

The asymptotic structure of $\mathcal{E}_{ee}(\mathbf{r})$ and its Hartree $\mathcal{E}_H(\mathbf{r})$, and Pauli-Coulomb $\mathcal{E}_{xc}(\mathbf{r})$ components is

$$\mathcal{E}_{ee}(r) \underset{r\to 0}{\sim} 0.137\,r - 0.0360\,r^3, \qquad \mathcal{E}_{ee}(r) \underset{r\to\infty}{\sim} \frac{1}{r^2} - \frac{0.0754}{r^3} - \frac{24.0}{r^4},$$
$$\mathcal{E}_H(r) \underset{r\to\infty}{\sim} \frac{2}{r^2} - \frac{0.0287}{r^3} + \frac{16.3}{r^4}, \qquad \mathcal{E}_{xc}(r) \underset{r\to\infty}{\sim} -\frac{1}{r^2} - \frac{0.0467}{r^3} - \frac{40.3}{r^4}. \quad \text{(B.17)}$$

Electron-interaction energy E_{ee}

$$E_{ee} = \pi^2\,N^2\{\tfrac{4}{\Omega^5}[24BC + 4(AB + C)\Omega + A\,\Omega^2] + [105\,C^2 +$$
$$15(B^2 + 2\,AC)\,\Omega + 3(A^2 + 2\,AB)\Omega^2 + \,\Omega^3]\} = 0.254158\,(a.u.)^\star. \quad \text{(B.18)}$$

Kinetic energy tensor $t_{\alpha\beta}[\mathbf{r};\gamma]$

For the derivation of the tensor $t_{\alpha\beta}[\mathbf{r};\gamma]$ see Appendix C.

$$t_{\alpha\beta}[\mathbf{r};\gamma] = \frac{r_\alpha\, r_\beta}{r^2}\, f(r) + \delta_{\alpha\beta}\, k(r), \tag{B.19}$$

where

$$f(r) = \pi\, N^2 e^{-2\Omega r^2}\left[\frac{r}{\Omega}\frac{\partial f_1(r)}{\partial r} - 2\, r\frac{\partial f_2(r)}{\partial r}\right] + \frac{\Omega^2}{2}\, r^2 \rho(r), \tag{B.20}$$

$$k(r) = \pi\, N^2\, e^{-2\Omega r^2}\left[\frac{f_1(r)}{\Omega} + 2 f_3(r)\right], \tag{B.21}$$

$$\begin{aligned} f_1(r) = \frac{e^{\Omega r^2/2}}{8\,\Omega^3}\{&4\, e^{\Omega^2 r^2/2}[\,(90\, C^2\,\Omega + (8B^2 + 14AC)\Omega^2)(1 + r^2) + 15\, C^2\,\Omega^2 r^4] + \\ &\sqrt{\pi\Omega}[(165BC + (30\, AB + 18\, C)\Omega + 4\, A\,\Omega^2) + r^2(198BC\,\Omega + (20AB + \\ &12\, C)\Omega^2) + 44BC\Omega^2 r^4]I_0(\Omega r^2/2) + \sqrt{\pi\Omega}\,[(33BC + (10\, AB + 6\, C)\Omega + \\ &4\, A\Omega^2) + r^2(154BC\Omega + (20AB + 12\, C)\Omega^2) + 44BC\Omega^2 r^4]I_1(\Omega r^2/2)\} \end{aligned} \tag{B.22}$$

$$\begin{aligned} f_2(r) = \frac{e^{\Omega r^2/2}}{16\Omega^4}\{&8e^{\Omega r^2/2}[(24\, C^2 + (6B^2 + 12\, AC)\Omega + (2\, A^2 + 4B)\Omega^2 + \Omega^3) + \\ &(72\, C^2\Omega + (12B^2 + 24\, AC)\Omega^2 + (2\, A^2 + 4B)\Omega^3)r^2 + (36\, C^2\Omega^2 + (3B^2 + \\ &6AC)\Omega^3)r^4 + 4\, C^2\Omega^3 r^6] + \sqrt{\pi\Omega}[(105BC + 30(AB + C)\Omega + 12\, A\Omega^2) + \\ &28BC\Omega^3 r^6 + (315BC\Omega + 60(AB + C)\Omega^2 + 12\, A\Omega^3)r^2 + (196BC\Omega^2 + \\ &20(AB + C)\Omega^3)r^4]I_0(\Omega r^2/2) + \sqrt{\pi\Omega}\,\Omega[(161BC + 40(AB + C)\Omega + \\ &12\, A\Omega^2)r^2 + (168BC\Omega + 20(AB + C)\Omega^2)r^4 \\ &+28BC\Omega^2 r^6]I_1(\Omega r^2/2)\} \end{aligned} \tag{B.23}$$

$$\begin{aligned} f_3(r) = \frac{e^{\Omega r^2/2}}{8\Omega^4}\{&4e^{\Omega r^2/2}[(6\, C^2 + 2(2B^2 + 4AC)\Omega + (A^2 + 2B)\Omega^2 + \Omega^3) + \\ &(18\, C^2\Omega + 4(B^2 + 2AC)\Omega^2 + (A^2 + 2B)\Omega^3)r^2 + (9\, C^2\Omega^2 + (B^2 + \\ &2AC)\Omega^3)r^4 + C^2\Omega^3 r^6] + \sqrt{\pi\Omega}[(15BC + 6(AB + C)\Omega + 4\, A\Omega^2) + \\ &(45BC\Omega + 12(AB + C)\Omega^2 + 4A\Omega^3)r^2 + (28BC\Omega^2 + 4(AB + C)\Omega^3)r^4 + \\ &4BC\Omega^3 r^6]I_0(\Omega r^2/2) + \sqrt{\pi\Omega}\,\Omega[(23BC + 8(AB + C)\Omega + 4\, A\Omega^2)r^2 + \\ &(24BC\Omega + 4(AB + C)\Omega^2)r^4 + 4BC\Omega^2 r^6]I_1(\Omega r^2/2)\}. \end{aligned} \tag{B.24}$$

Kinetic 'force' $z_\alpha[\mathbf{r};\gamma]$

$$z_\alpha[\mathbf{r};\gamma] = \frac{2\, r_\alpha}{r}\left\{\frac{\partial[f(r) + k(r)]}{\partial r} + \frac{f(r)}{r}\right\}, \tag{B.25}$$

where the functions $f(r)$ and $k(r)$ are given in (B.20) and (B.21), respectively. The asymptotic structures are

$$z(r) \underset{r\to 0}{\sim} 0.00174\, r + 0.00811\, r^3 + \cdots \tag{B.26}$$

$$z(r) \underset{r\to\infty}{\sim} e^{-\Omega r^2} 10^{-5}(-0.0116r^{11} - 0.0860r^{10} + 0.218r^9 - 0.0700r^8 + 5.55r^7 + \cdots). \tag{B.27}$$

Kinetic Energy T

$$T = \frac{N_R^2\,\pi}{4} + \frac{N_r^2\,\pi}{\Omega^4}\big(480C^2 + 12\,A^2\Omega^2 + 3\,A\sqrt{\pi/2}\,\Omega^{3/2}(13B + 3\Omega) + 1.5\,C(85B\sqrt{2\pi\Omega} + 64\,A\Omega + 5\sqrt{2\pi}\Omega^{3/2}) + 4\Omega(16B^2 + 4B\Omega + \Omega^2)\big) = 0.615577\ (a.u.)^\star, \tag{B.28}$$

where $N_r = 0.05431655771$, $N_R = 0.4136182782$.

Lorentz $\mathcal{L}(\mathbf{r})$ and Internal Magnetic $\mathcal{I}_m(\mathbf{r})$ Fields

$$\mathcal{L}(\mathbf{r}) = \frac{2\omega_L j(r)}{\rho(r)}\,\hat{\mathbf{i}}_r, \tag{B.29}$$

$$\mathcal{I}_m(\mathbf{r}) = \left[-\frac{2\omega_L j(r)}{\rho(r)} + \omega_L^2 r\right]\hat{\mathbf{i}}_r, \tag{B.30}$$

$$\mathcal{M}(\mathbf{r}) = -[\mathcal{L}(\mathbf{r}) + \mathcal{I}_m(\mathbf{r})] = -\omega_L^2 r\ \mathbf{i}_r, \tag{B.31}$$

where $j(r) = j_p(r) + j_d(r) + j_m(r)$ (see (B.10–B.12).)

External Electrostatic E_{es} and Magnetostatic E_{mag} Energies

$$\begin{aligned} E_{es} + E_{mag} &= \int \rho(r)\left[\frac{1}{2}k_{\text{eff}}r^2\right]d\mathbf{r} \\ &= k_{\text{eff}}\frac{\pi^2 N^2}{4\Omega^7}\big(64\Omega^2\,(A^2 + 2B) + 480\Omega(2\,AC + B^2) + 135\sqrt{2\pi}\,\Omega^{3/2}(AB + C) \\ &+ \sqrt{2\pi\Omega}\,(21\,A\Omega^2 + 1155BC) + 4608\,C^2 + 12\Omega^3\big) \\ &= 0.742657\ (a.u.)^\star \end{aligned} \tag{B.32}$$

Expectation values

With the complete elliptical integrals [2]

$$K(p) = \int_0^{\pi/2} \frac{d\theta}{\sqrt{1 - p^2 \sin^2 \theta}}, \tag{B.33}$$

$$E(p) = \int_0^{\pi/2} \sqrt{1 - p^2 \sin^2 \theta d\theta}, \tag{B.34}$$

the expressions and values of the various expectations are:

$$< r >= \int \rho(r)\, r\, d\mathbf{r}$$

$$= \frac{\pi^2 N^2}{2\Omega^{13/2}} \Big[\frac{47}{2} \sqrt{\pi}\, \Omega^2 (A^2 + 2B) + \frac{639}{4} \sqrt{\pi}\, \Omega (2AC + B^2) + \sqrt{2}\, \Omega^{3/2} \big(174 E(1/2) - 47\, K(1/2)\big)(AB + C) + 2\sqrt{2}\, A\Omega^{5/2} \big(15 E(1/2) - 4\, K(1/2)\big) + 5\sqrt{2\Omega}\, BC \big(273 E(1/2) - 74\, K(1/2)\big) + \frac{11313}{8} \sqrt{\pi}\, C^2 + 5\sqrt{\pi}\, \Omega^3 \Big] = 5.823553\ (a.u.)^{\star}, \tag{B.35}$$

$$< r^2 >= \int \rho(r)\, r^2\, d\mathbf{r}$$

$$= \frac{\pi^2 N^2}{2\Omega^7} \big(4608\, C^2 + 1155 BC \sqrt{2\pi\Omega} + 480(B^2 + 2\, AC)\Omega + 135(AB + C)\sqrt{2\pi}\Omega^{3/2} + 64(A^2 + 2B)\Omega^2 + 21\, A\sqrt{2\pi}\, \Omega^{5/2} + 12\Omega^3\big) = 20.567403\ (a.u.)^{\star}, \tag{B.36}$$

$$\left\langle \frac{1}{r} \right\rangle = \int \rho(r) \left(\frac{1}{r} \right) d\mathbf{r}$$

$$= \frac{\pi^2 N^2}{8\Omega^{11/2}} [76\sqrt{\pi}\, \Omega^2 (A^2 + 2B) + 378\sqrt{\pi}\, \Omega (2\, AC + B^2) + 48\sqrt{2}\, \Omega^{3/2} \big(9E(1/2) - 2\, K(1/2)\big) (AB + C) + 16\sqrt{2}\, A\Omega^{5/2} \big(6E(1/2) - K(1/2)\big) + 8\sqrt{2\Omega}\, BC \big(336 E(1/2) - 83\, K(1/2)\big) + 2601\sqrt{\pi}\, C^2 + 24\sqrt{\pi}\, \Omega^3] = 1.041717\ (a.u.)^{\star}, \tag{B.37}$$

$$< \delta(\mathbf{r}) >= \int \rho(r) \delta(\mathbf{r}) d\mathbf{r} = \rho(0)$$

$$= \frac{N^2 \pi}{4\Omega^5} [3\sqrt{\pi\Omega}\, \big(35 BC + 10(AB + C)\Omega + 4\, A\Omega^2\big) + 8\big(24\, C^2 + 6(B^2 + 2AC)\Omega + 2(A^2 + 2B)\Omega^2 + \Omega^3\big)\,] = 0.0555377\ (a.u.)^{\star}, \tag{B.38}$$

References

1. M. Slamet, V. Sahni, Chem. Phys. **546**, (2021) 111073
2. M. Abramowitz, I.A. Stegun, *Handbook of Mathematical Functions*, Dover, New York, 1972

Appendix C

Derivation of the Kinetic-Energy Tensor and Kinetic 'Force' for the 2^3S State of a Two-Electron Quantum Dot in a Magnetic Field

The spatial part $\Psi(\mathbf{r}_1\mathbf{r}_2)$ of the first excited triplet state wave function of a 2-electron quantum dot in a magnetic field is (see [1] and Chaps. 5, 9)

$$\Psi(\mathbf{r}_1, \mathbf{r}_2) = Ne^{i\theta}e^{-\Omega(r_1^2+r_2^2)/2}g_0(u), \tag{C.1}$$

$$g_0(u) = u + c_2u^2 + c_3u^3 + c_4u^4, \tag{C.2}$$

where the values of the coefficients N, Ω, c_2, c_3, c_4 are given in Sect. 5.1, and θ is the angle of the relative coordinate $\mathbf{u} = \mathbf{r}_2 - \mathbf{r}_1$.

The kinetic energy tensor $t_{\alpha\beta}(\mathbf{r}; \gamma)$ is defined as

$$t_{\alpha\beta}(\mathbf{r}; \gamma) = \frac{1}{4}\left[\frac{\partial^2}{\partial r_{p\alpha}\partial r_{q\beta}} + \frac{\partial^2}{\partial r_{p\beta}\partial r_{q\alpha}}\right]\gamma(\mathbf{r}_p, \mathbf{r}_q)\bigg|_{\mathbf{r}_p=\mathbf{r}_q=\mathbf{r}}, \tag{C.3}$$

where the single-particle density matrix $\gamma(\mathbf{r}_p\mathbf{r}_q)$ is

$$\gamma(\mathbf{r}_p, \mathbf{r}_q) = 2\int \Psi^\star(\mathbf{r}_p, \mathbf{r}_2)\Psi^\star(\mathbf{r}_q, \mathbf{r}_2)d\mathbf{r}_2. \tag{C.4}$$

Hence, the components of the tensor are

V. Sahni, *Schrödinger Theory of Electrons: Complementary Perspectives*,
Springer Tracts in Modern Physics 285,
https://doi.org/10.1007/978-3-030-97409-1

$$t_{xx} = \int \left(\frac{\partial \Psi^{\star}_{p,2}}{\partial x_p} \frac{\partial \Psi_{q,2}}{\partial x_q} \right) \Bigg|_{\mathbf{r}_p=\mathbf{r}_q=r} d\mathbf{r}_2, \tag{C.5}$$

$$t_{xy} = \frac{1}{2} \int \left(\frac{\partial \Psi^{\star}_{p,2}}{\partial x_p} \frac{\partial \Psi_{q,2}}{\partial y_q} + \frac{\partial \Psi^{\star}_{p,2}}{\partial y_p} \frac{\partial \Psi_{q,2}}{\partial x_q} \right) \Bigg|_{\mathbf{r}_p=\mathbf{r}_q=\mathbf{r}} d\mathbf{r}_2, \tag{C.6}$$

$$t_{yx} = t_{xy}, \tag{C.7}$$

$$t_{yy} = \int \left(\frac{\partial \Psi^{\star}_{p,2}}{\partial y_p} \frac{\partial \Psi_{q,2}}{\partial y_q} \right) \Bigg|_{\mathbf{r}_p=\mathbf{r}_q=\mathbf{r}} d\mathbf{r}_2. \tag{C.8}$$

We next determine the derivatives in the components of the tensor.

(i) Writing $\mathbf{r}_1 = \mathbf{r}_p$,

$$\frac{\partial}{\partial x_p} e^{-\Omega(r_p^2+r_2^2)/2} \Bigg|_{\mathbf{r}_p=r} = -\Omega x e^{-\Omega(r^2+r_2^2)/2}. \tag{C.9}$$

(ii) Writing $\mathbf{r}_p = \mathbf{r}$, and defining $\mathbf{r}_2 - \mathbf{r} = \mathbf{r}_3$,

$$\frac{\partial}{\partial x_p} |\mathbf{r}_2 - \mathbf{r}_p| = -\frac{x_3}{r_3}, \tag{C.10}$$

$$\frac{\partial}{\partial x_p} |\mathbf{r}_2 - \mathbf{r}_p|^2 = -2x_3, \tag{C.11}$$

$$\frac{\partial}{\partial x_p} |\mathbf{r}_2 - \mathbf{r}_p|^3 = -3r_3 x_3, \tag{C.12}$$

$$\frac{\partial}{\partial x_p} |\mathbf{r}_2 - \mathbf{r}_p|^4 = -4r_3^2 x_3. \tag{C.13}$$

Thus,

$$\frac{\partial}{\partial x_p} g_0(r_3) = -x_3 \left(\frac{1}{r_3} + 2c_2 + 3c_3 r_3 + 4c_4 r_3^2 \right) \tag{C.14}$$

$$= -x_3 g_1(r_3). \tag{C.15}$$

(iii) With

$$\theta_{p,2} = \tan^{-1} \left(\frac{y_2 - y_p}{x_2 - x_p} \right), \tag{C.16}$$

$$\frac{\partial}{\partial x_p} e^{-i\theta_{p,2}} = -i \frac{y_3}{r_3^2} e^{-i\theta_{p,2}}. \tag{C.17}$$

Hence, the first derivative of the integrand of t_{xx} of (C.5) is

$$\frac{\partial \Psi^{\star}_{p,2}}{\partial x_p} = N e^{-\frac{\Omega}{2}(2r^2+r_3^2+2\mathbf{r}\cdot\mathbf{r}_3)} \left[-\Omega x g_0 - x_3 g_1 - i \frac{y_3}{r_3^2} g_0 \right] e^{-i\theta_3}. \tag{C.18}$$

In a similar manner, the second derivative $\partial\Psi_{q,2}/\partial x_q$ is obtained, so that the integrand of t_{xx} of (C.5) is

$$\frac{\partial\Psi^\star_{p,2}}{\partial x_p}\cdot\frac{\partial\Psi_{q,2}}{\partial x_q}=N^2e^{-\Omega(2r^2+r_3^2+2\mathbf{r}\cdot\mathbf{r}_3)}\left[\Omega^2x^2g_0^2+x_3^2g_1^2+2\Omega xx_3g_0g_1+\frac{y_3^2}{r_3^4}g_0^2\right]. \tag{C.19}$$

Similarly, the integrand of t_{yy} of (C.8) is

$$\frac{\partial\Psi^\star_{p,2}}{\partial y_p}\cdot\frac{\partial\Psi_{q,2}}{\partial y_p}=N^2e^{-\Omega(2r^2+r_3^2+2\mathbf{r}\cdot\mathbf{r}_3)}\left[\Omega^2y^2g_0^2+y_3^2g_1^2+2\Omega yy_3g_0g_1+\frac{x_3^2}{r_3^4}g_0^2\right], \tag{C.20}$$

and that of t_{xy} of (C.6) is

$$\frac{1}{2}\left(\frac{\partial\Psi^\star_{p,2}}{\partial x_p}\cdot\frac{\partial\Psi_{q,2}}{\partial y_q}+\frac{\partial\Psi^\star_{p,2}}{\partial y_p}\cdot\frac{\partial\Psi_{q,2}}{\partial x_q}\right)=N^2e^{-\Omega(2r^2+r_3^2+2\mathbf{r}\cdot\mathbf{r}_3)}$$
$$\left[\Omega^2xyg_0^2+x_3y_3g_1^2+\Omega xy_3g_0g_1+\Omega yx_3g_0g_1-\tfrac{x_3y_3}{r_3^4}g_0^2\right]. \tag{C.21}$$

Let us first consider the off-diagonal component t_{xy}. In this component, consider the contribution of the first term of (C.21) in the square parentheses which is

$$N^2e^{-2\Omega r^2}xy\Omega^2\int g_0^2(r_3)e^{-\Omega(r_3^2+2\mathbf{r}\cdot\mathbf{r}_3)}d\mathbf{r}_3 \tag{C.22}$$

$$=(xy\Omega^2)\left[N^2e^{-2\Omega r^2}\int_0^\infty e^{-\Omega r_3^2}g_0^2(r_3)r_3dr_3\int_0^{2\pi}e^{-2\Omega rr_3\cos\theta_3}d\theta_3\right] \tag{C.23}$$

$$=(xy\Omega^2)\left[N^2e^{-2\Omega r^2}2\pi\int_0^\infty e^{-\Omega r_3^2}g_0^2(r_3)I_0(2\Omega rr_3)r_3dr_3\right], \tag{C.24}$$

$$=\frac{(xy\Omega^2)}{2}\rho(r), \tag{C.25}$$

where $I_0(x)$ is the zeroth-order modified Bessel function [2] (see Appendix B). (This term can be written more generally as $(r_\alpha r_\beta\Omega^2\rho(r)/2)$, where r_α, r_β represent either x or y.)

The vector components x_3 and y_3 of the second term in the square parentheses of (C.21) can be eliminated through the equalities

$$x_3e^{-2\Omega\mathbf{r}\cdot\mathbf{r}_3}=-\frac{1}{2\Omega}\frac{\partial}{\partial x}e^{-2\Omega\mathbf{r}\cdot\mathbf{r}_3}, \tag{C.26}$$

and

$$y_3e^{-2\Omega\mathbf{r}\cdot\mathbf{r}_3}=-\frac{1}{2\Omega}\frac{\partial}{\partial y}e^{-2\Omega\mathbf{r}\cdot\mathbf{r}_3}, \tag{C.27}$$

and then by evaluating the $d\theta_3$ integral of (C.6) first, the contribution of the second term of (C.21) to t_{xy} is

$$\frac{2\pi N^2}{4\Omega^2} e^{-2\Omega r^2} \frac{\partial^2}{\partial x \partial y} \int_0^\infty r_3 g_1^2(r_3) e^{-\Omega r_3^2} I_0(2\Omega r r_3) dr_3. \tag{C.28}$$

As the lowest-order of $g_1(r_3)$ is $1/r_3$, the integrand of (C.28) goes as $1/r_3$, which is singular at $r_3 = 0$. In order to eliminate the singularity, we employ

$$\frac{\partial}{\partial y} I_0(2\Omega r r_3) = \frac{2\Omega y r_3}{r} I_1(2\Omega r r_3), \tag{C.29}$$

where $I_1(x)$ is the first-order modified Bessel function [2] (see Appendix B).

The contribution of the fifth term of (C.21) to the integral of (C.6) also goes as $1/r_3$ to lowest-order, and the singularity is treated as above. Then by evaluating the dr_3 integral, and employing the equality for a general function $w(r)$ as follows:

$$\frac{\partial}{\partial r_\alpha}\left[r_\beta w(r)\right] = \delta_{\alpha\beta} w(r) + \frac{r_\alpha r_\beta}{r} \frac{\partial w(r)}{\partial r}, \tag{C.30}$$

the contribution of the combination of the second and fifth terms of (C.21) to (C.6) for t_{xy} may be written as

$$\frac{\pi N^2}{\Omega} e^{-2\Omega r^2}\left[\delta_{\alpha\beta} f_1(r) + \frac{r_\alpha r_\beta}{r} \frac{\partial f_1(r)}{\partial r}\right], \tag{C.31}$$

where

$$f_1(r) = \frac{1}{r} \int_0^\infty \left(g_1^2 - \frac{g_0^2}{r_3^4}\right) e^{-\Omega r_3^2} r_3^2 I_1(2\Omega r r_3) dr_3. \tag{C.32}$$

(See (B.22) for $f_1(r)$).

To evaluate contribution of the third and fourth cross-terms of (C.21) to (C.6), which are identical, we apply the equalities (C.26) and (C.27), evaluate the $d\theta_3$ integral first (no singularity in this case), then evaluate the dr_3 integral, and employ the following equality for any function $w(r)$

$$r_\beta \frac{\partial w(r)}{\partial r_\alpha} = \frac{r_\alpha r_\beta}{r} \frac{\partial w(r)}{\partial r}. \tag{C.33}$$

Then the sum of the cross-terms may be written as

$$-2\pi N^2 e^{-2\Omega r^2}\left(\frac{r_\alpha r_\beta}{r}\right) \frac{\partial f_2(r)}{\partial r}, \tag{C.34}$$

where

$$f_2(r) = \int_0^\infty r_3 g_1(r_3) g_0(r_3) e^{-\Omega r_3^2} I_0(2\Omega r r_3) dr_3. \tag{C.35}$$

(See (B.23) for $f_2(r)$).

Next consider the diagonal elements t_{xx} and t_{yy} of (C.5) and (C.8), respectively. The first three terms of the corresponding integrands given by (C.19) and (C.20) are evaluated in the same way as the first 3 terms of the off-diagonal element t_{xy} as described above.

Note that the contribution of the last term of (C.19) to t_{xx} is proportional to y_3^2 (instead of x_3^2), whereas that of the last term of (C.20) to t_{yy} is proportional to x_3^2 (instead of y_3^2). Since $y_3^2 = r_3^2 - x_3^2$, the last term of (C.19) may be written as $y_3^2 g_0^2/r_3^4 = (r_3^2 - x_3^2)g_0^2/r_3^4$. This term may be further generalized to include the corresponding term of the off-diagonal element t_{xy} by writing it as

$$\left(\delta_{\alpha\beta} r_3^2 - r_{3\alpha} r_{3\beta}\right)\frac{g_0^2}{r_3^4}. \tag{C.36}$$

Notice that (C.36) is identical to the fifth term in (C.21) for t_{xy}, because $\delta_{\alpha\beta} = 0$ when $\alpha \neq \beta$. (In this case $\alpha = x$, and $\beta = y$).

We next determine the contribution of the $\delta_{\alpha\beta} r_3^2$ term of (C.36) to t_{xx}. From (C.19), this contribution is

$$N^2 e^{-2\Omega r^2} \int_0^\infty e^{-\Omega r_3^2} \frac{g_0^2}{r_3^4} \cdot r_3^2 \cdot r_3 dr_3 \int_0^{2\pi} e^{-2\Omega r r_3 \cos\theta_3} d\theta_3 \tag{C.37}$$

$$= 2\pi N^2 e^{-2\Omega r^2} f_3(r) \tag{C.38}$$

where

$$f_3(r) = \int_0^\infty e^{-\Omega r_3^2} \frac{g_0^2}{r_3} I_0(2\Omega r r_3) dr_3. \tag{C.39}$$

(See (B.24) for $f_3(r)$).

The second term of (C.36) is the same as the last term of (C.21), and its contribution has been previously evaluated.

Thus, in summing all the requisite terms, the tensor $t_{\alpha\beta}$ may be written as

$$t_{\alpha\beta}(r) = \frac{r_\alpha r_\beta}{r^2} f(r) + \delta_{\alpha\beta} k(r) \tag{C.40}$$

where $f(r)$ and $k(r)$ are defined in (B.20) and (B.21).

The kinetic 'force' component is defined as

$$z_\alpha(r) = 2\sum_{\beta=1}^{2} \nabla_\beta t_{\alpha\beta}(r). \tag{C.41}$$

Upon substituting $t_{\alpha\beta}(r)$ of (C.40) into (C.41) we obtain

$$z_\alpha(r) = 2\sum_{\beta=1}^{2} \frac{\partial}{\partial r_\beta}\left[\frac{r_\alpha r_\beta}{r^2} f(r) + \delta_{\alpha\beta} k(r)\right], \tag{C.42}$$

where $f(r)$ and $k(r)$ are given in (B.20) and (B.21).

For the $2D$ coordinate system, it can be shown

$$\sum_{\beta=1}^{2} \frac{\partial}{\partial r_\beta}(r_\alpha r_\beta) = 3r_\alpha, \tag{C.43}$$

$$\sum_{\beta=1}^{2} r_\alpha r_\beta \frac{\partial}{\partial r_\beta}\Big[\frac{f(r)}{r^2}\Big] = r_\alpha\Big[\frac{1}{r}\frac{\partial f(r)}{\partial r} - \frac{2f(r)}{r^2}\Big], \tag{C.44}$$

$$\sum_{\beta=1}^{2} \frac{\partial}{\partial r_\beta}\big[\delta_{\alpha\beta}k(r)\big] = \frac{r_\alpha}{r}\frac{\partial k(r)}{\partial r}. \tag{C.45}$$

Finally, by substituting (C.43), (C.44), and (C.45) into (C.42), we obtain the components of the kinetic 'force' as given in (B.25).

References

1. M. Slamet, V. Sahni, Chem. Phys. **546**, 111073 (2021)
2. M. Abramowitz, I.A. Stegun, *Handbook of Mathematical Functions* (Dover, New York, 1972)

Appendix D

Quantal Density Functional Theory Expressions for the Singlet State 2^1S of a Two-Electron Quantum Dot in a Magnetic Field

In this Appendix the analytical and semi-analytical expressions of the Quantal Density Functional Theory mapping from a first *excited* singlet 2^1S state of a 2-electron quantum dot in a magnetic field to one of noninteracting fermions in a *ground* state possessing the same electronic density $\rho(\mathbf{r})$ and physical current density $\mathbf{j}(\mathbf{r})$ is provided [1]. The interacting system wave function is given in Chap. 6. (In the expressions below $C_{01} = C_1$, and $\Omega = \Omega_1$ of the expression for the wave function given in Chap. 6.)

Electron density $\rho_1(\mathbf{r})$

$$\rho_1(r) = \frac{C_{01}^2}{2\Omega^4}\big\{4\pi e^{-\Omega r^2}\big(K_1 + L_1 r^2 + M_1 r^4 + N_1 r^6\big) + \\ \pi\sqrt{\pi\Omega}\, e^{-\frac{3}{2}\Omega r^2}\big[K_2 + L_2 r^2 + M_2 r^4 + N_2 r^6\big)\, I_o\big(\Omega r^2/2\big) + \\ \big(L_3 r^2 + M_3 r^4 + N_3 r^6\big)\, I_1\big(\Omega r^2/2\big)\big]\big\}, \tag{D.1}$$

where

$$C_{01} = 0.1085631994,$$
$$\Omega = \frac{2}{2.91199} = 0.686816\,,$$
$$K_1 = 6B^2 + (2\,A^2 + 4B)\,\Omega + (1 + 2\,A)\,\Omega^2 + \Omega^3,$$
$$L_1 = 18\,B^2\,\Omega + 4(A^2 + 2B)\,\Omega^2 + (1 + 2\,A)\,\Omega^3,$$
$$M_1 = 9\,B^2\,\Omega^2 + (A^2 + 2B)\,\Omega^3,$$
$$N_1 = B^2\,\Omega^3,$$
$$K_2 = 15\,AB + 6(A + B)\,\Omega + 4\,\Omega^2,$$
$$L_2 = 45\,AB\,\Omega + 12(A + B)\,\Omega^2 + 4\,\Omega^3,$$

V. Sahni, *Schrödinger Theory of Electrons: Complementary Perspectives*, Springer Tracts in Modern Physics 285,
https://doi.org/10.1007/978-3-030-97409-1

$$
\begin{aligned}
M_2 &= 28\,AB\,\Omega^2 + 4(A + B\,\Omega^3,\\
N_2 &= 4\,AB\,\Omega^3,\\
L_3 &= 23\,AB\,\Omega + 8(A + B)\,\Omega^2 + 4\,\Omega^3,\\
M_3 &= 24\,AB\,\Omega^2 + 4(A + B\,\Omega^3,\\
N_3 &= 4\,AB\,\Omega^3,\\
A &= \frac{1}{4}(1 + \Omega - 2\,\epsilon_r) = -0.2651111137,\\
B &= \frac{1}{9}\left(\frac{1}{4} - \frac{5\epsilon_r}{2} + \frac{9}{4}\Omega\right) = -0.1820822248,\\
\epsilon_r &= 1.37363
\end{aligned}
\tag{D.2}
$$

and $I_0(x)$ and $I_1(x)$ are the zeroth- and first-order modified Bessel functions [2] (see Appendix B).
The asymptotic structure of $\rho(\mathbf{r})$ near the center of the quantum dot, and in the classically forbidden region, respectively, are as follows:

$$\rho_1(r) \underset{r\to 0}{\sim} 0.207 - 0.141\,r^2 + 0.0753\,r^4 + \cdots \tag{D.3}$$

with $\rho_1(0) = 0.207299\;a.u.$,

$$\rho_1(r) \underset{r\to\infty}{\sim} e^{-\Omega r^2}(-0.00171\,r^3 + 0.00252\,r^2 + 0.0549\,r + 0.113 + \tfrac{0.0398}{r} + \tfrac{0.00902}{r^3} + \cdots) \tag{D.4}$$

Pair-correlation density $g(\mathbf{r}\mathbf{r}')$

$$g(\mathbf{r}\mathbf{r}') = \frac{2\,C_{01}^2\,e^{-\Omega(r^2+r'^2)}\left[1 + |\mathbf{r}-\mathbf{r}'| + A|\mathbf{r}-\mathbf{r}'|^2 + B|\mathbf{r}-\mathbf{r}'|^3\right]^2}{\rho_1(\mathbf{r})}, \tag{D.5}$$

where the constants C_{01}, Ω, A and B are given in (D.2).

Single-particle density matrix $\gamma(\mathbf{r}\mathbf{r}')$

$$\gamma(\mathbf{r}\mathbf{r}') = 2\,C_{01}^2\,e^{-\Omega(r^2+r'^2)/2}\int e^{-\Omega\,y^2}\left[1 + |\mathbf{y}-\mathbf{r}| + A|\mathbf{y}-\mathbf{r}|^2 + B|\mathbf{y}-\mathbf{r}|^3\right]\left[1 + |\mathbf{y}-\mathbf{r}'| + A|\mathbf{y}-\mathbf{r}'|^2 + B|\mathbf{y}-\mathbf{r}'|^3\right]d\mathbf{y}, \tag{D.6}$$

where the constants C_{01}, Ω, A and B are given in (D.2).

Dirac density matrix $\gamma_s(\mathbf{r}\mathbf{r}')$

$$\gamma_s(\mathbf{r}\mathbf{r}') = \sqrt{\rho_1(\mathbf{r})\rho_1(\mathbf{r}')} \tag{D.7}$$

Electron-interaction field $\mathcal{E}_{ee}(\mathbf{r})$ and its components

$$\mathcal{E}_{ee}(\mathbf{r}) = \frac{\pi\, C_{01}^2\, e^{-2\Omega r^2}}{4\, r\, \Omega^{5/2} \rho_1(\mathbf{r})}\Big\{\sqrt{\pi}\, r^2\, e^{\Omega r^2/2}\big[I_1(\Omega\, r^2/2)\, \big(r^2\big(4\, A^2 \Omega^2 + 14 B^2 \Omega + 8 B \Omega^2\big) + 2\, A^2 \Omega + 8\, A \Omega^2 + 4 B^2 r^4 \Omega^2 + 4 B \Omega - 8 \Omega^3 + 4 \Omega^2 + 3 B^2\big) + I_0\big(\Omega r^2/2\big)\, \big(r^2\big(4\, A^2 \Omega^2 + 18 B^2 \Omega + 8 B \Omega^2\big) + 6\, A^2 \Omega + 8\, A \Omega^2 + 4 B^2 r^4 \Omega^2 + 12 B \Omega + 8 \Omega^3 + 4 \Omega^2 + 15 B^2\big)\big] + 16 \sqrt{\Omega}\big[e^{\Omega r^2}\big(A B r^4 \Omega + r^2 (A \Omega + B \Omega + 2\, A B) + \Omega\big) - \Omega\big]\Big\}, \quad \text{(D.8)}$$

where the constants C_{01}, Ω, A and B are given in (D.2).
The asymptotic structure of $\mathcal{E}_{\mathrm{ee}}(\mathbf{r})$ and its Hartree $\mathcal{E}_H(\mathbf{r})$, Pauli-Coulomb $\mathcal{E}_{xc}(\mathbf{r})$, Pauli $\mathcal{E}_x(\mathbf{r})$, and Coulomb $\mathcal{E}_c(\mathbf{r})$ components are

$$\begin{aligned}
&\mathcal{E}_{ee}(r) \underset{r\to 0}{\sim} 0.955\, r - 0.494\, r^3, && \mathcal{E}_{ee}(r) \underset{r\to\infty}{\sim} \frac{1}{r^2} - \frac{7.64}{r^4} + \frac{4.24}{r^5},\\
&\mathcal{E}_H(r) \underset{r\to\infty}{\sim} \frac{2}{r^2} + \frac{4.92}{r^4}, && \mathcal{E}_{xc}(r) \underset{r\to\infty}{\sim} -\frac{1}{r^2} - \frac{12.6}{r^4},\\
&\mathcal{E}_x(r) \underset{r\to\infty}{\sim} -\frac{1}{r^2} - \frac{2.45}{r^4}, && \mathcal{E}_c(r) \underset{r\to\infty}{\sim} -\frac{10.1}{r^4}.
\end{aligned} \quad \text{(D.9)}$$

Electron-interaction potential $W_{ee}(\mathbf{r})$ and its components

$$W_{ee}(\mathbf{r}) = -\int_{\infty}^{r} \mathcal{E}_{ee}(y)\, dy, \quad \text{(D.10)}$$

where $\mathcal{E}_{ee}(r)$ is given in (D.8). The Hartree potential $W_H(\mathbf{r})$ in two dimensions can be written for cylindrically symmetric densities $\rho(\mathbf{r})$ as [3]

$$\begin{aligned}
W_H(\mathbf{r}) &= \int \frac{\rho(\mathbf{r}')}{|\mathbf{r}-\mathbf{r}'|}\, d\mathbf{r}'\\
&= 4\int_0^r dr'\, \frac{r'}{r}\, \rho(r')\, K\left(\frac{r'^2}{r^2}\right) + 4\int_r^{\infty} dr' \rho(r')\, K\left(\frac{r^2}{r'^2}\right),
\end{aligned} \quad \text{(D.11)}$$

where $K(x)$ is the complete elliptic integral of the first kind [2]. The asymptotic structure of $W_{ee}(r)$ and its Hartree $W_H(r)$, Pauli-Coulomb $W_{xc}(r)$, Pauli $W_x(r)$, and Coulomb $W_c(r)$ components are

$$\begin{aligned}
&W_{ee}(r) \underset{r\to\infty}{\sim} \frac{1}{r} - \frac{2.55}{r^3}, && W_H(r) \underset{r\to\infty}{\sim} \frac{2}{r} + \frac{1.64}{r^3},\\
&W_{xc}(r) \underset{r\to\infty}{\sim} -\frac{1}{r} - \frac{4.18}{r^3}, && W_x(r) \underset{r\to\infty}{\sim} -\frac{1}{r} - \frac{0.818}{r^3},\\
&W_c(r) \underset{r\to\infty}{\sim} -\frac{3.37}{r^3},
\end{aligned} \quad \text{(D.12)}$$

Electron-interaction energy E_{ee}

$$E_{ee} = \pi^2 C_{01}^2 \Big[\sqrt{\tfrac{\pi}{2\Omega^9}}\big(\Omega\ (2\,A\Omega + \Omega^2 + \Omega + 3\,A^2 + 6B) + 15B^2\big) + 2\,\big(2\Omega(A+B) + \Omega^2 + 8\,AB\big)/\Omega^4\Big] = 0.600476\ a.u., \tag{D.13}$$

where the constants C_{01}, Ω, A and B are given in (D.2).

Kinetic energy tensor $t_{\alpha\beta}[\mathbf{r};\gamma]$

$$t_{\alpha\beta}[\mathbf{r};\gamma] = \frac{r_\alpha\, r_\beta}{r^2}\, f(r) + \delta_{\alpha\beta}\, k(r), \tag{D.14}$$

where

$$f(r) = \pi\, C_{01}^2 e^{-2\Omega r^2}\left[\frac{r}{\Omega}\frac{\partial f_1(r)}{\partial r} - 2\,r\frac{\partial f_2(r)}{\partial r}\right] + \frac{\Omega^2}{2}\, r^2 \rho_1(r), \tag{D.15}$$

$$k(r) = \frac{\pi\, C_{01}^2}{\Omega}\, e^{-2\Omega r^2} f_1(r), \tag{D.16}$$

$$f_1(r) = \tfrac{1}{2\,\Omega^2\, r^2}\Big\{ -\Omega + e^{\Omega r^2}\big[\Omega + (4\Omega A^2 + 6\Omega B + 18\,A^2 + 9\Omega B^2 r^2)\, r^2\big] + A\sqrt{\pi\Omega}\; r^2 e^{\Omega r^2/2}\big[(2\Omega + 9B + 6\Omega B r^2)\; I_0(\Omega r^2/2) + (2\Omega + 3B + 6\Omega B r^2)\; I_1(\Omega r^2/2)\big]\Big\}, \tag{D.17}$$

$$f_2(r) = \tfrac{e^{\Omega r^2/2}}{8\Omega^3}\Big\{4e^{\Omega r^2/2}\big[(6B^2 + 2\,A^2\Omega + 4B\Omega + \Omega^2 + 2\,A\Omega^2) + (12B^2\Omega + 2\,A^2\Omega^2 + 4B\Omega^2)\, r^2 + 3B^2\Omega^2 r^4\big] + \sqrt{\pi\Omega}\, \Big[\big(15\,AB + 6\,A\Omega + 6B\Omega + 4\Omega^2 + (30\,AB\Omega + 6(A+B)\,\Omega^2)\, r^2 + 10\,AB\Omega^2 r^4\big)\; I_0(\Omega r^2/2) + \big(2\Omega(10\,AB + 3\,A\Omega + 3B\Omega) r^2 + 10\,AB\Omega^2 r^4\big)\; I_1(\Omega r^2/2)\Big]\Big\}, \tag{D.18}$$

where the constants C_{01}, Ω, A and B are given in (D.2). The derivation of the tensor $t_{\alpha\beta}[\mathbf{r};\gamma]$ is similar to that for the triplet state given in Appendix C. For details of the derivation, (see [1]).

Kinetic energy tensor $t_{s,\alpha\beta}[\mathbf{r};\gamma_s]$

$$t_{s,\alpha\beta}[\mathbf{r};\gamma_s]=\frac{r_\alpha\, r_\beta}{8r^2\rho_1(r)}\left(\frac{\partial\rho_1(r)}{\partial r}\right)^2. \tag{D.19}$$

Kinetic 'force' $z_\alpha[\mathbf{r};\gamma_s]$

$$z_\alpha[\mathbf{r};\gamma_s]=\frac{2\,r_\alpha}{r}\left\{\frac{\partial[f(r)+k(r)]}{\partial r}+\frac{f(r)}{r}\right\}, \tag{D.20}$$

where the functions $f(r)$ and $k(r)$ are given in (D.15) and (D.16), respectively. The asymptotic structures are

$$z(r)\underset{r\to 0}{\sim}0.703\,r-r^3+0.459\,r^5+\cdots \tag{D.21}$$

$$\begin{aligned}z(r)\underset{r\to\infty}{\sim}\;&e^{-\Omega r^2}(-0.0417r^5-0.346r^4-0.215r^3+0.332r^2+0.202r\\&-0.0772+\frac{0.216}{r}+\frac{0.0343}{r^2}-\frac{0.0650}{r^4}+\cdots,\end{aligned} \tag{D.22}$$

where Ω is given in (D.2).

Kinetic 'force' $z_{s,\alpha}[\mathbf{r};\gamma_s]$

$$z_{s,\alpha}[\mathbf{r};\gamma_s]=2\sum_{\beta=1}^{2}{}^{'}\,\nabla_\beta\, t_{s,\alpha\beta}[\mathbf{r};\,{}_s]$$

$$=\frac{r_\alpha}{2r\rho_1(r)}\frac{\partial\rho_1(r)}{\partial r}\left[-\frac{1}{2\rho_1(r)}\left(\frac{\partial\rho_1(r)}{\partial r}\right)^2+\frac{\partial^2\rho_1(r)}{\partial r^2}+\frac{1}{2r}\frac{\partial\rho_1(r)}{\partial r}\right], \tag{D.23}$$

$$z_{s,\alpha}(r)\underset{r\to 0}{\sim}0.287\,r-0.699\,r^3+0.519\,r^5+\cdots \tag{D.24}$$

$$\begin{aligned}z_{s,\alpha}(r)\underset{r\to\infty}{\sim}\;&e^{-\Omega r^2}(-0.757r^5+2.46r^4-3.50r^3+3.90r^2+\\&32.2r-76.8+\frac{58.5}{r}+\frac{204}{r^2}-\frac{815}{r^3}+\cdots,\end{aligned} \tag{D.25}$$

where Ω is given in (D.2).

Correlation-kinetic field $\mathcal{Z}_{t_c}(r)$ and potential $\mathbf{W}_{t_c}(\mathbf{r})$

$$\mathcal{Z}_{t_c}(r) = \frac{z_s(r) - z(r)}{\rho_1(r)}, \qquad \mathcal{Z}_{t_c}(r) \underset{r\to\infty}{\sim} \frac{9}{r^3} - \frac{20}{r^4} + \cdots, \tag{D.26}$$

$$W_{t_c}(\mathbf{r}) = -\int_{\infty}^{r} \mathcal{Z}_{t_c}(y)\, dy, \qquad W_{t_c}(r) \underset{r\to\infty}{\sim} \frac{9}{2\,r^2} + \cdots . \tag{D.27}$$

Effective electron-interaction potential $v_{ee}(\mathbf{r})$

$$v_{ee}(r) = W_{ee}(r) + W_{t_c}(r), \tag{D.28}$$

$$v_{ee}(r) \underset{r\to 0}{\sim} 2.07 + 0.262\, r^2 + \cdots, \qquad v_{ee}(r) \underset{r\to\infty}{\sim} \frac{1}{r} + \frac{9}{2\,r^2} + \cdots . \tag{D.29}$$

Kinetic Energy T

$$\begin{aligned} T &= \int\int \Psi_1(\mathbf{rR})\big(\nabla_r^2 + \tfrac{1}{4}\nabla_R^2\big)\Psi_1(\mathbf{rR})\, d\mathbf{r}\, d\mathbf{R} \\ &= \tfrac{\pi^2\, C_{01}^2}{4\Omega^4}\big(16\Omega\ (2\,A^2 + 3B) + \sqrt{2\pi}\ \Omega^{3/2}(11\,A + 3B) + 57\sqrt{2\pi\Omega}\,AB \\ &\quad + 8(A+1)\Omega^2 + 3\sqrt{2\pi}\Omega^{5/2} + 4\Omega^3 + 192B^2\big) \\ &= 1.266683\ a.u., \end{aligned} \tag{D.30}$$

where the constants C_{01}, Ω, A and B are given in (D.2).

External Energy E_{ext}

$$\begin{aligned} E_{ext} &= \int \rho_1(\mathbf{r})\ \big(\tfrac{1}{2}k_{\text{eff}}r^2\big)d\mathbf{r} \\ &= \tfrac{\pi^2\, C_{01}^2}{4\Omega^4}\big(64\Omega\ (A^2 + 2B) + 21\sqrt{2\pi}\ \Omega^{3/2}(A + B) + 135\sqrt{2\pi\Omega}\,AB + \\ &\quad 12(2\,A + 1)\Omega^2 + 5\sqrt{2\pi}\Omega^{5/2} + 4\Omega^3 + 480B^2\big) \\ &= 1.566917\ a.u., \end{aligned} \tag{D.31}$$

where the constants C_{01}, Ω, A and B are given in (D.2), and $k_{\text{eff}} = \Omega^2 = 0.471716$.

Expectation values

$$< r >= \int \rho_1(r)\, r\, d\mathbf{r}$$

$$= \frac{\pi^{3/2}\, C_{01}^2}{8\Omega^{11/2}}\Bigg[\pi\big(94\Omega\ (A^2 + 2B) + 20(2A + 1)\Omega^2 + 8\Omega^3 + 639B^2\big) +$$

$$4\sqrt{2}\,\Gamma\left(\frac{3}{4}\right)^2\sqrt{\Omega}\big(15\Omega(A + B) + 4\Omega^2 + 87\,AB\big) +$$

$$\left.\frac{32\sqrt{2\Omega}\pi^2\left(7\Omega(A+B)+2\Omega^2+40AB\right)}{\Gamma\left(-\frac{1}{4}\right)^2}\right]$$

$$= 3.276784\ a.u., \tag{D.32}$$

where the constants C_{01}, Ω, A and B are given in (D.2), and $\Gamma(x)$ is the Gamma function [2].

$$< r^2 >= \int \rho_1(r)\, r^2\, d\mathbf{r}$$

$$= \frac{\pi^2\, C_{01}^2}{2\Omega^6}\Big(64\Omega\ (A^2+2B)+21\sqrt{2\pi}\ \Omega^{3/2}(A+B)+135\sqrt{2\pi\Omega}AB+$$

$$12(2A+1)\Omega^2+5\sqrt{2\pi}\Omega^{5/2}+4\Omega^3+480B^2\Big)$$

$$= 6.643480\ a.u., \tag{D.33}$$

where the constants C_{01}, Ω, A and B are given in (D.2).

$$\left\langle\frac{1}{r}\right\rangle = \int \rho_1(r)\left(\frac{1}{r}\right) d\mathbf{r}$$

$$= \frac{\pi^{3/2}\, C_{01}^2}{4\Omega^{9/2}}\left[\pi\left(\Omega\ (38\,A^2+76B)+4\Omega^2(2\Omega+6\,A+3)+189B^2\right)+\right.$$

$$4\sqrt{2}\,\Gamma\left(\frac{3}{4}\right)^2\sqrt{\Omega}\left(6\Omega(A+B)+2\Omega^2+27\,AB\right)+$$

$$\left.\frac{32\sqrt{2\Omega}\pi^2\left(4\Omega(A+B)+2\Omega^2+15AB\right)}{\Gamma\left(-\frac{1}{4}\right)^2}\right]$$

$$= 1.930015\ a.u., \tag{D.34}$$

where the constants C_{01}, Ω, A and B are given in (A.2), and $\Gamma(x)$ is the Gamma function [2].

$$< \delta(\mathbf{r}) >= \rho_1(0)$$

$$= \frac{C_{01}^2}{2\Omega^4}\Big[4\pi\left((2\,A^2+4B)\Omega+(2\,A+1)\Omega^2+\Omega^3+6B^2+\right.$$

$$\left.\pi^{3/2}\sqrt{\Omega}\left(6\Omega(A+B)+4\Omega^2+15\,AB\right)\right]$$

$$= 0.207299\ a.u., \tag{D.35}$$

where the constants C_{01}, Ω, A and B are given in (D.2).

References

1. M. Slamet, V. Sahni, Comp. Theor. Chem. **1114**, 125 (2017)
2. M. Abramowitz, I.A. Stegun, *Handbook of Mathematical Functions* (Dover, New York, 1972)
3. T. Yang, X.-Y. Pan, V. Sahni, Phys. Rev. A **83**, 042518 (2011)

Index

V. Sahni, *Schrödinger Theory of Electrons: Complementary Perspectives*,
Springer Tracts in Modern Physics 285,
https://doi.org/10.1007/978-3-030-97409-1